T0329099

LONDON MATHEMATICAL SOCIETY LECTURE NOTE SERIES

Managing Editor: Professor M. Reid, Mathematics Institute,
University of Warwick, Coventry CV4 7AL, United Kingdom

The titles below are available from booksellers, or from Cambridge University Press at
http://www.cambridge.org/mathematics

London Mathematical Society Lecture Note Series: 440

Surveys in Combinatorics 2017

Edited by

ANDERS CLAESSON
University of Iceland

MARK DUKES
University College Dublin

SERGEY KITAEV
University of Strathclyde

DAVID MANLOVE
University of Glasgow

KITTY MEEKS
University of Glasgow

CAMBRIDGE
UNIVERSITY PRESS

CAMBRIDGE
UNIVERSITY PRESS

University Printing House, Cambridge CB2 8BS, United Kingdom

One Liberty Plaza, 20th Floor, New York, NY 10006, USA

477 Williamstown Road, Port Melbourne, VIC 3207, Australia

314-321, 3rd Floor, Plot 3, Splendor Forum, Jasola District Centre, New Delhi - 110025, India

79 Anson Road, #06-04/06, Singapore 079906

Cambridge University Press is part of the University of Cambridge.

It furthers the University's mission by disseminating knowledge in the pursuit of education, learning and research at the highest international levels of excellence.

www.cambridge.org
Information on this title: www.cambridge.org/9781108413138
DOI: 10.1017/9781108332699

© Cambridge University Press 2017

First published 2017

A catalogue record for this publication is available from the British Library

ISBN 978-1-108-41313-8 Hardback

Contents

Preface

The Twenty-Sixth British Combinatorial Conference was held at the University of Strathclyde in Glasgow in July 2017. The British Combinatorial Committee had invited nine distinguished combinatorialists to give survey lectures in their areas of expertise, and this volume contains the survey articles on which these lectures were based.

In compiling this volume, we are very grateful to the authors for preparing their excellent surveys so professionally, and to the anonymous referees for their detailed and timely responses. We would also like to thank the team at Cambridge University Press, in particular Roger Astley, Clare Dennison and Abigail Walkington, for all their assistance and advice. Finally, we could not have completed this task without the experience of the editors of earlier volumes of Surveys and the guidance of the British Combinatorial Committee.

Relations among partitions

R. A. Bailey

Abstract

Combinatorialists often consider a balanced incomplete-block design to consist of a set of points, a set of blocks, and an incidence relation between them which satisfies certain conditions. To a statistician, such a design is a set of experimental units with two partitions, one into blocks and the other into treatments; it is the relation between these two partitions which gives the design its properties. The most common binary relations between partitions that occur in statistics are refinement, orthogonality and balance. When there are more than two partitions, the binary relations may not suffice to give all the properties of the system. I shall survey work in this area, including designs such as double Youden rectangles.

1 Introduction

Many combinatorialists think of a balanced incomplete-block design (BIBD) as a set $\mathcal{P}$ of points together with a collection $\mathcal{B}$ of subsets of $\mathcal{P}$, called *blocks*, which satisfy various conditions. For example, see [52]. Some papers, such as [16, 65, 201], call a BIBD simply a *design*. Others think of it as the pair of sets $\mathcal{P}$ and $\mathcal{B}$ with a binary incidence relation between their elements. These views are both rather different from that of a statistician who is involved in designing experiments. The following examples introduce the statistical point of view, as well as serving as a basis for the combinatorial ideas in this paper.

Example 1.1 A horticultural enthusiast wants to compare three varieties of lettuce for people to grow in their own gardens. He enlists twelve people in his neighbourhood. Each of these prepares three patches in their vegetable garden, and grows one of the lettuce varieties on each patch, so that each gardener grows all three varieties.

Here the patches of land are experimental units. There may be some differences between the gardeners, so the three patches in a single garden form what is called a *block*. Each variety occurs just once in each block, and so the blocks are said to be *complete*. Complete-block designs were advocated by Fisher in [78], and are frequently used in practice.

Example 1.2 Now suppose that the number of lettuce varieties is increased to nine. It is not reasonable to expect an amateur gardener to

A	D	G	A	B	C	A	B	C	A	B	C
B	E	H	D	E	F	E	F	D	F	D	E
C	F	I	G	H	I	I	G	H	H	I	G

Figure 1: Balanced incomplete-block design in Example 1.2: columns represent blocks and letters represent varieties

grow nine different varieties, so each gardener still uses only three patches of ground, and thus can grow only three varieties. The blocks are now *incomplete*, in the terminology of Yates [227].

One possible layout is shown in Figure 1. This incomplete-block design has the property that each pair of distinct varieties concur in the same number of blocks (here, exactly one). Yates originally called incomplete-block designs with this property *symmetrical*, but the adjective had been changed to *balanced* within a few years [46, 80].

To a statistician, the partition of the set of experimental units into blocks is inherent and is known before the decision is taken about which variety to allocate to each unit. This allocation gives another partition of the set of experimental units, and it is the relation between these two partitions that is regarded as *balance*. It is not a symmetric relation, in general. In Example 1.2 the varieties are balanced with respect to the blocks, but the blocks are not balanced with respect to the varieties because some pairs of blocks have one variety in common while others have none. This relation is discussed in more detail in Section 5.

In fact, statisticians usually call these partitions *factors*, because the names of the parts are relevant. In Example 1.2 the names of the varieties are not interchangeable; we probably want to find out which one does best. Thus a factor is typically regarded as a function from the set of experimental units to a finite set: if B and L denote the factors for blocks and lettuce varieties respectively and ω is a vegetable patch then $B(\omega)$ is the block (garden) containing ω and $L(\omega)$ is the variety grown on ω. Furthermore, $|B(\omega)|$ is the size of the block containing ω, while $|L(\omega)|$ is the number of patches with the same variety as that grown on ω.

A response Y_ω, such as total yield of edible lettuce in kilograms, is measured on each patch ω. It is usually assumed that Y_ω is a random variable and that there are constants τ_i and β_j such that

$$Y_\omega = \tau_{L(\omega)} + \beta_{B(\omega)} + \varepsilon_\omega, \tag{1.1}$$

where the final terms ε_ω are independent random variables with zero mean and the same variance σ^2; often they are assumed to be normally dis-

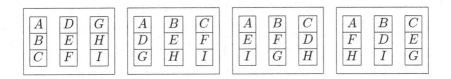

Figure 2: Resolved balanced incomplete-block design in Example 1.3: columns represent blocks, rectangles represent districts and letters represent varieties

tributed. The purpose of the experiment is to estimate the constants τ_i. Of course, this is impossible, because equation (1.1) is unchanged if a constant is added to every τ_i and subtracted from every β_j, but we aim to estimate differences such as $\tau_1 - \tau_2$, that is, to estimate the τ_i up to an additive constant.

Thus the two partitions have different roles. One (the partition B) is inherent, and we are usually not interested in the effects β_j of the different parts. The other (the partition L) has its parts allocated by the experimenter, and the purpose of the experiment is to find out what differences there are between its parts. Nonetheless, this paper will concentrate on the combinatorial relation between them. Before doing so, we give some examples with three partitions.

Example 1.3 Suppose that the twelve gardeners in Example 1.2 do not all live in the same neighbourhood. Instead, they are spread over four different districts, with three per district. If the first three blocks in Figure 1 represent the gardens in the first district, and so on, then each variety is grown once in each district, as shown in Figure 2. This is convenient if other people want to look at the different varieties during the course of the experiment.

Each block is contained within a single district, so the partition into blocks is a *refinement* of the partition into districts. Section 3 discusses refinement in more detail. On the other hand, the partitions into districts and into varieties have the property that each part of one (a district) meets each part of the other (a variety) in a single experimental unit. This is a special case of *strict orthogonality*, which is explained in Section 4.

The assumption about Y_ω might remain as in (1.1) or it might be

$$Y_\omega = \tau_{L(\omega)} + \beta_{B(\omega)} + \gamma_{D(\omega)} + \varepsilon_\omega, \qquad (1.2)$$

where $D(\omega)$ is the district containing ω. Of course, if the β_j and γ_k are all constants then they are not estimable, because we can add a constant

A	B	C	D	E	F	G
C	D	E	F	G	A	B
D	E	F	G	A	B	C
E	F	G	A	B	C	D

Figure 3: Row–column design in Example 1.4: rows represent months, columns represent people and letters represent exercise regimes

to γ_1 and subtract it from β_j for all blocks j in district 1. However, it is sometimes assumed that the β_j are independent random variables with zero mean and the same variance σ_B^2. Section 6.2 discusses further the potential difficulty in an assumption like (1.2) when one partition is a refinement of another.

An incomplete-block design whose blocks can be grouped into collections each of which contains each variety just once, as in Example 1.3, is called *resolvable*. Section 15 gives more information about such designs.

Example 1.4 In order to assess the benefits of different exercise regimes, a health scientist asks seven healthy people to participate in an experiment over four months. Each month each person will be allocated one of seven exercise regimes. At the end of each month, the change in some measure of fitness, such as heart rate, will be recorded for each person.

Now each experimental unit is one person for one month. The partitions into months and into people are inherent, but the scientist chooses the partition into exercise regimes. Figure 3 shows one possible design for this experiment. The partitions into months and into people are strictly orthogonal to each other, as are the partitions into months and into exercise regimes. The partitions into people and into exercise regimes are both balanced with respect to each other.

Example 1.5 A small modification of Example 1.4 has five months, six people and ten exercise regimes. One possible design is shown in Figure 4, where rows represent months, columns represent people and letters represent exercise regimes.

Example 1.6 A modification of Example 1.2 has ten gardens of three vegetable patches each, and six varieties of lettuce. In addition, there are are five possible watering regimes. Each patch must have one variety of lettuce and one watering regime. The design in Figure 4 can be used, but now rows represent watering regimes, columns represent lettuce varieties and letters represent gardens.

H	J	I	G	F	E
J	I	H	C	B	D
D	F	A	J	G	C
A	B	G	E	D	I
E	A	C	B	H	F

Figure 4: Combinatorial design used in Examples 1.5 and 1.6

Denote by R, C and L the partitions into rows, columns and letters in the design in Figure 4. From the point of view of the statistician, the uses of this design in Examples 1.5 and 1.6 are quite different. In the former, the partitions R and C are inherent while L is at the choice of the experimenter; in the latter, L is inherent while the experimenter chooses R and C. However, in both cases it may be assumed that

$$Y_\omega = \alpha_{R(\omega)} + \phi_{C(\omega)} + \tau_{L(\omega)} + \varepsilon_\omega. \tag{1.3}$$

From a combinatorial point of view, Figure 4 simply shows a set with three partitions. The partitions R and C are strictly orthogonal to each other, while each of R and C is balanced with respect to letters. In fact, there is a third property, called *adjusted orthogonality*, that will be defined in Section 8.

For further explanation of how combinatorial design problems arise from statistically designed experiments, see [22, 33, 177, 204].

The remainder of this paper treats a combinatorial design as a collection of partitions of a finite set. Section 2 establishes some notation for partitions and their associated matrices and subspaces. Sections 3–5 discuss the three most important binary relations between partitions, all of which have been seen in the examples so far. Section 6 explains more about the background to equations (1.1)–(1.3). Section 7 discusses the relations between the subspaces defined by partitions, and shows that sometimes there is a need for a ternary relation. Sections 8 and 9 give more details of two important non-binary relations. These are used in Section 10, which considers possibilities for three partitions. This leads to several different types of combinatorial design, considered in the remaining sections. Each type is defined by three partitions, or is a simple generalization with more partitions but no need for any further non-binary relations.

2   Partitions on a finite set

Let Ω be a finite set of size e, where $e > 1$. The elements of Ω will be called *experimental units*, or just *units*. The rest of this paper deals with partitions of Ω.

If F is such a partition, denote by n_F its number of parts. The $e \times n_F$ *incidence matrix* X_F has (ω, i)-entry equal to 1 if unit ω is in part i of F; otherwise, this entry is zero. Thus $X_F X_F^\top$ is the $e \times e$ *relation matrix* for F, with (ω_1, ω_2)-entry equal to 1 if ω_1 and ω_2 are in the same part of F, and equal to 0 otherwise. The $n_F \times n_F$ matrix $X_F^\top X_F$ is diagonal, with (i, i)-entry equal to the size of the i-th part of F.

Definition A partition is *uniform* if all of its parts have the same size.

Many statisticians, including Tjur [208, 209], call uniform partitions *balanced*, but this conflicts with the notion of balance introduced in Section 1. This terminology is discussed again in Section 9. Preece reviewed the overuse of the word *balance* in design of experiments in [155]. The adjectives *homogeneous* [44], *proper* [151] and *regular* [66] are also used.

If F is uniform, denote the size of all its parts by k_F. Then $n_F k_F = e$ and $X_F^\top X_F = k_F I_{n_F}$, where I_n is the identity matrix of order n.

Denote by $\mathbf{R}^\Omega$ the real vector space of dimension e whose coordinates are labelled by the elements of Ω, so that each vector may be regarded as a function from Ω to $\mathbf{R}$. If F is a partition of Ω, denote by V_F the subspace of $\mathbf{R}^\Omega$ consisting of vectors which are constant on each part of F. Then $\dim(V_F) = n_F$.

We assume the standard inner product on $\mathbf{R}^\Omega$. Denote by P_F the matrix of orthogonal projection onto V_F. Then P_F replaces the coordinate y_ω of any vector y by the average value of y_ν for ν in $F(\omega)$, which is the part of F containing ω. In fact, $P_F = X_F \left(X_F^\top X_F \right)^{-1} X_F^\top$. If F is uniform then $X_F X_F^\top = k_F P_F$.

Equations (1.1)–(1.3) all have the form

$$Y = \sum_{F \in \mathcal{F}} X_F \psi_F + \varepsilon, \tag{2.1}$$

where Y and ε are random vectors of length e, $\mathcal{F}$ is a set of partitions of Ω, and, for F in $\mathcal{F}$, ψ_F is a real vector of length n_F. Thus the expectation $\mathbb{E}(Y)$ of Y is in the subspace $\sum_{F \in \mathcal{F}} V_F$.

There are two trivial partitions on Ω, which are different when $e > 1$. The parts of the *equality* partition E are singletons, so $k_E = 1$, $n_E = e$ and $X_E = I_e = P_E$. At the other extreme, the *universal* partition U has a single part, so $n_U = 1$, $k_U = e$, $X_U X_U^\top = J_{ee}$ and $P_U = e^{-1} J_{ee}$, where J_{nm} denotes the $n \times m$ matrix with all entries equal to 1. Moreover, V_E is the whole space $\mathbf{R}^\Omega$, while V_U is the 1-dimensional subspace of constant vectors.

If F and G are two partitions of Ω, their $n_F \times n_G$ *incidence matrix* N_{FG} is defined by $N_{FG} = X_F^\top X_G$. The (i, j)-entry is the size of the intersection of the i-th part of F with the j-th part of G. In particular, $N_{EF} = X_F$.

Given a set $\mathcal{F}$ of partitions of Ω, denote by $\mathcal{A}_\mathcal{F}$ the algebra of $e \times e$ real matrices generated by the projection matrices P_F for F in $\mathcal{F}$, and denote by $\mathcal{J}_\mathcal{F}$ the algebra generated by the relation matrices $X_F X_F^\top$ for F in $\mathcal{F}$. These are the same if all partitions in $\mathcal{F}$ are uniform. James called $\mathcal{J}_\mathcal{F}$ the *relationship algebra* of $\mathcal{F}$ in [99], but it was shown in [100, 115] that $\mathcal{A}_\mathcal{F}$ is more useful for understanding the properties of $\mathcal{F}$ relevant to a designed experiment.

3   Refinement

Definition   If F and G are partitions of Ω, then F is *finer* than G (equivalently, G is *coarser* than F) if every part of F is contained in a single part of G but at least one part of G is not a part of F. This relation is denoted $F \prec G$ or $G \succ F$.

In Example 1.3, $B \prec D$. If $F \prec G$ then $n_F > n_G$ and $V_G < V_F$.

Write $F \preccurlyeq G$ (or $G \succcurlyeq F$) to mean that either $F \prec G$ or $F = G$. Then $\preccurlyeq$ is a partial order. For every partition F, it is true that $E \preccurlyeq F \preccurlyeq U$ and $V_U \leq V_F \leq V_E$.

Proposition 3.1   *Let F and G be partitions of Ω. If $F \preccurlyeq G$ then $P_F P_G = P_G P_F = P_G$.*

As with any partial order, there is a choice about which of the two objects should be considered 'smaller'. Some statisticians write the refinement partial order in the opposite way to that used here. For example, see [31, 208, 209].

Since there are only a finite number of partitions of Ω, there is no difficulty with the next definition.

Definition   Let F and G be partitions of Ω. The *infimum* $F \wedge G$ of F and G is the coarsest partition H satisfying $H \preccurlyeq F$ and $H \preccurlyeq G$; its parts are the non-empty intersections of a part of F and a part of G. Thus $F \wedge G = E$ if and only if no part of F intersects any part of G in more than one unit. The *supremum* $F \vee G$ of F and G is the finest partition K satisfying $F \preccurlyeq K$ and $G \preccurlyeq K$; its parts are the connected components of the graph with vertex-set Ω and an edge between ω_1 and ω_2 if $F(\omega_1) = F(\omega_2)$ or $G(\omega_1) = G(\omega_2)$.

Thus if $F \preccurlyeq G$ then $F \wedge G = F$ and $F \vee G = G$. In the design in Figure 4, $R \wedge C = R \wedge L = C \wedge L = E$ and $R \vee C = R \vee L = C \vee L = U$.

Proposition 3.2 *If F and G are partitions of Ω then $V_F \cap V_G = V_{F \vee G}$.*

4   Orthogonality

4.1   Definitions

As Preece noted in [154], the word *orthogonal* has many different meanings in the statistical literature. Here I use the terminology in [23, 25, 32, 208].

Proposition 3.2 shows that subspaces V_F and V_G can never be orthogonal to each other. This motivates the following definition, from [208].

Definition Let V and W be subspaces of $\mathbf{R}^\Omega$. Then V and W are *geometrically orthogonal* to each other if the subspaces $V \cap (V \cap W)^\perp$ and $W \cap (V \cap W)^\perp$ are orthogonal to each other.

Proposition 4.1 *Let F and G be partitions of Ω. The following statements are equivalent:*

(i) V_F is geometrically orthogonal to V_G;

(ii) $P_F P_G = P_G P_F$;

(iii) $P_F P_G = P_{F \vee G}$;

(iv) for every unit ω, we have $|F(\omega)|\,|G(\omega)| = |(F \wedge G)(\omega)|\,|(F \vee G)(\omega)|$.

The second statement above is sometimes called 'projectors commute', and the fourth 'proportional meeting within each class of the supremum'.

Definition Let F and G be partitions of Ω. Then F is *orthogonal* to G, written $F \perp G$, if $P_F P_G = P_G P_F$; and F is *strictly orthogonal* to G, written $F \underline{\perp} G$, if $P_F P_G = P_G P_F = P_U$.

Duquenne calls these two concepts *local orthogonality* and *orthogonality* respectively in [66]; the latter agrees with Gilliland's definition of orthogonality in [83]. Some authors split the definitions further according to whether or not $F \wedge G$ is uniform.

Proposition 3.1 shows that if $F \preccurlyeq G$ then $F \perp G$. In particular, all partitions are orthogonal to both E and U, and every partition is orthogonal to itself.

A	B	B	C	A	C
C		A		B	

Figure 5: A 2×3 row–column design with nine units and three letters, giving mutually orthogonal partitions into rows, columns and letters

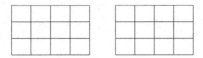

Figure 6: Two blocks, each of which is a 3×4 rectangle, so that there are 6 rows and 8 columns

In the design in Figure 4, $R \perp C$. If R, C and L denote the partitions into rows, columns and letters in Figure 5, then $R \perp C$, $R \perp L$ and $C \perp L$ even though R, $R \wedge C$ and $R \wedge L$ are not uniform, because the 'proportional meeting' condition in Proposition 4.1(iv) is satisfied for all pairs and all pairwise suprema are equal to U. If B, R and C denote the partitions into blocks, rows and columns in Figure 6, then $R \perp C$ but R is not strictly orthogonal to C because $R \vee C = B \neq U$.

Proposition 4.2 *Let F and G be partitions of Ω. Then $F \perp G$ if and only if $N_{FG} = e^{-1}(X_F^\top X_F) J_{n_F n_G}(X_G^\top X_G)$.*

4.2 Orthogonal arrays

Definition An *orthogonal array* of strength two on Ω is a collection $\mathcal{F}$ of at least two uniform partitions of Ω with the property that every pair of distinct partitions is strictly orthogonal. Inductively, for $m \geq 3$, a collection $\mathcal{F}$ of at least m partitions of Ω is an orthogonal array of strength m if it is an orthogonal array of strength $m - 1$ and, whenever $F_1, \ldots, F_m$ are distinct partitions in $\mathcal{F}$, the infimum $F_1 \wedge F_2 \wedge \cdots \wedge F_{m-1}$ is strictly orthogonal to F_m.

Figure 7 shows an orthogonal array of strength two with $e = 12$, $|\mathcal{F}| = 11$, and $n_F = 2$ for all F in $\mathcal{F}$. It is equivalent to that given by Plackett and Burman [142]. Replacing each 0 by -1 and adjoining a row of 1s gives a Hadamard matrix of order 12. The paper [142] inspired Rao to define orthogonal arrays and begin to develop a general theory of them in [181, 182].

F_1	0	0	1	0	0	0	1	1	1	0	1	1
F_2	1	0	0	1	0	0	0	1	1	1	0	1
F_3	0	1	0	0	1	0	0	0	1	1	1	1
F_4	1	0	1	0	0	1	0	0	0	1	1	1
F_5	1	1	0	1	0	0	1	0	0	0	1	1
F_6	1	1	1	0	1	0	0	1	0	0	0	1
F_7	0	1	1	1	0	1	0	0	1	0	0	1
F_8	0	0	1	1	1	0	1	0	0	1	0	1
F_9	0	0	0	1	1	1	0	1	0	0	1	1
F_{10}	1	0	0	0	1	1	1	0	1	0	0	1
F_{11}	0	1	0	0	0	1	1	1	0	1	0	1

Figure 7: Orthogonal array of strength two, consisting of 11 partitions of a set of size 12 into two parts: columns represent elements of the set, and each row shows one partition

For $n \geq 2$, the rows, columns and letters of any Latin square of order n give an orthogonal array of strength two on a set of size n^2, with three partitions into parts of size n. See [95] for many uses and constructions of orthogonal arrays, as well as more theory. Eendebak and Schoen maintain a catalogue on the web page [76].

From Finney [77] onwards, finite Abelian groups have been a fruitful source of orthogonal arrays, under the name *fractional factorial designs*. For $i = 1, \ldots, s$ let G_i be an Abelian group of order n_i, where $n_i \geq 2$. Let G be the product group $G_1 \times G_2 \times \cdots \times G_s$. Every complex irreducible character χ of G has the form $\chi = (\chi_1, \chi_2, \ldots, \chi_s)$ where χ_i is an irreducible character of G_i and $\chi(g_1, g_2, \ldots, g_s) = \chi_1(g_1)\chi_2(g_2) \cdots \chi_s(g_s)$. Let H be a subgroup of G, and let F_i be the partition of H defined by the values of the i-th coordinate. Then $\{F_1, \ldots, F_s\}$ forms an orthogonal array of strength m on H if and only if the only non-trivial characters χ of G whose restriction to H is trivial have non-trivial components χ_i for at least $m + 1$ values of i. For example, if $s = 3$, $n_1 = n_2 = n_3 = 7$ and G_i is $\mathbf{Z}_7$ written additively for $i = 1$, 2 and 3 then $\{F_1, F_2, F_3\}$ forms an orthogonal array of strength two on the subgroup $H = \{(g_1, g_2, g_3) : g_1 + g_2 + g_3 = 0\}$. Up to isotopism (permutations of the names of the parts of each partition), this is the Latin square obtained as the Cayley table of $\mathbf{Z}_7$.

Some papers, such as [61, 112, 141, 215], call an orthogonal array *regular* if and only if it is made from an Abelian group in this way. There are two problems with this. The first is that, in each experiment, the parts of F_i (such as varieties of lettuce) are unlikely to be labelled by the elements of a finite Abelian group. How is the statistician analysing the

A	B	C	D	E
E	A	B	C	D
D	E	A	B	C
C	D	E	A	B
B	C	D	E	A

A	B	C	D	E
B	A	D	E	C
E	D	A	C	B
C	E	B	A	D
D	C	E	B	A

(a)   (b)

Figure 8: Two Latin squares of order five: square (a) is isotopic to the Cayley table of $\mathbf{Z}_5$, but square (b) is not

data to know whether or not the orthogonal array was constructed from Abelian groups? In any case, this derivation makes no difference to the data analysis. The second problem is that, for an experiment designed using a Latin square, it is usually of no practical importance whether or not the square is isotopic to the Cayley table of an Abelian group. As n increases, so does the proportion of Latin squares of order n which are not isotopic to such Cayley tables. Figure 8 shows two Latin squares of order five: only one of them is isotopic to a Cayley table.

The definitions in this section show that, in an orthogonal array $\mathcal{F}$ of strength two, P_F commutes with P_G for all F and G in $\mathcal{F}$. Proposition 4.1 shows that commutativity cannot be destroyed by inclusion of suprema. Thus Grömping and Bailey, in their paper [86] giving some more lenient definitions of regularity, proposed calling an orthogonal array *geometrically regular* if $P_{G_1 \wedge \cdots \wedge G_r}$ commutes with $P_{H_1 \wedge \cdots \wedge H_s}$ for all subsets $\{G_1, \ldots, G_r\}$ and $\{H_1, \ldots, H_s\}$ of $\mathcal{F}$. Thus the two orthogonal arrays in Figure 8 are geometrically regular, while the one in Figure 7 is not, because $F_1 \wedge F_2$ is not orthogonal to F_3.

4.3 Tjur block structures and orthogonal block structures

Proposition 4.1 leads to these definitions, given in [23], building on the work in [208].

Definition Let $\mathcal{F}$ be a set of partitions on a finite set Ω. Then $\mathcal{F}$ is a *Tjur block structure* if $\mathcal{F}$ is closed under taking suprema, every pair of partitions in $\mathcal{F}$ is orthogonal, and $E \in \mathcal{F}$. If, in addition, $\mathcal{F}$ is closed under taking infima, every partition in $\mathcal{F}$ is uniform, and $U \in \mathcal{F}$, then $\mathcal{F}$ is an *orthogonal block structure*.

For example, if $\mathcal{F}$ is an orthogonal array of strength two then $\mathcal{F} \cup \{E, U\}$ is a Tjur block structure.

For F in $\mathcal{F}$, define a further subspace W_F of $\mathbf{R}^\Omega$ by

$$W_F = V_F \cap \bigcap_{F \prec G \in \mathcal{F}} V_G^\perp = V_F \cap \left(\sum_{F \prec G \in \mathcal{F}} V_G \right)^\perp.$$

Theorem 4.3 *If $\mathcal{F}$ is a Tjur block structure then the subspaces W_F, for F in $\mathcal{F}$, are mutually orthogonal and their sum is $\mathbf{R}^\Omega$.*

It follows that the dimensions of the subspaces W_F, and the matrices of orthogonal projection onto them, can be calculated recursively, starting with the coarsest partition in $\mathcal{F}$. Moreover, the algebra $\mathcal{A}_\mathcal{F}$ is commutative, and consists of all real linear combinations of the matrices P_F, for F in $\mathcal{F}$. The subspaces W_F are the mutual eigenspaces of $\mathcal{A}_\mathcal{F}$. For any partition in $\mathcal{F}$ which is uniform, its relation matrix is also in $\mathcal{A}_\mathcal{F}$.

Tjur block structures are used widely in statistics, in two different contexts, which are explained more in Section 6. One concerns covariance, and the other expectation.

The covariance $\mathrm{cov}(Y_\alpha, Y_\beta)$ of responses Y_α and Y_β is defined to be $\mathbb{E}[(Y_\alpha - \mathbb{E}(Y_\alpha))(Y_\beta - \mathbb{E}(Y_\beta))]$. The variance-covariance matrix $\mathrm{Cov}(Y)$ of the random vector Y in equation (2.1) is the $e \times e$ matrix whose entry in row α and column β is $\mathrm{cov}(Y_\alpha, Y_\beta)$. It is often assumed that $\mathrm{Cov}(Y)$ is an unknown matrix in $\mathcal{J}_\mathcal{H}$ for a specified Tjur block structure $\mathcal{H}$ with $\mathcal{H} \subseteq \mathcal{F}$. If the partitions are all uniform then $\mathcal{J}_\mathcal{H} = \mathcal{A}_\mathcal{H}$ and so the eigenspaces of $\mathrm{Cov}(Y)$ are known. Then closure under suprema ensures that there is no pre-determined linear dependence among the eigenvalues, which avoids complications in estimating their values: see [35].

The other use is to give a collection of models for the expectation $\mathbb{E}(Y)$ of Y. It is assumed, as in equations (1.1)–(1.3), that there is a subset $\mathcal{G}$ of $\mathcal{F}$ such that $\mathbb{E}(Y) \in \sum_{G \in \mathcal{G}} V_G$, and we would like to find the smallest such $\mathcal{G}$: see [28]. Closure under suprema is essential for the existence of such a smallest subset.

The set of partitions $\{E, R, C, L, U\}$ in any Latin square forms an orthogonal block structure. Apart from Latin squares, and sets of mutually orthogonal Latin squares, most orthogonal block structures in common use are poset block structures, described in the next subsection.

4.4   Poset block structures

Apart from Großmann's *AutomaticAnova* [87], and the package recently introduced by Bate and Chatfield [42, 43], most statistical software cannot currently recognise the $\prec$ relation, unless the names of the partitions contain a clue. For example, if R, C and L denote the partitions of a Latin square into rows, columns and letters respectively, then standard analysis of variance in the popular statistical software R [175] can recognise that $R \wedge C \preccurlyeq R$ and $R \wedge C \preccurlyeq C$ but not that $R \wedge C \preccurlyeq L$; nor can it recognise that $H \preccurlyeq R$ if H is another name for the partition $R \wedge C$. Each software has its own symbol for the binary operator "$\wedge$", usually something like "." or ":" that is available on standard keyboards. The usual rule seems to be that $A_1 \wedge A_2 \wedge \cdots \wedge A_n$ is recognised to be finer than $B_1 \wedge B_2 \wedge \cdots \wedge B_m$ if and only if $\{B_1, B_2, \ldots, B_m\}$ is a proper subset of $\{A_1, A_2, \ldots, A_n\}$. This rule is true for so-called poset block structures when their partitions are named canonically.

The definition of poset block structures needs another partial order, which I shall write as $\sqsubseteq$. If $(\mathcal{P}, \sqsubseteq)$ is a partially ordered set (*poset* for short), a subset $\mathcal{Q}$ of $\mathcal{P}$ is defined to be *ancestral*, or an *up-set*, if whenever $i \in \mathcal{Q}$ and $j \in \mathcal{P}$ with $i \sqsubset j$ then $j \in \mathcal{Q}$.

Definition Let $\mathcal{P} = \{1, \ldots, s\}$ be a finite set with a partial order $\sqsubseteq$. For $i = 1$, …, s, let Ω_i be a finite set of size n_i, where $n_i \geq 2$. Put $\Omega = \Omega_1 \times \Omega_2 \times \cdots \times \Omega_s$. Let F_i be the partition of Ω defined by the values of the i-th coordinate, for $i = 1$, …, s. If $\mathcal{Q} \subseteq \mathcal{P}$, define the partition $F_{\mathcal{Q}}$ of Ω by $F_{\mathcal{Q}} = \bigwedge_{i \in \mathcal{Q}} F_i$. The *poset block structure* on Ω defined by $(\mathcal{P}, \sqsubseteq)$ is $\{F_{\mathcal{Q}} : \mathcal{Q}$ is an ancestral subset of $\mathcal{P}\}$.

Example 4.4 If $\mathcal{P} = \{1, 2, 3\}$ with $1 \sqsupset 2$ and $1 \sqsupset 3$ then the corresponding poset block structure may be visualized as n_1 rectangles (parts of $F_{\{1\}}$) each defined by n_2 rows (parts of $F_{\{1,2\}}$) and n_3 columns (parts of $F_{\{1,3\}}$). Figure 6 shows an example with $n_1 = 2$, $n_2 = 3$ and $n_3 = 4$.

Example 4.5 Extend Example 1.4 so that there are 14 people, of whom seven are men and seven are women. If we ignore the partition into exercise regimes, we have the poset block structure defined by $\{1, 2, 3\}$ with $n_1 = 2$, $n_2 = 7$, $n_3 = 4$ and $2 \sqsubset 1$. The parts of $F_{\{1\}}$ are the genders; the parts of $F_{\{1,2\}}$ are the people; the parts of $F_{\{3\}}$ are the months; each part of $F_{\{1,3\}}$ is one gender for one month; and the parts of $F_{\{1,2,3\}}$ are the units.

Theorem 4.6 *The following statements hold for any poset block structure defined by a poset $(\mathcal{P}, \sqsubseteq)$.*

(i) *If $\mathcal{Q}$ is an ancestral subset of $\mathcal{P}$, then $F_{\mathcal{Q}}$ is uniform, with all parts of size $\prod_{i \in \mathcal{P} \setminus \mathcal{Q}} n_i$.*

(ii) *The subsets $\emptyset$ and $\mathcal{P}$ are both ancestral. Moreover, $F_{\emptyset} = U$ and $F_{\mathcal{P}} = E$.*

(iii) If Q and R are both ancestral subsets of P, then $F_Q \perp F_R$, $F_Q \vee F_R = F_{Q \cap R}$ and $F_Q \wedge F_R = F_{Q \cup R}$.

It follows that every poset block structure is an orthogonal block structure. Latin squares show that the converse is not true.

Poset block structures were investigated extensively (but not so named) by Yates [226] and later by Kempthorne and his colleagues [107, 207, 234], but these authors did not manage to completely distinguish between the two partial orders involved. By restricting himself to series-parallel posets, Nelder was able to provide recursive definitions and constructions in [126] for what he called *simple orthogonal block structures*. This approach led to algorithms that underlie many different programs used today for the analysis of variance. Speed and Bailey pointed out in [199, 200] that Nelder's approach can be used for arbitrary finite posets. Further details and examples are in [23, 26, 30].

5 Balance

In this section we denote by B and L two partitions of Ω whose parts will be called *blocks* and *letters* respectively.

Definition The relationship between L and B is *binary* if $L \wedge B = E$; this means that each letter occurs at most once in each block. It is *generalized binary* if no two intersections of a part of L with a part of B differ in size by more than one.

Confusion alert! The first of these really is a binary relation whose name is 'binary'.

An $n \times n$ matrix is called *completely symmetric* if it is a linear combination of I_n and J_{nn}.

5.1 Combinatorial notions of balance

Let i and j be two letters, not necessarily distinct. The number of ordered pairs (ω_1, ω_2) in $\Omega \times \Omega$ with the properties that $B(\omega_1) = B(\omega_2)$, $L(\omega_1) = i$ and $L(\omega_2) = j$ is equal to the (i, j)-entry of $N_{LB}N_{BL}$. It is called the *concurrence* of i and j in blocks.

The classical definition of balance, given by Yates in [227], follows.

Definition If the partition B is uniform and $L \wedge B = E$ then the partition L is *balanced* with respect to B if the off-diagonal elements of $N_{LB}N_{BL}$ are all the same but not zero.

For a binary design, a counting argument shows that if L is balanced with respect to B then L is also uniform and hence that $N_{LB}N_{BL}$ is completely symmetric. This definition of balance includes as a special case complete-block designs. In these, L is uniform, $n_B n_L = e$, $N_{LB}N_{BL} = n_B J_{n_L n_L}$ and $L \perp B$. For all other designs which are balanced according to this definition, the coefficient of I_{n_L} in $N_{LB}N_{BL}$ is non-zero and therefore $N_{LB}N_{BL}$ has rank n_L.

Fisher proved his famous inequality in [79]: if $L \wedge B = E$, B is uniform, and L is balanced with respect to B but not orthogonal to B, then $n_L \le n_B$. There are now many proofs of this result. One of the simplest is the observation that the rank of $N_{LB}N_{BL}$ cannot be greater than the number of columns of N_{LB}. Conversely, if $N_{LB}N_{BL}$ has rank n_L and $n_L = n_B$, it follows that N_{LB} is invertible: hence if $N_{LB}N_{BL}$ is completely symmetric then so is $N_{BL}N_{LB}$ and therefore B is also balanced with respect to L. See [52, Chapter 1] and [204, Chapter 2].

How should this definition be generalized if B is not uniform or the relationship between L and B is not binary? Relaxing the uniformity of B gives *pairwise balanced designs*, introduced in [110]. Now a counting argument shows that the entries on the diagonal of $N_{LB}N_{BL}$ are all strictly bigger than the common off-diagonal entry, and so $N_{LB}N_{BL}$ is positive definite; therefore it has rank n_L, and Fisher's inequality follows as before. As [225] shows, pairwise balanced designs have been a very fruitful field of research, which includes results about the existence of BIBDs. However, as we show in the next subsection, this notion of balance does not match what is needed from the statistical point of view.

5.2   Statistical notions of balance

The vector form of equation (1.1) is $Y = X_B\beta + X_L\tau + \varepsilon$. To estimate the vector τ up to an additive constant, it is necessary to project the data vector onto the subspace $(V_L + V_B) \cap V_B^\perp$. The $n_L \times n_L$ *information matrix* C_{LB} is defined by $C_{LB} = X_L^\top(I_e - P_B)X_L$. Note that $I_e - P_B$ is the matrix of orthogonal projection onto $V_B^\perp$. Also, if B is uniform then $X_L^\top P_B X_L = k_B^{-1} N_{LB}N_{BL}$.

The matrix C_{LB} is symmetric, with row-sums zero, so it is singular. If $B \preccurlyeq L$ then it is impossible to estimate any difference $\tau_i - \tau_j$. In this case, $C_{LB} = 0$, and the block design is not considered to be balanced. If C_{LB} has rank $n_L - 1$ then all differences $\tau_i - \tau_j$ can be estimated and C_{LB} has a Moore–Penrose generalized inverse C_{LB}^-. Under the assumption that $\mathrm{Cov}(Y) = \sigma^2 I_e$, standard linear model theory shows that the variance of the estimator of $\tau_i - \tau_j$ is

$$(C_{LB}^-(i,i) + C_{LB}^-(j,j) - C_{LB}^-(i,j) - C_{LB}^-(j,i))\sigma^2.$$

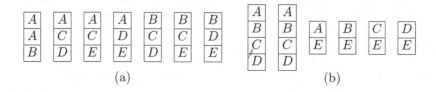

(a) (b)

Figure 9: Two variance-balanced block designs: columns represent blocks

See [33] for further explanation for cominatorialists. Thus a block design is called *variance-balanced* if C_{LB} is completely symmetric but not zero: this terminology was introduced by Tocher in [210]. In this case both C_{LB} and C_{LB}^- are scalar multiples of $n_L I_{n_L} - J_{n_L n_L}$.

Figure 9 shows two block designs which are variance-balanced but are not BIBDs. In the design in Figure 9(a), taken from [33, 210], B is uniform but the relationship between L and B is not binary, even though $k_B = 3 < 5 = n_L$. In the design in Figure 9(b), B is not uniform.

If a block design is variance-balanced then the off-diagonal entries of $X_L^\top P_B X_L$ are all equal. If L is not orthogonal to B, a counting argument similar to that used for pairwise balanced designs shows that every diagonal entry is strictly bigger than this common value, and so $X_L^\top P_B X_L$ has rank n_L. Since P_B has rank n_B, this shows that $n_B \geq n_L$, so that Fisher's inequality holds for variance-balanced designs. If, in a variance-balanced design, B is also uniform and $n_B = n_L$, then the argument in Section 5.1 shows that the design obtained by interchanging the roles of B and L is also variance-balanced.

Hedayat and Federer [92] gave examples to show that neither of pairwise balance and variance balance implies the other.

If the experimenter is more interested in estimating some differences of the form $\tau_i - \tau_j$ than others, the experiment may well be designed so that L is not uniform. If there are no blocks, then the information matrix is C_{LU}, defined by $C_{LU} = X_L^\top (I_e - P_U) X_L = X_L^\top X_L - e^{-1} X_L^\top J_{ee} X_L$. In a block design, the *efficiency* for the estimation of $\tau_i - \tau_j$ is the ratio of the variance in an unblocked design, in which each part of L has the same size as it does in the block design, to that in the block design, assuming that the value of σ^2 is unchanged: see [135]. The block design is said to be *efficiency-balanced* if C_{LB} is a scalar multiple of C_{LU}. This concept was introduced by Jones in [105], but not named until later. In the terminology of [100], the partition L has *first order balance* with respect to B.

One easy construction of an efficiency-balanced block design is to take

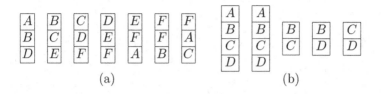

Figure 10: Two efficiency-balanced designs: columns represent blocks

a BIBD and identify two letters. For example, this gives the non-binary design in Figure 10(a). Figure 10(b) shows an binary efficiency-balanced block design where B is not uniform; it is taken from [221].

If L is not orthogonal to B, the information matrices C_{LB} and C_{LU} cannot be equal. Hence, if one is a scalar multiple of the other then $X_L^\top X_L$ is a linear combination of $X_L^\top P_B X_L$ and $X_L^\top P_U X_L$. The diagonal matrix $X_L^\top X_L$ has rank n_L, while the rank of any linear combination of $X_L^\top P_B X_L$ and $X_L^\top P_U X_L$ is bounded above by n_B, so, once again, Fisher's inequality holds.

5.3   Balance between partitions

As Figures 9 and 10 show, block designs which are variance-balanced or efficiency-balanced tend to have either one or both of the partitions L and B being non-uniform. In fact, if L is uniform then variance balance is equivalent to efficiency balance; otherwise, it is impossible for a design to have both properties. On the other hand, adjoining two BIBDs with different block sizes gives a design which is pairwise balanced, variance-balanced and efficiency-balanced but has non-uniform partition B.

These considerations motivate the following definition.

Definition Let L and B be uniform partitions of Ω. Then L is *balanced* with respect to B if $X_L^\top (I_e - P_B) X_L$ is completely symmetric but not zero. It is *strictly balanced* if it is balanced and the relationship between L and B is generalized binary.

The second part of this definition follows [25]. Strict balance is called *balance* by Kiefer, who showed in [108] that strictly balanced block designs are *optimal* in the sense, described more fully in [33, 193], that the average variance of the estimators of differences like $\tau_i - \tau_j$ is minimized. Binary balance is called *total balance* in [98, 134, 145, 147]. If $L \perp B$ and $L \wedge B$ is uniform then L is balanced with respect to B and B is balanced with respect to L.

A	B	C	D	E	F	G
A	A	A	A	A	A	A
B	B	B	B	B	B	B
C	C	C	C	C	C	C
D	D	D	D	D	D	D
E	E	E	E	E	E	E
F	F	F	F	F	F	F
G	G	G	G	G	G	G

(a)

A	B	C	D	E	F	G
A	B	C	D	E	F	G
C	D	E	F	G	A	B
C	D	E	F	G	A	B
D	E	F	G	A	B	C
D	E	F	G	A	B	C
E	F	G	A	B	C	D
E	F	G	A	B	C	D

(b)

Figure 11: Two block designs in which letters are balanced with respect to blocks, which are represented by columns

In general, balance is not a symmetric relation, unlike orthogonality. To emphasize this, here we write $L \blacktriangleright B$ or $B \blacktriangleleft L$ to indicate that L is balanced with respect to B but not strictly orthogonal to B, with $\blacktriangleright$ and $\blacktriangleleft$ replaced by $\triangleright$ and $\triangleleft$ for strict balance. If L and B are both strictly balanced with respect to each other but not orthogonal to each other, we write $L \bowtie B$.

Example 5.1 Suppose that $e = 56$ and $n_B = n_L = 7$. Figure 11 shows two block designs in which $L \blacktriangleright B$. Neither is binary. The one in Figure 11(a) has strict balance, but the one in Figure 11(b) does not.

The results in Section 5.2 show that if $L \blacktriangleright B$ then $n_L \leq n_B$, and that if, in addition, $n_L = n_B$, then $B \blacktriangleright L$.

6   Linear Models

6.1   Fixed effects and random effects

As explained briefly in equation (2.1), in an experiment whose design is defined by a set $\mathcal{F}$ of partitions, it is usually assumed that the response data, which form a vector in $\mathbf{R}^e$, give a realization of a random vector Y which satisfies

$$Y = \sum_{F \in \mathcal{F}} X_F \psi_F + \varepsilon. \tag{6.1}$$

Here ε, which can be regarded as ψ_E, is a random vector of length e with zero mean, and $\mathrm{Cov}(\varepsilon) = \sigma^2 I_e$. For some partitions F we assume that

ψ_F is a real vector of length n_F. Then F is said to have *fixed effects*. The other possibility is that ψ_F is a random vector with zero mean and $\text{Cov}(\psi_F) = \sigma_F^2 I_{n_F}$. Then F is said to have *random effects*.

Equation (6.1), together with the assumptions about fixed and random effects, are called the *linear model* for the data.

When all effects (apart from E) are fixed, the main step in estimating the vector ψ_F (up to an additive constant) is the projection of the data vector onto the subspace

$$\sum_{H \in \mathcal{F}} V_H \cap \left(\sum_{G \in \mathcal{F} \setminus \{F\}} V_G \right)^{\perp}.$$

Section 7 gives more details about the subspaces involved. Sections 8 and 9 discuss combinatorial conditions necessary for the projections to have good properties.

6.2   Partitions related by refinement

In the discussion of equation (1.2) in Section 1, we saw that if $F \prec G$ then there is a potential difficulty in including both $X_F \psi_F$ and $X_G \psi_G$ in the linear model, because $V_G < V_F$. Here is an explanation of how this is handled in three common situations.

6.2.1   Nested block designs   If F and G are both inherent and $F \prec G$ then we have a situation like that in Example 1.3. In the notation used there, $B \prec D$ and the parts of B and D can be thought of as small blocks and large blocks respectively. Some people say that the small blocks are *nested* in the large blocks. A third partition, L, is the one in which the experimenter is actually interested. Put $\tau = \psi_L$.

It is common to assume that L and D have fixed effects while B has random effects, because if B has fixed effects then the relation between L and D is immaterial. Then one simple estimate of τ, up to an additive constant, can be obtained by projecting the data vector onto $V_B^{\perp}$. If L is not orthogonal to B, another can be obtained by projecting the data onto $V_B \cap V_D^{\perp}$. The variances of these estimators are proportional to σ^2 and $\sigma^2 + k_B \sigma_B^2$ respectively. Once the quantities σ^2 and σ_B^2 have been estimated from the data, an appropriate linear combination of these two estimates of τ gives a better estimate: see [24, 128].

In a variant of this, D is also assumed to have random effects. If L is not orthogonal to D, a third estimate of τ is obtained by projecting the data onto V_D, and then all three estimates are combined.

6.2.2 Split-plot designs There are some circumstances where F is inherent and G is not, but practical constraints force $F \prec G$. In Example 1.6, the gardeners might object that it is too cumbersome for any of them to use more than one watering regime. If the parts of B and R are gardens and watering regimes respectively, then B is inherent, R is not, and $B \prec R$. Now we assume that B has random effects and R has fixed effects. The vector ψ_R is estimated from the projection of the data onto V_B.

To understand the common name for these designs, rename the gardens as plots. There is only one watering regime on each plot but, in Example 1.6, each plot is split up into three patches, and different lettuce varieties are grown on each patch.

6.2.3 Simpler models If F is allocated by the experimenter and G is an innate grouping of the parts of F, then there is no avoiding the relation $F \prec G$ even when F and G both have fixed effects. For example, suppose that the parts of F are the exercise regimes in Example 1.5. If five of these have individual exercise while five involve activity with other people, this gives a partition G which groups the parts of F into two groups of five.

Now the linear model which includes $X_G\psi_G$ but not $X_F\psi_F$ is a submodel of the one that includes $X_F\psi_F$ but not $X_G\psi_G$. Using the former gives an estimate of ψ_G from the data, while using the latter gives an estimate of ψ_F. Because $V_G < V_F$, the projection of the data vector onto V_F is no further from the original data than the projection onto V_G, and so it is clear that the model which includes $X_F\psi_F$ provides a better fit to the data. However, the improvement might be no more than could easily happen by chance. The subtle statistical business of *hypothesis testing* addresses the question "Can we attribute the different effects of different exercise regimes to the simple distinction between communal activity and solo activity?"

7 Subspaces

In this section we examine the subspaces derived from two or more subspaces of a real vector space with an inner product. The theory applies to subspaces of any sort, but we shall present it for subspaces defined by partitions as in Section 2.

7.1 Two subspaces

In [100], James and Wilkinson examined the further subspaces defined by V_F and V_G for any pair of partitions F and G of Ω, building on the algebra in [99]. The subscript notation that follows is my own responsibility. Put $V_{F-G} = V_F \cap V_G^\perp$. Since $V_{F\lor G} = V_F \cap V_G$ by Proposition 3.2, the subspaces V_{F-G} and $V_{F\lor G}$ are orthogonal to each other, and both are subspaces of V_F. Put $V_{FG} = V_F \cap (V_{F-G})^\perp \cap V_{F\lor G}^\perp$. Denote by $V_{F\vdash G}$ the image of the projection of V_F onto $V_G^\perp$, which is $(V_F + V_G) \cap V_G^\perp$. Define V_{G-F}, V_{GF} and $V_{G\vdash F}$ analogously.

If $\mathbb{E}(Y) = X_F\psi_F + X_G\psi_G$ then, in order to estimate the vector ψ_F (up to addition of a vector in $V_{F\lor G}$), it is necessary to project the data onto $V_{F\vdash G}$.

Put $Q_F = P_F - P_{F\lor G}$, which is the matrix of orthogonal projection onto $V_F \cap V_{F\lor G}^\perp$, and $Q_G = P_G - P_{F\lor G}$. The following results are in [25, 99, 100].

Theorem 7.1   *(i) V_F is the orthogonal direct sum of $V_{F\lor G}$, V_{F-G} and V_{FG}. Hence $n_F = \dim(V_F) = n_{F\lor G} + \dim(V_{F-G}) + \dim(V_{FG})$.*

(ii) The column space of $Q_F Q_G$ is V_{FG} and the column space of $Q_G Q_F$ is V_{GF}. Hence $\dim(V_{FG}) = \dim(V_{GF})$.

(iii) If $F \perp G$ then the subspaces V_{FG} and V_{GF} are both zero.

(iv) Let x be an eigenvector of $Q_F Q_G Q_F$ with non-zero eigenvalue λ, and let θ be the angle between x and $Q_G x$. Then $\cos^2 \theta = \lambda$. Moreover, the projection of x onto $V_{F\vdash G}$ is $x - Q_G x$, the angle between this and x is $\pi/2 - \theta$, and x is an eigenvector of $Q_F(I_e - Q_G)Q_F$ with eigenvalue $1 - \lambda$.

(v) If F is balanced with respect to G then $F \lor G = U$ and every vector in $V_F \cap V_U^\perp$ is an eigenvector of $Q_F Q_G Q_F$. Hence either $F\perp G$ and $V_{F-G} = V_F \cap V_U^\perp$ or F is not orthogonal to G and $V_{F-G} = \{0\}$. In the second case, the unique eigenvalue λ is in $(0,1)$, $Q_F Q_G Q_F = \lambda Q_F$, and the matrix of orthogonal projection onto $(V_F + V_G) \cap V_U^\perp$ is

$$Q_G + (1-\lambda)^{-1}(Q_F - Q_G Q_F - Q_F Q_G + Q_G Q_F Q_G), \qquad (7.1)$$

which simplifies to

$$(1-\lambda)^{-1}(Q_F + Q_G - Q_F Q_G - Q_G Q_F) \qquad (7.2)$$

if $n_F = n_G$; moreover, if $F \land G = E$ then $\lambda = (n_F - k_G)/[(n_F-1)k_G]$.

(vi) The column space of $(I_e - Q_G)Q_F$ is $V_{F\vdash G}$.

(vii) $V_{F-G} \le V_{F\vdash G}$, and the orthogonal complement of V_{F-G} in $V_{F\vdash G}$ is $(V_{FG} + V_{GF}) \cap V_G^\perp$, which has dimension equal to $\dim(V_{FG})$.

Parts (i), (ii) and (v) give yet another proof of Fisher's inequality. If $F \blacktriangleright G$ then $n_F - 1 = \dim(V_{FG}) = \dim(V_{GF}) \leq n_G - 1$.

7.2   Three subspaces but no refinement relation

When there are three or more partitions, the statistical issues are more affected by which are inherent and which are of interest. This has been discussed in Section 6.2 for the case that one partition is finer than another. Here we assume that the partitions are R, C and L, with no relation of refinement among them. For simplicity of exposition, assume that $R \vee C = L \vee R = L \vee C = U$, so that $V_R \cap V_C = V_L \cap V_R = V_L \cap V_C = V_U$. Then the design in Figure 4 provides a working example.

Now Equation (6.1) becomes

$$Y = X_R \alpha + X_C \phi + X_L \tau + \varepsilon. \tag{7.3}$$

Put $Q_R = P_R - P_U$, $Q_C = P_C - P_U$ and $Q_L = P_L - P_U$, which are the matrices of orthogonal projection onto $V_R \cap V_U^{\perp}$, $V_C \cap V_U^{\perp}$ and $V_L \cap V_U^{\perp}$.

7.2.1   Row–column designs   In the most common use of such a design in experiments, rows and columns are inherent, with $R \wedge C = E$ and $R \perp C$. The experimenter chooses the partition into letters and wants to estimate τ up to an additive constant.

If rows, columns and letters all have fixed effects and we want to estimate τ, then we have to project the data onto $(V_R + V_C)^{\perp}$. Since $R \perp C$, the matrix of this projection is $I_e - Q_R - Q_C - P_U$. If $L \blacktriangleright R$ and $L \blacktriangleright C$ then $V_{LR} = V_{LC} = V_L \cap V_U^{\perp}$ and $Q_L(I_e - Q_R - Q_C - P_U)Q_L$ is a scalar multiple of Q_L. Unless the scalar is zero, this implies that this design has variance balance for the estimation of τ.

On the other hand, if rows and columns both have random effects then further estimates of τ can be obtained by projecting the data onto $V_R \cap V_U^{\perp}$ and $V_C \cap V_U^{\perp}$. If $L \blacktriangleright R$ and $L \blacktriangleright C$ then, again, $Q_L Q_R Q_L$ and $Q_L Q_C Q_L$ are both non-zero scalar multiples of Q_L, and so the combined estimates of τ still have variance balance.

In this case, $Q_L Q_R Q_L$ commutes with $Q_L Q_C Q_L$. This property is called *general balance* by Nelder in [127]. There is not room here for a full discussion of general balance, but I will mention another special case that is very useful if it is not possible for L to be orthogonal to or balanced with respect to both R and C, for example if $n_L > n_R$ and $n_L > n_C$.

Suppose that $V_{LR} \perp V_{LC}$. If $x \in V_{LR}$ then $x \in V_C^{\perp}$ and so the linear combination $x^{\top} X_L \tau$ can be estimated in $(V_R + V_C)^{\perp}$ and in $V_R \cap V_U^{\perp}$ but not in $V_C \cap V_U^{\perp}$. There is an analogous conclusion if $x \in V_{LR}$. On the other hand, if $x \in V_L \cap (V_U + V_{LR} + V_{LC})^\perp$ then $x^\top X_L \tau$ is estimated only in $(V_R + V_C)^\perp$. Thus each linear combination is estimated by combining at most two simple estimates, which is sometimes considered an advantage. This property is discussed further in Section 8.

7.2.2 Block designs for two non-interacting sets of treatments In the other main use of a design like the one in Figure 4, the partition into letters is inherent (and letters are called 'blocks'), the experimenter chooses the other two partitions and wants to estimate α and ϕ up to additive constants. It is desirable that $R \perp C$, so that every part of R occurs with every part of C equally often.

Usually it is assumed that R, C and L all have fixed effects. If we ignore C, a simplistic method of estimating α is to project the data onto $V_{R \vdash L}$. If this subspace is not orthogonal to $V_{C \vdash L}$ then this estimate of α is contaminated by the actual value of ϕ. Thus it is desirable for these two subspaces to be orthogonal to each other. The formal definition is given in the next section.

8 Adjusted orthogonality

8.1 Definition and results

The definition of orthogonality in Section 4 seems surprising at first, because the vector subspaces V_F and V_G corresponding to two partitions F and G can never be orthogonal to each other. If F is strictly orthogonal to G then it is the projections of V_F and V_G onto the orthogonal complement of V_U that are orthogonal to each other; equivalently, $X_F^\top (I_e - P_U) X_G = 0$. It is tempting to say that F and G are orthogonal to each other after adjusting for U. More generally, F is orthogonal to G if and only if $X_F^\top (I_e - P_{F \vee G}) X_G = 0$, which could be considered to be orthogonality after adjusting for $F \vee G$.

These comments lead to the notion of two partitions having adjusted orthogonality with respect to a third partition. For continuity with Section 7, we define what it means for partitions R and C to have adjusted orthogonality with respect to partition L.

Definition Partitions R and C have *adjusted orthogonality* with respect to partition L if $X_R^\top (I_e - P_L) X_C = 0$.

Lemma 8.1 *Partitions R and C have adjusted orthogonality with respect to partition L if and only if*

$$N_{RL}(X_L^\top X_L)^{-1} N_{LC} = N_{RC}. \tag{8.1}$$

If L is uniform, this is equivalent to

$$N_{RL}N_{LC} = k_L N_{RC}. \tag{8.2}$$

Theorem 8.2 *Partitions R and C have adjusted orthogonality with respect to partition L if and only if*

$$Q_R Q_L Q_C = Q_R Q_C, \tag{8.3}$$

where $Q_F = P_F - P_U$ for F in $\{R, C, L\}$.

Proof Since $P_F P_U = P_U P_F = P_U$ for every partition F, equation (8.3) is equivalent to

$$P_R P_L P_C = P_R P_C. \tag{8.4}$$

Pre-multiplying both sides of this by $X_R^\top$ and post-multiplying both sides by X_C gives $X_R^\top P_L X_C = X_R^\top X_C$. Conversely, pre-multiplying both sides of equation (8.1) by $X_R(X_R^\top X_R)^{-1}$ and post-multiplying both sides by $(X_C^\top X_C)^{-1} X_C^\top$ gives equation (8.4). $\qquad\square$

Note that this result does not require any of R, C and L to be uniform, nor does it need orthogonality between R and C.

If $R \perp C$ and $R \wedge C$ is uniform then the entries in N_{RC} are all the same. Then condition (8.2) becomes

$$N_{RL}N_{LC} \text{ is a scalar multiple of } J_{n_R n_C}. \tag{8.5}$$

This has a clean combinatorial interpretation: the subset of letters in any row has a constant number of letters in common with every column (as usual, this needs a more precise explanation if the 'subset' is actually a multiset).

In the design in Figure 4, every row has three letters in common with every column. Therefore rows and columns have adjusted orthogonality with respect to letters.

Example 8.3 The design in Figure 12 is taken from [157], rearranged to show the partitions R and C as rows and columns. Here $R \wedge C$ is not uniform, because some parts of $R \wedge C$ have size one and others have size two. Moreover, R is not orthogonal to C, because $N_{RC} = I_5 + J_5$ when the rows and columns are labelled in the obvious way. However, $N_{RL}N_{LC} = 3N_{RC}$, so rows and columns have adjusted orthogonality with respect to letters even though rows are not orthogonal to columns. Expressed in another way, row i has six letters in common with column j if $i = j$ but only three letters in common if $i \neq j$.

A	F	D	G	J	C
D	B	G	E	H	F
G		E	C	H	A	I
J		H	A	D	I	B
C		F	I	B	E	J

Figure 12: A design for 30 units, with partitions into rows, columns and letters

8.2 Link with subspaces

Some observations from Section 7 are gathered here.

Proposition 8.4 *Let R, C and L be partitions of Ω. Put $Q_F = P_F - P_U$ for F in $\{R, C, L\}$. If $R \vee C = L \vee R = L \vee C = U$ then the following hold.*

(i) $V_{LR} \perp V_{LC}$ *if and only if* $Q_R Q_L Q_C = 0$.

(ii) *If* $V_{LR} \perp V_{LC}$ *then* $\dim(V_{LR}) + \dim(V_{LC}) \le n_L - 1$.

(iii) *If* $R \blacktriangleright L$, $C \blacktriangleright L$ *and* $V_{LR} \perp V_{LC}$ *then* $n_R + n_C - 1 \le n_L$.

(iv) $V_{R \vdash L} \perp V_{C \vdash L}$ *if and only if* $Q_R(I_e - Q_L)Q_C = 0$.

(v) *If* $R \perp C$ *then* $Q_R Q_C = 0$ *and therefore* $V_{LR} \perp V_{LC}$ *if and only if* $V_{R \vdash L} \perp V_{C \vdash L}$.

This shows that an alternative characterization of R and C having adjusted orthogonality with respect to L is that the subspaces $V_{R \vdash L}$ and $V_{C \vdash L}$ are orthogonal to each other. Moreover, if, in addition, $R \perp C$, then the subspaces V_{LR} and V_{LC} are orthogonal to each other. Then part (iii) gives the following result.

Theorem 8.5 *If* $R \perp C$, $R \blacktriangleright L$, $C \blacktriangleright L$, *and* R *and* C *have adjusted orthogonality with respect to* L *then* $n_R + n_C - 1 \le n_L$.

8.3 A little history

It seems that Potthoff [143] was the first to notice the importance of conditions like (8.2) for experiments with three partitions under the assumption of model (7.3). Preece also gave condition (8.2) in [145, 147] in the situation where $R \perp C$, $L \rhd R$ and $L \rhd C$. Since they both assumed that $R \perp C$ and $R \wedge C = E$, they stated the condition in the form (8.5).

Preece spent the year 1974–1975 in Australia: see [29]. At the end of his stay, he presented work on these designs at the Australian Conference on Combinatorial Mathematics in Adelaide: see [151]. This led Sterling and Wormald to give some constructions for such designs in [203] and Seberry and Street to take the ideas further in [189, 205].

Meanwhile, Eccleston and Russell had independently invented the idea of adjusted orthogonality, which they wrote as $R(L) \perp C(L)$ in [72], where they proved Lemma 8.1. They introduced the name 'adjusted orthogonality' in [73]. Papers [12, 19, 68, 69, 70, 74, 102, 186, 191, 192] followed, and the book [193] also had a section on adjusted orthogonality, but none of these mentioned the work of Preece.

Preece, Eccleston and Russell were all working in Statistics at universities in Sydney during the last four months of 1974. They met for discussions during this time, and Preece was external examiner for Russell's 1977 PhD thesis [185], whose Chapter 4 was devoted to adjusted orthogonality. Eccleston and Russell both report on Preece's very thorough reading of this[1]. Nonetheless, when Preece cited [72] in his 1977 paper [154] it was only to say that this concept of orthogonality was not related to anyone else's. However, his article [157] for the *Genstat Newsletter* did use the phrase 'adjusted orthogonality', and explained it very clearly in the context of Example 8.3. I have not found an instance of his using the phrase 'adjusted orthogonality' in the mainstream literature before [164].

Condition (8.5) was also used, but not named, by Raghavarao and co-authors in [178, 180], with no mention of Preece or any of the literature on adjusted orthogonality. As we show in Section 13, this condition also played an important role in the series of papers [3, 4, 5, 6] by Agrawal in 1966. This mulitple introduction of the concept in the 1960s and 1970s shows how active was research at this time into the designs for the situations described in Section 7.2.

Bagchi listed authors who had constructed designs with adjusted orthogonality in [17], again with no mention of the many designs given by Preece in [147], and proved Theorem 8.5 in [18].

Eccleston and McGilchrist extended the ideas of [100] to three subspaces in [71] and applied their results to row–column designs. In particular, they proved that the average variance of estimators of differences like $\tau_i - \tau_j$ is bounded below by a known function of the average variances in the two block designs obtained when one of the partitions into rows and columns is ignored, and that this bound is achieved if rows and columns have adjusted orthogonality with respect to letters. Bagchi and Shah generalized this to a stronger notion of optimality in [21], but with no mention

---

[1]Personal communications from JAE and KGR.

of [71].

Independently of Eccleston, Russell, and their co-authors, but building on the work of Preece in [145, 147, 151], Morgan and Uddin in [123] defined a block design for two non-interacting treatment factors F and G to be an *orthogonal BIBD* if $F \rhd B$, $G \rhd B$ and $k_B N_{FG} = N_{FB} N_{BG}$; this last condition is precisely adjusted orthogonality. The terminology OBIBD is also used in [1, 85, 120]; the first two of these state that the third condition is the same as adjusted orthogonality. Rees presented [1] at the British Combinatorial Conference in Sussex (2001).

8.4   Adjusting for more than one partition

In [72], Eccleston and Russell proposed a more general version of adjusted orthogonality. If $\mathcal{L}$ is a set of partitions of Ω, put $V_{\mathcal{L}} = \sum_{L \in \mathcal{L}} V_L$, and let $P_{\mathcal{L}}$ be the matrix of orthogonal projection onto $V_{\mathcal{L}}$. By convention, $P_{\emptyset} = P_{\{U\}} = P_U = e^{-1} J_{ee}$. In the notation of [72], $R(\mathcal{L}) \perp C(\mathcal{L})$ if $(V_R + V_{\mathcal{L}}) \cap V_{\mathcal{L}}^{\perp}$ is orthogonal to $(V_C + V_{\mathcal{L}}) \cap V_{\mathcal{L}}^{\perp}$; equivalently,

$$X_R^{\top}(I_e - P_{\mathcal{L}})X_C = 0.$$

In words, R and C have adjusted orthogonality with respect to $\mathcal{L}$.

In an important special case of this, $\mathcal{F}$ is a Tjur block structure and $\mathcal{L} = \mathcal{F} \setminus \{E\}$. Then R and C have adjusted orthogonality with respect to $\mathcal{L}$ if and only if the projections of V_R and V_C onto W_E are orthogonal to each other, where W_E is the subspace defined in Section 4.3.

9   Adjusted balance

9.1   Terminology

Section 8 discussed adjusted orthogonality, which is a possible relation between partitions F and G when everything is projected onto the orthogonal complement $V_H^{\perp}$ of V_H for some other partition H. What happens when $F = G$? In this case, for clarity, we write $F = G = L$ and $H = B$.

Recall from Section 2 that $X_L^{\top} X_L$ is a diagonal matrix whose diagonal entries are the sizes of the parts of L. Thus L is uniform if and only if $X_L^{\top} X_L$ is completely symmetric. Some authors say that L is balanced. Following on from Section 8, it would be natural to say that L has adjusted balance with respect to B if $X_L^{\top}(I_e - P_B)X_L$ is completely symmetric. This is always true when $n_L \leq 2$, because the $n_L \times n_L$ matrix $X_L^{\top}(I_e - P_B)X_L$ is symmetric and has zero row-sums.

The requirement that $X_L^{\top}(I_e - P_B)X_L$ be completely symmetric is the main part of the definition of balance in Section 5.3. For consistency

A	B	H	F		B	C	I	G		C	D	A	H		D	E	B	I		E	F	C	A
B	H	F	A		C	I	G	B		D	A	H	C		E	B	I	D		F	C	A	E

F	G	D	B		G	H	E	C		H	I	F	D		I	A	G	E
G	D	B	F		H	E	C	G		I	F	D	H		A	G	E	I

Figure 13: Design in Example 9.1: there are nine letters in nine blocks, each of which is a 2×4 rectangle

with Section 5, I shall continue to say 'is balanced with respect to' rather than 'has adjusted balance with respect to', but this discussion does show that the ideas in Sections 5 and 8 are closely related. However, there is one twist. In Section 5, we required $X_L^\top (I_e - P_B) X_L$ to be completely symmetric but not zero. Since $X_L^\top X_L$ itself cannot be zero, it is not unreasonable to include 'not zero' in the definition of balance.

This twist shows a difference between orthogonality and balance. If $F \preccurlyeq G$ then $F \perp G$, so we regard refinement as a special case of orthogonality. However, if $F \preccurlyeq G$ then $X_F^\top (I_e - P_G) X_F = 0$, so that F is not balanced with respect to G, as noted in Section 5.2.

9.2 General definition

We can now define balance with respect to a set of partitions in a way that is analogous to the more general definition of adjusted orthogonality in Section 8.4.

Definition Let $\mathcal{G}$ be a set of partitions of Ω, and let L be a partition of Ω. Then L is balanced with respect to $\mathcal{G}$ if $X_L^\top (I_e - P_{\mathcal{G}}) X_L$ is completely symmetric but not zero.

It is immediate that L cannot be balanced with respect to $\mathcal{G}$ if there is any G in $\mathcal{G}$ for which $G \preccurlyeq L$. As in Section 8.4, an important special case occurs when $\mathcal{G} = \mathcal{F} \setminus \{E\}$ for some Tjur block structure $\mathcal{F}$. In this case, L is balanced with respect to $\mathcal{G}$ if it is balanced with respect to G for all in $\mathcal{G}$ and $X_L^\top (I_e - P_{\mathcal{G}}) X_L$ is not zero. The following example shows that it is possible to achieve balance with respect to $\mathcal{G}$ without having balance with respect to all G in $\mathcal{G}$.

Example 9.1 The design in Figure 13 has 72 units, in nine blocks, each of which is a 2×4 rectangle. Nine letters have been allocated to the units.

D	E	F	G	H	I
H	I	G	C	A	B
C	A	B	E	F	D
B	H	E	I	C	F
F	C	I	D	G	A
G	D	A	B	E	H

Figure 14: Row–column design in which letters are balanced with respect to $\{R, C\}$ while rows and columns have adjusted orthogonality with respect to letters

Denote by B, R, C and L the partitions into blocks, rows, columns and letters respectively. Then $P_{R,C,B} = P_R + P_C - P_B$ and so

$$X_L^\top(I_e - P_{R,C,B})X_L = X_L^\top(I_e - P_C)X_L - X_L^\top(P_R - P_B)X_L.$$

In this design, the two rows in each block have exactly the same set of letters, with the result that $X_L^\top(P_R - P_B)X_L = 0$. It follows that $X_L^\top(I_e - P_{R,C,B})X_L = X_L^\top(I_e - P_C)X_L$, which is completely symmetric and nonzero, because $L \rhd C$. Hence L is balanced with respect to $\{R, C, B\}$, even though it is not balanced with respect to either R or B. In fact, in this example $P_{R,C,B} = P_{R,C}$ because $R \prec B$, and so L is also balanced with respect to $\{R, C\}$.

9.3   Balance with respect to a pair of partitions

The most common use of this more general concept of balance is for the case that $\mathcal{G} = \{R, C\}$ and $R \vee C = U$, so that we are interested in the projection of the data onto $(V_R + V_C)^\perp$. Let Q_{RC} be the matrix of orthogonal projection onto $(V_R + V_C) \cap V_U^\perp$; and put $Q_R = P_R - P_U$ and $Q_C = P_C - P_U$. Then L is balanced with respect to the pair $\{R, C\}$ if $X_L^\top(I_e - Q_{RC} - P_U)X_L$ is completely symmetric but not zero.

This terminology is consistent with that in [25], but it is not ideal, because 'L is balanced with respect to R and C' might mean '$L \blacktriangleright R$ and $L \blacktriangleright C$' or it might mean 'L is balanced with respect to $\{R, C\}$'. In [151], Preece calls it 'L has overall total balance with respect to the rest of the design'; in later papers this becomes 'L is fully balanced ...'.

If $R \blacktriangleright C$ then equation (7.1) gives $X_L^\top Q_{RC} X_L$. If, additionally, either $L \perp R$ or $L \perp C$ then $X_L^\top Q_{RC} X_L$ is a scalar multiple of one of $X_L^\top Q_R X_L$ and $X_L^\top Q_C X_L$, so the properties of $X_L^\top Q_{RC} X_L$ follow from those of the binary relations between L and R and between L and C.

If $R \perp C$ then $Q_{RC} = Q_R + Q_C$ and so the properties of $X_L^\top Q_{RC} X_L$ are derivable from those of $X_L^\top Q_R X_L$ and $X_L^\top Q_C X_L$ considered together. It may be possible for their sum to be completely symmetric even though neither is. For example, in a resolvable BIBD in which $n_B = 2k_B$ it may be possible to allocate the letters to the cells of a $k_B \times k_B$ square in such a way that the rows form half of the original blocks and the columns form the others. Figure 14 shows an example with $n_L = 9$, $n_B = 12$ and $k_B = 6$ (the blocks are the complements of those in Figure 1). In this design, it is also true that rows and columns have adjusted orthogonality with respect to letters. Many more examples are given in [113, 134]. Preece found 345 species (that is, merging isomorphism classes obtainable from each other by interchanging rows and columns) of designs with these parameters and properties in [149, 153], while McSorley and Phillips completed the enumeration to 348 by a computer search reported in [117].

9.4   Three-way balance with pairwise balance

Suppose that $C \blacktriangleright R$. Equation (7.1) shows that the condition for L to be balanced with respect to $\{R, C\}$ is that the matrix

$$(1 - \lambda) X_L^\top X_L - (1 - \lambda) X_L^\top P_U X_L - (1 - \lambda) X_L^\top Q_R X_L - X_L^\top Q_C X_L$$
$$+ X_L^\top Q_R Q_C X_L + X_L^\top Q_C Q_R X_L - X_L^\top Q_R Q_C Q_R X_L$$

is completely symmetric but not zero. If $L \blacktriangleright R$ and $L \blacktriangleright C$ then the first four terms are completely symmetric, and so the first part of this condition becomes

$$k_R (N_{LR} N_{RC} N_{CL} + N_{LC} N_{CR} N_{RL}) - N_{LR} N_{RC} N_{CR} N_{RL}$$

$$\text{is completely symmetric.} \quad (9.1)$$

Condition (9.1) is given explicitly in [151].

If, in addition, C and L have adjusted orthogonality with respect to R, then Lemma 8.1 shows that $N_{CR} N_{RL}$ is a scalar multiple of N_{CL}. Therefore the matrix in (9.1) is a multiple of $N_{LC} N_{CL}$, which is completely symmetric because $L \blacktriangleright C$. Thus adjusted orthogonality gives a special case of this type of three-way balance.

In [147], Preece gave 59 designs with three partitions R, C and L satisfying $n_R = n_C < n_L$, $R \triangleright L$, $C \triangleright L$, $R \bowtie C$ and $N_{RL} N_{LC} = k_L N_{RC}$. It follows that R and C have adjusted orthogonality with respect to letters, that R is balanced with respect to $\{C, L\}$, and that C is balanced with respect to $\{R, L\}$. For all these designs, $N_{RC} = J_{n_R n_R} \pm I_{n_R}$. Figures 12 and 15 show examples. Street generalized his constructions in [205] to give infinite families of designs, and widened the scope by relaxing the condition

	L	I	F	J	G	D	A
E		M	J	G	K	A	B
B	F		N	K	A	L	C
M	C	G		H	L	B	D
C	N	D	A		I	M	E
N	D	H	E	B		J	F
K	H	E	I	F	C		G
H	I	J	K	L	M	N	

Figure 15: A design for 56 units, with partitions into rows, columns and letters: blank cells indicate empty row-column intersections

that the relation between R and C is generalized binary. Agrawal and Sharma gave further designs of this type in [10].

On the other hand, if $n_L \leq n_C = n_R$ then equation (7.2) gives the following (ignoring the possibility that $X_L^\top(I_e - Q_{RC} - P_U)X_L$ might be zero).

Proposition 9.2 *Suppose that $R \bowtie C$, $L \blacktriangleright R$ and $L \blacktriangleright C$. Then L is balanced with respect to $\{R, C\}$ if and only if*

$$N_{LR}N_{RC}N_{CL} + N_{LC}N_{CR}N_{RL} \text{ is completely symmetric.} \qquad (9.2)$$

A stronger condition is

$$N_{LR}N_{RC} \text{ is a linear combination of } N_{LC} \text{ and } J_{n_L n_C}. \qquad (9.3)$$

Proofs of the following are in [25].

Proposition 9.3 *If $R \bowtie C$, $L \blacktriangleright R$ and $L \blacktriangleright C$ then the following hold.*

(i) *Condition (9.3) implies condition (9.2).*

(ii) *If $n_L = n_C$ and condition (9.2) is satisfied for the ordered triple (L, R, C) then it is satisfied for any permutation of $\{L, R, C\}$.*

(iii) *If $n_L = n_C$ and condition (9.3) is satisfied for the ordered triple (L, R, C) then it is satisfied for any permutation of $\{L, R, C\}$.*

Figure 16 shows two designs for 28 units with three partitions having seven parts of size four. In both designs, all the pairwise relations between partitions are strict balance in both directions. However, in the design in Figure 16(a), taken from [98], none of these partitions is balanced with

A	B	C		D		
E	A				B	F
	D	A	F			G
		C		E	A	G
B		G	D			E
F			C	G		B
		E		F	D	C

A	E	B		C		
	B	F	C		D	
		C	G	D		E
F			D	A	E	
	G			E	B	F
G		A			F	C
D	A		B			G

(a)   (b)

Figure 16: Two designs for 28 units, with partitions into rows, columns and letters: blank cells indicate empty row-column intersections

respect to the other pair, whereas in the design in Figure 16(b), taken from [145], every one of these partitions is balanced with respect to the other pair.

10   Three partitions

10.1   Supreme sets of uniform partitions

For simplicity, from now on we confine our interest to uniform partitions on Ω. The following definition is taken from [32].

Definition   A set $\mathcal{F}$ of partitions on Ω is *supreme* if $\mathcal{F} \cup \{U\}$ is closed under taking suprema.

We shall examine supreme sets $\mathcal{F}$ of uniform non-trivial partitions of Ω with the property that if F and G are in $\mathcal{F}$ then at least one of the following holds: (i) F is orthogonal to G (this includes $F \prec G$ and $G \prec F$); (ii) at least one of F and G is strictly balanced with respect to the other.

Pearce and co-authors discussed sets of (usually) uniform partitions in [98, 134] and introduced notation for various binary relations between them. Preece augmented the notation in [145] and displayed the relations in a matrix whose rows and columns are labelled by the partitions. The diagonal is empty. If $F \neq G$ then the (F, G)-entry is O (for 'orthogonal') if F is strictly orthogonal to G and $F \wedge G$ is uniform; it is T (for 'total balance') if F is binary balanced with respect to G but not orthogonal to G; it is T' (with the connotation that the transpose indicates the reverse relationship) if F is neither orthogonal nor binary balanced with respect to G but G is binary balanced with respect to F.

For example, denote by R, C and L the partitions into rows, columns and letters in Figures 3 and 4. In this notation, the matrices for these row–column designs are

$$
\begin{array}{c} \\ R \\ C \\ L \end{array}
\begin{array}{ccc} R & C & L \end{array}
\left[\begin{array}{ccc} - & O & O \\ O & - & T \\ O & T & - \end{array} \right]
\quad \text{and} \quad
\begin{array}{c} \\ R \\ C \\ L \end{array}
\begin{array}{ccc} R & C & L \end{array}
\left[\begin{array}{ccc} - & O & T \\ O & - & T \\ T' & T' & - \end{array} \right]
$$

respectively.

In [32], Bailey and Cameron convey the same information by showing each partition as a vertex of a directed graph and labelling the edges with symbols to show the relationships. Here we shall simply use the symbols $\perp$, $\underline{\perp}$, $\prec$, $\succ$, $\lhd$, $\rhd$ and $\bowtie$.

10.2   Two partitions

Suppose that $\mathcal{F} = \{F, G\}$, where F and G are distinct uniform non-trivial partitions. If $\mathcal{F}$ is supreme, then, up to renaming, either $F \prec G$ or $F \vee G = U$. The first case gives a poset block structure with $\mathcal{P} = \{1, 2, 3\}$ and $3 \sqsubset 2 \sqsubset 1$, where $F = F_{\{1,2\}}$ and $G = F_{\{1\}}$.

If $F \vee G = U$ and $F \perp G$ then $F \underline{\perp} G$ and all parts of $F \wedge G$ have the same size, by Proposition 4.1. Then F and G can be regarded as the partitions of a rectangle into rows and columns, with each row-column intersection containing the same number of units. If $F \wedge G = E$ this is the poset block structure defined by $\mathcal{P} = \{1, 2\}$ with trivial partial order.

If $F \vee G = U$ but F is not orthogonal to G then, up to renaming, $F \rhd G$. Altogether, we have these possible structures.

A.1 The poset block structure defined by $\mathcal{P} = \{1, 2, 3\}$ and $3 \sqsubset 2 \sqsubset 1$, with $\mathcal{F} = \{F_{\{1\}}, F_{\{1,2\}}\}$.

A.2 The poset block structure defined by $\mathcal{P} = \{1, 2\}$ with trivial partial order, with $\mathcal{F} = \{F_{\{1\}}, F_{\{2\}}\}$.

A.3 The poset block structure defined by $\mathcal{P} = \{1, 2, 3\}$, $3 \sqsubset 1$ and $3 \sqsubset 2$, with $\mathcal{F} = \{F_{\{1\}}, F_{\{2\}}\}$.

A.4 $F \rhd G$ but G is not balanced with respect to F, so that $n_F < n_G$.

A.5 $F \bowtie G$, so that $n_F = n_G$.

10.3 Three partitions: three orthogonal relations

Suppose that $\mathcal{F} = \{F, G, H\}$, where F, G and H are distinct uniform non-trivial partitions and all the pairwise relations are orthogonality. If $\mathcal{F}$ is supreme but no supremum is U then, up to renaming, either $F \prec G \prec H$ or $F \vee G = H$. The first case gives a poset block structure with $\mathcal{P} = \{1, 2, 3, 4\}$ and $4 \sqsubset 3 \sqsubset 2 \sqsubset 1$, where $F = F_{\{1,2,3\}}$, $G = F_{\{1,2\}}$ and $H = F_{\{1\}}$. In the second case, the parts of H can be considered as rectangles, each of which is partitioned into rows and columns as in structures A.2 or A.3: the first of these gives Example 4.4. Because F, G and H are all uniform, Proposition 4.1 shows that $F \wedge G$ is uniform.

If only one supremum is U, suppose that it is $G \vee H$. Then $F \prec G$ and $F \prec H$, so $F \preccurlyeq G \wedge H$. This gives two cases: $F = G \wedge H$ and $F \prec G \wedge H$. If precisely two suprema are U, suppose that they are $F \vee H$ and $G \vee H$. Then either $F \prec G$ or $G \prec F$, and we obtain a poset block structure like the one in Example 4.5. If all three suprema are U then we have an orthogonal array of strength two. This is not necessarily derived from a poset block structure: for example, F, G and H could be three of the partitions in Figure 7.

This gives the following possible structures.

B.1 The poset block structure defined by $\mathcal{P} = \{1, 2, 3, 4\}$ and $4 \sqsubset 3 \sqsubset 2 \sqsubset 1$, with $\mathcal{F} = \{F_{\{1\}}, F_{\{1,2\}}, F_{\{1,2,3\}}\}$.

B.2 The poset block structure defined by $\mathcal{P} = \{1, 2, 3\}$, $2 \sqsubset 1$ and $3 \sqsubset 1$, with $\mathcal{F} = \{F_{\{1\}}, F_{\{1,2\}}, F_{\{1,3\}}\}$.

B.3 The poset block structure defined by $\mathcal{P} = \{1, 2, 3, 4\}$, $4 \sqsubset 3 \sqsubset 1$ and $4 \sqsubset 2 \sqsubset 1$, with $\mathcal{F} = \{F_{\{1\}}, F_{\{1,2\}}, F_{\{1,3\}}\}$.

B.4 The poset block structure defined by $\mathcal{P} = \{1, 2, 3\}$, $3 \sqsubset 1$ and $3 \sqsubset 2$, with $\mathcal{F} = \{F_{\{1\}}, F_{\{2\}}, F_{\{1,2\}}\}$.

B.5 The poset block structure defined by $\mathcal{P} = \{1, 2, 3, 4\}$, $4 \sqsubset 3 \sqsubset 1$ and $4 \sqsubset 3 \sqsubset 2$, with $\mathcal{F} = \{F_{\{1\}}, F_{\{2\}}, F_{\{1,2,3\}}\}$.

B.6 The poset block structure defined by $\mathcal{P} = \{1, 2, 3\}$ and $2 \sqsubset 1$, with $\mathcal{F} = \{F_{\{1\}}, F_{\{1,2\}}, F_{\{3\}}\}$.

B.7 An orthogonal array of strength two containing three partitions.

10.4 Three partitions: two relations of orthogonality and one of balance

Suppose that $\mathcal{F} = \{F, G, H\}$, where F, G and H are distinct uniform non-trivial partitions, $\mathcal{F}$ is supreme, $F \vartriangleright G$, $F \perp H$ and $G \perp H$. Since

$F \vee G = U$, we cannot have $H \succ F$ and $H \succ G$, because that implies that $H \not\succeq F \vee G$.

If $H \prec F$ and $H \prec G$ then $E \neq H \preceq F \wedge G$ and so the relationship between F and G cannot be binary. If it is generalized binary then the parts of $F \wedge G$ have size differing by one, but this cannot happen, because H is uniform.

Because F and G are not related by $\prec$, H cannot be finer than one and coarser than the other. If $H \perp G$ then every part of H meets every part of G in $k_H k_G / e$ units. If, in addition, $H \prec F$, then every part of F meets every part of G in a constant number of units, so $F \perp G$, contrary to the assumptions. We get a similar contradiction if $H \perp F$ and $H \prec G$.

If $G \prec H$ and $H \perp F$ then the parts of G are grouped into parts of H, each of which has m units in common with every part of F, where $m = k_H k_F / e$. With the parts of F and G considered as letters and blocks respectively, this is called an *m-resolvable* design. When $m = 1$, this is just a resolvable design, as in Example 1.3. Bose proved a generalization of Fisher's inequality in [47]: in such a design, $n_G - n_H \geq n_F - 1$. This can be proved by using parts (i), (ii) and (v) of Theorem 7.1 and noting that $V_{G-F} \geq V_H \cap V_U^\perp$. Hence G cannot be balanced with respect to F.

On the other hand, if $F \prec H$ and $H \perp G$ then a short counting argument shows that the average concurrence (in parts of G) of parts of F within the same part of H is $k_F(k_G - n_H)/(n_F - n_H)$ while the average concurrence between other parts of F is $k_F k_G / n_F$. These cannot be the same unless $n_F = k_G$, which is impossible when $F \rhd G$.

In the remaining case, $H \perp F$ and $H \perp G$. To aid thought, rename H as R, G as C and F as L, so that the parts of R and C are the rows and columns of a rectangle. Either $R \wedge C = E$ or $E \prec R \wedge C$. Moreover, $L \perp R$, $L \rhd C$, and either $L \bowtie C$ or not.

First suppose that $R \wedge C = E$. If $n_L = n_C$ then Hall's Marriage Theorem shows that any strictly balanced block design can have the letters allocated to the units in the rectangle in such a way that the columns are blocks and every letter comes exactly once in each row. When $n_R < n_L$ this is called a *Youden square*. Figure 3 shows an example.

The usage of *generalized Youden design* in [4, 194, 206] relaxes the condition that $n_L = n_C$. There seems little to lose if the condition that $n_R < n_L$ is also relaxed. For example, putting two copies of the design in Figure 3 side by side and putting two Latin squares of order 7 underneath them gives a design with $n_R = 11$, $n_C = 14$ and $n_L = 7$ in which $R \perp C$, $R \perp L$, $R \wedge C = E$ and $L \rhd C$. This is not the only method of construction. For example, start with the design in Figure 1, glue the blocks together to make a 3×12 rectangle, make two further copies by permuting whole rows by a cycle of order 3, and then place all of these side by side to obtain a

A	B	C	D	E	F	G
B	C	D	E	F	G	A
G	A	B	C	D	E	F
C	D	E	F	G	A	B
D	E	F	G	A	B	C
F	G	A	B	C	D	E

A	B	C	D	E	F	G
B	C	D	E	F	G	A
C	D	E	F	G	A	B
D	E	F	G	A	B	C
E	F	G	A	B	C	D
F	G	A	B	C	D	E

(a)   (b)

Figure 17: Two 3×7 row–column designs in which letters and columns are strictly balanced with respect to each other and strictly orthogonal to rows

A	A	A	A	A	B	C	D	E	F	C	D	E	F	B
B	C	D	E	F	C	D	E	F	B	E	F	B	C	D
C	D	E	F	B	A	A	A	A	A	D	E	F	B	C
D	E	F	B	C	E	F	B	C	D	F	B	C	D	E

Figure 18: A 2×15 row–column design in which letters are strictly balanced with respect columns, but not vice versa, and both are strictly orthogonal to rows

design with $n_R = 3$, $n_C = 36$ and $n_L = 9$.

Finally suppose that $E \prec R \wedge C$. Because $R \wedge C$ is not in $\mathcal{F}$, there is no constraint on the relationship between L and $R \wedge C$. For example, in the designs in Figures 17(a) and 18 letters are balanced with respect to $R \wedge C$, whereas in the designs in Figures 17(b) and 19 they are not. As with orthogonal arrays of strength two, we ignore this distinction here. In both designs in Figure 17, and in the design in Figure 19, $L \bowtie C$. This is not true in the design in Figure 18.

Thus we have the following possible structures.

C.1 An m-resolved strictly balanced block design, for $m \geq 1$, with partitions into letters (L), blocks (B) and districts (D) as in Example 1.3. Then $\mathcal{F} = \{L, B, D\}$. See Section 15.

C.2 A Youden square: a row–column design in which $R \wedge C = C \wedge L = E$, letters and columns are both strictly orthogonal to rows and $L \bowtie C$, or the generalization that does not demand that $C \wedge L = E$. Here $\mathcal{F} = \{R, C, L\}$. See Section 11.

C.3 A row–column design in which $R \wedge C = E$, rows are strictly or-

B	A	D	C	F	E	H	G	J	I	L	K	N	M	P	O
E	F	G	H	A	B	C	D	M	N	O	P	I	J	K	L
C	D	A	B	G	H	E	F	K	L	I	J	P	O	N	M
I	J	K	L	M	N	O	P	E	F	G	H	A	B	C	D
D	C	B	A	H	G	F	E	L	K	J	I	O	P	M	N
M	N	O	P	I	J	K	L	A	B	C	D	E	F	G	H

Figure 19: A 3×16 row–column design in which letters and columns are strictly balanced with respect to each other and strictly orthogonal to rows

thogonal to both columns and letters, $L \rhd C$ and $n_C > n_L$. Here $\mathcal{F} = \{R, C, L\}$.

C.4 A row–column design in which $R \wedge C \neq E$, rows are strictly orthogonal to both columns and letters, and $L \bowtie C$. Here $\mathcal{F} = \{R, C, L\}$.

C.5 A row–column design in which $R \wedge C \neq E$, rows are strictly orthogonal to both columns and letters, $L \rhd C$ and $n_C > n_L$. Here $\mathcal{F} = \{R, C, L\}$.

10.5 Three partitions: one relation of orthogonality and two of balance

Suppose that $\mathcal{F} = \{F, G, H\}$ where $F \perp G$ and the relation between H and each of the others is non-orthogonal strict balance in at least one direction. If $\mathcal{F}$ is supreme then, up to renaming, either $F \prec G$ or $F \perp G$.

If $F \prec G$ it is convenient to rename F, G and H as B, D and L respectively, as in Example 1.3. If L is strictly balanced with respect to both B and D then $\mathcal{F}$ is a *nested balanced block design*. If, in addition, $L \wedge D = E$, then $\mathcal{F}$ is a *nested BIBD*: see Section 12. If $L \wedge D \neq E$ then $L \wedge B$ may or may not be E. Possible linear models, and consequent methods of estimation, are given in Section 6.2.1.

In this case, Fisher's inequality gives $n_B > n_D \geq n_L$. Thus it is possible to have $L \bowtie D$ but not to have $L \bowtie B$.

On the other hand, if $n_L \geq n_B > n_D$ then it may be possible to have both $B \rhd L$ and $D \rhd L$. Figure 20 shows an example. Then every pair of blocks concur in at least one part of L, and so $L \wedge D \neq E$. In Figure 20, $n_B = n_L$ and so $B \bowtie L$. If each letter in Figure 20 is replaced by two letters, so that $k_B = 6$ and $n_L = 8$, then we have an example with $n_L > n_B > n_D$.

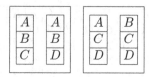

Figure 20: A nested block design in which blocks (shown as columns) and districts (shown as rectangles) are both strictly balanced with respect to letters

A	A	B	B	C		A	A	A	B	D
B	E	D	C	D		B	C	C	C	E
F	F	E	E	F		D	D	E	F	F

Figure 21: A nested block design in which districts (shown as rectangles) are strictly balanced with respect to letters and letters are strictly balanced with respect to blocks (shown as columns)

The third possibility is $n_B > n_L > n_D$ with $D \triangleright L \triangleright B$. Figure 21 shows such a design with $n_D = 2$, $n_L = 6$ and $n_B = 10$.

If $F \perp G$ then, for clarity, rename F, G and H as R, C and L so that the parts of R and C can be visualized as the rows and columns of a rectangle. It may or may not be the case that $R \wedge C = E$.

Since $k_R = n_C k_{R \wedge C}$, if n_L divides n_C then the relationship between L and R cannot be generalized binary unless $L \perp R$, which is contrary to our assumptions. Hence $n_L \neq n_C$. Similarly, $n_L \neq n_R$.

Since both relations involving L are not orthogonality, Section 7.2 shows that we need to consider adjusted orthogonality. If $L \triangleright R$ then $V_{LR} = V_L \cap V_U^\perp$, so Proposition 8.4 shows that it is impossible for R and C to have adjusted orthogonality with respect to L. The same is true if $L \triangleright C$. Thus adjusted orthogonality requires $R \triangleright L$, $C \triangleright L$ and $n_R + n_C - 1 \leq n_L$.

If $n_L < n_R$ and $n_L < n_C$ then it may be possible to have $L \triangleright R$ and $L \triangleright C$. Figure 22 shows three non-isomorphic such designs with $R \wedge C = E$, $n_L = 4$ and $n_R = n_C = 6$. The one in Figure 22(a) was given in [136]; that in Figure 22(b), which was used for an experiment on strawberries, was given in [134]. A complete enumeration of species for such designs of this size is in [166], which was presented by Côté at the BCC in Surrey (1991). In such a design, every vector in $V_L \cap V_U^\perp$ makes the same angle with $(V_R + V_C)^\perp$, with the result that estimates of

A	B	C	D	A	B
B	C	D	A	A	D
C	D	A	B	B	C
D	A	B	C	C	A
A	B	C	D	C	D
B	C	D	A	D	B

A	D	C	B	A	D
B	B	D	A	C	C
D	C	B	C	D	A
C	A	D	A	B	B
C	C	B	D	A	A
B	D	A	D	B	C

A	B	C	D	A	B
B	A	D	C	B	C
C	D	A	B	C	D
D	C	B	A	D	A
A	C	B	D	A	C
C	B	D	A	B	D

(a)   (b)   (c)

Figure 22: Three non-isomorphic row–column designs in which letters are strictly balanced with respect to both rows and columns

A	H	G	F	J	E
F	A	I	C	H	D
I	B	J	G	D	A
G	F	B	D	E	C
E	J	C	H	I	B

Figure 23: Combinatorial design in which rows and columns are both balanced with respect to letters but do not have adjusted orthogonality with respect to letters

all differences like $\tau_i - \tau_j$ have the same variance. Such a design cannot have adjusted orthogonality. Some authors include designs like this in *generalized Youden designs* if $R \wedge C = E$: see [108, 109].

If $n_L > n_R$ and $n_L > n_C$ then it may be possible to have $R \rhd L$ and $C \rhd L$. This occurs in the design in Figure 4, which does have adjusted orthogonality. Figure 23 shows a design with all the same properties except that it does not have adjusted orthogonality. These designs are called *double arrays* in [116, 118].

The third possibility is, up to interchanging rows and columns, that $n_R < n_L < n_C$ and $R \rhd L$ and $L \rhd C$. Again, adjusted orthogonality is impossible. Figure 24 shows an example.

Therefore, we have the following possibilities.

D.1 A nested balanced incomplete-block design with small blocks (B) nested in large blocks (D), in which letters (L) satisfy $L \rhd B$, $L \rhd D$ and $L \wedge D = E$, where $n_D > n_L$ so that D is not balanced with respect to L. Here $\mathcal{F} = \{B, D, L\}$.

D.2 A nested balanced incomplete-block design with small blocks (B)

C	A	B	B	A	C	F	D	E	B	A	C	D	E	F
E	F	C	D	F	A	E	C	B	D	F	A	E	C	B
D	E	A	F	B	E	C	A	D	F	E	B	C	A	D
F	B	D	E	C	D	B	F	A	C	D	E	B	F	A

Figure 24: A 4×15 row–column design in which rows are strictly balanced with respect to letters, which are in turn strictly balanced with respect to columns

nested in large blocks (D), in which letters (L) satisfy $L \rhd B$, $L \rhd D$ and $L \wedge D = E$, where $n_D = n_L$ so that $L \bowtie D$. Here $\mathcal{F} = \{B, D, L\}$.

D.3 A nested balanced block design with small blocks (B) nested in large blocks (D), in which letters (L) satisfy $L \rhd B$, $L \rhd D$, $L \wedge D \neq E$ but $L \wedge B = E$, where $n_D > n_L$ so that D is not balanced with respect to L. Here $\mathcal{F} = \{B, D, L\}$.

D.4 A nested balanced block design with small blocks (B) nested in large blocks (D), in which letters (L) satisfy $L \rhd B$, $L \rhd D$ and $L \wedge B \neq E$, where $n_D > n_L$ so that D is not balanced with respect to L. Here $\mathcal{F} = \{B, D, L\}$.

D.5 A nested balanced block design with small blocks (B) nested in large blocks (D), in which letters (L) satisfy $L \rhd B$, $L \rhd D$, $L \wedge D \neq E$ but $L \wedge B = E$, where $n_D = n_L$ so that $L \bowtie D$. Here $\mathcal{F} = \{B, D, L\}$.

D.6 A nested balanced block design with small blocks (B) nested in large blocks (D), in which letters (L) satisfy $L \rhd B$, $L \rhd D$ and $L \wedge B \neq E$, where $n_D = n_L$ so that $L \bowtie D$. Here $\mathcal{F} = \{B, D, L\}$.

D.7 Small blocks (B) are nested in large blocks (D), and letters (L) are arranged so that $B \rhd L$ and $D \rhd L$. Also, $n_L = n_B$, so that $B \bowtie L$. Here $\mathcal{F} = \{B, D, L\}$.

D.8 Small blocks (B) are nested in large blocks (D), and letters (L) are arranged so that $B \rhd L$ and $D \rhd L$. Also, $n_L > n_B$, so that L is not balanced with respect to B. Here $\mathcal{F} = \{B, D, L\}$.

D.9 Small blocks (B) are nested in large blocks (D), and letters (L) are arranged so that $D \rhd L \rhd B$, where $n_B > n_L > n_D$, so that B is not balanced with respect to L, and L is not balanced with respect to D. Here $\mathcal{F} = \{B, D, L\}$.

D.10 A row–column design in which letters (L) are arranged so that $L \rhd R$ and $L \rhd C$, $n_L < n_R$ and $n_L < n_C$. Here $\mathcal{F} = \{R, C, L\}$.

D.11 A row–column design in which letters (L) are arranged so that $R \rhd L$, $C \rhd L$, and adjusted orthogonality is achieved. Here $\mathcal{F} = \{R, C, L\}$. This is called *triple array* when $R \wedge C = L \wedge C = L \wedge R = E$. See Section 13.

D.12 A row–column design in which letters (L) are arranged so that $R \rhd L$, $C \rhd L$, and adjusted orthogonality is not achieved, even though $n_L \geq n_R + n_C - 1$. Here $\mathcal{F} = \{R, C, L\}$.

D.13 A row–column design in which letters (L) are arranged so that $R \rhd L$, $C \rhd L$, and adjusted orthogonality cannot be achieved, because $n_L < n_R + n_C - 1$. Here $\mathcal{F} = \{R, C, L\}$.

D.14 A row–column design in which letters (L) are arranged so that $R \rhd L \rhd C$. Here $\mathcal{F} = \{R, C, L\}$.

10.6   Three partitions: three relations of balance

Suppose that $\mathcal{F} = \{F, G, H\}$ and all the pairwise relations are non-orthogonal strict balance in at least one direction. Then $\mathcal{F}$ is supreme. There is no loss of generality in assuming that $n_F \leq n_G \leq n_H$.

If $n_F = n_G = n_H$ then $F \bowtie G$, $F \bowtie H$ and $G \bowtie H$. As Section 9.4 shows, either each partition is balanced with respect to the other pair, or none is.

If $n_F = n_G < n_H$ then it can happen that F and G have adjusted orthogonality with respect to H, in which case each is balanced with respect to the other pair, as shown in Section 9.4. Figure 15 shows an example. Such designs were called *pergolas* in [183], which was presented by Preece at the 1997 BCC at Queen Mary University of London. It may be possible to have such balance without adjusted orthogonality.

If $n_F < n_G = n_H$ then F may or may not be balanced with respect to $\{G, H\}$. Figure 25 shows an example in which $N_{FG}N_{GH}N_{HF}$ is completely symmetric because N_{GH} is completely symmetric and $N_{FG} = N_{FH}$. In fact, the triple (F, G, H) satisfies condition (9.3). However, neither G nor H is balanced with respect to the other pair.

In a real experiment, it is unlikely that two inherent partitions would have a non-orthogonal relation of strict balance. If partition L is of interest and partition R is not then the desirability of variance balance means that it is more usual to have $L \rhd R$ than $R \rhd L$. If R is inherent then there is usually no need for it to be balanced with respect to any subset of other partitions. Thus statisticians have not investigated the other possibilities.

	A	B	A	B	C
A		C	B	C	A
B	C		C	A	B
A	B	C		A	B
C	A	B	A		C
B	C	A	B	C	

Figure 25: A design for 30 units, with partitions F, G and H into letters, rows and columns, in which $G \bowtie H$, $F \rhd G$, $F \rhd H$ and F is balanced with respect to $\{G, H\}$: blank cells denote empty row-column intersections

Therefore, we have the following list of possible structures, which is not divided as finely as the lists in previous subsections. In every case, $\mathcal{F} = \{F, G, H\}$.

E.1 $F \bowtie G$, $F \bowtie H$ and $G \bowtie H$, and each partition is balanced with respect to the other pair. See Section 14.

E.2 $F \bowtie G$, $F \bowtie H$ and $G \bowtie H$, but no partition is balanced with respect to the other pair.

E.3 $F \bowtie G$, $F \rhd H$, $G \rhd H$, $n_F < n_H$, and F and G have adjusted orthogonality with respect to H, which forces each of F and G to be balanced with respect to the other pair.

E.4 $F \bowtie G$, $F \rhd H$, $G \rhd H$, $n_F < n_H$, and F and G do not have adjusted orthogonality with respect to H, but nonetheless each of F and G is balanced with respect to the other pair.

E.5 $F \bowtie G$, $F \rhd H$, $G \rhd H$, $n_F < n_H$, and neither of F and G is balanced with respect to the other pair.

E.6 $F \rhd G$, $F \rhd H$, $G \bowtie H$, $n_F < n_H$ and F is balanced with respect to $\{G, H\}$.

E.7 $F \rhd G$, $F \rhd H$, $G \bowtie H$, $n_F < n_H$ and F is not balanced with respect to $\{G, H\}$.

E.8 $F \rhd G \rhd H$ and $n_F < n_G < n_H$.

A	B	C	D	E	F	G
B	D	F	E	G	A	C
C	F	E	A	B	G	D

1			3		2	
2	1			3		
3		1				2
	2		1			3
			3	2	1	
		3	2			1
				2	3	1

(a)   (b)

Figure 26: Different representations of the same 3×7 Youden square

11   Youden squares and their generalizations

11.1   A rectangle with one set of letters

Definition An incomplete-block design is *symmetric* if both partitions are uniform and they have the same number of parts.

This word 'symmetric' is standard, but it is unfortunate, because it does not imply that the incidence matrix is symmetric or that there is an automorphism interchanging the two partitions. The adjective *square* is also used: see [45, 52].

Definition An $n \times m$ *Youden square* is an $n \times m$ rectangle in which $R \wedge C = E$ and $R \perp C$, so that there is exactly one cell in each row-column intersection, with one of m letters allocated to each cell, in such a way that (i) each letter occurs once in each row and (ii) the allocation of letters to columns forms a balanced binary incomplete-block design. Thus $n < m$.

These designs were introduced by Youden in [232]. His first example, with $n = 3$ and $m = 7$, is shown in Figure 26(a). If the roles of letters and rows are interchanged, the design can be represented as an $m \times m$ square with $m(m - n)$ empty cells, as in Figure 26(b). This representation led to the confusing name 'Youden square', which has been used since [80].

Deletion of any row of an $m \times m$ Latin square gives an $(m - 1) \times m$ Youden square. Yates had recommended these in [228]. How can they be constructed if $n < m - 1$?

In [232], Youden wrote that 'some patience is required to obtain' the rectangular layout from the incomplete-block design. In [198], Smith and Hartley proved that this is always possible, whether or not the incomplete-block design is balanced. In fact, this follows from Hall's Marriage Theorem [89], as shown in [28, Chapter 11]. There is a greedy algorithm, where

A	C	C	B	D	B	D
D	A	B	C	C	D	A
B	D	D	A	A	C	B
C	B	A	D	B	A	C

Figure 27: A regular generalized Youden square, relaxing condition (ii), shown with rows and columns interchanged

one row is constructed at a time as a 'system of distinct representatives' of the columns; then induction is used. This argument generalizes easily to situations where condition (i) is weakened to 'all letters appear equally often in each row' and/or the words 'binary incomplete' in condition (ii) are weakened to 'generalized binary but not orthogonal', in which case n may be greater than m.

Relaxing condition (ii) allows a design such as that in Figure 27, given in [98]. Condition (i) was weakened as above in [4, 194]. Doing both gives what Kiefer called *regular generalized Youden designs* in [109].

However, when both conditions are weakened to 'strictly balanced but not orthogonal' then the method of constructing transversals by using Hall's Marriage Theorem is no longer guaranteed to work, and so Kiefer had to give many explicit constructions in [109]. Further non-regular generalized Youden designs are in [15, 190].

Preece gave many direct constructions of Youden squares, and classified them up to various notions of isomorphism, in [146, 158, 160, 165]. The review in [158] was presented to the 1989 BCC in Norwich.

11.2 Double Youden rectangles

If a Youden square is a poor man's Latin square, what should a poor man's Graeco-Latin square be? We already have partitions R, C and L with $n_R < n_C = n_L$. Should the new partition G have $n_G = n_R$ or $n_G = n_C$? If $n_G = n_C$ then we have three partitions with pairwise strict balance. In this case, as we saw in Section 9, we need to consider their ternary relation as well, so we defer this to Section 14.3.

Definition An $n \times m$ *double Youden rectangle* is an $n \times m$ rectangle with partitions R (rows), C (columns), L (Latin letters) and G (Greek letters) such that $n_R = n_G = n < m$, $n_C = n_L = m$, $R \perp C$, $R \perp L$, $G \perp C$, $G \perp L$, $R \bowtie G$ and $C \bowtie L$, and $R \wedge C = R \wedge L = G \wedge C = G \wedge L = E$. This means that every row meets every column and every Latin letter in exactly one cell, every Greek letter meets every Latin letter and every

B α	E α	D δ	A β	C γ
C β	A γ	E β	B δ	D α
D γ	C δ	A α	E γ	B β
E δ	D β	B γ	C α	A δ

Figure 28: A 4 × 5 double Youden rectangle

A ♠	3 ♣	4 ♡	7 ♡	8 ♣	2 ♣	10 ◇	J ♠	5 ♠	6 ◇	Q ◇	K ♠	9 ♡
2 ◇	5 ♡	3 ◇	4 ♠	6 ♠	7 ◇	8 ♠	9 ♣	10 ♣	K ♡	J ♡	Q ♣	A ◇
4 ♣	J ◇	6 ♣	K ◇	5 ◇	9 ♠	7 ♣	8 ♡	Q ♡	10 ♠	A ♣	2 ♡	3 ♠
10 ♡	2 ♠	Q ♠	5 ♣	A ♡	6 ♡	3 ♡	4 ◇	9 ◇	J ♣	7 ♠	8 ◇	K ♣

Figure 29: A 4 × 13 double Youden rectangle

column in exactly one cell, the relation between Latin letters and columns is a symmetric binary balanced block design, while the relation between Greek letters and rows is a symmetric generalized binary balanced block design.

Thus removal of G leaves a Youden square, while removal of L leaves one of the weak generalizations of Youden squares. This name was given in [22], but the concept was not new. Clarke had given the 4 × 5 double Youden rectangle in Figure 28 in [59], and Preece had given the 4 × 13 double Youden rectangle in Figure 29 in [156]. Clarke also gave a 5 × 6 double Youden rectangle in [59] and a 4 × 7 one in [60], while Freeman gave some designs in [82] which Preece was able to interpret in [150] as double Youden rectangles, including one of size 6 × 7. This is useful because this size is excluded from the construction by Hedayat, Parker and Federer in [93], which uses an $(m-1) \times (m-1)$ Graeco-Latin square with a transversal to build an $(m - 1) \times m$ double Youden rectangle. Saha and Das gave a construction for size $(m - 1) \times m$ when m is odd in [188]. Some of these small double Youden rectangles are also given in [145]. Several of size 5 × 11 are in [152].

The review paper [158] seems to have spurred Preece and his collaborators to find further double Youden rectangles. One of size 6 × 11 is given in [159]. Preece gave several 7 × 15 double Youden rectangles in [160], including the interesting one in Figure 30. Here the Youden square given by ignoring the Greek letters can be obtained from projective geometry using the methods of [75] and [97, Section 17.5]: it admits the alternating

A α	H α	K δ	M η	E γ	J ζ	C β	B ε	I α	O ζ	N δ	F γ	L η	G β	D ε
B β	C ζ	H β	L ε	N α	F δ	K η	D γ	E ζ	J β	I η	O ε	G δ	M α	A γ
C γ	E δ	D η	H γ	M ζ	O β	G ε	L α	B δ	F η	K γ	J α	I ζ	A ε	N β
D δ	M β	F ε	E α	H δ	N η	I γ	A ζ	O γ	C ε	G α	L δ	K β	J η	B ζ
E ε	B η	N γ	G ζ	F β	H ε	O α	J δ	C η	I δ	D ζ	A β	M ε	L γ	K α
F ζ	K ε	C α	O δ	A η	G γ	H ζ	I β	L β	D α	J ε	E η	B γ	N ζ	M δ
G η	J γ	L ζ	D β	I ε	B α	A δ	H η	N ε	M γ	E β	K ζ	F α	C δ	O η

Figure 30: A 7×15 double Youden rectangle

group A_7 as a group of automorphisms. An interesting connection with Kirkman's schoolgirls problem is in [88].

Christofi enumerated 4×5 and 5×6 double Youden rectangles in [57], which was presented at the BCC in Surrey (1991), and those of size 6×7 in [58]. Preece gave some double Youden rectangles of sizes 8×15, 5×11 and 6×11 in [161], [162] and [165] respectively. The last two were presented at the BCCs in Surrey (1991) and Stirling (1995).

In [216], Vowden gave a construction for $p \times (2p + 1)$ double Youden rectangles for all primes p congruent to 3 modulo 4, excluding the size 3×7, for which there is no such rectangle. This was augmented in [172] by the results of a computer search to give double Youden rectangles of sizes $p \times (2p+1)$ and $(p+1) \times (2p+1)$ for p in $\{5, 7, 9, 11\}$. Vowden adapted his construction to give $p \times (2p + 1)$ double Youden rectangles for prime powers p congruent to 1 modulo 4, excluding $p = 5$, in [217]. Further computer searches, described in [137], gave positive results for sizes 5×21, 6×31 and 8×57 in [173], and size 9×37 in [138].

12   Nested balanced incomplete-block designs

12.1   Small blocks inside large blocks

In the original idea for nested BIBDs, there are two inherent uniform partitions B and D of Ω. The parts of B are small blocks, and the parts of D are large blocks, each comprising k_D/k_B small blocks. Given a set of n_L letters, the aim is to allocate letters to experimental units in such a way that L and B form one BIBD and L and D form another.

Definition Uniform partitions B, D and L on a set Ω form a *nested balanced-incomplete block design* if $B \prec D$ and L is binary balanced with respect to each of B and D separately but not orthogonal to either.

Preece defined these designs in [148], where he cited two previous uses [106, 111] in scientific experiments, and, as was usual for him, gave a table of designs for small sizes, many of them made by cyclic development from one or more initial small blocks (parts of B) and large blocks (parts of D).

Constructions based on finite fields were given in [64, 101]. Designs with $n_D = n_L$, $k_D = n_L - 1$ and $k_B = 2$ were given for all odd n_L in [36].

Morgan included such designs in his survey chapter [119]. Preece talked about them at the 1999 BCC in Kent, attended by Morgan. As a result, they teamed up, and together with Rees, generalized the concept, first to *doubly nested BIBDs*, and then to *multiply nested BIBDs*. The former have $B \prec D_1 \prec D_2$ with L binary balanced with respect to each of B, D_1 and D_2, while the latter is the obvious generalization. Paper [170] is about doubly nested BIBDS, while [121] is a much expanded and updated survey, including the observation that most designs for whist tournaments can be considered as nested BIBDS: see [14].

As discussed in Section 6.2.1, if B has fixed effects then there is no benefit from having L balanced with respect to D. Preece was clear in [148] that he was assuming random effects of small blocks. As the ratio σ_B^2/σ^2 increases, there is less need to have L balanced with respect to D.

12.2   Nested row–column designs

These ideas can be extended to the poset block structure in Example 4.4, with partitions B, R and C into blocks, rows and columns. If B, R and C all have fixed effects, then it is desirable to allocate letters in such a way that L is balanced with respect to $\{R, C, B\}$. As shown in Section 9.2, this is equivalent to L being balanced with respect to $\{R, C\}$.

Some nested row–column designs with this property and with L being binary with respect to B are given in [8, 9, 56, 124, 196, 202, 205, 211, 212, 213, 214]. The lack of specified binary relations between L and B, L and R, and L and C takes us somewhat outside the scope of this paper, but there is an excellent survey in [119]. The square in Figure 14 gives an example with $n_B = 1$ (except that L is not binary in respect to B).

If $n_L < k_B$ then it seems intuitive that the best designs should be among those for which L is binary with respect to blocks. However, Bagchi, Mukhopadhyay and Sinha [20] and Chang and Notz [55] showed independently in 1990 that, under fixed effects, a design like the one in Figure 13, in which, in every block, each row has the same set of letters, gives the smallest average variance of the estimators of differences $\tau_i - \tau_j$.

On the other hand, if rows and columns have random effects with small variances then the arguments in [24, 56] show that there is a reason to allocate the parts of L in such a way that $L \rhd G$ for all G in $\{R, C, B\}$.

That is, letters should be allocated so that they form a BIBD in rectangles, in rows and in columns separately. Several constructions for this are given in [148]. One with seven letters in seven 2×3 rectangles is obtained by developing

2	3	5
0	6	4

modulo 7. In [122], Morgan and Uddin gave an inequality involving σ^2, σ_R^2, σ_C^2 and σ_B^2 which determines exactly when such a design is better than one like that in Figure 13.

13 Triple arrays

13.1 Definitions and examples

Definition Consider an $n \times m$ rectangle in which one of v letters is allocated to each cell. This is a *double array* if the partitions R (rows), C (columns) and L (letters) are all uniform with pairwise binary relations such that $R \perp C$, $R \triangleright L$ and $C \triangleright L$. It is a *triple array* if it is a double array and, additionally, R and C have adjusted orthogonality with respect to L. Thus we have an $n \times m$ rectangle with precisely one unit in each row-column intersection. Letters are allocated to units, in such a way that no letter is repeated in any row or column. It is a double array if it satisfies the two conditions (i) every pair of rows has the same number of letters in common, and (ii) every pair of columns has the same number of letters in common. If, in addition, (iii) every row has nm/v letters in common with every column, then it is a triple array.

Figure 4 shows a triple array with $n = 5$, $m = 6$ and $v = 10$.

These definitions were given by McSorley, Phillips, Wallis and Yucas in [118], but without mentioning 'adjusted orthogonality'. Instead, they gave condition (8.5), as Preece had done in [145, 147]. They mentioned one design given in [143] and one in [145], and referred to a table of design parameters in [151]. Being unaware of the literature on adjusted orthogonality, they did not realise that Anderson and Eccleston [12] and Bagchi and Shah [21] had also written about such designs, and they proved Theorem 8.5, which Bagchi had published in [18].

Preece became aware of [118] before publication. He began joint research with its authors, adopted the name 'triple arrays' with enthusiasm, and talked about them in the 2003 BCC in Bangor: see [139]. McSorley also talked about related work there.

Triple arrays which meet the bound in Theorem 8.5 are very interesting. They were called *extremal* in [37], which, being unpublished, had not

34	45	35	25	24	23
45	35	34	14	13	15
15	24	12	45	25	14
12	13	25	23	15	35
23	12	14	13	34	24

Figure 31: Relabelled version of Figure 4

been seen by McSorley and his co-authors. The following two examples of extremal triple arrays are taken from [37].

Example 13.1 Replacing the letters A–I in Figure 4 lexicographically by the unordered pairs from $\{1, 2, 3, 4, 5\}$ gives the design in Figure 31. The rows can be labelled 1–5 and the columns labelled by six pentagons with vertex-set $\{1, 2, 3, 4, 5\}$ which form a single orbit under the action of the alternating group A_5. Then the unordered pair in row i and column j is the edge opposite to vertex i in pentagon j.

Thus A_5 acts as a group of automorphisms of the array in Figure 31 fixing each of the three partitions. In the action on ordered pairs from $\{1, 2, 3, 4, 5\}$, the irreducible subspaces of this action have dimensions 1, 4 and 5. Because rows and columns are both balanced with respect to letters, $\dim(V_{LR}) = n - 1 = 4$ and $\dim(V_{LC}) = m - 1 = 5$. Both subspaces are fixed by the automorphisms, so they must be the two non-trivial irreducible subspaces and so they are orthogonal to each other. The discussion in Section 8.2 shows that this array has adjusted orthogonality. It is an extremal triple array because $n + m - 1 = 10 = v$.

Example 13.2 Figure 32 shows an extremal triple array with $n = 6$, $m = 10$ and $v = 15$. The letters are shown as unordered pairs from $\{1, 2, 3, 4, 5, 6\}$; the rows are labelled 1–6 in natural order, and the columns are labelled by the partitions of $\{1, 2, 3, 4, 5, 6\}$ into two parts of size three. The unordered pair in row i and column j consists of the two numbers in the same part as i in partition j. The symmetric group S_6 acts a group of automorphisms of this design, and so a similar argument to that in Example 13.1 shows that this is a triple array.

Sterling and Wormald gave extremal triple arrays of sizes 4×9, 5×16 and 6×25 in [203].

Example 13.3 The non-extremal triple array shown in Figure 33 was given by McSorley et al. in [118], thereby answering a question posed by

| 123 | 124 | 125 | 126 | 134 | 135 | 136 | 145 | 146 | 156 |
| 456 | 356 | 346 | 345 | 256 | 246 | 245 | 236 | 235 | 234 |

23	24	25	26	34	35	36	45	46	56
13	14	15	16	56	46	45	36	35	34
12	56	46	45	14	15	16	26	25	24
56	12	36	35	13	26	25	15	16	23
46	36	12	34	26	13	24	14	23	16
45	35	34	12	25	24	13	23	14	15

Figure 32: An extremal triple array with $n = 6$, $m = 10$ and $v = 15$

e	A	R	P	G	J	E	C	D	B	g	N	S	O	L
Z	f	A	B	c	d	b	T	a	K	E	h	C	H	D
A	Q	M	I	C	D	U	V	F	i	Y	E	X	B	W
F	a	g	b	P	M	i	d	O	J	I	Z	L	Q	c
P	L	W	f	h	U	O	g	X	V	K	J	H	Y	T
U	V	b	X	Y	S	G	N	R	c	a	W	Z	d	e
K	G	H	N	M	f	T	F	h	R	S	Q	i	e	I

Figure 33: A non-extremal triple array with $n = 7$, $m = 15$ and $v = 35$

Preece at the 1973 BCC in Aberystwyth and listed in [151] as unresolved. Here $n = 7$, $m = 15$ and $v = 35$. Phillips and Wallis described in [140] how the design had been found by a computer search. Yucas showed in [233] how it can be obtained from projective 3-space over GF(2), which suggests an intriguing connection with the design in Figure 30.

13.2 The connection with symmetric BIBDs

Given an extremal triple array, construct a symmetric incomplete-block design (in the usual combinatorialist's sense) with $v + 1$ blocks of size n as follows. The points are $\{u_1, \ldots, u_n, w_1, \ldots, w_m\}$. Considered as a set of points, block i is

$$\{w_j : \text{letter } i \text{ occurs in column } j\} \cup \{u_\ell : \text{letter } i \text{ does not occur in row } \ell\}$$

for $i = 1, \ldots, v$, and block $v + 1$ is $\{u_1, \ldots, u_n\}$.

The following, proved in [118] and [37], and was implicit in [6].

Theorem 13.4 *The symmetric incomplete-block design made from an extremal triple array by the foregoing construction is balanced.*

The common concurrence in this block design is $n(v-m)/v$, and so the difference between the block size and this is nm/v, which is the number of times that each letter occurs in the triple array, which I have been writing as k_L. As proved in [116], this is also the number of letters which any row and column have in common.

Can we reverse the foregoing construction, and make the extremal triple array from the symmetric BIBD? Here is another piece of interesting history. In 1966, Agrawal published several papers in this area. Paper [6] is particularly relevant: it constructs some of the same designs that Preece was constructing, and it gives condition (8.5). It is not surprising that Preece and Agrawal were unaware of each other's work published in the same year, but Preece became aware of Agrawal's work and cited it in [151].

Given the symmetric BIBD, we label the rows of the $n \times m$ rectangle by the points in the first block and label the columns by the other points. Then ignore that block. For row i, write down the set of other blocks that do not contain point i; for column j, write down the set of other blocks that do contain point j. Put the block-names in column j in any order: then it is simply a matter of re-arranging the names in each column in such a way that each row has the correct set of names.

'Simply'? This runs into exactly the same problem that we met for non-regular generalized Youden designs in Section 11.1: the method of selecting a system of distinct representatives used in Hall's Marriage Theorem is no longer guaranteed to work. In fact, when $n = 3$, $m = 4$ and $v = 6$ (so that $k_L = 2$) there is a symmetric BIBD but no triple array. This leads to the following conjecture in [118].

Conjecture 13.5 *If $k_L > 2$ then the foregoing construction can be reversed to give an extremal triple array from the symmetric BIBD.*

Agrawal wrote in [6] that 'In the examples tried by the author it was found that if $k_L > 2$ such arrangement is always possible (we have no mathematical proof for the above observation).' (His notation has been changed to match that used here.) In [125], Nageswara Rao says that the fact that this 'arrangement of symbols is always possible can be shown with the help of generalized systems of distinct representatives', yet does not cite the earlier paper [179], of which he was apparently a co-author, which claims to give a proof, not excluding the case $k_L = 2$. The procedure given in [179] finds a set of distinct representatives for each row, with no

E	H	J	F	I	J	G	H	I	E	G	J	F	G	H	E	F	I
D	H	J	B	I	J	C	H	I	B	C	J	B	D	H	C	D	I
A	D	J	A	F	J	A	C	G	C	G	J	D	F	G	C	D	F
A	D	E	A	B	I	A	G	I	B	E	G	B	D	G	D	E	I
A	E	H	A	B	F	A	C	H	B	C	E	B	F	H	C	E	F

Figure 34: Array of subsets to illustrate the problem of choosing distinct representatives

check to ensure that there are no repeats in any column. Moreover, the final paragraph gives a different short proof (with acknowledgement to the referee) which gives a design which does not have adjusted orthogonality. Wallis and Yucas pointed out flaws in this proof in [220].

Meanwhile, Anderson and Eccleston claimed in [12] that algorithms implemented in software described in [104] and [187] always succeed in creating the row–column design. More detail about the first is in [103].

Here is a way of thinking about the problem. We know the set $\mathcal{R}_i$ of m letters that should go in row i, and we know the set $\mathcal{C}_j$ of n letters that should go in column j. Write the set $\mathcal{R}_i \cap \mathcal{C}_j$ in cell (i, j) to give an array like that in Figure 34. Can we choose just one letter in each cell (an *array of distinct representatives*) in such a way that the chosen letters in each row are distinct (and hence make up $\mathcal{R}_i$) and similarly for columns? In this particular case, the design in Figure 4 gives a solution.

Fon-Der-Flaass considered a more general version of this problem. Given an $n \times m$ array, with a set $\mathcal{S}_{ij}$ of letters in cell (i, j), can we choose one letter from each cell in such a way that the chosen letters in each row are all distinct, as are the chosen letters in each column? In [81], he showed that this decision problem is NP-complete.

13.3   Paley triple arrays

If q is an odd prime power then there is a Hadamard matrix of order $2(q + 1)$. Any Hadamard matrix can be normalized so that one row has all of its entries equal to $+1$. If one column is also normalized, and then that row and column are removed, the positions of the $+1$ entries in the remaining $(2q + 1) \times (2q + 1)$ matrix form the incidence matrix of a symmetric BIBD for $2q + 1$ points in blocks of size q. These designs are sometimes called *Hadamard 2-designs*: see [52, 96].

If the construction in Section 13.2 can be reversed, this design gives an extremal $q \times (q + 1)$ triple array. The triple arrays in Figures 4 and 31 are like this, as are many in [6, 143, 145, 151]. Preece pointed out in [151] that a method given by Agrawal and Mishra in [7] could be used to give an extremal 11 × 12 triple array.

After Preece's Adelaide talk in 1975, Seberry was motivated to find such arrays. Her paper [189] originally showed Preece as a co-author, but he was so furious at her submitting it without consulting him that he wrote to the journal editor and asked for his name to be removed.[2] That is why the paper starts with Preece's explanation of the problem, including condition (8.5), followed by the sentence 'We thank D. A. Preece for writing this introduction and now proceed to give our construction and some examples.' The remainder of the paper gives constructions which purport to prove the existence of an extremal $q \times (q + 1)$ triple array whenever q is an odd prime power.

In [205], Street also proved this for odd prime powers q with $q \equiv 3$ (mod 4), constructing two series of designs not isomorphic to those given in [189]. It was independently proved again for prime powers $q \equiv 3$ (mod 4) in [37], this time explicitly excluding $q = 3$, for which it is not true.

In [18], Bagchi proved the result for odd prime powers q strictly greater than 3, unless $q \equiv 1$ (mod 8).

Preece, Wallis and Yucas named these *Paley triple arrays* in [174], and proved the following.

Theorem 13.6 *If q is an odd prime power and $q > 3$, then there exists an extremal $q \times (q + 1)$ triple array.*

This is a very nice result, with explicit constructions given and verified. However, the paper does not mention the results in [18] or [205], which independently prove this for some congruence classes modulo 8. It does cite [37, 189], but neither states their results nor finds fault with their proofs. It does credit [10] with giving some 'partial results'. Recently Nilson showed in [129] that some of the constructions in [189] do not give adjusted orthogonality, and Cameron[3] has independently found similar problems in [189].

13.4 Recent work

In [131], Nilson and Heidtmann followed up on the connection between extremal triple arrays and symmetric BIBDs. They defined a property of symmetric BIBDs called *inner balance*, which is achieved by the symmetric BIBDs with 11 points in blocks of either size five or size six. The ensuing

---

[2]Personal communication from JS.
[3]Personal communication from PJC.

theory enables them to prove that there are no extremal triple arrays for
the parameter sets left as undecided in [118]. They gave an infinite family of parameter sets for potential symmetric BIBDs with inner balance.
Broughton [48] proved that there are no other possibilities.

Nilson and Öhman show in [132] how some Youden squares may be
used to construct extremal triple arrays. In particular, all Paley triple
arrays may be constructed like this. They also give a construction using
difference sets. Further constructions from difference sets are in [130].

The recent survey paper [219] largely consists of material copied ver-
batim from [118, 174]. It mentions none of the literature on adjusted
orthogonality.

14   Universal balance

14.1   Definitions and results

This section picks up the ideas of Section 9, while specializing them
to insist that every partition in the set is balanced with respect to every
subset of the others. It introduces the term 'universal balance' as well as
Theorem 14.2 and Conjecture 14.4, in an attempt to pull together some
different pieces of work into a single framework.

Recall that if $\mathcal{L}$ is a set of partitions of Ω then $V_{\mathcal{L}} = \sum_{L \in \mathcal{L}} V_L$ and $P_{\mathcal{L}}$
denotes the matrix of orthogonal projection onto $V_{\mathcal{L}}$. Moreover, partition
F is balanced with respect to $\mathcal{L}$ if $X_F^{\top}(I_e - P_{\mathcal{L}})X_F$ is completely symmetric
but not zero. The insistence on 'not zero' ensures that $\dim(V_{\mathcal{L} \cup \{F\}}) =
\dim(V_{\mathcal{L}}) + n_F - 1$.

Definition   Let $\mathcal{F}$ be a set of partitions of Ω. Then $\mathcal{F}$ has *universal
balance* if all pairwise relations between distinct partitions in $\mathcal{F}$ are gener-
alized binary and, whenever $F \in \mathcal{F}$ and $\mathcal{L} \subseteq \mathcal{F} \setminus \{F\}$, then F is balanced
with respect to $\mathcal{L}$ but V_F is not geometrically orthogonal to $V_{\mathcal{L}}$.

Taking $\mathcal{L}$ to be $\emptyset$ in the above definition shows that if $\mathcal{F}$ has universal
balance and $F \in \mathcal{F}$ then F is uniform. Likewise, if $F \in \mathcal{F}$, $G \in \mathcal{F}$ and
$F \neq G$ then $F \bowtie G$ and so $n_F = n_G$. Write m for the common value of
n_F for F in $\mathcal{F}$.

These remarks give an inductive proof of the following result, which is
stated in [13].

Theorem 14.1   *If $\mathcal{F}$ has universal balance, $|\mathcal{F}| = s$, and every partition
in $\mathcal{F}$ has m parts of size k, then $1 + s(m-1) \le e = mk$.*

Write $Q_{\mathcal{L}}$ for $P_{\mathcal{L}} - P_U$, simplifying $Q_{\{R\}}$ to Q_R and $Q_{\{R,C\}}$ to Q_{RC} as in Section 9. Then the arguments used in Section 8.1 show that the following are equivalent:

(i) F is balanced with respect to $\mathcal{L}$ but V_F is not geometrically orthogonal to $V_{\mathcal{L}}$;

(ii) there is some scalar μ in $(0, 1)$ such that $Q_F Q_{\mathcal{L}} Q_F = \mu Q_F$.   (14.1)

When condition (14.1) is satisfied, the proof underlying Theorem 7.1(v) shows that

$$Q_{\mathcal{L}\cup\{F\}} = Q_{\mathcal{L}} + (1-\mu)^{-1}(Q_F - Q_{\mathcal{L}}Q_F - Q_F Q_{\mathcal{L}} + Q_{\mathcal{L}}Q_F Q_{\mathcal{L}}).   (14.2)$$

When $\mathcal{L} = \{G\}$ and F and G are partitions into m parts of size k whose pairwise relation is generalized binary then the value of μ in condition (14.1) is determined by the values of m and k. Writing this common value as μ_1, expression (7.2) shows that

$$(1 - \mu_1)Q_{RC} = Q_R + Q_C - Q_R Q_C - Q_C Q_R   \qquad (14.3)$$

whenever $\mathcal{F}$ has universal balance and R and C are distinct partitions in $\mathcal{F}$.

Let L be a third such partition. Then there is a scalar μ_2 in $(0, 1)$ such that

$$(1 - \mu_1)^{-1}Q_L(Q_R + Q_C - Q_R Q_C - Q_C Q_R)Q_L = \mu_2 Q_L,$$

which implies that

$$2\mu_1 Q_L - Q_L Q_R Q_C Q_L - Q_L Q_C Q_R Q_L = \mu_2(1 - \mu_1)Q_L$$

and hence that

$$Q_L Q_R Q_C Q_L + Q_L Q_C Q_R Q_L = (2\mu_1 - \mu_2 + \mu_1\mu_2)Q_L.   (14.4)$$

Pre- and post-multiplication of equation (14.4) by $X_L^\top$ and X_L respectively gives condition (9.2); pre- and post-multiplication by Q_R or Q_C prove a stronger version of Proposition 9.3(ii) which includes the fact that the value of μ_2 does not depend on which one of $\{R, C, L\}$ is distinguished.

However, when $|\mathcal{L} \cup \{F\}| \geq 4$ then the value of μ in condition (14.1) can be different for different choices of the distinguished partition. Examples of this were given in [11, 63, 145, 157].

After some manipulation, expression (7.2) and equations (14.1), (14.2), (14.3) and (14.4) show that if $\{R, C, L\}$ has universal balance then

$$
\begin{aligned}
(1 - \mu_1)(1 - \mu_2)Q_{RCL} = {} & (1 - \mu_1)(Q_R + Q_C + Q_L) \\
& - (Q_R Q_C + Q_C Q_R + Q_R Q_L + Q_L Q_R + Q_C Q_L + Q_L Q_C) \\
& + (Q_R Q_C Q_L + Q_L Q_C Q_R + Q_C Q_L Q_R + Q_R Q_L Q_C \\
& \quad + Q_L Q_R Q_C + Q_C Q_R Q_L).
\end{aligned}
$$

Hence induction gives the following.

Theorem 14.2 *If $\mathcal{F}$ has universal balance then $Q_{\mathcal{F}}$ is a linear combination of products of the matrices Q_F for F in $\mathcal{F}$.*

Corollary 14.3 *If $\mathcal{F}$ has universal balance, $\mathcal{L} \subset \mathcal{F}$ and $F \in \mathcal{F} \setminus \mathcal{L}$ then $X_F^\top Q_{\mathcal{L}} X_F$ is a sum of matrices of the form*

$$
N_{FL_1} N_{L_1 L_2} \cdots N_{L_r F}
$$

where $(L_1, L_2, \ldots, L_r)$ is a sequence of partitions in $\mathcal{L}$, possibly having repeated entries.

Conjecture 14.4 *If $\mathcal{F}$ has universal balance then $Q_{\mathcal{F}}$ is a linear combination of matrices of the form*

$$
Q_{F_1} Q_{F_2} \cdots Q_{F_r} + Q_{F_r} \cdots Q_{F_2} Q_{F_1}
$$

for sequences $(F_1, F_2, \ldots, F_r)$, with no repeated entries, of partitions in $\mathcal{F}$.

Apart from the 'non-zero' check, this would imply that $\mathcal{F}$ has universal balance if, whenever $\mathcal{L} \subset \mathcal{F}$ and $F \in \mathcal{F} \setminus \mathcal{L}$ and $(L_1, L_2, \ldots, L_r)$ is a sequence of distinct partitions in $\mathcal{L}$, then the matrix

$$
N_{FL_1} N_{L_1 L_2} \cdots N_{L_r F} + N_{FL_r} \cdots N_{L_2 L_1} N_{L_1 F}
$$

is completely symmetric.

14.2 Sets of partitions with universal balance

Assume that $s \geq 3$, so that condition (9.2) must be satisfied for all sets of three partitions in $\mathcal{F}$. We know only three families of incidence matrices which satisfy this and the more general condition given at the end of Section 14.1. Those in Section 14.2.1 have every pairwise incidence matrix equal to $J_{mm} - I_m$, and there is a direct construction of the partitions on the set Ω. For those in Section 14.2.2, each pairwise incidence matrix is one of only two possibilities, which are the transposes of each other and commute with each other. The pairwise incidence matrices of those in Section 14.2.3 satisfy the stronger condition (9.3). For these last two, we defer the construction of the designs until Section 14.3.

14.2.1 Constructions from orthogonal Latin squares

Probably the earliest example of universal balance to be identified has $n_F = m = k_F + 1$ for all F in $\mathcal{F}$. The starting point is a set of $s - 2$ mutually orthogonal $m \times m$ Latin squares with a common transversal, where $s \geq 2$. Use the common transversal to label the rows, columns and letters of each square. Then Ω consists of the $m(m - 1)$ cells not in that transversal, and $\mathcal{F} = \{R, C, L_1, \ldots, L_{s-2}\}$, where the parts of R, C and L_i are rows, columns and the letters of the i-th square. The common labelling ensures that all of the incidence matrices N_{FG} for $F \neq G$ with $\{F, G\} \subseteq \mathcal{F}$ satisfy $N_{FG} = J_{mm} - I_m$. See [5, 13, 94, 176, 195].

14.2.2 Constructions from doubly regular tournaments

Many of the designs in [11, 145, 147, 205] belong to another infinite family. The design in Figure 16(b) gives an example. Here $N_{RC} = I_7 + A$ and $N_{CL} = N_{LR} = I_7 + A^\top$, where $A(i, j) = 1$ if $j - i$ is a non-zero square in GF(7) and $A(i, j) = 0$ otherwise. These matrices satisfy condition (9.2) but not the stronger condition (9.3), or the intermediate one that $N_{LR}N_{RC}N_{CL}$ is completely symmetric.

Given a directed graph with m vertices, its *adjacency matrix* A is the $m \times m$ matrix with $A(i, j) = 1$ if there is an edge from vertex i to vertex j and $A(i, j) = 0$ otherwise. The directed graph is called a *doubly regular tournament* if $A + A^\top = J_{mm} - I_m$ and

$$AA^\top = \frac{m + 1}{4} I_m + \frac{m - 3}{4} J_{mm}, \tag{14.5}$$

from which it follows that

$$A^2 = \frac{m + 1}{4} (J_{mm} - I_m) - A \tag{14.6}$$

and $m \equiv 3 \pmod 4$. See [184] for more details.

Many doubly regular tournaments come from finite fields. If m is power of a prime and $m \equiv 3 \pmod 4$ then make a directed graph whose vertices are the elements of GF(m) with an edge from i to j if $j - i$ is a non-zero square.

Suppose that $e = mk$ and every partition in $\mathcal{F}$ has m parts of size k. If there is a doubly regular tournament of size m and the parts of every partition in $\mathcal{F}$ can be labelled by its vertices in such a way that either

(i) $k = (m + 1)/2$ and every $m \times m$ incidence matrix is equal to either $I + A$ or $I + A^\top$ or (ii) $k = (m - 1)/2$ and every $m \times m$ incidence matrix is equal to either A or $A^\top$, then equations (14.5) and (14.6) show that if M is any product of a sequence of such incidence matrices then $M + M^\top$ is completely symmetric. Thus Theorem 14.2 shows that $\mathcal{F}$ has universal balance unless there is some F in $\mathcal{F}$ and subset $\mathcal{L}$ of $\mathcal{F} \setminus \{F\}$ such that $X_F^\top(I_e - P_{\mathcal{L}})X_F$ is zero.

Unlike the straightforward situation in Section 14.2.1, the scalar μ in condition (14.1) is not determined by the values of m and k when $s \geq 3$. Draw a directed graph with s vertices, one for each partition, and an edge from F to G if N_{FG} includes A rather than $A^\top$. The value of μ depends on the isomorphism class of this directed graph. When $s = 3$ it may be a cycle or a total order. There are more possibilities as s increases. See [3, 5, 13, 40, 63, 145, 169].

14.2.3 Constructions for powers of four

As noted in Section 1, many combinatorialists think of block designs as incidence relations between different sets. Working in this mode, Cameron and Seidel developed *systems of linked symmetric designs* in [50, 53], using permutation groups, Steiner systems and quadratic forms. The pairwise incidence matrices satisfy condition (9.3), and so Theorem 14.2 shows that if these are the incidence matrices of $\mathcal{F}$ then $\mathcal{F}$ has universal balance.

Here is one way of producing the incidence matrices. Start with the Steiner system $\mathfrak{S}(5, 8, 24)$: see [52, 114]. Let $\mathcal{Q}$ be an octad in this, and let $\mathcal{P}$ be the remaining set of 16 points. If u, w are distinct points in $\mathcal{Q}$ then there are precisely 16 octads whose intersection with $\mathcal{Q}$ is $\{u, w\}$: their intersections with $\mathcal{P}$ give the blocks of a symmetric BIBD with 16 blocks of size six. Call this set of blocks $\mathcal{Q}_{uw}$, and denote the incidence matrix between $\mathcal{P}$ and $\mathcal{Q}_{uw}$ as $N_{0,uw}$.

Let x be a point in $\mathcal{Q} \setminus \{u, w\}$. Then each block of $\mathcal{Q}_{ux}$ intersects six blocks in $\mathcal{Q}_{uw}$ in one point, the remainder in three points. Define a block in $\mathcal{Q}_{uw}$ to be incident with one in $\mathcal{Q}_{ux}$ if they intersect in one point. Then

$$N_{uw,ux} \text{ is a linear combination of } N_{uw,0}N_{0,ux} \text{ and } J_{16,16}. \qquad (14.7)$$

Hence $N_{uw,ux}N_{ux,uw}$ is a linear combination of $N_{uw,0}N_{0,ux}N_{ux,0}N_{0,uw}$ and $J_{16,16}$, which is completely symmetric because both $N_{0,ux}N_{ux,0}$ and $N_{uw,0}N_{0,uw}$ are. Therefore the relation between $\mathcal{Q}_{ux}$ and $\mathcal{Q}_{uw}$ is another symmetric BIBD with 16 blocks of size six.

Moreover, equation (14.7) shows that the triple $\mathcal{P}$, $\mathcal{Q}_{ux}$, $\mathcal{Q}_{uw}$ satisfies the strong condition (9.3), as does the triple $\mathcal{Q}_{ux}$, $\mathcal{Q}_{uw}$, $\mathcal{Q}_{xw}$ and any triple of the form $\mathcal{Q}_{ux}$, $\mathcal{Q}_{uw}$, $\mathcal{Q}_{uv}$ with v in $\mathcal{Q} \setminus \{u, w, x\}$. Thus the relations among $\{\mathcal{P}, \mathcal{Q}_{ux}, \mathcal{Q}_{uw}, \mathcal{Q}_{xw}\}$ have universal balance, as have those among $\mathcal{P}$ and any collection of $\mathcal{Q}_{uz}$ for z in $\mathcal{Q} \setminus \{u\}$.

Can any of these systems of linked symmetric designs be realised as partitions of a single set of size 96? Theorem 14.1 shows that $s \leq 6$, so the whole of the second one cannot be realised without linear dependence. For $s = 2$, the pair of partitions simply gives a BIBD. Here we show how to deal with the case $s = 3$, and defer the more general case until Section 14.3.

If Φ is a block of $\mathcal{Q}_{uw}$ then there are six blocks of $\mathcal{Q}_{ux}$ which intersect it in a single point; conversely each point in Φ occurs in exactly one such intersection. Hence the set of triples (p, Φ, Ψ) for which $p \in \mathcal{P}$, Φ is a block of $\mathcal{Q}_{uw}$, Ψ is a block of $\mathcal{Q}_{ux}$ and $\Phi \cap \Psi = \{p\}$ has size 96 and admits $\{\mathcal{P}, \mathcal{Q}_{ux}, \mathcal{Q}_{uw}\}$ as a set of partitions with universal balance.

This construction for $m = 16$ is rather special and does not generalize. However, this family of incidence matrices can also be constructed by using quadratic forms, and this approach does generalize to higher powers of 4: see Section 14.3.

It is interesting to note that the approach of Cameron and Seidel was based on the idea of equal angles between subspaces, very much as in Section 7. In fact, the definition of universal balance at the start of this section can be interpreted as saying that every vector in $V_F \cap V_U^{\perp}$ makes the same angle with $V_{\mathcal{L}} \cap V_U^{\perp}$ (and this angle is neither zero nor $\pi/2$). There is more about the link with equal angles in [49], while [67] shows the interest of this topic to workers in quantum information.

14.3 Multi-stage Youden rectangles

Double Youden rectangles provide one generalization of Youden squares. A different generalization, given in [2, 10, 13, 40, 51, 54, 144, 145, 147, 151, 169, 171, 197, 218], has a set $\mathcal{F}$ of partitions with universal balance, and another partition G orthogonal to all of these. Vowden presented [171] at the Keele BCC in 1993 and [218] at the Stirling BCC in 1995, but the authors later admitted[4] that the second paper has errors. Many different names have been used: multi-stage or multi-letter, combined with Youden designs or squares or rectangles, and even Freeman–Youden rectangles. Here I choose a name that allows me to present this subsection and the next in a unified way.

Definition Let $n < m$. A *multi-stage $n \times m$ Youden rectangle* consists of a set $\mathcal{F}$ of at least two partitions of a set Ω of size nm into m parts of size n, such that $\mathcal{F}$ has universal balance, together with a partition G of Ω into n parts of size m such that $F \wedge G = E$ and $F \perp G$ for all F in $\mathcal{F}$.

---

[4]See Mathematical Reviews 1675100

	0			1			2			3			4			5			6		
0	0	0	0	1	1	1	2	2	2	3	3	3	4	4	4	5	5	5	6	6	6
1	1	2	4	2	3	5	3	4	6	4	5	0	5	6	1	6	0	2	0	1	3
2	2	4	1	3	5	2	4	6	3	5	0	4	6	1	5	0	2	6	1	3	0
4	4	1	2	5	2	3	6	3	4	0	4	5	1	5	6	2	6	0	3	0	1

Figure 35: A multi-stage 4×7 Youden rectangle with four stages: the columns give one stage, the first number in each cell gives the second stage, and the other two numbers in each cell give the remaining two stages

Put $s = |\mathcal{F}|$. If $s = 2$ then this is just a Youden square. Most other examples in the literature have $s = 3$.

Proposition 14.5 *If $n < m$ and there exists a multi-stage $n \times m$ Youden rectangle with s partitions into m parts then $s \leq n$.*

The proof is similar to the proof of Theorem 14.1.

For the case that $n = m - 1$, if there exists a set of $s - 1$ mutually orthogonal $m \times m$ Latin squares then use $s - 2$ of these in the construction in Section 14.2.1, taking the common transversal to be the positions of one letter in the remaining Latin square. Then the n other letters of that Latin square give the parts of a partition G orthogonal to all the other partitions.

For most other known sets of partitions with universal balance, the irony is that the simplest way to construct them is to make a multi-stage Youden rectangle and then ignore the distinguished partition G.

Here is a construction for $n \times m$ multi-stage Youden rectangles when m is an odd prime power q, $q \equiv 3 \pmod 4$ and n is either $(q-1)/2$ or $(q+1)/2$. It uses the doubly regular tournament derived from the set $\mathcal{S}$ of non-zero squares in $GF(q)$. For $n = (q+1)/2$, label the rows of a rectangle by the elements of $\{0\} \cup \mathcal{S}$, and label the columns by the elements of $GF(q)$. The rows form the parts of G. For $u \in \{0\} \cup \mathcal{S}$, define the partition F_u to be the kernel of the function mapping cell (i, j) to $ui + j$. Thus F_0 is just the partition into columns. If we use all $(q+1)/2$ such partitions, the bound in Proposition 14.5 is achieved.

For $n = (q-1)/2$, remove row 0 from the previous construction. To avoid linear dependence, at most $(q-1)/2$ of the partitions F_u can be used.

Many constructions of multi-stage Youden rectangles given in the literature use a version of the above method. Figure 35 shows the result when $m = q = 7$ and $n = 4 = (q+1)/2$.

A	B	C	D	E	F	G	H	I	J	K	L	M	N	O	P
A	B	C	D	E	F	G	H	I	J	K	L	M	N	O	P
A	B	C	D	E	F	G	H	I	J	K	L	M	N	O	P
H	G	F	E	D	C	B	A	P	O	N	M	L	K	J	I
G	H	E	F	C	D	A	B	O	P	M	N	K	L	I	J
B	A	D	C	F	E	H	G	J	I	L	K	N	M	P	O
K	L	I	J	O	P	M	N	C	D	A	B	G	H	E	F
J	I	L	K	N	M	P	O	B	A	D	C	F	E	H	G
D	C	B	A	H	G	F	E	L	K	J	I	P	O	N	M
M	N	O	P	I	J	K	L	E	F	G	H	A	B	C	D
I	J	K	L	M	N	O	P	A	B	C	D	E	F	G	H
E	F	G	H	A	B	C	D	M	N	O	P	I	J	K	L
O	P	M	N	K	L	I	J	G	H	E	F	C	D	A	B
F	E	H	G	B	A	D	C	N	M	P	O	J	I	L	K
L	K	J	I	P	O	N	M	D	C	B	A	H	G	F	E
P	O	N	M	L	K	J	I	H	G	F	E	D	C	B	A
C	D	A	B	G	H	E	F	K	L	I	J	O	P	M	N
N	M	P	O	J	I	L	K	F	E	H	G	B	A	D	C

Figure 36: A multi-stage 6×16 Youden rectangle with four stages: the columns give one stage, the first letter in each cell gives the second stage, the second and third letters give the other two stages

What about the constructions for powers of 4? After Preece's talk to the 1973 BCC in Aberystwyth, Cameron told him about the results in Section 14.2.3, and they pooled their ideas to produce the designs with $m = 16$ and $n = 6$ or 10 in [167]. These include the 6×16 design in Figure 36, where the columns and the letters in each of the three positions in each cell give four partitions with the incidence structure of $\{\mathcal{P}, \mathcal{Q}_{ux}, \mathcal{Q}_{uw}, \mathcal{Q}_{xw}\}$. The rows form another partition orthogonal to all of these.

It turns out that the Cameron–Seidel construction also has a link with coding theory. A set like $\{\mathcal{P}, \mathcal{Q}_{ux}, \mathcal{Q}_{uw}, \mathcal{Q}_{xw}\}$ gives a linear code and can be realised as a set of partitions. One like $\{\mathcal{P}, \mathcal{Q}_{ux}, \mathcal{Q}_{uw}, \mathcal{Q}_{uv}, \ldots\}$ gives a non-linear code (a Kerdock code) and cannot.

In spite of being a co-author of [167], Cameron was still not comfortable thinking in terms of different partitions of a single set. When he finally adjusted to this mindset, he generalized 16 to arbitrary powers 4^t of 4, with blocks of size $2^{2t-1} - 2^{t-1}$ or $2^{2t-1} + 2^{t-1}$, and produced the designs in both forms—as a set of incidence relations satisfying (9.3) and as multi-stage Youden rectangles. He spoke about these results at the 2001 BCC in Sussex: see [51].

There is also a link with another nice combinatorial object. The complete graph K_{16} on 16 vertices can be decomposed into three strongly regular graphs with valency five. Their edge-sets are disjoint, so their adjacency matrices A_1, A_2 and A_3 sum to $J_{16,16} - I_{16}$. If $\{i, j, \ell\} = \{1, 2, 3\}$ then $A_i^2 = 5I_{16} + 2(A_j + A_\ell)$ and $(I_{16} + A_i)(I_{16} + A_j) = 3J_{16,16} - 2(I_{16} + A_\ell)$. The elements of each of $\mathcal{P}$, $\mathcal{Q}_{ux}$, $\mathcal{Q}_{uw}$ and $\mathcal{Q}_{xw}$ can be labelled by the same set of size 16 in such a way that the three incidence matrices involving any one are $I_{16} + A_1$, $I_{16} + A_2$ and $I_{16} + A_3$. Each of these strongly regular graphs contains no triangles, and so together they give a 3-colouring of the edges of K_{16} with no monochromatic triangles. This shows that the corresponding Ramsey number is at least 17, as was first observed by Greenwood and Gleason in [84].

There is more! Replacing the 0 entries of A_i by 1 and the 1 entries by -1 gives a Hadamard matrix H_i, and $H_i H_j = 4H_\ell$. Therefore $\{I_{16}, 4^{-1}H_1, 4^{-1}H_2, 4^{-1}H_3\}$ forms an elementary Abelian 2-group. Higher even powers of 2 give a similar group formed from Hadamard matrices, and these lead to the multi-stage Youden rectangles in [51] and to the results in [49].

14.4   Multi-layered Youden rectangles

An $n \times m$ double Youden rectangle has two sets, $\mathcal{F}$ and $\mathcal{G}$, of partitions of Ω. Those in $\mathcal{F}$ have m parts of size n, while those in $\mathcal{G}$ have n parts of size m. If $F \in \mathcal{F}$ and $G \in \mathcal{G}$ then $F \perp G$. Each of $\mathcal{F}$ and $\mathcal{G}$ has universal balance, and $|\mathcal{F}| = |\mathcal{G}| = 2$. Choose the labelling so that $n < m$.

Multi-stage Youden rectangles satisfy all the same conditions, except that $|\mathcal{F}| \geq 2$ and $|\mathcal{G}| = 1$. When $|\mathcal{F}| = 2$ this is just a Youden square.

Preece introduced *triple Youden rectangles* in [163]. The naive reader might think that these would be like double Youden rectangles with $|\mathcal{F}| = |\mathcal{G}| = 3$. In fact, they have $|\mathcal{F}| = 2$ and $|\mathcal{G}| = 3$. He gave examples of size 4×13 and 7×15.

Preece and Morgan generalized this concept in [168] by allowing the size of $\mathcal{G}$ to vary. They changed the name to *multi-layered Youden rectangle*. Thus the partitions in $\mathcal{F}$ are regarded as stages, while those in $\mathcal{G}$ are regarded as layers.

Definition An $n \times m$ *multi-layered Youden rectangle* with s layers has two sets, $\mathcal{F}$ and $\mathcal{G}$, of partitions of a set Ω of size nm. Each partition in $\mathcal{F}$ has m parts of size n, and each partition in $\mathcal{G}$ has n parts of size m, where $n < m$. If $F \in \mathcal{F}$ and $G \in \mathcal{G}$ then $F \perp G$. Each of $\mathcal{F}$ and $\mathcal{G}$ has universal balance, $|\mathcal{F}| = 2$ and $|\mathcal{G}| = s$.

If $s = 1$ this is just a Youden square; if $s = 2$ it is a double Youden rectangle.

Because $|\mathcal{F}| = 2$, a multi-layered Youden rectangle can be shown as an $m \times m$ square with $m - n$ empty cells in each row and column, as in Figure 26(b). Each non-empty cell has s letters, one for each layer. When $s = 3$, these are called *triple Youden arrays*, not to be confused with the triple arrays in Section 13.

In fact, the definitions in [163, 168] do not specify universal balance. For $\mathcal{F}$, this is equivalent to pairwise balance, because $|\mathcal{F}| = 2$. In all the examples in [163, 168], $m - 1$ is a multiple of n. Therefore pairwise balance among the layers implies that they can all have their parts labelled by the same set in such a way that each incidence matrix between layers is equal to $I_n + [(m - 1)/n]J_{nn}$. Thus universal balance is guaranteed.

Preece and Morgan give an explicit construction which proves the following.

Theorem 14.6 *If q is an odd prime power, $m = 2q+1$ and $s \leq q-4$ then there exists a $q \times (2q + 1)$ multi-layered Youden rectangle with s layers.*

15   Resolved designs

So far, all designs have been described as collections of partitions of the set Ω of experimental units. Now we change that viewpoint.

Statisticians usually call each tuple of values of the non-inherent partitions a *treatment*. Denote by Γ the set of treatments.

Definition A design is *resolved* is there is a partition D of Ω such that every inherent non-trivial partition H of Ω satisfies $H \preccurlyeq D$ and every part of D contains each treatment exactly once.

In a resolved design, it is useful to look at the partitions of Γ induced within each part of D. In the next two subsections we do this for block designs and for row–column designs.

15.1   Block designs

Let B be the partition of Ω into blocks in a resolved block design. For $i = 1, \ldots, n_D$, let B_i be the partition of Γ induced by B in the i-th part of D. Properties of $\{B_1, \ldots, B_{n_D}\}$ give information about the block design.

Among the earliest examples of incomplete-block designs were the *square lattice* designs of Yates [229]. In these, $|\Gamma| = k_B^2$ and B_1, B_2,

..., B_{n_D} form the partitions of a $k_B \times k_B$ square into rows, columns and the letters of $n_D - 2$ mutually orthogonal Latin squares. The design in Figure 2 has this form, as do those made from it by removing one or two parts of D.

In a square lattice design, the set of partitions $\{B_1, \ldots, B_{n_D}\}$ forms an orthogonal array of strength two on Γ, with all parts of the same size. In [47], Bose defined *affine-resolvable* designs. Orthogonal arrays had not yet been invented, but now we can say that affine-resolved designs are precisely those resolved incomplete-block designs for which $\{B_1, \ldots, B_{n_D}\}$ forms an orthogonal array of strength two on Γ with all parts of the same size.

For example, the first four rows of the orthogonal array in Figure 7 give an affine-resolved incomplete-block design for 12 treatments in eight blocks of size six. More generally, if any row of a Hadamard matrix is normalized to have all its entries the same, then the remaining rows form an orthogonal array of strength two. If an affine-resolved design is made by using each of these rows once, then the block design is balanced in all the senses described in Section 5. Such designs are sometimes called *Hadamard 3-designs*: see [52, 96].

Many constructions of affine-resolved incomplete-block designs appear in [38], where it is proved that they are optimal among resolved designs.

If k_B^2 does not divide $|\Gamma|$ then an affine-resolvable design is not possible. In work [133] presented at the 1975 BCC in Aberdeen, Patterson and Williams observed that if $n_D = 2$ then $\{B_1, B_2\}$ may be regarded as the partitions B and L of an incomplete-block design with $n_B = n_L$. For example, the partitions C and L of the set of size 28 in Figure 3 can be used to give a resolved incomplete-block design for 28 treatments in two districts of seven blocks of size four. It was proved in [223] that the resolved design is optimal among resolved designs if and only if the symmetric incomplete-block design is optimal among IBDs of that size. This result gives a technique for finding other optimal resolved incomplete-block designs when $n_D = 2$: see [224].

If $n_D \geq 3$ and k_B^2 does not divide $|\Gamma|$ then we run into the problem discussed in Section 9: the pairwise relations among $B_1, B_2, \ldots, B_{n_D}$ do not suffice to give the overall properties of the design. Sets of partitions with universal balance have been recommended. When $|\Gamma| = k_B(k_B + 1)$, the construction in Section 14.2.1 gives resolved IBDs called *rectangular lattice* designs: see [41, 62, 90, 91, 222].

For other values of $|\Gamma|$, it may be possible to use a set of n_D partitions having a different type of universal balance. When $n_D = 3$ these give the families of resolved incomplete-block designs in [27].

15.2   Nested row–column designs

In a nested row–column design each part of D is a rectangle with partitions R and C into rows and columns, and $R \wedge C = E$. If it is resolved then $|\Gamma| = k_R k_C$ and we may consider the partitions $R_1, R_2, \ldots,$ R_{n_D} and $C_1, C_2, \ldots, C_{n_D}$ induced on Γ. Then $R_i \perp C_i$ for $i = 1, \ldots, n_D$.

If $k_R = k_C$ and n_D is small enough then we may be able to arrange for all of these partitions to be pairwise strictly orthogonal by using the rows, columns and letters of $2(n_D - 1)$ mutually orthogonal $k_R \times k_R$ Latin squares. These were called *quasi-Latin squares* when introduced by Yates in [230] but the name soon changed to *lattice square* designs in [231].

Otherwise, we may be able to choose the partitions so that $R_i \perp C_j$, $R_i \bowtie R_j$ and $C_i \bowtie C_j$ for $\{i, j\} \subseteq \{1, \ldots, n_D\}$ with $i \neq j$. When $n_D = 2$ this gives a different use for double Youden rectangles, as explained in [39], where it is proved that these are optimal among resolved nested row–column designs. When $n_D \geq 3$ we need to consider the multi-way relations among $\{R_1, \ldots, R_{n_D}\}$ and $\{C_1, \ldots, C_{n_D}\}$, so some of the designs in Section 14.3 may be suitable.

16   Factorial designs in blocks

An experiment is called *factorial* if more than one partition is of interest and can have its parts allocated by the experimenter, as in Example 1.6 and Section 7.2.2. If the only inherent partition is the partition B into blocks, and there are two treatment factors F and G, then it may be useful to consider the set Δ of all triples of values of F, G and B that occur. If we show the parts of F and G as rows and columns, then usually $F \wedge G \neq E$, unlike in classical row–column designs.

Here we give two examples, considering B, F and G as partitions of Δ, to show the effect of different practical constraints.

Example 16.1 Modify Example 1.6 so that there are still six varieties of lettuce and five watering regimes, but now the ten gardens each have room for six vegetable patches and there are no practical constraints. Using a resolved design, as in Section 15.1, it is natural to think of the watering regimes and lettuce varieties as the rows and columns of a 5×6 rectangle, with five of the blocks corresponding to letters of a Latin square with the last column repeated, and the remaining five blocks corresponding to the letters of a second Latin square, orthogonal to the first, also with the last column repeated. This can give the design in Figure 37.

Here Δ is the same as Ω. The design has $F \perp G$, $B \perp G$, and $F \rhd B$. Thus it is of type C.5 from Section 10.4.

A	F	B	G	C	H	D	I	E	J	E	J
E	I	A	J	B	F	C	G	D	H	D	H
D	G	E	H	A	I	B	J	C	F	C	F
C	J	D	F	E	G	A	H	B	I	B	I
B	H	C	I	D	J	E	F	A	G	A	G

Figure 37: Design in Example 16.1: rows represent watering regimes, columns represent lettuce varieties, letters A–E represent the first five blocks and letters F–J represent the other five blocks

A	B	A	F	A	I	F	I	B	F	B	I
B	C	G	J	C	G	C	J	B	G	B	J
C	D	D	F	C	H	C	F	F	H	D	H
D	E	D	G	G	I	E	I	E	G	D	I
A	E	A	J	A	H	E	J	E	H	H	J

Figure 38: Design in Example 16.2: rows represent drugs, columns represent types of cancer and letters represent medical centres (blocks)

Example 16.2 Valerii Fedorov[5] posed the problem of assigning cancer types (C) and cancer drugs (R) to medical centres (B) in such a way that no centre deals with very many types of drug or cancer, each centre allocates drugs to patients in such a way that each of its drugs is tested on each of its types of cancer, cancer types and drugs are both balanced with respect to medical centres, and each drug is paired with each cancer type at the same number of medical centres. Since we cannot foresee how many patients will be included in the trial, it make sense to consider the set Δ of all triples of values of R, C and B that occur.

Figure 38 shows such a design for five drugs, six cancer types and ten medical centres: rows represent drugs, columns represent cancer types, and letters represent medical centres (also called 'blocks'). It was made by assigning a subset of two drugs and a subset of three cancer types to each medical centre, and then giving each medical centre all six combinations of its assigned drugs and cancer types. This method of construction ensures that R and C have adjusted orthogonality with respect to blocks, in addition to the requirements $R \perp C$, $R \blacktriangleright B$ and $C \blacktriangleright B$. However, the relationships between R and B, and C and B, are not generalized binary, so the balance is not strict. More designs for this problem are in [34].

---

[5] Meeting at the Isaac Newton Institute in July 2015

17 Some open problems

1. Divide the cases in list E in Section 10.6 as finely as those in the other lists in Section 10. Prove that the resulting list gives all possibilities, and give an example of each.

2. Is there an interesting connection, using finite projective geometry or group theory, between the double Youden rectangle in Figure 30 and the non-extremal triple array in Figure 33?

3. Prove Conjecture 13.5 or find a counter-example.

4. The (hopefully) more unified approach to adjusted balance presented in Sections 9 and 14 is new, so may have errors or things that can be done better. Improve it and take it further.

5. Are there any other families of square matrices, apart from those given in Section 14.2, which can occur as the incidence matrices of a set of partitions with universal balance?

6. Are there designs like those in Sections 14.3–14.4 in which both $|\mathcal{F}|$ and $|\mathcal{G}|$ are greater than two?

Acknowledgements I am very grateful to Donald Preece (R.I.P.) for many illuminating conversations about the topics in this paper over many years. I also thank Chris Brien, Peter Cameron, John Eccleston, Ulrike Grömping, Jonathan Hall, Tomas Nilson, Gordon Royle, Ken Russell and Jennifer Seberry for their very helpful inputs while I was writing this paper.

References

[1] R. Julian R. Abel, Malcolm Grieg & D. H. Rees, Existence of OBIBDs with $k = 4$ with and without nesting, *Discrete Mathematics* **266** (2003), 3–36.

[2] K. Afsarinejad & A. Hedayat, Some contributions to the theory of multistage Youden design, *Annals of Statistics* **3** (1975), 707–711.

[3] Hira Lal Agrawal, Two way elimination of heterogeneity, *Calcutta Statistical Association Bulletin* **15** (1966), 32–38.

[4] Hira Lal Agrawal, Some generalizations of distinct representatives with applications to statistical design, *Annals of Mathematical Statistics* **37** (1966), 525–528.

[5] Hira Lal Agrawal, Some systematic methods of construction for two-way elimination of heterogeneity, *Calcutta Statistical Association Bulletin* **15** (1966), 93–108.

[6] Hira Lal Agrawal, Some methods of construction of designs for two-way elimination of heterogeneity—1, *Journal of the American Statistical Association* **61** (1966), 1153–1171.

[7] H. L. Agrawal & R. I. Mishra, Some methods of constructing 4DIB designs, *Calcutta Statistical Association Bulletin* **20** (1971), 89–92.

[8] H. L. Agrawal & J. Prasad, Some methods of construction of balanced incomplete block designs with nested rows and columns, *Biometrika* **69** (1982), 481–483.

[9] H. L. Agrawal & J. Prasad, On construction of balanced incomplete block designs with nested rows and columns, *Sankhyā, Series B* **45** (1983), 345–350.

[10] H. L. Agrawal & K. L. Sharma, On construction of two-way designs, *Journal of the Indian Statistical Association* **13** (1975), 1–31.

[11] H. L. Agrawal & K. L. Sharma, On construction of four-dimensional incomplete block designs, *Journal of the American Statistical Association* **73** (1978), 844–849.

[12] D. A. Anderson & J. A. Eccleston, On the construction of a class of efficient row–column designs, *Journal of Statistical Planning and Inference* **11** (1985), 131–134.

[13] D. A. Anderson & W. T. Federer, Multidimensional balanced designs, *Communications in Statistics – Theory and Methods* **5** (1976), 1193–1204.

[14] I. Anderson, A hundred years of whist tournaments, *Journal of Combinatorial Mathematics and Combinatorial Computation* **19** (1995), 129–150.

[15] A. Ash, Generalized Youden designs: constructions and tables, *Journal of Statistical Planning and Inference* **5** (1981), 1–25.

[16] E. F. Assmus, Jr, On the theory of designs, in *Surveys in Combinatorics, 1989* (ed. J. Siemons), *London Mathematical Society Lecture Notes Series*, 141, Cambridge University Press, Cambridge (1989), pp. 1–21.

[17] Sunanda Bagchi, An infinite series of adjusted orthogonal designs with replication two, *Statistica Sinica* **6** (1996), 975–987.

[18] Sunanda Bagchi, On two–way designs, *Graphs and Combinatorics* **14** (1998), 313–319.

[19] S. Bagchi & E. E. M. van Berkum, On the optimality of a new class of adjusted orthogonal designs, *Journal of Statistical Planning and Inference* **28** (1991), 61–65.

[20] Sunanda Bagchi, A. C. Mukhopadhyay & Bikas K. Sinha, A search for optimal nested row–column designs, *Sankhyā, Series B* **52** (1990), 93–104.

[21] S. Bagchi & K. R. Shah, On the optimality of a class of row–column designs, *Journal of Statistical Planning and Inference* **23** (1989), 397–402.

[22] R. A. Bailey, Designs: mappings between structured sets, in *Surveys in Combinatorics, 1989* (ed. J. Siemons), *London Mathematical Society Lecture Notes Series*, 141, Cambridge University Press, Cambridge (1989), pp. 22–51.

[23] R. A. Bailey, Orthogonal partitions in designed experiments, *Designs, Codes and Cryptography* **8** (1996), 45–77.

[24] R. A. Bailey, Choosing designs for nested blocks, *Listy Biometryczne* **36** (1999), 85–126.

[25] R. A. Bailey, Resolved designs viewed as sets of partitions, in *Combinatorial designs and their applications* (eds. Fred C. Holroyd, Kathleen A. S. Quinn, Chris Rowley & Bridget S. Webb), *Research Notes in Mathematics*, 403, Chapman & Hall, Boca Raton (1999), pp. 17–47.

[26] R. A. Bailey, *Association Schemes: Designed Experiments, Algebra and Combinatorics, Cambridge Studies in Advanced Mathematics*, 84, Cambridge University Press, Cambridge (2004).

[27] R. A. Bailey, Six families of efficient resolvable designs in three replicates, *Metrika* **62** (2005), 161–173.

[28] R. A. Bailey, *Design of Comparative Experiments, Cambridge Series in Statistical and Probabilistic Mathematics*, Cambridge University Press, Cambridge (2008).

[29] R. A. Bailey, Donald Arthur Preece: A life in statistics, mathematics and music, arXiv 1402.2220.

[30] R. A. Bailey, Structures defined by factors, in *Handbook of Design and Analysis of Experiments* (eds. Angela Dean, Max Morris, John Stufken & Derek Bingham), *Chapman and Hall/ CRC Handbooks of Modern Statistical Methods*, Chapman and Hall/ CRC, Boca Raton (2015), pp. 371–414.

[31] R. A. Bailey & C. J. Brien, Randomization-based models for multi-tiered experiments: I. A chain of randomizations, *Annals of Statistics* **44** (2016), 1131–1164.

[32] R. A. Bailey & Peter J. Cameron, What is a design? How should we classify them?, *Designs, Codes and Cryptography* **44** (2007), 223–238.

[33] R. A. Bailey & Peter J. Cameron, Combinatorics of optimal designs, in *Surveys in Combinatorics 2009* (eds. Sophie Huczynska, James D. Mitchell & Colva Roney-Dougal), *London Mathematical Society Lecture Note Series*, 365, (2009), pp. 19–73.

[34] R. A. Bailey & Peter J. Cameron, Designs which allow each medical centre to treat only a limited number of cancer types with only a limited number of drugs, preprint, University of St Andrews, 2016.

[35] R. A. Bailey, Sandra S. Ferreira, Dário Ferreia & Célia Nunes, Estimability of variance components when all model matrices commute, *Linear Algebra and its Applications* **492** (2016), 144–160.

[36] R. A. Bailey, D. C. Goldrei & D. F. Holt, Block designs with block size two, *Journal of Statistical Planning and Inference* **10** (1984), 257–263.

[37] R. A. Bailey & P. Heidtmann, Extremal row–column designs with maximal balance and adjusted orthogonality, preprint, Goldsmiths' College, University of London, 1994.

[38] R. A. Bailey, H. Monod & J. P. Morgan, Construction and optimality of affine-resolvable designs, *Biometrika* **82** (1995), 187–200.

[39] R. A. Bailey & H. D. Patterson, A note on the construction of row-and-column designs with two replicates, *Journal of the Royal Statistical Society, Series B* **53** (1991), 645–648.

[40] R. A. Bailey, D. A. Preece & C. A. Rowley, Randomization for a balanced superimposition of one Youden square on another, *Journal of the Royal Statistical Society, Series B* **57** (1995), 459–469.

[41] R. A. Bailey & T. P. Speed, Rectangular lattice designs: efficiency factors and analysis, *Annals of Statistics* **14** (1986), 874–895.

[42] Simon T. Bate & Marion J. Chatfield, Identifying the structure of the experimental design, *Journal of Quality Technology* **48** (2016), 343–364.

[43] Simon T. Bate & Marion J. Chatfield, Using the structure of the experimental design and the randomization to construct a mixed model, *Journal of Quality Technology* **48** (2016), 365–387.

[44] Gerhard Behrendt, Equivalence systems with finitely many relations, *Monatshefte für Mathematik* **103** (1987), 77–83.

[45] Thomas Beth, Dieter Jungnickel & Hanfried Lenz, *Design Theory*, Volume I, second edition, *Encyclopedia of Mathematics and its Applications*, 69, Cambridge University Press, Cambridge (1999).

[46] R. C. Bose, On the construction of balanced incomplete block designs, *Annals of Eugenics* **9** (1939), 353–399.

[47] R. C. Bose, A note on the resolvability of balanced incomplete block designs, *Sankhā* **6** (1942), 105–110.

[48] Wayne Broughton, Admissible parameters of symmetric designs satisfying $v = 4(k - \lambda) + 2$ and symmetric designs with inner balance, *Designs, Codes and Cryptography* **73** (2014), 77–83.

[49] A. R. Calderbank, P. J. Cameron, W. M. Kantor & J. J. Seidel, $\mathbb{Z}_4$-Kerdock codes, orthogonal spreads, and extremal Euclidean line-sets, *Proceedings of the London Mathematical Society, Series 3* **75** (1997), 436–480.

[50] Peter J. Cameron, On groups with several doubly-transitive permutation representations, *Mathematische Zeitschrift* **128** (1972), 1–14.

[51] P. J. Cameron, Multi-letter Youden rectangles from quadratic forms, *Discrete Mathematics* **266** (2003), 143–151.

[52] P. J. Cameron & J. H. van Lint, *Designs, Graphs, Codes and their Links, London Mathematical Society Student Texts*, 22, Cambridge University Press, Cambridge (1991).

[53] P. J. Cameron & J. J. Seidel, Quadratic forms over GF(2), *Proceedings of the Koninklijke Nederlandse Akademie van Wetenschappen, Series A* **76** (1973), 1–8.

[54] B. D. Causey, Some examples of multi-dimensional incomplete block designs, *Annals of Mathematical Statistics* **39** (1968), 1577–1590.

[55] J. Y. Chang & W. I. Notz, A method for constructing universally optimal block designs with nested rows and columns, *Utilitas Mathematica* **38** (1990), 263–276.

[56] C.-S. Cheng, A method for constructing balanced incomplete-block designs with nested rows and columns, *Biometrika* **73** (1986), 695–700.

[57] C. Christofi, Enumerating 4×5 and 5×6 double Youden rectangles, *Discrete Mathematics* **125** (1994), 129–135.

[58] C. Christofi, On the number of 6×7 double Youden rectangles, *Ars Combinatoria* **47** (1997), 223–241.

[59] G. M. Clarke, A second set of treatments in a Youden square design, *Biometrics* **19** (1963), 98–104.

[60] G. M. Clarke, Four-way balanced designs based on Youden squares with 5, 6, or 7 treatments, *Biometrics* **23** (1967), 803–812.

[61] Dominique Collombier, *Plans d'expérience factoriels, Mathématiques & Applications*, 21, Springer-Verlag, Berlin (1996).

[62] L. C. A. Corsten, Rectangular lattices revisited, in *Linear Statistical Inference* (eds. T. Caliński & W. Klonecki), *Lecture Notes in Statistics*, 35, Springer, Berlin (1985), pp. 29–38.

[63] G. Dall'Aglio, Blocs incomplets éqilibrés orthogonaux, *Colloques Internationaux du Centre National de la Recherche Scientifique* **110** (1963), 105–214.

[64] A. Dey, U. S. Das & A. K. Banerjee, Construction of nested balanced incomplete block designs, *Calcutta Statistical Association Bulletin* **35** (1986), 161–167.

[65] Jean Doyen, Designs and automorphism groups, in *Surveys in Combinatorics, 1989* (ed. J. Siemons), *London Mathematical Society Lecture Notes Series*, 141, Cambridge University Press, Cambridge (1989), pp. 75–83.

[66] V. Duquenne, What can lattices do for experimental designs?, *Mathematical Social Sciences* **11** (1986), 243–281.

[67] Thomas Durt, Berthold-Georg Englert, Ingemar Bengtsson & Karol Życzkowski, On mutually unbiased bases, *International Journal of Quantum Information* **8** (2010), 535–640.

[68] J. A. Eccleston & J. A. John, Recovery of row and column information in row-column designs with adjusted orthogonality, *Journal of the Royal Statistical Society, Series B* **48** (1986), 238–243.

[69] J. A. Eccleston, J. A. John & D. Whitaker, Some row–column designs with adjusted orthogonality, *Journal of Statistical Planning and Inference* **36** (1993), 323–330.

[70] J. A. Eccleston & J. Kiefer, Relationships of optimality for individual factors of a design, *Journal of Statistical Planning and Inference* **5** (1981), 213–219.

[71] J. A. Eccleston & C. A. McGilchrist, Algebra of a row–column design, *Journal of Statistical Planning and Inference* **12** (1985), 305–310.

[72] J. Eccleston & K. Russell, Connectedness and orthogonality in multifactor designs, *Biometrika* **62** (1975), 341–345.

[73] J. A. Eccleston & K. G. Russell, Adjusted orthogonality in nonorthogonal designs, *Biometrika* **64** (1977), 339–345.

[74] J. A. Eccleston & A. Street, Construction for adjusted orthogonal designs, *Ars Combinatoria* **28** (1990), 117–128.

[75] W. L. Edge, The geometry of the linear fractional group LF(4, 2), *Proceedings of the London Mathematical Society, Series 3* **4** (1954), 317–342.

[76] P. Eendebak & E. Schoen, Complete series of non-isomorphic orthogonal arrays, http://pietereendebak.nl/oapage/

[77] D. J. Finney, The fractional replication of factorial experiments, *Annals of Eugenics* **12** (1945), 291–301.

[78] R. A. Fisher, *Design of Experiments*, Oliver and Boyd, Edinburgh (1935).

[79] R. A. Fisher, An examination of the different possible solutions of a problem in incomplete blocks, *Annals of Eugenics* **10** (1940), 52–75.

[80] R. A. Fisher & F. Yates, *Statistical Tables for Biological, Agricultural and Medical Research*, Oliver and Boyd, Edinburgh (1938).

[81] Dmitri G. Fon-Der-Flaass, Arrays of distinct representatives — a very simple NP-complete problem, *Discrete Mathematics* **171** (1997), 295–298.

[82] G. H. Freeman, Some non-orthogonal partitions of 4×4, 5×5 and 6×6 Latin squares, *Annals of Mathematical Statistics* **37** (1966), 666–681.

[83] Dennis C. Gilliland, A note on orthogonal partitions and some well-known structures in design of experiments, *Annals of Statistics* **5** (1977), 565–570.

[84] R. E. Greenwood & A. M. Gleason, Combinatorial relations and chromatic graphs, *Canadian Journal of Mathematics* **7** (1955), 1–7.

[85] M. Greig & D. H. Rees, Existence of balanced incomplete block designs for many sets of treatments, *Discrete Mathematics* **261** (2003), 299–324.

[86] Ulrike Grömping & R. A. Bailey, Regular fractions of factorial arrays, in *mODa 11—Advances in Model-Oriented Design and Analysis* (eds. Joachim Kunert, Christine H. Müller & Anthony C. Atkinson), Springer International Publishing, Switzerland (2016), pp. 143–151.

[87] Heiko Großmann, Automating the analysis of variance of orthogonal designs, *Computational Statistics and Data Analysis* **70** (2014), 1–18.

[88] J. I. Hall, On identifying $PG(3, 2)$ and the complete 3-design on seven points, *Annals of Discrete Mathematics* **7** (1980), 131–141.

[89] P. Hall, On representatives of subsets, *Journal of the London Mathematical Society* **10** (1935), 26–30.

[90] B. Harshbarger, Preliminary report on the rectangular lattices, *Biometrics* **2** (1946), 115–119.

[91] B. Harshbarger, Triple rectangular lattices, *Biometrics* **5** (1949), 1–13.

[92] A. Hedayat & W. T. Federer, Pairwise and variance balanced incomplete block designs, *Annals of the Institute of Statistical Mathematics* **26** (1974), 331–338.

[93] A. Hedayat, E. T. Parker & W. T. Federer, The existence and construction of two families of designs for two successive experiments, *Biometrika* **57** (1970), 351–355.

[94] A. Hedayat, E. Seiden & W. T. Federer, Some families of designs for multistage experiments: mutually balanced Youden designs when the number of treatments is prime power or twin primes. I., *Annals of Mathematical Statistics* **43** (1972), 1517–1527.

[95] A. S. Hedayat, N. J. A. Sloan & J. Stufken, *Orthogonal Arrays*, Springer-Verlag, New York (1999).

[96] A. Hedayat & W. D. Wallis, Hadamard matrices and their applications, *Annals of Statistics* **6** (1978), 1184–1238.

[97] J. W. P. Hirschfeld, *Finite Projective Spaces of Three Dimensions*, *Oxford Mathematical Monographs*, Oxford University Press, Oxford (1985).

[98] T. N. Hoblyn, S. C. Pearce & G. H. Freeman, Some considerations in the design of successive experiments in fruit plantations, *Biometrics* **10** (1954), 503–515.

[99] A. T. James, The relationship algebra of an experimental design, *Annals of Mathematical Statistics* **28** (1957), 993–1002.

[100] A. T. James & G. N. Wilkinson, Factorization of the residual operator and canonical decomposition of nonorthogonal factors in the analysis of variance, *Biometrika* **58** (1971), 279–294.

[101] M. Jimbo & S. Kuriki, Construction of nested designs, *Ars Combinatoria* **16** (1983), 275–285.

[102] J. A. John & J. A Eccleston, Row–column α-designs, *Biometrika* **73** (1986), 301–306.

[103] Byron Jones, Algorithms to search for optimal row-and-column designs, *Journal of the Royal Statistical Society, Series B* **41** (1979), 210–216.

[104] Byron Jones, Algorithm AS156: Combining two component designs to form a row-and-column design, *Applied Statistics* **29** (1980), 334–345.

[105] R. Morley Jones, On a property of incomplete blocks, *Journal of the Royal Statistical Society, Series B* **21** (1959), 172–179.

[106] B. Kassanis & A. Kleczkowski, Inactivation of a strain of tobacco necrosis virus and of the RNA isolated from it, *Photochemistry and Photobiology* **4** (1965), 209–214.

[107] O. Kempthorne, G. Zyskind, S. Addelman, T. N. Throckmorton & R. N. White, *Analysis of Variance Procedures*, Aeronautical Research Laboratory, Ohio, Report No. 149, 1961.

[108] J. Kiefer, On the nonrandomized optimality and randomized nonoptimality of symmetrical designs, *Annals of Mathematical Statistics* **29** (1958), 675–699.

[109] J. Kiefer, Balanced block designs and generalized Youden designs, I. Construction (patchwork), *Annals of Statistics* **3** (1975), 109–118.

[110] K. Kishen, Symmetrical unequal block arrangements, *Sankhyā* **5** (1940–1941), 329–344.

[111] A. Kleczkowski, Interpreting relationships between the concentrations of plant viruses and numbers of local lesions, *Journal of General Microbiology* **4** (1950), 53–69.

[112] A. Kobilinsky & H. Monod, Experimental designs generated by group morphisms, *Scandinavian Journal of Statistics* **18** (1991), 119–134.

[113] A. M. Kshirsagar, On balancing in designs in which heterogeneity is eliminated in two directions, *Calcutta Statistical Assocation Bulletin* **7** (1957), 161–166.

[114] Heinz Lüneberg, *Transitive Erweiterungen endlicher Permutationsgruppen*, Lecture Notes in Mathematics, 4, Springer-Verlag, Berlin (1969).

[115] H. B. Mann, The algebra of a linear hypothesis, *Annals of Mathematical Statistics* **31** (1960), 1–15.

[116] John P. McSorley, Double arrays, triple arrays, and balanced grids with $v = r + c - 1$, *Designs, Codes and Cryptography* **37** (2005), 313–318.

[117] John P. McSorley & Nicholas C. Phillips, Complete enumeration and properties of binary pseudo-Youden designs PYD$(9, 6, 6)$, *Journal of Statistical Planning and Inference* **137** (2007), 1464–1473.

[118] John P. McSorley, N. C. K. Phillips, W. D. Wallis & J. L. Yucas, Double arrays, triple arrays and balanced grids, *Designs, Codes and Cryptography* **35** (2005), 21–45.

[119] J. P. Morgan, Nested designs, in *Design and Analysis of Experiments* (eds. S. Ghosh & C. R. Rao), *Handbook of Statistics*, 13, Elsevier Science, Amsterdam (1996), pp. 939–976.

[120] J. P. Morgan, Properties of superimposed BIBDs, *Journal of Statistical Planning and Inference* **73** (1998), 135–148.

[121] J. P. Morgan, D. A. Preece & D. H. Rees, Nested balanced incomplete block designs, *Discrete Mathematics* **231** (2001), 351–389.

[122] John P. Morgan & Nizam Uddin, Optimality and construction of nested row and column designs, *Journal of Statistical Planning and Inference* **37** (1993), 81–93.

[123] J. P. Morgan & Nizam Uddin, Optimal blocked main effects plans with nested rows and columns and related designs, *Annals of Statistics* **24** (1996), 1185–1208.

[124] Rahul Mukerjee & Sudhir Gupta, Geometric construction of balanced block designs with nested rows and columns, *Discrete Mathematics* **91** (1991), 105–108.

[125] G. Nageswara Rao, Further contributions to balanced generalized two-way elimination of heterogeneity designs, *Sankhyā* **38** (1976), 72–79.

[126] J. A. Nelder, The analysis of randomized experiments with orthogonal block structure. I. Block structure and the null analysis of variance, *Proceedings of the Royal Society of London, Series A* **283** (1965), 147–162.

[127] J. A. Nelder, The analysis of randomized experiments with orthogonal block structure. II. Treatment structure and the general analysis of variance, *Proceedings of the Royal Society of London, Series A* **283** (1965), 163–178.

[128] J. A. Nelder, The combination of information in generally balanced designs, *Journal of the Royal Statistical Society, Series B* **30** (1968), 303–311.

[129] Tomas Nilson, Row-column designs with adjusted orthogonality, Masters thesis, Mid Sweden University, Department of Engineering, Physics and Mathematics, 2007.

[130] Tomas Nilson & Peter J. Cameron, Triple arrays from difference sets, arXiv 1609.00152.

[131] Tomas Nilson & Pia Heidtmann, Inner balance of symmetric designs, *Designs, Codes and Cryptography* **71** (2014), 247–260.

[132] Tomas Nilson & Lars-Daniel Öhman, Triple arrays and Youden squares, *Designs, Codes and Cryptography* **75** (2015), 429–451.

[133] H. D. Patterson & E. R. Williams, Some theoretical results on general block designs, *Congressus Numerantium* **15** (1976), 489–496.

[134] S. C. Pearce, The use and classification of non-orthogonal designs, *Journal of the Royal Statistical Society, Series A* **126** (1963), 353–377.

[135] S. C. Pearce, The efficiency of block designs in general, *Biometrika* **57** (1970), 339–346.

[136] S. C. Pearce & J. Taylor, The changing of treatments in a long-term trial, *Journal of Agricultural Science* **38** (1948), 402–410.

[137] N. C. K. Phillips & D. A. Preece, Finding double Youden rectangles, in *Designs 2002: Further Computational and Constructive Design Theory* (ed. W. D. Wallis), *Mathematics and its Applications*, 563, Kluwer Academic Publishing, Boston MA (2003), pp. 301–315.

[138] N. C. K. Phillips, D. A. Preece & D. H. Rees, Double Youden rectangles for the four biplanes with $k = 9$, *Journal of Combinatorial Mathematics and Combinatorial Computing* **44** (2003), 169–176.

[139] N. C. K. Phillips, D. A. Preece & W. D. Wallis, The seven classes of 5×6 triple arrays, *Discrete Mathematics* **293** (2005), 213–218.

[140] N. C. K. Phillips & W. D. Wallis, An elusive array, *Bulletin of the Institute of Combinatorics and its Applications* **39** (2003), 39–40.

[141] Giovanni Pistone & Maria-Piera Rogantin, Indicator function and complex coding for mixed fractional factorial designs, *Journal of Statistical Planning and Inference* **138** (2008), 787–802.

[142] R. L. Plackett & J. P. Burman, The design of optimum multifactorial experiments, *Biometrika* **33** (1946), 305–325.

[143] Richard F. Potthoff, Three-factor additive designs more general than the Latin square, *Technometrics* **4** (1962), 187–208.

[144] Richard F. Potthoff, Some illustrations of four-dimensional incomplete block constructions, *Calcutta Statistical Association Bulletin* **12** (1963), 19–30.

[145] D. A. Preece, Some row and column designs for two sets of treatments, *Biometrics* **22** (1966), 1–25.

[146] D. A. Preece, Classifying Youden rectangles, *Journal of the Royal Statistical Society, Series B* **28** (1966), 118–130.

[147] D. A. Preece, Some balanced incomplete block designs for two sets of treatments, *Biometrika* **53** (1966), 497–506.

[148] D. A. Preece, Nested balanced incomplete block designs, *Biometrika* **54** (1967), 479–486.

[149] D. A. Preece, Balanced 6 × 6 designs for 9 treatments, *Sankhyā, Series B* **30** (1968), 443–446.

[150] D. A. Preece, Some new balanced row-and-column designs for two non-interacting sets of treatments, *Biometrics* **27** (1971), 426–430.

[151] D. A. Preece, Non-orthogonal Graeco-Latin designs, in *Combinatorial Mathematics IV* (eds. Louis R. A. Casse & Walter D. Wallis), *Lecture Notes in Mathematics*, 560, Springer-Verlag, Berlin (1976), pp. 7–26.

[152] D. A. Preece, Some designs based on 11×5 Youden 'squares', *Utilitas Mathematica* **9** (1976), 139–146.

[153] D. A. Preece, A second domain of balanced 6 × 6 designs for nine equally-replicated treatments, *Sankhyā, Series B* **38** (1976), 192–194.

[154] D. A. Preece, Orthogonality and designs: a terminological muddle, *Utilitas Mathematica* **12** (1977), 201–223.

[155] D. A. Preece, Balance and designs: another terminological tangle, *Utilitas Mathematica* **21C** (1982), 85–186.

[156] D. A. Preece, Some partly cyclic 13 × 4 Youden 'squares' and a balanced arrangement for a pack of cards, *Utilitas Mathematica* **22** (1982), 255–263.

[157] D. A. Preece, Genstat analyses for complex balanced designs with non-interacting factors, *Genstat Newsletter* **21** (March 1988), 33–45.

[158] D. A. Preece, Fifty years of Youden squares: a review, *Bulletin of the Institute of Mathematics and its Applications* **26** (1990), 65–75.

[159] D. A. Preece, Double Youden rectangles of size 6 × 11, *Mathematical Scientist* **16** (1991), 41–45.

[160] D. A. Preece, Enumeration of some 7 × 15 Youden squares and construction of some 7 × 15 double Youden rectangles, *Utilitas Mathematica* **41** (1992), 51–62.

[161] D. A. Preece, A set of double Youden rectangles of size 8 × 15, *Ars Combinatoria* **36** (1993), 215–219.

[162] D. A. Preece, Double Youden rectangles—an update with examples of size 5 × 11, *Discrete Mathematics* **125** (1994), 309–317.

[163] D. A. Preece, Triple Youden rectangles: A new class of fully balanced combinatorial arrangements, *Ars Combinatoria* **37** (1994), 175–182.

[164] D. A. Preece, Multi-factor balanced block designs with complete adjusted orthogonality for all pairs of treatment factors, *Australian Journal of Statistics* **38** (1996), 223–230.

[165] D. A. Preece, Some 6 × 11 Youden squares and double Youden rectangles, *Discrete Mathematics* **167/168** (1997), 527–541.

[166] D. A. Preece, P. W. Brading, C. W. H. Lam & M. Côté, Balanced 6 × 6 designs for 4 equally replicated treatments, *Discrete Mathematics* **125** (1994), 319–327.

[167] D. A. Preece & P. J. Cameron, Some new fully-balanced Graeco-Latin Youden 'squares', *Utilitas Mathematica* **8** (1975), 193–204.

[168] D. A. Preece & J. P. Morgan, Multi-layered Youden rectangles, *Journal of Combinatorial Designs*, **25** (2017), 75–84. doi:10.1002/jcd.21252

[169] D. A. Preece & N. C. K. Phillips, A new type of Freeman-Youden rectangle, *Journal of Combinatorial Mathematics and Combinatorial Computing* **25** (1997), 65–78.

[170] D. A. Preece, D. H. Rees & J. P. Morgan, Doubly nested balanced incomplete block designs, *Congressus Numerantium* **137** (1999), 5–18.

[171] D. A. Preece & B. J. Vowden, Graeco-Latin squares with embedded balanced superpositions of Youden squares, *Discrete Mathematics* **138** (1995), 353–363.

[172] D. A. Preece, B. J. Vowden & N. C. K. Phillips, Double Youden rectangles of sizes $p \times (2p+1)$ and $(p+1) \times (2p+1)$, *Ars Combnatoria* **51** (1999), 161–171.

[173] D. A. Preece, B. J. Vowden & N. C. K. Phillips, Double Youden rectangles of sizes $(p + 1) \times (p^2 + p + 1)$, *Utilitas Mathematica* **59** (2001), 139–154.

[174] D. A. Preece, W. D. Wallis & J. L. Yucas, Paley triple arrays, *Australasian Journal of Combinatorics* **33** (2005), 237–246.

[175] The R Project for Statistical Computing, https://www.r-project.org.

[176] D. Raghavarao, A note on some balanced generalized two-way elimination of heterogeneity designs, *Journal of the Indian Society of Agricultural Statistics* **22** (1970), 49–52.

[177] D. Raghavarao, *Constructions and Combinatorial Problems in Design of Experiments*, John Wiley and Sons, New York (1971).

[178] D. Raghavarao & W. T. Federer, On connectedness in two-way elimination of heterogeneity designs, *Annals of Statistics* **3** (1975), 730–735.

[179] D. Raghavarao & G. Nageswararao, A note on a method of construction of designs for two-way elimination of heterogeneity, *Communications in Statistics* **3** (1974), 197–199.

[180] D. Raghavarao & K. R. Shah, A class of D_0 designs for two-way elimination of heterogeneity, *Communications in Statistics—Theory and Methods* **9** (1980), 75–80.

[181] C. R. Rao, Factorial arrangements derivable from combinatorial arrangements of arrays, *Journal of the Royal Statistical Society, Supplement* **9** (1947), 128–139.

[182] C. R. Rao, On a class of arrangements, *Proceedings of the Edinburgh Mathematical Society* **8** (1949), 119–125.

[183] D. H. Rees & D. A. Preece, Perfect Graeco-Latin balanced incomplete block designs (pergolas), *Discrete Mathematics* **197/198** (1999), 691–712.

[184] K. B. Reid & E. Brown, Doubly regular tournaments are equivalent to skew Hadamard matrices, *Journal of Combinatorial Theory, Series A* **12** (1972), 332–338.

[185] Kenneth Graham Russell, *On the theory of row-column designs*, PhD thesis, University of New South Wales, 1977.

[186] K. G. Russell, A comparison of six methods of analysing row-column designs with inter-block information, *Statistics and Computing* **9** (1999), 239–246.

[187] K. G. Russell, J. A. Eccleston & G. J. Knudsen, Algorithms for the construction of (M, S) - optimal block designs and row–column designs, *Journal of Statistical Computation and Simulation* **12** (1981), 93–105.

[188] G. M. Saha & A. D. Das, A note on construction of mutually balanced Youden designs, *Utilitas Mathematica* **33** (1988), 5–8.

[189] Jennifer Seberry, A note on orthogonal Graeco-Latin designs, *Ars Combinatoria* **8** (1979), 85–94.

[190] E. Seiden & C. Y. Wu, A geometric construction of generalized Youden designs for v a power of a prime, *Annals of Statistics* **6** (1978), 452–460.

[191] K. R. Shah, On uniformly better combined estimates in row-column designs with adjusted orthogonality, *Communications in Statistics— Theory and Methods* **17** (1988), 3121–3124.

[192] K. R. Shah & J. A. Eccleston, On some aspects of row–column designs, *Journal of Statistical Planning and Inference* **15** (1986), 87–95.

[193] Kirti R. Shah & Bikas K. Sinha, *Theory of Optimal Designs, Lecture Notes in Statistics*, 54, Springer-Verlag, New York (1989).

[194] S. S. Shrikande, Designs for two-way elimination of heterogeneity, *Annals of Mathematical Statistics* **22** (1951), 235–247.

[195] M. Singh & A. Dey, Two-way elimination of heterogeneity, *Journal of the Royal Statistical Society, Series B* **40** (1978), 58–63.

[196] M. Singh & A. Dey, Block designs with nested rows and columns, *Biometrika* **66** (1979), 321–326.

[197] N. P. Singh & Gular Singh, Analysis of row-column experiments involving several non-interacting-sets of treatments and multistage Youden square designs, *Biometrical Journal* **26** (1984), 893–899.

[198] C. A. B. Smith & H. O. Hartley, The construction of Youden squares, *Journal of the Royal Statistical Society, Series B* **10** (1948), 262–263.

[199] T. P. Speed & R. A. Bailey, On a class of association schemes derived from lattices of equivalence relations, in *Algebraic Structures and Applications* (eds. P. Schulz, C. E. Praeger & R. P. Sullivan), Marcel Dekker, New York (1982), pp. 55–74.

[200] T. P. Speed & R. A. Bailey, Factorial dispersion models, *International Statistical Review* **55** (1987), 261–277.

[201] Edward Spence, Construction and classification of combinatorial designs, in *Surveys in Combinatorics, 1995* (ed. Peter Rowlinson), *London Mathematical Society Lecture Notes Series*, 218, Cambridge University Press, Cambridge (1995), pp. 191–213.

[202] S. K. Srivastav & J. P. Morgan, On the class of 2×2 balanced incomplete block designs with nested rows and columns, *Communications in Statistics—Theory and Methods* **25** (1996), 1859–1870.

[203] Leon S. Sterling & Nicholas Wormald, A remark on the construction of designs for two-way elimination of heterogeneity, *Bulletin of the Australian Mathematical Society* **14** (1976), 383–388.

[204] Anne Penfold Street & Deborah J. Street, *Combinatorics of Experimental Design*, Oxford University Press, Oxford (1987).

[205] Deborah J. Street, Graeco-Latin and nested row and column designs, in *Combinatorial Mathematics VIII* (ed. Kevin L. McAvaney), *Lecture Notes in Mathematics*, 884, Springer-Verlag, Berlin (1981), pp. 304–313.

[206] W. B. Taylor, Incomplete block designs with row balance and recovery of inter-block information, *Biometrics* **13** (1957), 1–12.

[207] T. N. Throckmorton, *Structures of classification data*, PhD thesis, Ames, Iowa, 1961.

[208] Tue Tjur, Analysis of variance models in orthogonal designs, *International Statistical Review* **52** (1984), 33–65.

[209] Tue Tjur, Analysis of variance and design of experiments, *Scandinavian Journal of Statistics* **18** (1991), 273–322.

[210] K. D. Tocher, The design and analysis of block experiments, *Journal of the Royal Statistical Society, Series B* **14** (1952), 45–91.

[211] Nizam Uddin, Constructions for some balanced incomplete block designs with nested rows and columns, *Journal of Statistical Planning and Inference* **31** (1992), 253–261.

[212] Nizam Uddin, On recursive construction for balanced incomplete block designs with nested rows and columns, *Metrika* **42** (1995), 341–345.

[213] Nizam Uddin & John P. Morgan, Some constructions for balanced incomplete block designs with nested rows and columns, *Biometrika* **77** (1990), 193–202.

[214] Nizam Uddin & John P. Morgan, Two constructions for balanced incomplete block designs with nested rows and columns, *Statistica Sinica* **1** (1991), 229–232.

[215] P. M. van de Ven & A. Di Bucchianico, On the equivalence of definitions for regular fractions of mixed-level factorial designs, *Journal of Statistical Planning and Inference* **139** (2009), 2351–2361.

[216] Barry Vowden, Infinite series of double Youden rectangles, *Discrete Mathematics* **125** (1994), 385–391.

[217] B. J. Vowden, A new infinite series of double Youden rectangles, *Ars Combinatoria* **56** (2000), 133–145.

[218] B. J. Vowden & D. A. Preece, Some new infinite series of Freeman-Youden rectangles, *Ars Combinatoria* **51** (1999), 49–63.

[219] W. D. Wallis, Triple arrays and related designs, *Discrete Applied Mathematics* **163** (2014), 220–236.

[220] W. D. Wallis & J. L. Yucas, Note on Agrawal's "Designs for Two-way Elimination of Heterogeneity", *Journal of Combinatorial Mathematics and Combinatorial Computation* **46** (2003), 155–160.

[221] E. R. Williams, Efficiency-balanced designs, *Biometrika* **62** (1975), 686–689.

[222] E. R. Williams, A note on rectangular lattice designs, *Biometrics* **33** (1977), 410–414.

[223] E. R. Williams, H. D. Patterson & J. A. John, Resolvable designs with two replications, *Journal of the Royal Statistical Society, Series B* **38** (1976), 296–301.

[224] E. R. Williams, H. D. Patterson & J. A. John, Efficient two-replicate resolvable designs, *Biometrics* **33** (1977), 713–717.

[225] R. M. Wilson, An existence theorem for pairwise balanced designs. III. Proof of the existence conjecture, *Journal of Combinatorial Theory, Series A* **18** (1975), 71–79.

[226] F. Yates, Complex experiments, *Journal of the Royal Statistical Society, Supplement* **2** (1935), 181–247.

[227] F. Yates, Incomplete randomized blocks, *Annals of Eugenics* **7** (1936), 121–140.

[228] F. Yates, Incomplete Latin squares, *Journal of Agricultural Science* **26** (1936), 301–315.

[229] F. Yates, A new method of arranging variety trials involving a large number of varieties, *Journal of Agricultural Science* **26** (1936), 424–455.

[230] F. Yates, A further note on the arrangement of variety trials: quasi-Latin squares, *Annals of Eugenics* **7** (1937), 319–332.

[231] F. Yates, Lattice squares, *Journal of Agricultural Science* **30** (1940), 672–787.

[232] W. J. Youden, Use of incomplete block replications in estimating tobacco-mosaic virus, *Contributions from Boyce Thompson Institute* **9** (1937), 41–48.

[233] Joseph L. Yucas, The structure of a 7 × 15 triple array, *Congressus Numerantium* **154** (2002), 43–47.

[234] G. Zyskind, On structure, relation, sigma, and expectation of mean squares, *Sankhyā, Series A* **24** (1962), 115–148.

School of Mathematics and Statistics
University of St Andrews
North Haugh
St Andrews
Fife KY16 9SS, U.K.
rab24@st-andrews.ac.uk

Large-scale structures in random graphs

Julia Böttcher

Abstract

In recent years there has been much progress in graph theory on questions of the following type. What is the threshold for a certain large substructure to appear in a random graph? When does a random graph contain all structures from a given family? And when does it contain them so robustly that even an adversary who is allowed to perturb the graph cannot destroy all of them? I will survey this progress, and highlight the vital role played by some newly developed methods, such as the sparse regularity method, the absorbing method, and the container method. I will also mention many open questions that remain in this area.

1 Introduction

Erdős and Rényi introduced the notion of a random graph in their seminal paper [49]. They thus initiated the study of which type of property typical graphs of a certain density have or do not have, which turned out to be immensely influential in graph theory as well as in other related mathematical areas. The books [26, 63, 79] provide an excellent and extensive overview of the theory of random graphs and its applications.

This survey is concerned with a particular type of properties of random graphs, namely the appearance of given large-scale subgraphs. In the past two decades, the theory of large-scale structures in random graphs $G(n, p)$ underwent swift development and originated powerful new tools. The following three main directions of research can be distinguished in this area.

Firstly and naturally, one may study for which edge probability p the random graph $G(n, p)$ is likely to possess *one particular* spanning (or large) structure. This structure could for example be a perfect matching, a Hamilton cycle, or a disjoint collection of triangles covering as many vertices of $G(n, p)$ as possible. More generally, for any sequence (H_n) of graphs one can ask when H_n is a subgraph of $G(n, p)$. Questions of this type were pursued since the early days of the theory of random graphs, and some turned out to be extremely challenging.

Secondly and more generally, instead of considering a single sequence (H_n) of subgraphs one may ask for $G(n, p)$ to be *universal* for a a given sequence of families $(\mathcal{H}_n)$ of graphs, that is, to simultaneously contain a copy of each graph in $\mathcal{H}_n$. A typical example of a question of this type is for which p the random graph $G(n,p)$ is likely to contain every binary tree on n vertices. Since the number of such trees is huge, it is clear that an answer to this question is not trivially entailed by a result on the appearance of any fixed spanning binary tree in $G(n,p)$. Hence, such universality questions are in general harder than the questions for single subgraph sequences. Universality questions were originally motivated by problems in circuit design, data representation, and parallel computing (see [23] for relevant references, and more history concerning universality). Their study in random graphs is more recent and it is often observed or conjectured that when $G(n,p)$ is likely to contain any fixed graph from $\mathcal{H}_n$ then it is already universal for $\mathcal{H}_n$.

Finally, one may ask how *resiliently* $G(n,p)$ possesses certain structures. In other words, if $G(n,p)$ is known to contain a subgraph H, but an adversary is allowed to delete edges from $G(n,p)$ under certain restrictions, when is the adversary likely able to destroy all copies of H. As it turns out the random graph is very robust towards such adversarial edge deletions. Another way of motivating resilience-type questions is from the perspective of extremal graph theory. Two main directions of research in extremal graph theory are the investigation of Turán-type questions, and of Dirac-type questions. Turán's theorem [132] states that K_r is a subgraph of any graph G on n vertices with more edges than the balanced complete $(r-1)$-partite graph on n vertices contains K_r as a subgraph, that is, graphs G with edge density at least $\frac{r-2}{r-1}+o(1)$ contain K_r. Dirac's theorem [47], on the other hand, asserts that any graph G with minimum degree $\delta(G) \geq \frac{1}{2}v(G)$ contains a Hamilton cycle. Resilience-type questions then ask for the transference of such results to sparse random graphs. For example, when does any subgraph of $G(n,p)$ with sufficiently many edges contain K_r, and when does any subgraph of $G(n,p)$ with sufficiently high minimum degree contain a Hamilton cycle? The former of these two questions proved to be surprisingly deep and both questions and their generalisations inspired much recent work in the area.

In this survey I attempt to give an overview of the progress in these three main directions. Let me stress that there is no material covered here that does not appear elsewhere. Instead, I try to outline the exciting developments in the area, and also give credit to the important new methods that allowed this progress. In some cases I will give simple examples of how these methods can be applied. These necessarily have to be brief, but pointers to further literature will be given.

What is not covered? There are several other important topics which recently received much attention and are closely connected to the the developments described in this survey in that progress in these areas influenced or was influenced by the methods and results provided in the following, but which are, to limit scope, not covered here. These topics include Ramsey theoretic results in random graphs, packing results in random graphs, and embedding results in various types of pseudorandom graphs. I also omit analogous results in random directed graphs and random hypergraphs, and embedding results for induced subgraphs in random graphs.

Organisation. The survey is structured as follows. Section 2 provides basic definitions and the relevant concepts from the theory of random graphs. Section 3 then collects, mainly for comparison, results on the appearance of fixed graphs H in $G(n, p)$. Section 4 reviews results on the appearance of a fixed sequence (H_n) in $G(n, p)$, where the graphs in (H_n) grow with n, while Section 5 considers corresponding universality results. Section 6 surveys progress on resilience results for large subgraphs of $G(n, p)$, and Section 7 discusses an important tool for this type of problem, the sparse blow-up lemma in random graphs.

2 Basic definitions and notation

For easy reference, this section collects the basic definitions we need in this survey. Throughout, we use the natural logarithm $\log x = \log_e x$. The set of the first n natural numbers is denoted by $[n] = \{1, \ldots, n\}$, and $(n)_k = n \cdot (n-1) \cdot \ldots \cdot (n-k)$ is the falling factorial. As is common in the area, ceilings and floors are omitted whenever they are not essential.

For a graph $G = (V, E)$ we denote by $v(G)$ the number of its vertices $|V|$ and by $e(G)$ the number of its edges $|E|$. The *minimum degree* of G is $\delta(G)$, while the *maximum degree* is $\Delta(G)$. The chromatic number of G is denoted by $\chi(G)$. The *girth* of a graph is the length of its shortest cycle. If H is a (not necessarily induced) *subgraph* of G we write $H \subseteq G$. An H-*copy* in G is a (not necessarily induced) copy of H in G. The *automorphism group* of G is denoted by $\mathrm{Aut}(G)$.

For a vertex $v \in V$ we write $N_G(v)$ for the *neighbourhood* of v in G, and $\deg_G(v) = |N_G(v)|$ for its *degree*. Similarly, if $U \subseteq V$ then $N_G(v; U)$ is the neighbourhood in G of v in the set U and $\deg_G(v; U) = |N_G(v; U)|$. When the graph G is clear from the context we often omit the subscript G in this notation.

2.1 Graph classes

The graph properties considered in this survey mainly concern the existence of certain subgraphs, and hence are monotone increasing. A *monotone increasing graph property* is a family $\mathcal{P}$ of graphs such that for any $G \in \mathcal{P}$ we have that a graph G' obtained from G by adding any edge is also in $\mathcal{P}$. A monotone increasing property is *non-trivial* if K_n is in $\mathcal{P}$ but the complement of K_n not, where K_n denotes the *complete graph* on n vertices.

A *balanced r-partite* graph is an r-partite graph whose partition classes are as equal as possible; it is *complete* if all the edges between all partition classes are present. The cycle on n vertices is denoted by C_n, and P_n is the n-vertex path. A *Hamilton cycle* (or *path*) of a graph G is a cycle (or path) containing all the vertices of G. A graph is called *Hamiltonian* if it has a Hamilton cycle. Let H be a fixed graph and G be a graph on n vertices. Then an *H-factor* in G is a collection of $\lfloor n/v(H) \rfloor$ vertex disjoint copies of H. In particular, when $v(H)$ divides $v(G)$ then an H-factor is a spanning subgraph of G. The *d-dimensional cube* Q_d is the graph on vertex set $\{0,1\}^d$ with edges uv whenever u and v differ in exactly one coordinate. The *$k \times k$-square grid* L_k is the graph on vertex set $[k] \times [k]$, with edges uv whenever u and v differ in exactly one coordinate by exactly one.

The *k-th power* of a graph H is the graph obtained from H by adding all edges between vertices of distance at most k. The 2-nd power of H is also called the *square* of H. We also denote the k-th power of H by H^k. In particular, C_n^k is the k-th power of a cycle C_n on n vertices. A graph H is *d-degenerate* if every subgraph of H contains a vertex of degree at most d. Equivalently, the vertices of H can be ordered in such a way that each vertex v sends at most d edges to vertices preceding v in this order. The *bandwidth* bw(H) of a graph H is the smallest integer b such that there is a labelling of $V(H)$ using all the integers $[v(H)]$ for which $|u - v| \leq b$ for each edge $uv \in E(H)$.

Further, the following classes of graphs are considered. Let $\mathcal{H}(n, \Delta)$ be the family of all graphs on n vertices with maximum degree at most Δ, and $\mathcal{H}(n, n, \Delta)$ be the class of all bipartite graphs with partition classes of order n each, and with maximum degree Δ. The class $\mathcal{T}(n, \Delta)$ contains all trees on n vertices with maximum degree Δ. Let me remark that sometimes these graph classes will be used to refer to small linear sized graphs, such as the class $\mathcal{H}(\gamma n, \Delta)$ for some small $\gamma > 0$, where one really should write $\mathcal{H}(\lfloor \gamma n \rfloor, \Delta)$, but I omit the floors and ceilings for simplicity.

2.2   Random graphs

The *binomial random graph* $G(n, p)$ is obtained by pairwise independently including each of the possible $\binom{n}{2}$ edges on n vertices with probability $p = p(n)$.[1] The *uniform random graph* $G(n, m)$, on the other hand,

---

[1]This model is also often called the Erdős–Rényi model, though this is objected to by part of the community because the model that Erdős and Rényi used in their papers

assigns each graph on vertex set $[n]$ with m edges probability $1/\binom{\binom{n}{2}}{m}$. An event holds *asymptotically almost surely* (abbreviated a.a.s.) in $G(n,p)$ (or in $G(n,m)$) if its probability tends to 1 as n tends to infinity. For a monotone increasing graph property $\mathcal{P}$ we say that $\tilde{p} = \tilde{p}(n)$ is a *threshold* for $\mathcal{P}$ if

$$\mathbb{P}\big(G(n,p) \in \mathcal{P}\big) \to \begin{cases} 0 & \text{if} \quad p/\tilde{p} \to 0, \\ 1 & \text{if} \quad p/\tilde{p} \to \infty. \end{cases}$$

As is common, in this case $\tilde{p}$ will also be called *the* threshold, even though it is not unique. Bollobás and Thomason [31] proved that every non-trivial monotone increasing property has a threshold. Moreover if the threshold is of the form $\log^a n/n^b$ with $a, b > 0$ fixed reals, as will be encountered frequently in this survey, then there is a *sharp* threshold $\tilde{p}$, that is, for any $\varepsilon > 0$

$$\mathbb{P}\big(G(n,p) \in \mathcal{P}\big) \to \begin{cases} 0 & \text{if} \quad p \le (1 - \varepsilon)\tilde{p}, \\ 1 & \text{if} \quad p \ge (1 + \varepsilon)\tilde{p}. \end{cases}$$

As explained in [61] this follows from the celebrated characterisation of sharp thresholds by Friedgut [60].

In this survey many results are considered that concern *spanning* subgraphs of $G(n,p)$ as n tends to infinity. These results therefore do not concern a single fixed subgraph, but rather a *sequence* of subgraphs, one for each value of n. Sometimes this fact is implicitly assumed when stating a result, but usually it is stressed by stating that we are given a sequence $H = (H_n)$ of graphs and that $G(n,p)$ contains H_n under certain conditions a.a.s.

2.3 Density parameters

In the results on subgraphs H of $G(n,p)$ that we will discuss, different density parameters are used, which will be defined next.[2] The first of these parameters is called *maximum 0-density*, and is given by

$$m_0(H) = \max_{H' \subseteq H} \frac{e(H')}{v(H')}.$$

The maximum 0-density is usually simply called maximum density in the literature. Let H^* be a subgraph of H realising the maximum in $m_0(H)$, and let X^* be the random variable counting the number of unlabelled

---

pioneering the area is the model $G(n,m)$.

[2]It is not true that these parameters are densities in the sense of being between 0 and 1. Rather, they are variations on the average degree of a graph.

copies of H^* in $G(n,p)$. Then

$$\mathbb{E}(X^*) = \big((n)_{v(H^*)}/|\operatorname{Aut}(H^*)|\big)p^{e(H^*)} \approx n^{v(H^*)}p^{e(H^*)},$$

which tends to infinity if $p \cdot n^{1/m_0(H)} \to \infty$. So, informally, we can say that in expectation the densest subgraph of H appears around $p = n^{-1/m_0(H)}$ in $G(n,p)$. Hence, it is natural to guess that this probability is the threshold for the appearance of H-copies in $G(n,p)$, which is indeed the case (see Theorem 3.1).

The other density parameters are slight variations on this first definition (which can, however, have an important influence on the resulting values of these parameters). These variations have similarly natural motivations as the 0-density. The *maximum 1-density* of a graph H with at least two vertices is

$$m_1(H) = \max_{\substack{H' \subseteq H \\ v(H')>1}} \frac{e(H')}{v(H')-1}.$$

This parameter is also called fractional arboricity in [12]. Again, let H^* be a subgraph of H realising the maximum in $m_1(H)$ and let v be a fixed vertex in $G(n,p)$. Then the expected number of H^*-copies in $G(n,p)$ containing v tends to infinity if $p \cdot n^{1/m_1(H)} \to \infty$. The threshold for the property that *each* vertex of $G(n,p)$ is contained in an H-copy is related to, but is not precisely equal to, $n^{-1/m_1(H)}$ (see also the explanations in Section 3).

The *maximum 2-density* of a graph H with at least one edge is

$$m_2(H) = \max_{\substack{H' \subseteq H \\ e(H')>0}} \begin{cases} \frac{e(H')-1}{v(H')-2}, & \text{if } v(H') > 2 \\ \frac{1}{2}, & \text{if } v(H') = 2. \end{cases}$$

If $p \cdot n^{1/m_2(H^*)} \to \infty$, in expectation a fixed edge of $G(n,p)$ is contained in many H^*-copies, where H^* realises the maximum in $m_2(H)$.

A graph H is called 0-*balanced* (or 1-*balanced*, or 2-*balanced*) if H is a maximiser in $m_0(H)$ (or $m_1(H)$, or $m_2(H)$, respectively). If H is the unique maximiser, then it is called *strictly* 0-balanced (or 1-balanced, or 2-balanced, respectively).

Riordan [120] defined a different density parameter $m_R(H)$ for H on at least 3 vertices, which I will call the *maximum Riordan-density* here and which is given by

$$m_R(H) = \max_{\substack{H' \subseteq H \\ v(H')>2}} \frac{e(H')}{v(H')-2}.$$

Note that, again, if H is a fixed graph and the maximum in $m_{\mathrm{R}}(H)$ is realised by H^*, then a fixed pair of vertices in $G(n,p)$ is in expectation contained in many H^*-copies if $p \cdot n^{1/m_{\mathrm{R}}(H^*)} \to \infty$. Riordan, however, uses this density in a result about copies of spanning graphs $H = (H_n)$ in $G(n,p)$. For such a graph H, the maximum in $m_{\mathrm{R}}(H)$ may well be realised by some $H^* = (H_n^*)$ with $v(H_n^*) \to \infty$, in which case $m_{\mathrm{R}}(H)$ is asymptotically equal to the maximum 0-density (or maximum 1-density) of H.

3 Small subgraphs

This survey focuses on large subgraphs of $G(n,p)$. Before we turn to the many results in this area, we will briefly review what is known for small, that is, fixed subgraphs H. Some of the relevant results will turn out useful for later comparison.

3.1 The appearance of small subgraphs

There are a number of natural questions that one may ask concerning the existence of a fixed subgraph H in $G(n,p)$:

1. What is the threshold for the appearance of an H-copy?
2. How many H-copies are there in $G(n,p)$?
3. How are the H-copies distributed in $G(n,p)$?

The second question is beyond the scope of this survey, though important and strong results were obtained in this direction (see, e.g., [79, Chapter 6] or [63, Chapter 5]). We shall concentrate on the other two, starting with the first. A classical result by Bollobás [24] in the theory of random graphs states that the threshold for the appearance of an H-copy in $G(n,p)$ is determined by its maximum 0-density.

Theorem 3.1 (see, e.g., Theorem 3.4 in [79]) *Let H be a graph with (and at least one edge). The threshold for $G(n,p)$ to contain a copy of H is*

$$n^{-1/m_0(H)}.$$

This answers the first question. Note that for 0-balanced graphs this threshold was already established by Erdős and Rényi [50].

So let us turn to the third question, which is phrased rather vaguely. In fact there are two meaningful interpretations which will play a more prominent role in this survey. One the one hand, one could ask: When do we find many vertex disjoint copies of H, or possibly even an H-factor? The latter is a difficult question, and we shall return to it in Section 4.3.

But a related question, considering a property which is clearly necessary for an H-factor, is much easier: When is every vertex of H contained in an H-copy? This question was answered by Ruciński [123] and Spencer [128]. For strictly 1-balanced graphs H the threshold is mainly influenced by the maximum 1-density of H.

Theorem 3.2 (see, e.g., Theorem 3.22 in [79]) *Let H be a strictly 1-balanced graph (with at least 2 vertices) and let COV_H be the event that every vertex of $G(n, p)$ is contained in a copy of H. The threshold for COV_H is*

$$\frac{(\log n)^{1/e(H)}}{n^{1/m_1(H)}}.$$

Similar results for non-strictly 1-balanced graphs exist (see [79, Theorem 3.22]). But these are more complicated: they need to take into account all the different ways of rooting the graph H at some vertex and all the different subgraphs H' of H containing this vertex. It is true, however, that the threshold for the event COV_H of Theorem 3.2 is $\Omega\left(\frac{(\log n)^{1/e(H)}}{n^{(v(H)-1)/e(H)}}\right)$ for every H.

The appearance of the log-factor in the threshold is not surprising. Recall that the expected number of H-copies in $G(n, p)$ containing a fixed vertex v is of order $n^{v(H)-1}p^{e(H)}$. Since we are asking for an H copy at *every* vertex of $G(n, p)$ it is natural to require that this quantity grows at least like $\log n$ (to allow for concentration), which is precisely the case for $p = \frac{(\log n)^{1/e(H)}}{n^{1/m_1(H)}}$ if H is strictly 1-balanced.

On the other hand, one could interpret the third question above as asking if $G(n, p)$ has a large subgraph without any H-copies. It is easy to show that this is the case below the 2-density-threshold.

Proposition 3.3 (see, e.g., Proposition 8.9 in [79]) *For all $0 < a < 1$ and all H with $\Delta(H) \geq 2$ there is a constant $c > 0$ such that the following holds. If $p \leq cn^{1/m_2(H)}$ then $G(n, p)$ a.a.s. has an H-free subgraph G with $e(G) \geq a \cdot e(G_{n,p})$.*

So H-copies in $G(n, p)$ are easy to delete once we are below the 2-density threshold. The reason for this is that the likely number of H-copies is comparable to the likely number of edges at this threshold. Above the threshold this changes, which is addressed in the following section.

3.2   The Erdős–Stone theorem in random graphs

What is the maximum number of edges in an H-free subgraph of $G(n, p)$? This question has inspired much research in the theory of random graphs. To understand what the answer to this question could reasonably be, let us first turn to dense graphs.

The Erdős–Stone theorem, one of the cornerstones of extremal graph theory, is a Turán-type theorem which states that the crucial property of a fixed graph H for determining the maximum number of edges in an H-free graph is its chromatic number.

Theorem 3.4 (Erdős, Stone [53]) *For each fixed graph H and every $\varepsilon > 0$ there is an n_0 such that for all $n \geq n_0$ the following holds. Any n-vertex graph G with at least $\left(\frac{\chi(H)-2}{\chi(H)-1} + \varepsilon\right)\binom{n}{2}$ edges contains H as a subgraph.*

As a balanced complete $(\chi(H)-1)$-partite graph has about $\frac{\chi(H)-2}{\chi(H)-1}\binom{n}{2}$ edges and is obviously H-free this is tight up to lower order terms. Similarly, a $(\chi(H)-1)$-partite subgraph of $G(n,p)$ with (roughly) equal sized random partition classes and all $G(n,p)$-edges between the partition classes contains $\left(\frac{\chi(H)-2}{\chi(H)-1} + o(1)\right) e(G(n,p))$ edges. Hence, when we ask for the maximum number of edges in an H-free subgraph of $G(n,p)$ we cannot go below this quantity. Moreover, as explained at the end of the last section, below the 2-density threshold the answer becomes (almost) trivial as all H-copies in $G(n,p)$ can be destroyed by deleting just a tiny fraction of the edges.

The question then is if these two observations already tell the whole story. The following breakthrough result on the transference of the Erdős–Stone theorem to sparse random graphs confirms that this is indeed the case, and was obtained independently by Conlon and Gowers [39] (for 2-balanced H) and Schacht [125] (for general H).

Theorem 3.5 (Schacht [125], Conlon, Gowers [39])
For every fixed graph H and every $\varepsilon > 0$ there are constants $0 < c < C$ such that the following holds. Let $\mathcal{A}$ be the property that the maximum number of edges in an H-free subgraph of $G(n,p)$ is at most $\left(\frac{\chi(H)-2}{\chi(H)-1} + \varepsilon\right) e(G(n,p))$. Then

$$\mathbb{P}[\mathcal{A}] \to \begin{cases} 0 & \text{if } p \leq cn^{-1/m_2(H)}, \\ 1 & \text{if } p \geq Cn^{-1/m_2(H)}. \end{cases}$$

Earlier results in this direction were obtained for special graphs H in [59, 65, 66, 69, 70, 74, 75, 89, 90], and for larger lower bounds on p in [93, 130]. In fact, the results of Conlon and Gowers [39] and of Schacht [125] are much more general statements, allowing the transference of a variety of extremal results on graphs, hypergraphs and sets of integers to sparse random structures. Both proofs reduce these problems to the analysis of random vertex subsets in certain auxiliary hypergraphs. In the case of Theorem 3.5 the vertices of the auxiliary hypergraph $\mathcal{H}$ are the edges of K_n, and the hyperedges are all $e(H)$-tuples that form H-copies. Hence, the random graph $G(n, p)$ corresponds to a random subset S of $V(\mathcal{H})$, and an H-free subgraph of $G(n, p)$ to an independent set in $\mathcal{H}[S]$.

Recently, a very general approach has been developed to analyse such independent sets in hypergraphs, the so-called container method developed independently by Balogh, Morris and Samotij [19], and Saxton and Thomasson [124], which has already proved tremendously useful for solving a variety of other problems as well.

The idea, in the language of the Erdős–Stone theorem in random graphs, is as follows. One naive approach to prove that $\mathbb{P}[\mathcal{A}] \to 1$ if $p \geq Cn^{-1/m_2(H)}$ in Theorem 3.5 is to first fix an H-free graph G on vertex set $[n]$, to calculate the probability that $G(n, p)$ contains more than $\left(\frac{\chi(H)-2}{\chi(H)-1} + \varepsilon \right) p \binom{n}{2}$ edges of G, and then use a union bound over all choices of G to conclude that a.a.s. $G(n, p)$ does not contain $\left(\frac{\chi(H)-2}{\chi(H)-1} + \varepsilon \right) p \binom{n}{2}$ edges of any H-free graph, so that any subgraph of $G(n, p)$ with that many edges cannot be H-free. This, of course, does not work because there are too many choices for G (and we did not use anything about the structure of H-free graphs). The crucial idea of the container method is to show that the set $\mathcal{G}$ of H-free graphs can be "approximated" by a much smaller set of good containers $\mathcal{C}$, that is, for each $G \in \mathcal{G}$ there is $C \in \mathcal{C}$ such that $G \subseteq C$ and $e(C) \leq \left(\frac{\chi(H)-2}{\chi(H)-1} + \frac{1}{2}\varepsilon \right) \binom{n}{2}$. This then basically allows us to run the union bound argument over $\mathcal{C}$. In reality, things are not quite so simple, and more properties are required of $\mathcal{C}$ (see also the excellent explanations in [124, Section 2.2]).

Restricting to the case when $H = K_r$, the analogous structural question to Theorem 3.5 of when the K_r-free subgraph of $G(n, p)$ with the most edges is $(r - 1)$-partite was first considered by Babai, Simonovits and Spencer [14], whose result was improved on by Brightwell, Panagiotou and Steger [35]. Finally, DeMarco and Kahn [46, 45] showed that this is a.a.s. the case when $p \geq C(\log n)^{1/(e(K_r)-1)}/n^{m_2(K_r)}$, which is optimal up to the value of C. For other graphs H a corresponding structural result has not yet been established. Observe that, in general, this is a difficult problem since we do not even know precise structural results in dense graphs for all H.

Question 3.6 *For some fixed H different from a complete graph, what is the structure of an H-free subgraph of $G(n, p)$ with the most edges?*

This question is already interesting when $H = C_5$, for example, when this subgraph should be bipartite (for p sufficiently large).

4 Large subgraphs

In this section we shall consider the question when the random graph $G(n, p)$ contains a fixed sequence of spanning graphs $H = (H_n)$ as subgraphs. Answers to this question come in various levels of accuracy. For some classes of graphs H we only know non-matching lower and upper bounds on the threshold probability, while for others the threshold has been established. Even stronger hitting time results could so far only rarely be obtained.

We start this section with the most classical subgraphs of $G(n, p)$ to be considered: matchings and Hamilton cycles. In Section 4.2 we present a theorem of Alon and Füredi concerning more general spanning subgraphs and the powerful improvement on this result by Riordan. We also analyse the bounds on the threshold for various classes of graphs that Riordan's result gives. Section 4.3 turns to the deep Johansson–Kahn–Vu theorem which establishes, among others, the threshold for K_r-factors, while Section 4.4 considers the threshold for the containment of spanning bounded degree trees. The question of when a bounded degree graph H appears in $G(n, p)$ is addressed in Section 4.5. In Section 4.6 we discuss a very general conjecture of Kahn and Kalai concerning the form a threshold for the containment of some sequence (H_n) can take in $G(n, p)$. In Section 4.7, finally, we consider the question of algorithmically finding a spanning H-copy in $G(n, p)$.

4.1 Matchings and Hamilton cycles

Two of the most natural questions concerning spanning substructures of random graphs asks for the threshold of $G(n, p)$ to contain a perfect matching or to be Hamiltonian. These questions are as well understood as one can hope for.

Already Erdős and Rényi [51, 52] showed that the threshold for containing a perfect matching is $\log n/n$. Bollobás and Thomason [30] established a hitting time result, which considers $G(n, m)$ as a graph process, where we start from the empty graph on n vertices and randomly add edges one-by-one. The hitting time result then states that a.a.s. precisely the edge in this process which eliminates the last isolated vertex creates a perfect matching (if n is even). In other words, avoiding the most trivial obstacle for containing a perfect matching in fact guarantees a perfect matching (see also [28] for an alternative proof and related results). Luczak and

Ruciński [111] extended these results, showing that the same hitting time result is true for T-factors for any non-trivial tree T.

Turning to the Hamiltonicity problem, Pósa [119] and Korshunov [103, 104] showed that also the threshold for a Hamilton cycle (as well as for a Hamilton path) is $\log n/n$. Improving on this result, Komlós and Szemerédi [102] determined an exact formula for the probability of the existence of a Hamilton cycle. Bollobás [25] established the corresponding hitting time result, stating that as soon as $G(n,p)$ gets minimum degree 2, it also contains a Hamilton cycle. Hence, if $\phi(n)$ is any function tending to infinity, then $G(n,p)$ a.a.s. is Hamiltonian if $p \geq (\log n + \log\log n + \phi(n))/n$ and not Hamiltonian if $p \leq (\log n + \log\log n - \phi(n))/n$.

Algorithmic results for finding Hamilton cycles – a problem which is NP-hard in general – in random graphs above the threshold probability were also obtained. Gurevich and Shelah [71] and Thomason [131] obtained linear expected time algorithms when p is well above the threshold. Improving on polynomial time randomised algorithms by Angluin and Valiant [13] and Shamir [127], Bollobás, Fenner and Frieze [27] gave a deterministic polynomial time algorithm with a success probability that matches the probability that a Hamilton cycle exists given by Komlós and Szemerédi [102].

4.2   The Alon–Füredi theorem and Riordan's theorem

Let us now turn to results concerning more general results on spanning subgraphs of $G(n,p)$. Motivated by a question of Bollobás asking for a non-trivial probability p such that $G(n,p)$ with $n = 2^d$ a.a.s. contains a copy of the d-dimensional hypercube Q_d, Alon and Füredi [10] established the following result, providing an upper bound on the threshold for the appearance of a spanning graph with a given maximum degree. Their theorem can be seen as a first general result concerning the appearance of spanning subgraphs in $G(n,p)$, thus stimulating research in the area.

Theorem 4.1 (Alon, Füredi [10]) *Let $H = (H_n)$ be a fixed sequence of graphs on n vertices with maximum degree $\Delta(H) \leq \sqrt{\sqrt{n}-1}$. If*

$$p \geq \left(\frac{20\Delta(H)^2 \log n}{n}\right)^{1/\Delta(H)}$$

then $G(n,p)$ a.a.s. contains a copy of H.

Their proof uses the following simple strategy, which is based on a multi-round exposure of $G(n,p)$. Apply the Hajnal–Szemerédi theorem [72] to the square H^2 of H to obtain an equitable $(\Delta^2 + 1)$-colouring of H^2, that is, a partition of $V(H^2) = V(H)$ into $\Delta^2 + 1$ parts $X_1 \dot\cup \ldots \dot\cup X_{\Delta^2+1}$ which are as equal in size as possible and form independent sets in H^2. Observe that this implies that between each pair of these parts H induces a matching. Partition the vertices of $G(n,p)$ into sets $V_1 \dot\cup \ldots \dot\cup V_{\Delta^2+1}$ of sizes equal to these parts. Then embed X_i into V_i one by one, revealing the edges between V_i and $\bigcup_{j<i} V_j$, and showing that the partial embedding from the previous round can be extended. This is possible because for any $x \in X_i$ the set $N^-(x)$ of already embedded neighbours is of size at most Δ and disjoint from any $N^-(x')$ with $x \neq x' \in X_i$, and a random bipartite graph with edge probability p^Δ and partition classes of size $n/(\Delta^2 + 1)$ contains a perfect matching. Ideas from this basic strategy were re-used in many of the results on universality and local resilience we will mention later.

The theorem of Alon and Füredi was improved on by Riordan. He proved the following surprisingly powerful result.

Theorem 4.2 (Riordan's theorem [120]) *Let $H = (H_n)$ be a fixed sequence of graphs with $v(H) = n$ and $e(H) > n/2$ and let $p = p(n) < 1$ satisfy*

$$\frac{np^{m_R(H)}}{\Delta(H)^4} \to \infty.$$

Then a.a.s. $G(n,p)$ contains a copy of H.

This result can be found in this form in [118] (where it is in addition verified that this result also remains true for H with fewer edges but $\delta(H) \geq 2$). Observe, that the condition on p in Riordan's theorem implies that $\Delta(H)$ grows slower than $n^{1/4}$. In most of this survey, however, we will consider bounded degree graphs only, for which Theorem 4.2 requires that p grows faster than $n^{-1/m_R(H)}$.

Let me mention that in [120] this result is stated for $G(n,m)$ instead of $G(n,p)$ and it is in addition required that $p\binom{n}{2} \to \infty$, $\binom{n}{2} - 2e(H) \to \infty$ and $(1-p)\sqrt{n} \to \infty$. However, the result for $G(n,p)$ follows from a standard argument (e.g. [26, Theorem 2.2]) and the first additional requirement on p follows from the requirement in Theorem 4.2 since $m_R(H) \geq \frac{n/2}{n-2} > \frac{1}{2}$ because $e(H) > n/2$. The second and third additional requirements are satisfied if we take p as small as possible while still satisfying the conditions in Theorem 4.2 because $\Delta(H)$ grows slower than $n^{1/4}$ and $m_R(H) \leq \Delta(H)$. The conclusion then still remains true for larger p because the property of containing H is monotone increasing.

The heart of the proof of Riordan's theorem is an elegant second moment argument in the $G(n,m)$ model, which shows that the variance of the number of H-copies is small by bounding from above how much one H-copy in $G(n, m)$ can make another H-copy more likely. Using the same approach in $G(n, p)$ is not possible because if H contains many edges and one conditions on the appearance of a fixed H-copy in $G(n, p)$, then this boosts the number of edges in $G(n, p)$ sufficiently to make other H-copies significantly more likely.

To illustrate the power of Riordan's theorem a few straightforward consequences are collected in the following. The first two of these were already given by Riordan [120], and the third was observed by Kühn and Osthus [108].

Hypercubes. If $n = 2^d$ and

$$p \geq \frac{1}{4} + 6\frac{\log d}{d}$$

then a.a.s. $G(n, p)$ contains a copy of the d-dimensional cube Q_d, because $m_{\mathrm{R}}(Q_d) = \frac{dn}{2(n-2)}$ (that is, Q_d is the maximiser in $m_{\mathrm{R}}(Q_d)$). This results is close to best possible since for $p = \frac{1}{4}$ the expected number of Q_d-copies is $(n!/|\operatorname{Aut}(Q_d)|)(\frac{1}{4})^{\frac{1}{2}n\log n} \leq (n!/|\operatorname{Aut}(Q_d)|) \cdot n^{-n}$, which tends to zero as n tends to infinity.

Square grids. If $n = k^2$ and

$$p \cdot n^{1/2} \to \infty$$

then a.a.s. $G(n, p)$ contains a copy of the $k \times k$-square grid L_k, because $m_{\mathrm{R}}(L_k) = 2$ (that is, C_4 is the maximiser in $m_{\mathrm{R}}(L_k)$). Again, an easy first moment calculation show that for $p = n^{-1/2}$ the probability that $G(n, p)$ contains L_k tends to 0.

Powers of Hamilton cycles. If $k \geq 3$ and

$$p \cdot n^{1/k} \to \infty$$

then G contains the k-th power of a Hamilton cycle C_n^k, because $m_{\mathrm{R}}(C_n^k) \leq k + \frac{(k+1)k^2}{n}$ as shown in [108]. For $p \leq ((1-\varepsilon)e/n)^{1/k}$ the probability that $G(n, p)$ contains the k-th power of a Hamilton cycle tends to 0 (using again the first moment).

For $k = 2$ Riordan's theorem does not provide a (close to) optimal result, because $m_{\mathrm{R}}(C_n^k) = m_{\mathrm{R}}(K_3) = 3$. An approximately tight result has been obtained by Kühn and Osthus [108] though, who showed that $G(n, p)$ a.a.s. contains C_n^2 if $p \geq n^{\varepsilon - 1/2}$ for any fixed $\varepsilon > 0$. This was improved on by Nenadov and Škorić [116] who require $p \geq C\log^4 n/n^{1/2}$. Both the result of Kühn and Osthus and the result of Nenadov and Škorić use an absorbing-type method. Very recently, using a second moment argument again, Bennett, Dudek, and Frieze [22] announced a proof showing that $p = \sqrt{1/n}$ is the threshold for $G(n, p)$ to contain the square of a Hamilton cycle.

Trees. For trees T on at least 3 vertices we have $m_{\mathrm{R}}(T) = 2$, where the path on 3 vertices is the maximiser in $m_{\mathrm{R}}(T)$. It follows that if $T = (T_n)$ is a fixed sequence of bounded degree trees then $G(n, p)$ a.a.s. contains T if $p \cdot n^{1/2} \to \infty$. This is far from the best known upper bound of $\log^5 n/n$ for the threshold for containing such trees [113], to which we shall return in Section 4.4. Riordan's theorem allows to also consider trees with growing maximum degrees. However, the resulting threshold bounds are again far from the best known bounds (see Section 4.4).

Planar graphs. For a planar graph H' we have $e(H')/(v(H') - 2) \leq 3$, and hence any n-vertex planar graph H satisfies $m_{\mathrm{R}}(H) \leq 3$, with equality when H is a triangulation. Hence, if H has bounded degree, then a.a.s. $G(n, p)$ contains H if

$$p \cdot n^{1/3} \to \infty.$$

As was observed by Bollobás and Frieze [29] for $p = c/n^{1/3}$ with $c = (27e/256)^{1/3}$ the random graph $G(n, p)$ a.a.s. contains no spanning triangulation.

A planar graph H drawn uniformly from all planar graphs on n vertices a.a.s. has maximum degree less than $3 \log n$ [112, 48]. It follows that for such graphs H the random graph $G(n, p)$ a.a.s. contains H if $p \cdot \dfrac{n^{1/3}}{\log^{4/3} n} \to \infty$.

K_r-factors. For K_r-factors H we have $m_{\mathrm{R}}(H) = m_{\mathrm{R}}(K_r) = \frac{1}{2}r(r - 1)/(r - 2)$ and hence $G(n, p)$ a.a.s. has a K_r-factor when

$$p \cdot n^{\frac{2}{r} - \frac{2}{r(r-1)}} \to \infty.$$

The power in the exponent of n is surprisingly close to the right one, which is $-1/m_1(K_r) = -2/r$ (ignoring log-factors), as given by Theorem 4.3 in the next section.

Bounded degree graphs. Graphs H' with maximum degree $\Delta(H') \leq \Delta$ satisfy $e(H')/(v(H') - 2) \leq \frac{1}{2}\Delta + \Delta/(v(H') - 2)$. To maximise this quantity we should set $v(H') = \Delta + 1$ (since for smaller $v(H')$ an even better bound on $e(H')/(v(H') - 2)$ holds). Hence, for a maximum degree Δ graph H we have $m_{\mathrm{R}}(H) \leq \frac{1}{2}(\Delta + 1)\Delta/(\Delta - 1)$ and thus $G(n, p)$ a.a.s. contains H when

$$p \cdot n^{\frac{2}{\Delta+1} - \frac{2}{\Delta(\Delta+1)}} \to \infty.$$

This again is close to the lower bound, which is given by the lower bound for containing a $K_{\Delta+1}$-factor.

D-degenerate graphs. For D-degenerate graphs H' we have $e(H') \leq (v(H') - D)D + \binom{D}{2} \leq v(H')D - 2D$ for $D \geq 3$. It follows that a D-degenerate graph H satisfies $m_{\mathrm{R}}(H) \leq D$ for $D \geq 3$. So, if further the maximum degree of H is bounded by a constant (potentially much larger than D) then $G(n,p)$ a.a.s. contains H when

$$p \cdot n^{1/D} \to \infty.$$

As mentioned earlier for $p \leq ((1 - \varepsilon)e/n)^{1/D}$ the probability that $G(n,p)$ contains the D-th power of a Hamilton cycle tends to 0. Since the D-th power of a Hamilton path is D-degenerate this shows that the bound given by Riordan's theorem is close to best possible. Observe also that this bound is much better than the known bounds in universality results for D-degenerate graphs discussed in Section 5.1.

These examples illustrate that Riordan's theorem often, though not always, gives optimal or close to optimal bounds. As indicated, for K_r-factors and bounded degree trees better bounds have been obtained in recent years, and I shall discuss these in the following sections.

For spanning bounded degree graphs H the gap between lower bounds and the bound given by Riordan's theorem remains, though very recently near-optimal bounds have been obtained for almost spanning H and we shall return to this topic in Section 4.5.

4.3   The Johansson–Kahn–Vu Theorem

It is not too difficult to prove (see, e.g., Theorem 4.9 of [79], or [123]) that the threshold in $G(n,p)$ for an *almost spanning* H-factor, that is, a collection of vertex disjoint copies of H covering all but at most εn vertices, is $n^{-1/m_1(H)}$. For obtaining a spanning H-factor we need to go above this threshold by at least some (power of a) logarithmic factor in some cases: For strictly 1-balanced H, if p grows slower than $(\log n)^{1/e(H)}/n^{1/m_1(H)}$ then by Theorem 3.2 a.a.s. not every vertex of $G(n,p)$ is covered by a copy of H, hence $G(n,p)$ contains no spanning H-factor.

Ruciński [123] showed that if $np^{\delta^*(H)} - \log n \to \infty$, where $\delta^*(H) = \max\{\delta(H'): H' \subseteq H\}$, then $G(n,p)$ a.a.s. contains an H-factor. This implies that the threshold for a K_r-factor is at most $(\log n/n)^{1/(r-1)}$. This was improved on by Krivelevich [105], who proved that for each r there is a constant $C = C(r)$ such that if $p \geq Cn^{-2r/((r-1)(r+2))}$ then $G(n,p)$ a.a.s. contains a K_r-factor (see [79, Section 4.3] for a short exposition of the interesting proof of this result in the case $r = 3$). Observe that this bound on the threshold is also better than the one implied by Riordan's theorem (Theorem 4.2).

Finally, in a celebrated result, Johansson, Kahn, and Vu [81] proved that for strictly 1-balanced H the threshold for an H-factor does indeed coincide with the H-*cover threshold*, that is, the threshold for every vertex of $G(n,p)$ to be contained in an H-copy.

Theorem 4.3 (Johansson, Kahn, Vu [81])
For a strictly 1-balanced graph H the threshold for $G(n,p)$ to contain an H-factor is

$$\frac{(\log n)^{1/e(H)}}{n^{1/m_1(H)}}.$$

Johansson, Kahn, and Vu prove this theorem more generally for hypergraphs in [81]. When H is a single edge, that is, we are asking for a perfect hypergraph matching, it thus solves the famous Shamir problem. A good exposition of the proof in this case is given in [15].

In their proof Johansson, Kahn and Vu work (for some part of the argument) in $G(n,m)$. The basic idea is to think of $G(n,m)$ as a random graph obtained from K_n by successively *deleting* random edges until only m edges remain. They then show with the help of a martingale argument and certain entropy results that in each deletion step not too many H-factors get destroyed, implying that the number of H-factors in $G(n,m)$ is close to expectation.

Already Ruciński [123] and Alon and Yuster [12] observed that not for every H the H-factor threshold is the same as the H-cover threshold. Indeed, it was shown in [12, 123] that for graphs H with $\delta(H) < m_1(H)$ the H-factor threshold is at least $n^{-1/m_1(H)}$, while the H-cover threshold is of lower order of magnitude. It is not surprising that the thresholds for these two properties do not always coincide since there may be some vertex $x \in V(H)$ such that among all H-copies in G the vertex x is only mapped to few vertices u of G. Alon and Yuster [12] conjectured, however, that for each graph H with $e(H) > 0$ the threshold for an H-factor is

$$n^{-(1/m_1(H))+o(1)}.$$

Johansson, Kahn and Vu [81] prove this conjecture as well. Further, they conjecture that the obstacle identified in the last paragraph is the only one, that is, that the H-factor threshold coincides with the threshold for the property $LCOV_H$ that in an n-vertex graph G

1. each vertex of G is contained in an H-copy, and
2. for each $x \in V(H)$ there are at least $n/v(H)$ vertices $u \in V(G)$ such that some H-copy in G maps x to u.

Conjecture 4.4 (Johansson, Kahn, Vu [81]) *The threshold for containing an H-factor is the same as that for $LCOV_H$.*

A related conjecture appears also already in [123]. Johansson, Kahn, and Vu [81] think it even possible that a hitting time version of conjecture 4.4 is true. Further, they state that the threshold of $LCOV_H$ is as follows (for a proof see the arXiv version of [68, Lemma 2.5]). The *local 1-density* of H at $x \in V(H)$ is

$$m_1(x, H) = \max_{\substack{H' \subseteq H \\ x \in V(H')}} \frac{e(H')}{v(H') - 1}.$$

We call H *vertex-1-balanced* if $m_1(x, H) = m_1(H)$ for all $x \in V(H)$. Let $s(x, H)$ denote the minimum number of edges of a maximiser H' in $m_1(x, H)$, and let $s(H)$ be the maximum among all $s(x, H)$. The threshold of $LCOV_H$, which, following [81], we denote by $\text{th}^{[2]}(n)$, then satisfies

$$\text{th}^{[2]}(n) = \begin{cases} \frac{(\log n)^{1/s(H)}}{n^{1/m_1(H)}} & \text{if } H \text{ is vertex-1-balanced}, \\ n^{-1/m_1(H)} & \text{otherwise}. \end{cases}$$

Gerke and McDowell [68] proved Conjecture 4.4 for graphs which are not vertex-1-balanced. Hence, the only open case now is that of vertex-1-balanced graphs which are not strictly 1-balanced.

Theorem 4.5 (Gerke, McDowell [68]) *For a graph H which is not vertex-1-balanced the threshold for an H-factor in $G(n, p)$ is $n^{-1/m_1(H)}$.*

The idea of [68] is to identify dense subgraphs H' of H (which do not cover all vertices because H is non-vertex-1-balanced) and first embed a corresponding non-spanning H'-factor into $G(n, p)$. They then use a variant of Theorem 4.3 to complete the embedding. For obtaining this variant they verify that a partite version of the Johansson–Kahn–Vu theorem holds, which is also useful in other applications.

In fact, the method of Gerke and McDowell allows a proof of Conjecture 4.4 also in the case of many H which are vertex-1-balanced and not strictly 1-balanced. Moreover, for all other H (as for example a triangle and a C_4 glued along one edge) the upper bound given by their method is within a constant log-power of the conjectured bound (see the discussions in the concluding remarks of [68]).

4.4   Trees

The appearance of long paths in $G(n,p)$ was another topic considered early on in the theory of random graphs. As explained in Section 4.1 the threshold in $G(n,p)$ for a Hamilton path is $\log n/n$, where the lower bound follows from the fact that for $p < \log n/n$ there are a.a.s. isolated vertices in $G(n,p)$. Many related results were obtained in the sequel. To give an example, in [2, 57] paths of length cn in $G(n,p)$ for $0 < c < 1$ are considered. But one very natural question, which turned out to be difficult, is if the threshold result for Hamilton paths extends to other spanning trees with bounded maximum degree. The following conjecture, which claims that this is indeed the case and has prompted much recent work, is attributed to Kahn (see [84]), but also appears in [11].

Conjecture 4.6 *For every fixed Δ there is some constant C such that if $T = (T_n)$ is a fixed sequence of trees on n vertices with $\Delta(T) \leq \Delta$ then $G(n,p)$ a.a.s. contains T if $p \geq C \log n/n$.*

In the following I will summarise the progress that has been made towards proving this conjecture. Trees of small linear size were considered by Fernandez de la Vega [58], who proved that there are (large) constants C, C' such that for any fixed Δ and any fixed sequence $T = (T_n)$ of trees with $v(T) \leq n/C$ and $\Delta(T) \leq \Delta$, if $p \geq C'\Delta/n$ then $G(n,p)$ a.a.s. contains T. Alon, Krivelevich and Sudakov [11] improved on this and showed that the threshold in $G(n,p)$ for any sequence of almost spanning trees of bounded degree is $1/n$.

Theorem 4.7 (Alon, Krivelevich, Sudakov [11]) *Given $\Delta \geq 2$ and $0 < \varepsilon < \frac{1}{2}$, let $C = 10^6 \Delta^3 \varepsilon^{-1} \log \Delta \log^2(2/\varepsilon)$. If $p \geq C/n$ then $G(n,p)$ a.a.s. contains all trees T with $\Delta(T) \leq \Delta$ and $v(T) \leq (1 - \varepsilon)n$.*

Observe that Theorem 4.7 is a universality result, stating that $G(n,p)$ contains all these trees *simultaneously*. We shall discuss universality results in $G(n,p)$ in more detail in Section 5. Obtaining such a universality result is possible for Alon, Krivelevich, and Sudakov because they do not prove their result directly for $G(n,p)$, but instead for any graph satisfying certain degree and expansion properties. Their proof uses the well-known embedding result for small (linear sized) trees by Friedman and Pippenger [62]. Balogh, Csaba, Pei and Samotij [16] showed that using instead a related tree embedding result of Haxell [73], which works for larger trees, one can improve the constant in Theorem 4.7 to $C = \max\{1000\Delta \log(20\Delta), 30\Delta\varepsilon^{-1} \log(4e\varepsilon^{-1})\}$. This was further improved by

Montgomery [114] to $C = 30\Delta\varepsilon^{-1}\log(4e\varepsilon^{-1}))$, which comes close to the $C = \Theta(\Delta\log\varepsilon^{-1})$ believed possible in [11].

Alon, Krivelevich and Sudakov also observed in [11] that for every $\varepsilon > 0$ Theorem 4.7 immediately implies Conjecture 4.6 for trees T with εn leaves (for $p \geq C(\varepsilon, \Delta)\log n/n$), by using a two-round exposure of $G(n, p)$, finding in the first round a copy of T minus $(\varepsilon n/\Delta)$ leaves with distinct parents, and then embedding these leaves in the second round, which is easy because all it requires is to find a certain matching. Hefetz, Krivelevich, and Szabó observe in [77] that a similar strategy can be used for embedding trees T with a linearly sized *bare path*, that is a path whose inner vertices have degree 2 in T, also for $p \geq C(\varepsilon, \Delta)\log n/n$.

This leaves the case of trees with few leaves (and no long bare path) of Conjecture 4.6. Since each tree has average degree less than 2, however, these trees have many vertices of degree 2, and hence a linear number of (arbitrarily long) constant length bare paths. Krivelevich [106] used this fact and showed that the same strategy as outlined for trees with many leaves in the previous paragraph can be used for trees with many bare paths by replacing the matching argument by a partite version of the Johansson–Kahn–Vu theorem for embedding the bare paths. Krivelevich's strategy leads to the following result.

Theorem 4.8 (Krivelevich [106]) *For every $\varepsilon > 0$ and every sequence $T = (T_n)$ of trees with $v(T) \leq n$ the random graph $G(n, p)$ a.a.s. contains T if*

$$p \geq \frac{40\Delta(T)\varepsilon^{-1}\log n + n^{\varepsilon}}{n}.$$

In this result $\Delta(T)$ is allowed to grow with n (in particular, a different strategy than Theorem 4.7 is used for obtaining an almost spanning embedding).

Further progress on various classes of trees has been obtained by various groups. Hefetz, Krivelevich, and Szabó [77] show that trees with linearly many leaves and trees with linear sized bare paths, and Montgomery [114] that trees with $\alpha n/\log^9 n$ bare paths of length $\log^9 n$ for any $\alpha > 0$, are already a.a.s. contained in $G(n, p)$ for

$$p = (1 + \varepsilon)\log n/n.$$

Hefetz, Krivelevich, and Szabó [77] also argue that for the same p the random graph $G(n, p)$ a.a.s. contains any typical random tree T, that is, a tree with maximum degree $(1 + o(1))\log n/\log\log n$ as shown in [115].

Investigating a class of special trees called combs was suggested by Kahn (see [84]). A *comb* is a tree consisting of a path on n/k vertices with disjoint k-paths beginning at each of its vertices. Observe that, for example for $k = \sqrt{n}$, combs neither have linearly many leaves nor linear sized bare paths. Kahn, Lubetzky, and Wormald [84, 83] established Conjecture 4.6 for combs. This was improved on and generalised by Montgomery [114] who proved the following result. A *tooth* of length k in a tree is a bare path of length k where one end-vertex is a leaf. Montgomery showed that for any fixed $\alpha > 0$ a tree T with at least $\alpha n/k$ teeth of length k is contained a.a.s. in $G(n,p)$ for $p = (1 + \varepsilon)\log n/n$.

Finally, a result for general bounded degree trees has recently been established by Montgomery [113], which comes very close to the conjectured threshold.

Theorem 4.9 (Montgomery [113]) *If $T = (T_n)$ is a fixed sequence of trees on n vertices with maximum degree $\Delta = \Delta(n)$ then $G(n,p)$ a.a.s. contains T if $p \geq \Delta \log^5 n/n$.*

Montgomery also announced in [113] further work in progress leading to the proof of Conjecture 4.6. For proving Theorem 4.9 Montgomery follows the basic strategy outlined above of first finding an almost spanning subtree of T, leaving some bare paths to be embedded in a second stage (since the case of trees with many leaves is solved already). For embedding these bare paths, however, Montgomery uses an absorbing-type method.

4.5 Bounded degree graphs

Now we turn to the question of when $G(n,p)$ contains given spanning graphs of bounded maximum degree. Let Δ be a constant and $H = (H_n)$ be sequence of graphs with $\Delta(H) \leq \Delta$ and $v(H) \leq n$. Recall that the Theorem of Alon and Füredi (Theorem 4.1) implies that $G(n,p)$ a.a.s. contains H if $p \geq (C(\Delta)\log n/n)^{1/\Delta}$, and Riordan's theorem (Theorem 4.2) implies the same if $p \cdot n^{\frac{2}{\Delta+1} - \frac{2}{\Delta(\Delta+1)}} \to \infty$. This is unlikely to be optimal, though it cannot be far off. The optimum is widely believed to be as follows (see, e.g., [55]).

Conjecture 4.10 *Let $H = (H_n)$ be a sequence of graphs with $\Delta(H) \leq \Delta$ and $v(H) \leq n$. Then $G(n,p)$ a.a.s. contains H if*

$$p \cdot \frac{n^{2/(\Delta+1)}}{(\log n)^{1/\binom{\Delta+1}{2}}} \to \infty. \tag{4.1}$$

In other words, the conjecture states that $G(n,p)$ contains H from above the threshold for a $K_{\Delta+1}$-factor. Ferber, Luh and Nguyen [55] prove Conjecture 4.10 for almost spanning H.

Theorem 4.11 (Ferber, Luh, Nguyen [55]) *Let $\varepsilon > 0$ and Δ be fixed. Let $H = (H_n)$ be a fixed sequence of graphs with $\Delta(H) \leq \Delta$ and $v(H) \leq (1 - \varepsilon)n$. Then $G(n,p)$ a.a.s. contains H if p satisfies (4.1).*

The strategy for the proof of Theorem 4.11 is as follows. Ferber, Luh and Nguyen show that H can be partitioned into a sparse part H', which is sparse enough to be embedded with the help of Riordan's theorem, and a dense part which consists of a collection of induced subgraphs, each of constant size. Given an embedding of H' they then in constantly many rounds extend this embedding successively to embed also the constant size dense bits of H by finding a matching in a suitable auxiliary hypergraph, using a hypergraph Hall-type theorem of Aharoni and Haxell [1] (a similar idea was already used in [38]).

In fact, it is widely believed that even a universality version of Conjecture 4.10 is true (see Conjecture 5.2). Further recent advances were made in this direction, which we shall return to in Section 5.1.

4.6   The Kahn–Kalai conjecture

Let us round off the results presented in the previous sections with a far-reaching and appealing conjecture of Kahn and Kalai. We first need some motivation and definitions. Theorem 3.1 states that for fixed graphs H the threshold for the appearance of H in $G(n,p)$ coincides with what Kahn and Kalai [82] call the *expectation threshold* for H, written $p_{\mathbb{E}}(H,n)$, which is the least $p = p(n)$ such that for each subgraph H' of H the expected number of H' in $G(n,p)$ is at least 1. The expectation threshold can be defined analogously for sequences $H = (H_n)$ of graphs. In particular, for any $(H_n) = H$ we have that $p_{\mathbb{E}}(H,n)$ is the least $p = p(n)$ such that for every subgraph H' of H we have

$$\frac{(n)_{v(H')}}{|\operatorname{Aut}(H')|} p^{e(H')} \geq 1 \, .$$

For example, if H is an F-factor and F' is any subgraph of F then let H' be the vertex disjoint union of $\ell = n/v(F)$ copies of F'. Then the condition above requires that

$$\frac{(n)_{\ell v(F')}}{\ell! \, |\operatorname{Aut}(F')|} p^{\ell e(F')} \geq 1 \, ,$$

which can easily be calculated to be equivalent to $p \geq C n^{-(v(F')-1)/e(F')}$ for some constant C, and hence $p_{\mathbb{E}}(H,n)$ is of the order $n^{-m_1(F)}$ for F-factors. So, by Theorem 4.3, in this case $p_{\mathbb{E}}(H,n)$ is different from the threshold for the appearance of H if F is strictly balanced – but only by less than a $\log n$ factor. Kahn and Kalai [82] conjectured that this is the case for every H.

Conjecture 4.12 (Kahn, Kalai [82]) *There is a universal constant C such that for any sequence $H = (H_n)$ of graphs the threshold for $G(n,p)$ to contain H is at most $C p_{\mathbb{E}}(H, n) \log n$.*

Conjecture 4.6 on trees is a special case of Conjecture 4.12 because $p_{\mathbb{E}}(T, n)$ is of order $1/n$ for bounded degree trees T. Conjecture 4.4 on H-factors and Conjecture 4.10 on bounded degree graphs, on the other hand, are somewhat stronger than what is implied by Conjecture 4.12 because they specify a smaller log-power.

4.7 Constructive proofs

One question we have only occasionally taken up in the preceding sections is if the results on the various structures that exist in $G(n,p)$ a.a.s. for certain probabilities have constructive proofs, allowing for a deterministic or randomised algorithm which finds the desired structure. This question is important for two reasons:

1. Such constructive proofs often lead to polynomial time algorithms, making it possible to find the structures efficiently.
2. Constructive proofs often allow the identification of certain pseudo-random properties, that is, properties which $G(n,p)$ a.a.s. enjoys, which are sufficient for the construction to work. In this case universality results may become possible.

In particular, two prominent results we discussed, whose proofs were not constructive but used the second moment method, were Riordan's theorem and the Johansson–Kahn–Vu theorem. As outlined, these were also used as tools in the proof of other results, such as Theorem 4.5, Theorem 4.8, or Theorem 4.11. This motivates the following problem.

Problem 4.13 *Give a constructive proof of Riordan's theorem (Theorem 4.2) or the Johansson–Kahn–Vu theorem (Theorem 4.3).*

As I shall explain in Sections 5 and 6, many constructive proofs for embedding classes of spanning or almost spanning graphs H in $G(n,p)$ (or in subgraphs of $G(n,p)$) we know of follow a greedy-type paradigm: They embed H (or a suitable subgraph of H) vertex by vertex (or class of vertices by class of vertices), aiming at guaranteeing that unembedded common H-neighbours of already embedded H-vertices can still be embedded in the future. In this sense they crucially rely on the fact that all common neighbourhoods in $G(n,p)$ of $\Delta(H)$ vertices (or of D vertices if H is D-degenerate) are large, which fails to be true for $p \leq n^{-1/\Delta(H)}$. Hence, in Sections 5 and 6 probability bounds of this order shall often form a natural barrier not yet overcome in many instances, though they are not believed to be the right bounds.

5   Universality of random graphs

In this section we consider the question of when the random graph is a.a.s. universal for certain classes of graphs. More precisely, a graph G on n vertices is said to be *universal* for a class $\mathcal{H}$ of graphs, if it contains a copy of every graph $H \in \mathcal{H}$. The crucial difference for $G(n,p)$ to contain some $H \in \mathcal{H}$ a.a.s. and to be a.a.s. universal for $\mathcal{H}$ (if $\mathcal{H}$ is large) is that in the latter case we require a typical graph from $G(n,p)$ to contain all these $H \in \mathcal{H}$ *simultaneously*.

The graph classes for which universality results have been established, and which we shall consider in this section are bounded degree graphs, bounded degree graphs which further have (smaller) bounded maximum 0-density or bounded degeneracy, and bounded degree trees. Let me stress that none of the results presented in this section is believed to be optimal, indicating that the methods we have at hand for proving universality are still limited. Moreover, there are many other natural graph classes still to be considered. The following is just one example.

Question 5.1 *When is $G(n,p)$ a.a.s. universal for the class of all planar graphs with maximum degree Δ; or more generally for all maximum degree Δ graphs which are F-minor free for some fixed F?*

As an aside, n-vertex universal graphs with $O(n \log n)$ edges for n-vertex planar graphs with maximum degree Δ were constructed in [23], and graphs G with $v(G) + e(G) = O(n)$ that are universal for this class of graphs in [36]. For more background on constructions of universal graphs see the survey of Alon [6].

5.1   Universality for bounded degree graphs

Before we turn to results concerning the universality of $G(n,p)$ for the family $\mathcal{H}(n,\Delta)$ of all n-vertex graphs with maximum degree at most Δ, let us first briefly recall some lower bounds. A counting argument shows that any graph G that is universal for $\mathcal{H}(n,\Delta)$ must have edge density at least $\Omega(n^{-2/\Delta})$. This was observed in [9], and follows from the fact that $\sum_{i \leq \Delta n/2} \binom{e(G)}{i} \geq |\mathcal{H}(\Delta,n)|$ and well-known estimates of the number of Δ-regular graphs (for details see [9]). It is interesting to observe that this lower bound was matched by constructive results: Alon and Capalbo constructed graphs that are universal for $\mathcal{H}(n, \Delta)$ and have n vertices and $C(\Delta)n^{2-2/\Delta} \log^{4/\Delta} n$ edges in [7], and $(1+\varepsilon)n$ vertices and $C_2(\Delta, \varepsilon)n^{2-2/\Delta}$ edges for every $\varepsilon > 0$ in [8] (see also [6]).

For $G(n,p)$ the only better lower bound we know is the following, which is only slightly better and only appeals to one particular graph in $\mathcal{H}(n, \Delta)$ instead of universality. By Theorem 3.2, If p grows slower than $(\log n)^{1/\binom{\Delta+1}{2}}/n^{2/(\Delta+1)}$ then a.a.s. $G(n,p)$ contains no spanning $K_{\Delta+1}$-factor. If one turns to universality for smaller, but linearly sized graphs H, the known lower bound is not much smaller. Indeed, if $p \leq cn^{-2/(\Delta+1)}$ for some sufficiently small $c = c(\eta) > 0$ then $G(n,p)$ is not universal even for $\mathcal{H}(\eta n, \Delta)$ as it does not contain a vertex disjoint union of $K_{\Delta+1}$ covering ηn vertices because the expected number of $K_{\Delta+1}$ in $G(n,p)$ is at most $n^{\Delta+1}p^{(\Delta+1)\Delta/2} \leq c^{(\Delta+1)\Delta/2}n$.

As mentioned earlier, it is widely believed (see, e.g., [44, 55]) that the lower bound above reflects the truth, that is, when $G(n,p)$ starts containing every fixed sequence (H_n) of graphs from $\mathcal{H}(n, \Delta)$ a.a.s. then it is already universal for $\mathcal{H}(n, \Delta)$ (cf. Conjecture 4.10).

Conjecture 5.2 $G(n,p)$ *is a.a.s. universal for* $\mathcal{H}(n, \Delta)$ *if*

$$p \cdot \frac{n^{2/(\Delta+1)}}{(\log n)^{1/\binom{\Delta+1}{2}}} \to \infty.$$

At present we are still far from verifying Conjecture 5.2, though this problem attracted considerable attention since the turn of the millennium. Alon, Capalbo, Kohayakawa, Rödl, Ruciński and Szemerédi [9] considered almost spanning graphs and showed that for every $\varepsilon > 0$ and Δ there is C such that for $p \geq C(\log n/n)^{1/\Delta}$ the random graph $G(n,p)$ is a.a.s. universal for $\mathcal{H}((1 - \varepsilon)n, \Delta)$. After improvements in [43], Dellamonica, Kohayakawa, Rödl, and Ruciński [44] showed that for this probability $G(n,p)$ is also universal for spanning bounded degree graphs.

Theorem 5.3 (Dellamonica, Kohayakawa, Rödl, Ruciński [44])
For each $\Delta \geq 3$ *there is* C *such that* $G(n,p)$ *is a.a.s. universal for the family* $\mathcal{H}(n, \Delta)$ *if*

$$p \geq C\Big(\frac{\log n}{n}\Big)^{1/\Delta}.$$

Using a simpler argument (but the same basic strategy), Kim and Lee [85] showed that this result also holds for $\Delta = 2$. For proving their theorem Dellamonica, Kohayakawa, Rödl, and Ruciński present a randomised algorithm that uses a certain set of pseudorandom properties which $G(n,p)$ has a.a.s. and embeds every $H \in \mathcal{H}(n, \Delta)$ a.a.s. in every graph G with these pseudorandom properties. This algorithm is inspired by the various known techniques for proving the blow-up lemma in dense graphs [97, 98, 121, 122], and the underlying idea of using an embedding strategy based on matchings goes back to the proof of the theorem of Alon and Füredi (Theorem 4.1) outlined in Section 4.2.

As mentioned in Section 4.7 the exponent $1/\Delta$ forms a natural barrier to further improvement. So far, this barrier was broken only for almost spanning subgraphs and in the case $\Delta = 2$.

Theorem 5.4 (Conlon, Ferber, Nenadov, Škorić [38])
For every $\varepsilon > 0$ and $\Delta \geq 3$ the random graph $G(n,p)$ is a.a.s. universal for $\mathcal{H}\big((1 - \varepsilon)n, \Delta\big)$ if

$$p \cdot \frac{n^{1/(\Delta-1)}}{\log^5 n} \to \infty.$$

For the case of maximum degree $\Delta = 2$, Conlon, Ferber, Nenadov, and Škorić [38] also state that similar arguments as those used for showing this theorem show that $G(n,p)$ is a.a.s. universal for $\mathcal{H}\big((1 - \varepsilon)n, 2\big)$ if $p \geq Cn^{-2/3}$, which is best possible up to the value of C. Moreover, Ferber, Kronenberg, and Luh [54] very recently showed that $G(n,p)$ is a.a.s. universal for $\mathcal{H}(n, 2)$ if $p \geq C(\log n/n^2)^{1/3}$, which is again best possible up to the value of C. Their proof combines the Johansson–Kahn–Vu Theorem with arguments from Montgomery's [113] proof of Theorem 4.9.

The strategy of the proof of Theorem 5.4 is as follows. Each graph H under consideration is partitioned into a set of (small) components with at most $\log^4 n$ vertices, a set of induced cycles of length at most $2 \log n$, and the graph H' induced on the remaining vertices. They then show that any induced subgraph of $G(n,p)$ on $\frac{1}{2}\varepsilon n$ vertices is universal for $\mathcal{H}(\log^4 n, \Delta)$ and can thus be used for embedding the small components, that H' has a structure suitable for a technical embedding result of Ferber, Nenadov and Peter [56], and that the remaining short cycles can be embedded with the help of the hypergraph matching criterion of Aharoni and Haxell [1].

In [56] Ferber, Nenadov and Peter use the technical result just mentioned for a spanning universality result under additional constraints. More precisely, they consider graphs in $\mathcal{H}(n, \Delta)$ with maximum 0-density at most m_0, and provide a better bound than Theorem 5.3 for $m_0 < \Delta/4$. Note that for any H we have $m_0(H) \leq \Delta(H)/2$.

Theorem 5.5 (Ferber, Nenadov, Peter [56]) *For $\Delta = \Delta(n) > 1$ and $m_0 = m_0(n) \geq 1$ the random graph $G(n,p)$ is a.a.s. universal*

(a) for all $H \in \mathcal{H}(n, \Delta)$ with $m_0(H) \leq m_0$ if

$$p \cdot \frac{\Delta^{12} n^{1/(4m_0)}}{\log^3 n} \to \infty, \text{ and}$$

(b) for all $H \in \mathcal{H}(n, \Delta)$ with $m_0(H) \leq m_0$ and girth at least 7 if

$$p \cdot \frac{\Delta^{12} n^{1/(2m_0)}}{\log^3 n} \to \infty.$$

Ferber, Nenadov and Peter prove this result by using a similar embedding strategy (and a similar decomposition of the graphs H) as Dellamonica, Kohayakawa, Rödl, Ruciński [44] and Kim and Lee [85].

A related result is proven in [5], where D-degenerate graphs H in $\mathcal{H}(n, \Delta)$ are considered. It is not difficult to see that the degeneracy $D(H)$ of any graph H satisfies $m_0(H) \leq D(H) \leq 2m_0(H)$. The bound in the first part of the following result is better than that in the first part of Theorem 5.5 if $D(H) < 2m_0(H) - \frac{1}{2}$. The bound in the second part is better than that in Theorem 5.4 if $D(H) < (\Delta(H) - 1)/2$.

Theorem 5.6 (Allen, Böttcher, Hàn, Kohayakawa, Person [5])
For every $\varepsilon > 0$, $\Delta \geq 1$ and $D \geq 1$ there is C such that the random graph $G(n,p)$ a.a.s. is universal
(a) for all D-degenerate $H \in \mathcal{H}(n, \Delta)$ if $p \geq C(\frac{\log n}{n})^{1/(2D+1)}$, and
(b) for all D-degenerate $H \in \mathcal{H}((1 - \varepsilon)n, \Delta)$ if $p \geq C(\frac{\log n}{n})^{1/(2D)}$.

This result is a direct consequence of a sparse blow-up lemma for graphs with bounded degeneracy (and maximum degree) established in [5], which we shall return to in Section 7.

5.2 Universality for bounded degree trees

Recall that the result of Alon, Krivelevich and Sudakov [11] (Theorem 4.7) states that already for $p = C(\Delta, \varepsilon)/n$ the random graph $G(n,p)$ is a.a.s. universal for the family $\mathcal{T}((1 - \varepsilon)n, \Delta)$ of (almost spanning) trees on $(1 - \varepsilon)n$ vertices and maximum degree at most Δ.

For spanning trees the situation is less well understood. Hefetz, Krivelevich, and Szabó [77] showed that spanning trees with linearly long bare paths are universally a.a.s. contained in $G(n,p)$ for $p = (1 + \varepsilon) \log n/n$. The first universality result in $G(n,p)$ for the entire class $\mathcal{T}(n, \Delta)$ was obtained by Johannsen, Krivelevich, Samotij [80]. This is a consequence of the following universality result for graphs with certain natural expansion properties. The proof of this result relies on the embedding result of Haxell [73] for large trees in graphs with suitable expansion properties and a result of Hefetz, Krivelevich, and Szabó [76] on Hamilton paths between any pair of vertices in graphs with certain different expansion properties.

Theorem 5.7 (Johannsen, Krivelevich, Samotij [80])
There is a constant c such that for any n and Δ with $\log n \leq \Delta \leq cn^{1/3}$ every graph G on n vertices with

(i) $|N_G(X)| \geq 7\Delta n^{2/3}|X|$ *for all* $X \subseteq V(G)$ *with* $1 \leq |X| < \frac{n^{1/3}}{14\Delta}$, *and*

(ii) $e_G(X,Y) > 0$ *for all disjoint* $X, Y \subseteq V(G)$ *with* $|X| = |Y| = \lceil \frac{n^{1/3}}{14\Delta} \rceil$

is universal for $\mathcal{T}(n,\Delta)$.

This directly implies that if $\Delta \geq \log n$ then $G(n,p)$ is a.a.s. universal for $\mathcal{T}(n,\Delta)$ if $p \geq C\Delta \log n/n^{1/3}$, and hence universality for $\mathcal{T}(n,\Delta)$ with constant Δ if $p \geq C\log^2 n/n^{1/3}$.

The result of Ferber, Nenadov, and Peter [56] discussed in the previous section improved on this when Δ grows slower than $n^{1/66}/(\log n)^{1/22}$. Indeed, it follows from the second part of Theorem 5.5 that $G(n,p)$ is a.a.s. universal for $\mathcal{T}(n,\Delta)$ if

$$p \cdot n^{1/2}/(\Delta^{12}\log^3 n) \to \infty.$$

Further, Montgomery announced in [113] that, using refinements of his method for proving Theorem 4.9, establishing universality of $G(n,p)$ for $\mathcal{T}(n,\Delta)$ with $p = C(\Delta)\log^2 n/n$ is now within reach.

Finally, let us remark that, again, $G(n,p)$ has no chance in giving the sparsest graph that is universal for $\mathcal{T}(n,\Delta)$. Indeed, Bhatt, Chung, Leighton, and Rosenberg [23] constructed n-vertex graphs which are universal for $\mathcal{T}(n,\Delta)$ with constant maximum degree $C(\Delta)$. See the references in [6] for earlier constructions.

6   Resilience of random graphs

In this section we study the question of how easily an adversary can destroy copies of a graph H in $G(n,p)$. Questions of this type date back (at least[3]) to [9] where this phenomenon was dubbed *fault tolerance* (which also appears in [87]), but lately the term *resilience* has come into vogue, following Sudakov and Vu [129].

---

[3]Of course Turán-type problems in random graphs also fall in this category and were studied even earlier (cf. Section 3.2).

Let $\mathcal{P}$ be a monotone increasing graph property and Γ be a graph. The *global resilience* of Γ with respect to $\mathcal{P}$ is the minimum $\eta \in \mathbb{R}$ such that deleting a suitable set of $\eta e(\Gamma)$ edges from Γ results in a graph not in $\mathcal{P}$. In other words, whenever an adversary deletes less than a η-fraction of the edges of Γ, the resulting graph will still be in $\mathcal{P}$. Similarly, in the definition of local resilience the adversary is allowed to destroy a certain fraction of the edges incident to each vertex. Formally, the *local resilience* of Γ with respect to $\mathcal{P}$ is the minimum $\eta \in \mathbb{R}$ such that deleting a suitable set of edges, while respecting the restriction that for every vertex $v \in V(\Gamma)$ at most $\eta \deg_\Gamma(v)$ edges containing v are removed, results in a graph not in $\mathcal{P}$. For $\Gamma = G(n,p)$ with $p \geq C \log n/n$ for C sufficiently large (where we have degree concentration) this means that for any $\eta' > \eta$ any subgraph G of Γ with minimum degree at least $(1 - \eta')pn$ is in $\mathcal{P}$.

For the random graph $G(n,p)$ we may then ask what is the local or global resilience of $G(n,p)$ a.a.s. with respect to a property $\mathcal{P}$ for a given p? It turns out that the answer to this question usually is either trivial, that is, basically 0 or 1, or provided by some extremal result in dense graphs (in other words, it is as in $G(n,p)$ with $p = 1$). It is thus not surprising that the local resilience is heavily influenced by the chromatic number of the graphs under study. To the best of my knowledge, at present we do not know of any (subgraph) property which does not follow the pattern just described.

Question 6.1 *Let $\pi(\mathcal{H}_n)$ be the local resilience of $G(n,1)$ with respect to containing all graphs from $\mathcal{H}_n$. Is there any (interesting) family $\mathcal{H} = (\mathcal{H}_n)$ of graphs such that $\pi(\mathcal{H}) = \lim_{n \to \infty} \pi(\mathcal{H}_n)$ exists, and the limit as n tends to infinity of the local resilience of $G(n,p)$ with respect to containing all graphs in $\mathcal{H}_n$ exists but is not in $\{0, 1 - \pi(\mathcal{H})\}$?*

Let me remark that resilience and universality are orthogonal properties in the following sense. We might ask for which probabilities $G(n,p)$ has a.a.s. a certain resilience with respect to containing any fixed graph sequence $H = (H_n)$ from a family $\mathcal{H}$, or with respect to being universal for $\mathcal{H}$ and there is a priori no reason why the answers should turn out the same (though we typically expect them to be). However, in contrast to some results discussed in the previous two sections, at present the methods available for proving resilience generally are constructive and hence allow for universality results. On the other hand, a side effect of this is that many of the probability bounds obtained are far from best-possible.

I will start this section with a global resilience result for small linear sized bounded degree bipartite graphs in Section 6.1, which I also use to outline one approach often used for obtaining resilience results that relies on the sparse regularity lemma. I then review local resilience results for cycles in Section 6.2, for trees in Section 6.3, for triangle factors in Section 6.4, and for graphs of low bandwidth in Section 6.5.

6.1 Global resilience

Obviously, any graph must have trivial global resilience with respect to the containment of any spanning graph H, since an adversary can delete all copies of H by simply deleting all edges at some vertex. For small linearly sized bipartite graphs H, however, Alon, Capalbo, Kohayakawa, Ruciński [9] and Szemerédi, in a paper initiating research into the area of the resilience of random graphs, proved the following result.

Theorem 6.2 ([9]) *For every $\Delta \geq 2$ and $\gamma > 0$ there exist $\eta > 0$ and C such that if $p \geq C(\frac{\log n}{n})^{1/\Delta}$ then $G(n,p)$ a.a.s. has global resilience at least $1 - \gamma$ with respect to universality for the family $\mathcal{H}(\eta n, \eta n, \Delta)$ of all bipartite graphs with partition classes of size $\lfloor \eta n \rfloor$ and maximum degree at most Δ.*

Note that this shows that $G(n,p)$ contains many copies of all graphs in $\mathcal{H}(\eta n, \eta n, \Delta)$ everywhere. It is clear that such a result cannot hold for non-bipartite H because, as any other graph, $G(n,p)$ can be made bipartite by deleting half of its edges. The lower bound on p though is unlikely to be optimal.

Problem 6.3 *Improve the lower bound on p in Theorem 6.2.*

The proof in [9] of Theorem 6.2 uses the sparse regularity lemma, which I will present and explain in more detail in Section 7. The strategy is as follows. First, the sparse regularity lemma is applied to the graph G to obtain a sparse ε-regular partition of $V(G)$. It is then easy to show that some pair of clusters in this partition forms a sparse ε-regular pair (V_1, V_2) with sufficient density. The authors of [9] then develop an embedding result for bounded degree bipartite graphs with partition classes of size $\eta'|V_1|$ and $\eta'|V_2|$ in such a pair.[4]

Most other resilience results (with the exception of the results on cycles in the next section) mentioned in the following use proof strategies which are variations on this basic strategy: They use the sparse regularity lemma to obtain a regular partition, then use a result from dense extremal

---

[4]This result is only stated for $p \geq C(\log n/n)^{1/2\Delta}$ in [9] though, for example, with the bipartite sparse blow-up lemma inferred in [32] from their techniques and from newer regularity inheritance results, one easily obtains from their proof the probability bound claimed in Theorem 6.2.

graph theory on the so-called reduced graph to obtain a suitable structure of regular pairs in this partition, and then use or develop a suitable embedding lemma in such structures of regular pairs, which allows one (often with substantial extra work) to embed the desired graphs.

6.2 Local resilience for cycles

In the language of local resilience, Dirac's theorem [47] states that K_n has local resilience $1/2 - o(1)$ with respect to containing a Hamilton cycle. In this section we shall consider sparse analogues of this result in $G(n, p)$.

Clearly, the local resilience of $G(n, p)$ with respect to containing any graph on more than $n/2$ vertices is at most $\frac{1}{2} - o(1)$, since by deleting the edges of $G(n, p)$ in a random balanced cut we obtain a disconnected graph with components of size at most $\frac{1}{2}n$, and it can easily be shown that each vertex loses at most $(\frac{1}{2} - o(1))pn$ of its edges. Sudakov and Vu [129] then showed a corresponding lower bound. They proved that for every $\gamma > 0$ the local resilience of $G(n, p)$ with respect to containing a Hamilton cycle is a.a.s. at least $\frac{1}{2} - \gamma$ if $p > \log^4 n/n$.

Smaller probabilities were first considered by Frieze, Krivelevich [64], who proved that there are C and η such that for $p \geq C \log n/n$ the local resilience of $G(n, p)$ for containing a Hamilton cycle is a.a.s. at least η. Ben-Shimon, Krivelevich, Sudakov [20] then were able to replace η with $\frac{1}{6}(1 - \gamma)$, and then in [21] with $\frac{1}{3}(1 - \gamma)$. Finally Lee and Sudakov [110] showed that also for this range of p the local resilience is $\frac{1}{2} - o(1)$.

Theorem 6.4 (Lee, Sudakov [110]) *For every $\gamma > 0$ there is a constant C such that the local resilience of $G(n, p)$ with respect to containing a Hamilton cycle is a.a.s. at least $\frac{1}{2} - \gamma$ if $p > C \log n/n$.*

In [21] probabilities as close as possible to the threshold for Hamiltonicity, that is, $p \geq (\log n + \log \log n + \omega(1))/n$, are investigated, at which point the results need to be of a different form, because vertices of degree 2 may exist in $G(n, p)$. That Hamilton cycles are so well understood is connected to the fact that with the Pósa rotation-extension technique (see, e.g., [119]), which is used in the proof of all the aforementioned results, we have a powerful tool at hand for finding Hamilton cycles.

Even smaller probabilities, where we cannot hope for Hamilton cycles any longer, were considered by Dellamonica, Kohayakawa, Marciniszyn and Steger [42]. They show that a.a.s. the local resilience of $G(n, p)$ with respect to containing a cycle of length at least $(1 - \alpha)n$ is $\frac{1}{2} - o(1)$ for any $0 < \alpha < \frac{1}{2}$ if $p \cdot n \to \infty$.

Finally, Krivelevich, Lee and Sudakov [107] proved that if $p \cdot n^{1/2} \to \infty$ then the local resilience of $G(n, p)$ with respect to being *pancyclic*, that is, having cycles of all lengths between 3 and n, is a.a.s. $\frac{1}{2} - o(1)$. Here the probability required is higher than in the results on Hamilton cycles, which is necessary for ensuring the adversary cannot delete all triangles (see also the remarks in Section 6.4). An even stronger result was proved by Lee and Samotij in [109] who show that for the same probability a.a.s. every Hamiltonian subgraph of $G(n, p)$ containing at least $\left(\frac{1}{2} + o(1)\right)pn$ edges is pancyclic.

6.3   Local resilience for trees

Komlós, Sárközy, and Szemerédi [96] showed that for every $\gamma > 0$ and every Δ every sufficiently large n-vertex graph G with minimum degree at least $(\frac{1}{2} + \gamma)n$ contains a copy of any spanning tree T with maximum degree at most Δ. In [99] they then extended this result to trees with maximum degree at most $cn/\log n$. An analogue of the former result for random graphs in the case that T is almost spanning was obtained by Balogh, Csaba, and Samotij [17]. Recall that $\mathcal{T}(n, \Delta)$ is the family of all n-vertex trees with maximum degree at most Δ.

Theorem 6.5 (Balogh, Csaba, Samotij [17]) *For all $\Delta \geq 2$ and $\gamma > 0$ there is a constant C such that for $p \geq C/n$ the local resilience of $G(n, p)$ with respect to being universal for $\mathcal{T}((1 - \gamma)n, \Delta)$ a.a.s. is at least $\frac{1}{2} - \gamma$.*

The surprising aspect about this theorem is the small probability for which it was proven to hold. Clearly, this bound on p is sharp up to the value of C, since for smaller p the biggest component of $G(n, p)$ gets too small to contain a tree on $(1 - o(1))n$ vertices. Moreover, as argued in [17], at this probability we cannot ask for, say, balanced $D(n)$-ary trees on $(1 - o(1))n$ vertices for $D(n) \to \infty$, since we do not have enough vertices of degree $D(n)$. Further, the factor $\frac{1}{2}$ in this result is best possible by the discussion in the second paragraph of the previous section. To prove their result Balogh, Csaba, and Samotij [17] use an approach based on the regularity lemma and an embedding result for trees which is a suitable modification of the tree embedding result by Friedman and Pippenger [62].

The only local resilience result for spanning trees that I am aware of follows from Theorem 6.9 on the resilience of $G(n, p)$ for low-bandwidth graphs, which is presented in Section 6.5. It was proven by Chung [37] that trees with constant maximum degree have bandwidth at most $O(n/\log n)$.

Theorem 6.6 (Allen, Böttcher, Ehrenmüller, Taraz [3])
For all $\Delta \geq 2$ and $\gamma > 0$ there is C such that for $p \geq C\left(\frac{\log n}{n}\right)^{1/3}$ the local resilience of $G(n, p)$ with respect to being universal for $\mathcal{T}(n, \Delta)$ a.a.s. is at least $\frac{1}{2} - \gamma$.

This probability is not believed to be optimal. Indeed, it is conceivable that this result remains true down to the conjectured universality threshold for $\mathcal{T}(n, \Delta)$.

Conjecture 6.7 *The conclusion of Theorem 6.6 is true for $p \geq C \log n/n$.*

6.4 Local resilience for triangle factors

Corrádi and Hajnal [41] proved that any graph G with $\delta(G) \geq \frac{2}{3}n$ contains a triangle factor. One could then ask if this result can be transferred to $G(n, p)$ for p sufficiently large, that is, if the local resilience of $G(n, p)$ with respect to containing a triangle factor a.a.s. is $\frac{1}{3} - o(1)$. Huang, Lee, and Sudakov [78] observed that this is not the case even for constant p. Indeed, every vertex v in $G(n, p)$ has a.a.s. a neighbourhood $N(v)$ of roughly size pn, and every $w \in N(v)$ has $\deg\big(w; N(v)\big) \approx p^2 n$ neighbours in $N(v)$. Therefore, we can delete all triangles containing v by removing at most roughly $p^2 n < \gamma p n$ edges at each w if p is small compared to γ and hence obtain a graph without a triangle factor. With a more careful analysis it is possible to show that we can actually choose $O(p^{-2})$ vertices and delete all triangles containing any of these vertices by removing less that $\gamma p n$ edges at each vertex (for the details see [78, Proposition 6.3]).

So the question above should be refined to ask for an *almost spanning* triangle factor, covering all but $O(p^{-2})$ vertices. Balogh, Lee and Samotij [18] showed that this is indeed true if $p \geq C\big(\frac{\log n}{n}\big)^{1/2}$. Observe that this probability is larger than the threshold $\log^{1/3} n/n^{2/3}$ for a triangle factor as given by Theorem 4.3. If p grows slower than $n^{-1/2}$, however, the $O(p^{-2})$ term becomes trivial.

Theorem 6.8 (Balogh, Lee, Samotij [18]) *For every $\gamma > 0$ there are constants C and D such that for $p \geq C\big(\frac{\log n}{n}\big)^{1/2}$ the local resilience of $G(n, p)$ with respect to the containment of an almost spanning triangle factor covering all but at most Dp^{-2} vertices is a.a.s. at least $\frac{1}{3} - \gamma$.*

It should be remarked that a corresponding result with Dp^{-2} replaced by εn follows easily from the conjecture of Kohayakawa, Łuczak, and Rödl [90, Conjecture 23], which has long been known for triangles and was proved in full generality in [19, 40, 124]. This argument will be sketched for the purpose of illustrating the sparse regularity lemma in Section 7.1.

For proving their result Balogh, Lee, Samotij [18] develop a sparse analogue of the blow-up lemma for the special case of triangle factors. We shall discuss (more general) blow-up lemmas in Section 7.

Analogous questions concerning H-factors for general H were considered for constant p in [78], but the currently best bounds follow from Theorem 6.9, which we discuss in the next section.

6.5   The bandwidth theorem in random graphs

In [34] it was shown that for every Δ, r and $\gamma > 0$ there is $\beta > 0$ such that any sufficiently large n-vertex graph G with $\delta(G) \geq (\frac{r-1}{r} + \gamma)n$ contains any r-colourable $H \in \mathcal{H}(n, \Delta)$ with bandwidth $\mathrm{bw}(H) \leq \beta n$. This proved a conjecture of Bollobás and Komlós and is often referred to as the bandwidth theorem. It is easy to argue that some restriction like the bandwidth restriction in this result is necessary, and also that the $\frac{r-1}{r}$ in the minimum degree is best possible; it is also known that we cannot have $\gamma = 0$ (for details see the discussions in [34]). Further, as shown in [33], the bandwidth condition does not excessively restrict the class of embeddable graphs. Indeed, requiring the bandwidth of a bounded degree n-vertex graph to be $o(n)$ is equivalent to requiring the treewidth to be $o(n)$ or to have no large expanding subgraphs. This implies that bounded degree planar graphs, and more generally bounded degree graphs defined by some (or several) forbidden minor, have bandwidth $o(n)$.

A transference of the bandwidth theorem to $G(n, p)$ for constant p was obtained by Huang, Lee and Sudakov [78]. As discussed in the last section in such a result we cannot hope to cover all the graphs H embedded by the bandwidth theorem. More precisely we have to ask for at least $O(p^{-2})$ vertices of H not to be contained in a triangle. A result for smaller p in the special case of almost spanning bipartite graphs in $\mathcal{H}((1 - o(1))n, \Delta)$ with bandwidth at most βn was obtained in [32] for $p \geq C(\log n/n)^{1/\Delta}$. Recently a general sparse analogue of the bandwidth theorem became possible with the help of the sparse blow-up lemma (see Section 7). For a concise statement, let $\mathcal{H}(n, \Delta, r, \beta)$ be the class of all r-colourable n-vertex graphs with maximum degree Δ and bandwidth at most βn.

Theorem 6.9 (Allen, Böttcher, Ehrenmüller, Taraz [3])
For all Δ, D, r and $\gamma > 0$ there are $\beta > 0$ and C such that $(\frac{1}{r} - \gamma)$ is a.a.s. a lower bound on the local resilience of $G(n, p)$ with respect to universality for all $H \in \mathcal{H}(n, \Delta, r, \beta)$ such that either

(a) *at least $C \max\{p^{-2}, p^{-1} \log n\}$ vertices of H are not in triangles, and $p \geq C(\log n/n)^{1/\Delta}$, or*

(b) *at least $C \max\{p^{-2}, p^{-1} \log n\}$ vertices of H are in neither triangles nor $C_4 s$, and H is D-degenerate, and $p \geq C(\log n/n)^{1/(2D+1)}$.*

Here, the term $p^{-1} \log n$ in the bound on the vertices not in triangles is only relevant for relatively large probabilities $p > 1/\log n$. It is an artefact of our proof and we do not believe it is necessary. Similarly, the requirement on vertices not being contained in C_4 in (b) can probably be removed, but we need it for our proof.

Observe that this theorem provides two different lower bounds on the probability, where the second one is better if the degeneracy of H is much smaller than its maximum degree (note though that even in this case we require a constant bound on the maximum degree). We do not believe these bounds to be optimal, but the bound in (a) matches the corresponding currently known universality bound in Theorem 5.3 and is thus well justified. Hence, the following problem is hard.

Problem 6.10 *Improve the bounds on p in Theorem 6.9.*

The exponent of n in p cannot be improved beyond $1/m_2(K_{\Delta+1}) = 2/(\Delta + 2)$. Indeed, if p grows slower than $n^{-1/m_2(H)}$ then in $G(n,p)$ the expected number of H-copies containing any fixed vertex is $o(pn)$ and one can show, using a concentration inequality of Kim and Vu [86], that in fact a.a.s. every vertex of $G(n,p)$ lies in at most γpn copies of H (for the details see, e.g., [4, Lemma 3.3]). Hence, in this case an adversary can even easily delete all H-copies without removing more than a 2γ-fraction of the edges at each vertex.

It is possible that $2/(\Delta + 2)$ is indeed the correct exponent. A more precise conjecture is offered in the concluding remarks of [3].

A better probability bound than that in Theorem 6.9 was very recently obtained by Noever and Steger [117] for the special case of almost spanning squares of Hamilton cycles, which is approximately optimal.

Theorem 6.11 (Noever, Steger [117]) *For all $\gamma > 0$ and $p \geq n^{\gamma-1/2}$ the local resilience of $G(n,p)$ with respect to containing the square of a cycle on at least $(1 - \gamma)n$ vertices is a.a.s. at least $\frac{1}{3} - \gamma$.*

7 The blow-up lemma for sparse graphs

Szemerédi's regularity lemma proved extremely important for much of the progress in extremal graph theory (and other areas) over the past few decades. Together with the blow-up lemma it also allowed for a wealth of results on spanning substructures of dense graphs. For sparse graphs, such as sparse random graphs or their subgraphs, the error terms appearing in the regularity lemma though are too coarse. This inspired the development of sparse analogues of this machinery – which turned out to be a difficult task. In this section these sparse analogues are surveyed and some very simple example applications are provided to demonstrate how they are used. Section 7.1 introduces the sparse regularity lemma and explains how it is used for obtaining resilience results. Section 7.2 states so-called inheritance lemmas for sparse regular pairs, which are needed to work with the sparse blow-up lemma. Section 7.3 provides the sparse blow-up lemma for random graphs, and Section 7.4 outlines how it is applied.

To a certain degree I assume familiarity of the reader with the dense regularity lemma and blow-up lemma, and refer to the surveys [95, 100, 101] for the relevant background.

7.1 The sparse regularity lemma

In sparse versions of the regularity lemma, all edge densities are taken relative to an ambient density p. In our applications here, where we are interested in subgraphs G of some random graph, we may always take the edge probability of the random graph as the ambient density p. In order to state a sparse regularity lemma we need some definitions.

Let $G = (V, E)$ be a graph, and suppose $p \in (0, 1]$ and $\varepsilon > 0$ are reals. For disjoint nonempty sets $U, W \subseteq V$ the p-density of the pair (U, W) is defined as $d_{G,p}(U, W) = e_G(U, W)/(p|U||W|)$. The pair (U, W) is (ε, d, p)-regular (or (ε, d, p)-lower-regular) if there is $d' \geq d$ such that $d_{G,p}(U', W') = d' \pm \varepsilon$ (or if $d_{G,p}(U', W') \geq d - \varepsilon$, respectively) for all $U' \subseteq U$ and $W' \subseteq W$ with $|U'| \geq \varepsilon|U|$ and $|W'| \geq \varepsilon|W|$. We say that (U, W) is (ε, p)-regular (or (ε, p)-lower-regular), if it is (ε, d, p)-regular (or (ε, d, p)-lower-regular) for some $d \geq d_{G,p}(U, W) - \varepsilon$.

An ε-equipartition of V is a partition $V = V_0 \dot\cup V_1 \dot\cup \ldots \dot\cup V_r$ with $|V_0| \leq \varepsilon|V|$ and $|V_1| = \cdots = |V_r|$. An (ε, p)-regular partition (or an (ε, p)-lower-regular partition) of $G = (V, E)$ is an ε-equipartition $V_0 \dot\cup V_1 \dot\cup \ldots \dot\cup V_r$ of V such that (V_i, V_j) is an (ε, p)-regular pair (or an (ε, p)-lower-regular pair) in G for all but at most $\varepsilon\binom{r}{2}$ pairs $ij \in \binom{[r]}{2}$. The partition classes V_i with $i \in [r]$ are called the clusters of the partition and V_0 is the exceptional set.

The sparse regularity lemma by Kohayakawa and Rödl [88, 92] and Scott [126] asserts the existence of (ε, p)-regular partitions for sparse graphs G. In applications of this sparse regularity lemma one often only makes use of sufficiently dense regular pairs in the regular partition, and the reduced graph of the partition captures where these dense pairs are. Formally, an ε-equipartition $V_0 \dot\cup V_1 \dot\cup \ldots \dot\cup V_r$ of a graph $G = (V, E)$ is an (ε, d, p)-regular partition (or (ε, d, p)-lower-regular partition) with reduced graph R if $V(R) = [r]$ and the pair (V_i, V_j) is (ε, d, p)-regular (or (ε, d, p)-lower-regular) in G whenever $ij \in E(R)$. Observe that, given $d > 0$, an (ε, p)-regular partition gives rise to an (ε, d, p)-regular partition of G with reduced graph R, where R contains exactly the edges ij such that (V_i, V_j)

is (ε, p)-regular and $d_{G,p}(V_i, V_j) \geq d - \varepsilon$.

It then is a consequence of the sparse regularity lemma that graphs G with sufficiently large minimum degree relative to the ambient density p (and which do not have linear sized subgraphs of density much above p) allow for (ε, d, p)-regular partitions with a reduced graph R of high minimum degree. In this sense R inherits the minimum degree of G. The following lemma, which can be found e.g. in [3], makes this precise.

Lemma 7.1 (sparse regularity lemma, min. degree version)
For each $\varepsilon > 0$, $\alpha \in [0, 1]$, and $r_0 \geq 1$ there exists $r_1 \geq 1$ with the following property. For any $d \in [0, 1]$, any $p > 0$, and any n-vertex graph G with $\delta(G) \geq \alpha \cdot pn$ such that for any disjoint $X, Y \subseteq V(G)$ with $|X|, |Y| \geq \varepsilon \frac{n}{r_1}$ we have $e(X, Y) \leq (1 + \frac{\varepsilon^2}{10^3})p|X||Y|$, there is an (ε, d, p)-regular partition of $V(G)$ with reduced graph R with $\delta(R) \geq (\alpha - d - \varepsilon)|V(R)|$ and $r_0 \leq |V(R)| \leq r_1$.

The crucial point is that the reduced graph R in this lemma is a *dense* graph, which means that we can apply extremal graph theory results for dense graphs to R. It should be noted that analogous lemmas can easily be formulated where other properties are inherited by the reduced graph, such as the (relative) density of G.

The regularity lemma then becomes useful in conjunction with suitable embedding lemmas. These come in different flavours. Embedding constant sized graphs H in systems of regular pairs in $G(n, p)$ is allowed by the so-called *counting lemma*, which even allows to give good estimates on the number of H-copies. In a major breakthrough such counting lemmas were recently established for the correct p (that is, the threshold was established) in [19, 40, 124], verifying a conjecture of Kohayakawa, Łuczak, and Rödl [90, Conjecture 23]. An embedding lemma for H of small linear size, on the other hand, was provided in [94] for $p \geq C(\log n/n)^{-1/\Delta}$. This range of p is not believed to be best possible, but again matches the natural barrier. Finally, the blow-up lemma, which is stated in Section 7.3, handles spanning graphs (for the same edge probability p).

To illustrate how the sparse regularity lemma and the embedding lemmas interact, let us briefly sketch how to show that for $p \geq C(\log n/n)^{1/2}$ a.a.s. a subgraph G of $G(n, p)$ with $\delta(G) \geq (\frac{2}{3} + \gamma)pn$ has a triangle factor covering at least $(1 - \gamma)n$ vertices for every $\gamma > 0$ and C sufficiently large. Indeed, if we apply the minimum degree version of the sparse regularity lemma (Lemma 7.1) to G, with $\varepsilon \ll d$ sufficiently small and $r_0 = 3$, we obtain a reduced graph R with $\delta(R) \geq (\frac{2}{3} + \frac{\gamma}{2})v(R)$, which thus contains a (spanning) triangle factor by the theorem of Corrádi and Hajnal [41]. One triangle in this triangle factor corresponds to three (ε, d, p)-regular pairs in G, in which, according to the sparse counting lemma, we find one triangle. After removing the three vertices of this triangle, what remains of the three pairs is still (ε', d, p)-regular for ε' almost as big as ε. Hence, we can apply the counting lemma again to find another triangle. In fact, we can repeat this process until, say, a $\frac{1}{2}\gamma$-fraction of the original three pairs is left. Repeating this for each triangle in the triangle factor of R, we obtain an almost spanning triangle factor in G covering all but at most the εn vertices of the exceptional set V_0 and a $\frac{1}{2}\gamma$-fraction of $V \setminus V_0$.

7.2 Regularity inheritance in $G(n,p)$

In the dense setting, when embedding graphs H in systems of regular pairs one often proceeds in rounds, and for later rounds crucially relies on the following fact (and a two-sided version thereof). Assume (X, Y) is a regular pair into which we want to embed an edge $x'y'$ of H. Assume further that some neighbour z' of x' was embedded in previous rounds in a pair Z. Then the setup of the blow-up lemma will be such that (Z, X) is also a regular pair, and we will have chosen the image z of z' carefully enough so that z is "typical" in the pair (Z, X) in the sense that $N(z; X)$ will be of size $d|X| \gg \varepsilon|X|$ for a suitable constant d. It then easily follows from the definition of ε-regularity that the pair $(N(z; X), Y)$ is still a regular pair (with reduced regularity parameter), that is $(N(z; X), Y)$ *inherits* regularity from (X, Y). This then makes it easy to embed the edge $x'y'$ in (X, Y) such that x' is embedded into $N(z; X)$.

Trying to use a similar approach in sparse graphs we encounter the following problem: If (X, Y) and (Y, Z) are (ε, d, p)-regular pairs then a "typical" vertex $z \in Z$ has a neighbourhood of size about $dp|X|$ in X, which is much smaller than $\varepsilon|X|$ if p goes to 0. Hence, it is not clear any more that $(N(z; X), Y)$ inherits regularity from (X, Y) – in fact, this is false in general. Fortunately, however, if we consider regular pairs (X, Y) and (Y, Z) in a subgraph G of $G(n, p)$, then it is true for most $z \in Z$ that $(N_G(z; X), Y)$ inherits regularity from (X, Y). This phenomenon was observed in [67, 91, 94]. Based on the techniques developed in these papers, the following regularity inheritance lemmas are shown in [5].

Lemma 7.2 (One-sided regularity inheritance [5]) *For each $\varepsilon', d > 0$ there are $\varepsilon_0 > 0$ and C such that for all $0 < \varepsilon < \varepsilon_0$ and $0 < p < 1$, a.a.s. $\Gamma = G(n, p)$ has the following property. Let $G \subseteq \Gamma$ be a graph and X, Y be disjoint subsets of $V(\Gamma)$. If (X, Y) is (ε, d, p)-lower-regular in G and*

$$|X| \geq C \max\left(p^{-2}, p^{-1}\log n\right) \quad and \quad |Y| \geq Cp^{-1}\log n\,,$$

then the pair $\left(N_\Gamma(z; X), Y\right)$ is not (ε', d, p)-lower-regular in G for at most $Cp^{-1} \log n$ vertices $z \in V(\Gamma)$.

Observe that this lemma consider neighbourhoods in $\Gamma = G(n, p)$, rather than directly in G. More specifically, Lemma 7.2 establishes lower-regularity of $\left(N_\Gamma(z; X), Y\right)$. However, since for most vertices $z \in Z$ the order of magnitude of $\deg_G(z; X)$ and $\deg_\Gamma(z; X)$ differs by a factor of at most $2d$, the pair $\left(N_G(z; X), Y\right)$ then easily inherits regularity from $\left(N_\Gamma(z; X), Y\right)$.

Lemma 7.2 is complemented by the following two-sided version, which guarantees lower-regularity of the pair $\left(N_\Gamma(z; X), N_\Gamma(z; Y)\right)$. This plays an important role when we want to embed triangles.

Lemma 7.3 (Two-sided regularity inheritance [5]) *For each $\varepsilon', d > 0$ there are $\varepsilon_0 > 0$ and C such that for all $0 < \varepsilon < \varepsilon_0$ and $0 < p < 1$, a.a.s. $\Gamma = G(n, p)$ has the following property. Let $G \subseteq \Gamma$ be a graph and X, Y be disjoint subsets of $V(\Gamma)$. If (X, Y) is (ε, d, p)-lower-regular in G and*

$$|X|, |Y| \geq C \max\left(p^{-2}, p^{-1} \log n\right),$$

then the pair $\left(N_\Gamma(z; X), N_\Gamma(z; Y)\right)$ is not (ε', d, p)-lower-regular in G for at most $C \max(p^{-2}, p^{-1} \log n)$ vertices $z \in V(\Gamma)$.

These two lemmas are similar to [94, Proposition 15], and in fact equivalent when $p = \Theta\left((\log n/n)^{1/\Delta}\right)$, but not for larger p, when the bounds on $|X|$ and $|Y|$ and the number of vertices z are different, which is sometimes useful in applications.

They are moreover proved for lower-regular pairs rather than for regular pairs (which leads to a less strong assumption, but also to a weaker conclusion). In fact, it would be interesting to obtain analogous lemmas for sparse regular pairs.

Problem 7.4 *Prove analogues of Lemmas 7.2 and 7.3 for (ε, d, p)-regular pairs in subgraphs of $G(n, p)$.*

Lemmas 7.2 and 7.3 state that most vertices in Z satisfy regularity inheritance properties. In the sparse blow-up lemma, however, we will require this property from all vertices in a cluster. More precisely, let X, Y and Z be vertex sets in $G \subseteq \Gamma$, where X and Y are disjoint and X and Z are disjoint, but we do allow $Y = Z$. We say that (Z, X, Y) has *one-sided (ε, d, p)-inheritance* if for each $z \in Z$ the pair $\left(N_\Gamma(z, X), Y\right)$ is (ε, d, p)-lower-regular. If in addition X and Z are disjoint, then we say that (Z, X, Y) has *two-sided (ε, d, p)-inheritance* if for each $z \in Z$ the pair

$(N_\Gamma(z, X), N_\Gamma(z, Y))$ is (ε, d, p)-lower-regular. When applying the sparse blow-up lemma, our approach will be to simply remove the few vertices from each cluster whose neighbourhoods in certain other clusters do not inherit lower-regularity (and deal with them separately).

7.3 The random graphs blow-up lemma

The purpose of this section is to state a slightly simplified version of the blow-up lemma for random graphs proven in [5]. The setup in this blow-up lemma is as follows. We are given two graphs G and H on the same number of vertices, where G is a subgraph of the random graph $\Gamma = G(n, p)$. The graphs G and H are endowed with partitions $\mathcal{V} = \{V_i\}_{i \in [r]}$ and $\mathcal{X} = \{X_i\}_{i \in [r]}$ of their respective vertex sets, of which we require certain properties. Firstly, the partitions $\mathcal{V}$ and $\mathcal{X}$ need to be *size-compatible*, that is, $|V_i| = |X_i|$ for all $i \in [r]$. Secondly, $(G, \mathcal{V})$ needs to be κ-*balanced*, that is, there exists m such that $m \leq |V_i| \leq \kappa m$ for all $i, j \in [r]$.

Further, we will have two reduced graphs R and $R' \subseteq R$ on r vertices, where R represents the regular pairs of $(G, \mathcal{V})$. In fact, we work with lower-regularity instead of regularity, because that is what the inheritance lemmas discussed in the last section provide. Hence, we say that $(G, \mathcal{V})$ is an (ε, d, p)-*lower-regular R-partition* if for each edge $ij \in R$ the pair (V_i, V_j) is (ε, d, p)-lower-regular.[5] We require that H has edges only along lower-regular pairs of this partition. Formally, $(H, \mathcal{X})$ is an R-*partition* if each part of $\mathcal{X}$ is empty, and whenever there are edges of H between X_i and X_j, the pair ij is an edge of R.

As in the dense blow-up lemma, we cannot hope to embed a spanning graph solely in systems of regular pairs, as these may contain isolated vertices. Therefore, we will require certain pairs to be super-regular, that is to additionally satisfy a minimum degree condition. Where these super-regular pairs are is captured by the second reduced graph R'. A pair (X, Y) in $G \subseteq \Gamma$ is called (ε, d, p)-*super-regular* (in G) if it is (ε, d, p)-lower-regular and for every $x \in X$ and $y \in Y$ we have

$$\deg_G(x; Y) > (d - \varepsilon) \max\{p|Y|, \deg_\Gamma(x; Y)/2\},$$
$$\deg_G(y; X) > (d - \varepsilon) \max\{p|X|, \deg_\Gamma(y; X)/2\}.$$

The second term in these maxima is technically necessary to treat vertices x of exceptionally high Γ-degree into Y, but can be ignored for most

---

[5]Observe that this differs from an (ε, d, p)-lower-regular partition with reduced graph R in that we do not require the partition to be an ε-equipartition. In fact, in the partitions referred to in the blow-up lemma the exceptional set is omitted.

purposes (see also the discussion in [5]). The partition $(G, \mathcal{V})$ is (ε, d, p)-*super-regular on R'* if for every $ij \in E(R')$ the pair (V_i, V_j) is (ε, d, p)-super-regular.

But even requiring super-regularity is not enough, as for super-regular pairs (X, Y), (Y, Z), (X, Z) in $G(n, p)$ there may be vertices $z \in Z$ with no edge in $\big(N_G(z; X), N_G(z; Y)\big)$, which prevents us for example from embedding a triangle factor in (Z, X, Y). However, as argued in the previous section, lower-regularity does not get inherited on neighbourhoods for only a few vertices in (Z, X, Y). Hence, omitting these we can circumvent this problem. In the blow-up lemma we will thus require regularity inheritance along R'. Formally, $(G, \mathcal{V})$ has *one-sided inheritance* on R' if (V_i, V_j, V_k) has one-sided (ε, d, p)-inheritance for every $ij, jk \in E(R')$, where we do allow $i = k$. Similarly, $(G, \mathcal{V})$ has *two-sided inheritance* on R' if (V_i, V_j, V_k) has two-sided (ε, d, p)-inheritance for every $ij, jk, ik \in R'$.

It remains to describe which of the edges of H are required to go along the super-regular pairs captured by R'. It turns out that we only need to restrict a small linear fraction of the vertices of each X_i to having their neighbours and second neighbours along R'. We collect these special vertices in a so-called buffer. Formally, a family $\tilde{\mathcal{X}} = \{\tilde{X}_i\}_{i \in [r]}$ of subsets $\tilde{X}_i \subseteq X_i$ is an (α, R')-*buffer* for H if for each $i \in [r]$ we have $|\tilde{X}_i| \geq \alpha|X_i|$ and for each $x \in \tilde{X}_i$ and each $xy, yz \in E(H)$ with $y \in X_j$ and $z \in X_k$ we have $ij \in R'$ and $jk \in R'$. The buffer sets can be chosen by the user of the blow-up lemma, which asserts that for any graphs H and G with the setup as just described we can embed H into G (if p is sufficiently large).

Lemma 7.5 (Blow-up lemma for $G_{n,p}$ [5]) *For all $\Delta \geq 2$, $\Delta_{R'} \geq 1$, $\kappa \geq 1$, and $\alpha, d > 0$ there exists $\varepsilon > 0$ such that for all r_1 there is a C such that for $p \geq C(\log n/n)^{1/\Delta}$ the random graph $\Gamma = G(n, p)$ a.a.s. satisfies the following. Let R be a graph on $r \leq r_1$ vertices and let $R' \subseteq R$ be a spanning subgraph with $\Delta(R') \leq \Delta_{R'}$. Let H and $G \subseteq \Gamma$ be graphs with κ-balanced size-compatible vertex partitions $\mathcal{X} = \{X_i\}_{i \in [r]}$ and $\mathcal{V} = \{V_i\}_{i \in [r]}$, respectively, which have parts of size at least $m \geq n/(\kappa r_1)$. Let $\tilde{\mathcal{X}} = \{\tilde{X}_i\}_{i \in [r]}$ be a family of subsets of $V(H)$ and suppose that*

   (i) *$\Delta(H) \leq \Delta$, $(H, \mathcal{X})$ is an R-partition, and $\tilde{\mathcal{X}}$ an (α, R')-buffer for H,*

   (ii) *$(G, \mathcal{V})$ is an (ε, d, p)-lower-regular R-partition, which is (ε, d, p)-super-regular on R', and has one- and two-sided inheritance on R'.*

Then there is an embedding of H into G.

A number of remarks are in place. Firstly, one of the advantages in this formulation of the blow-up lemma, compared to that of the dense blow-up lemma, is that the required regularity constant ε does not depend on the number of clusters r of the reduced graph R, but only on the maximum degree of R'. This makes it possible to apply this blow-up lemma to the whole reduced graph of a regular partition given by the sparse regularity lemma, instead of the repeated applications of the blow-up lemma to small parts of the reduced graph together with a technique for "glueing" the different so-obtained subgraphs together that were the norm when applying the dense blow-up lemma. Since for $p = 1$ we recover the dense setting, this technique can now also be used for dense graphs G.

Secondly, the version of this blow-up lemma given in [5] is stronger in the following senses. One difference is that in [5] we only require two-sided inheritance on triangles of R' in which we want to embed some triangle of H containing a vertex in the buffer. In particular, we do not need two-sided inheritance at all if H has no triangles. This is useful in some applications, as explained in [5]. The other difference is that in [5] so-called image restrictions are allowed, that is, for some vertices x of H we are allowed to specify a relatively small set of vertices in G into which x is to be embedded. These image restrictions have somewhat more complex requirements than in the dense case, hence we omit them here, but the basic philosophy is that the requirements are those needed to guarantee compatibility with super-regularity and regularity inheritance in the remainder of the partition of G (and they are generalisations of the image restrictions in the dense case).

But why do we need image restrictions? In the dense case such image restrictions were usually used for "glueing" different blow-up lemma applications together, which is now no longer needed, as described above. However, as we will describe in the next section, when we want to apply the blow-up lemma to a partition obtained from the sparse regularity lemma, we will need to exclude a number of vertices from each cluster to guarantee super-regularity and regularity inheritance. In the dense case, usually all of these vertices can be redistributed to other clusters without destroying these properties, but in the sparse case this is not necessarily possible. Hence, if we want to obtain a spanning embedding result we will first need to embed certain H-vertices on these exceptional vertices of G by hand, which lead to image restrictions, before we can apply the sparse blow-up lemma to embed the remainder of H (see, e.g., [3] for more details).

Finally, again, the lower bound on p is unlikely to be best possible.

Problem 7.6 *Improve the exponent of n in the lower bound on p in Lemma 7.5.*

Even in a version of this lemma for only small linear sized graphs H

this would for example directly lead to an improvement on the known bounds on so-called size Ramsey numbers (cf. [94]), among many others.

Let me remark that in [5] additionally a version of the blow-up lemma for D-degenerate graphs H with maximum degree Δ is given. In this version the exponent in the power of n in the bound on p depends only on D (but the constant C still depends on Δ). Often we can choose this exponent to be $2D+1$, and in some cases even smaller, but the details depend on the choice of a suitable buffer and are more involved. In applications image restrictions are often needed in addition, which complicate the statement of a corresponding blow-up lemma even further. It is this version of the sparse blow-up lemma which is used to prove Theorem 6.9(b).

7.4 Applying the blow-up lemma

As an example application of the sparse blow-up lemma presented in the last section, let us briefly sketch how it can be used to show that for every $\gamma > 0$, if C is sufficiently large and $p \geq C\big(\log n/n\big)^{1/4}$, a.a.s. any subgraph G of $\Gamma = G(n,p)$ with $\delta(G) \geq (\frac{1}{2}+\gamma)pn$ contains a copy of the $k \times k$ square grid $H = L_k$ with $k = (1-\gamma)\sqrt{n}$.

We start by preparing G for the sparse blow-up lemma. To this end, we first apply the minimum degree version of the sparse regularity lemma (Lemma 7.1) to G and obtain an (ε, d, p)-regular partition $V_0 \dot\cup V_1 \dot\cup \ldots \dot\cup V_r$ with reduced graph R of minimum degree bigger than $\frac{1}{2}v(R)$. Hence, R has a Hamilton cycle C, which contains a perfect matching if $v(R)$ is even (otherwise first add one cluster to the exceptional set V_0). This matching is our second reduced graph R'. We assume without loss of generality that the Hamilton cycle is $1, 2, \ldots, r$, and that the matching $R' \subseteq C$ is $\{1,2\}, \{3,4\}, \ldots, \{r-1, r\}$.

We then have to transform the regular-partition of G into a super-regular partition with regularity inheritance. Hence, we remove from each cluster all those vertices violating super-regularity on R', which are at most $\varepsilon|V_i|$ vertices per cluster V_i, and all those vertices violating (one-sided) regularity inheritance on R', which by Lemmas 7.2 are at most $Cp^{-1}\log n$ vertices per cluster V_i.

Next we prepare H. For embedding the grid H we want to use the linear cycle structure of the Hamilton cycle C in the reduced graph R. Therefore, let us first show that we can cut H into roughly equal pieces along a linear structure. Indeed, any diagonal of H has at most $\sqrt{n}$ vertices, hence by choosing appropriate diagonals as cuts (that is, we "cut" along the diagonal) we can partition H into $\frac{r}{2}$ sets $Y_1, \ldots, Y_{r/2}$ of size $(2n/r) \pm \sqrt{n}$. A C-partition $X_1, \ldots, X_r$ of H is then obtained by letting X_{i-1} and X_i be the two colour-classes of $H[Y_{i/2}]$ for every even $i \in [r]$. Observe that most edges of H then go along the matching edges of R'. Since vertices of the buffer $\tilde{X}_i$ for $i \in [r]$ should have their first and second neighbourhood along R', we simply choose αn vertices in X_i as $\tilde{X}_i$ which are on diagonals of distance at least 3 to any of the cut diagonals.

It is easy to check that we have $|V_i| \geq |X_i|$ for each i, so we can add isolated vertices to each part X_i of H to ensure size-compatibility, and then we can apply the sparse blow-up lemma, Lemma 7.5, to embed H into the remainder of G.

In fact, for embedding the almost spanning H into G we could have chosen a much simpler setup: We could have added αn isolated vertices to each X_i, and used these for the buffer $\tilde{X}_i$. Then we could have set R' to be the empty graph. We chose to describe the more complicated setting here though, because it is this setting which is necessary for generalising this approach to obtain a spanning H-copy in G.

The idea of how to generalise the above proof is roughly as follows. We would like to "redistribute" the vertices v of G we deleted to different clusters of G where they do not violate the required properties. Because of the minimum degree condition on G this can easily be shown to be possible for vertices v which satisfy regularity inheritance for any pair of clusters (X_i, X_j) with $ij \in R$, and which further have roughly the expected Γ-degree in each X_i with $i \in [r]$. The latter condition is necessary because of the second term in the maximum in the definition of sparse super-regular pairs, but it can be shown that a.a.s. all but at most $r \cdot Cp^{-1} \log n$ vertices satisfy this condition. The remaining $r \cdot Cp^{-1} \log n + r^3 \cdot Cp^{-1} \log n$ vertices cannot be redistributed, and need to be dealt with "by hand". The sparse blow-up lemma with image restrictions can then be used to complete the embedding. The details are more complicated, because the redistribution process is iterative. In particular, we need to ensure that during the redistribution no new violations of other vertices are created. Moreover, we also have to adapt the sizes of the clusters of G to match the actual sizes of the X_i. The details are omitted (see [3] for more explanations).

Acknowledgment

I would like to thank the anonymous referee, Peter Allen, and Yury Person for helpful comments and corrections.

References

[1] R. Aharoni and P. Haxell, *Hall's theorem for hypergraphs*, J. Graph Theory **35** (2000), no. 2, 83–88.

[2] M. Ajtai, J. Komlós, and E. Szemerédi, *The longest path in a random graph*, Combinatorica **1** (1981), no. 1, 1–12.

[3] P. Allen, J. Böttcher, J. Ehrenmüller, and A. Taraz, *The bandwidth theorem in sparse graphs*, arXiv:1612.00661.

[4] P. Allen, J. Böttcher, S. Griffiths, Y. Kohayakawa, and R. Morris, *Chromatic thresholds in sparse random graphs*, Random Structures Algorithms, accepted, arXiv:1508.03875.

[5] P. Allen, J. Böttcher, H. Hàn, Y. Kohayakawa, and Y. Person, *Blow-up lemmas for sparse graphs*, arXiv:1612.00622.

[6] N. Alon, *Universality, tolerance, chaos and order*, An irregular mind, Bolyai Soc. Math. Stud., vol. 21, János Bolyai Math. Soc., Budapest, 2010, pp. 21–37.

[7] N. Alon and M. Capalbo, *Sparse universal graphs for bounded-degree graphs*, Random Structures Algorithms **31** (2007), no. 2, 123–133.

[8] ———, *Optimal universal graphs with deterministic embedding*, Proceedings of the Nineteenth Annual ACM-SIAM Symposium on Discrete Algorithms, ACM, New York, 2008, pp. 373–378.

[9] N. Alon, M. Capalbo, Y. Kohayakawa, V. Rödl, A. Ruciński, and E. Szemerédi, *Universality and tolerance (extended abstract)*, 41st Annual Symposium on Foundations of Computer Science (Redondo Beach, CA, 2000), IEEE Comput. Soc. Press, Los Alamitos, CA, 2000, pp. 14–21.

[10] N. Alon and Z. Füredi, *Spanning subgraphs of random graphs*, Graphs Combin. **8** (1992), no. 1, 91–94.

[11] N. Alon, M. Krivelevich, and B. Sudakov, *Embedding nearly-spanning bounded degree trees*, Combinatorica **27** (2007), no. 6, 629–644.

[12] N. Alon and R. Yuster, *Threshold functions for H-factors*, Combin. Probab. Comput. **2** (1993), no. 2, 137–144.

[13] D. Angluin and L. G. Valiant, *Fast probabilistic algorithms for Hamiltonian circuits and matchings*, J. Comput. System Sci. **18** (1979), no. 2, 155–193.

[14] L. Babai, M. Simonovits, and J. Spencer, *Extremal subgraphs of random graphs*, J. Graph Theory **14** (1990), no. 5, 599–622.

[15] D. Bal and A. Frieze, *The Johansson-Kahn-Vu solution of the Shamir problem*, https://www.math.cmu.edu/~af1p/Teaching/ATIRS/Papers/FRH/Shamir.pdf.

[16] J. Balogh, B. Csaba, M. Pei, and W. Samotij, *Large bounded degree trees in expanding graphs*, Electron. J. Combin. **17** (2010), no. 1, Research Paper 6, 9.

[17] J. Balogh, B. Csaba, and W. Samotij, *Local resilience of almost spanning trees in random graphs*, Random Structures Algorithms **38** (2011), no. 1-2, 121–139.

[18] J. Balogh, C. Lee, and W. Samotij, *Corrádi and Hajnal's theorem for sparse random graphs*, Combin. Probab. Comput. **21** (2012), no. 1-2, 23–55.

[19] J. Balogh, R. Morris, and W. Samotij, *Independent sets in hypergraphs*, J. Amer. Math. Soc. **28** (2015), no. 3, 669–709.

[20] S. Ben-Shimon, M. Krivelevich, and B. Sudakov, *Local resilience and Hamiltonicity maker-breaker games in random regular graphs*, Combin. Probab. Comput. **20** (2011), no. 2, 173–211.

[21] _____, *On the resilience of Hamiltonicity and optimal packing of Hamilton cycles in random graphs*, SIAM J. Discrete Math. **25** (2011), no. 3, 1176–1193.

[22] P. Bennett, A. Dudek, and A. Frieze, *Square of a Hamilton cycle in a random graph*, arXiv:1611.06570.

[23] S. N. Bhatt, F. R. K. Chung, F. T. Leighton, and A. L. Rosenberg, *Universal graphs for bounded-degree trees and planar graphs*, SIAM J. Discrete Math. **2** (1989), no. 2, 145–155.

[24] B. Bollobás, *Threshold functions for small subgraphs*, Math. Proc. Cambridge Philos. Soc. **90** (1981), no. 2, 197–206.

[25] _____, *The evolution of sparse graphs*, Graph theory and combinatorics (Cambridge, 1983), Academic Press, London, 1984, pp. 35–57.

[26] _____, *Random graphs*, second ed., Cambridge Studies in Advanced Mathematics, vol. 73, Cambridge University Press, Cambridge, 2001.

[27] B. Bollobás, T. I. Fenner, and A. M. Frieze, *An algorithm for finding Hamilton paths and cycles in random graphs*, Combinatorica **7** (1987), no. 4, 327–341.

[28] B. Bollobás and A. M. Frieze, *On matchings and Hamiltonian cycles in random graphs*, Ann. Discrete Math. **28** (1985), 23–46.

[29] _____, *Spanning maximal planar subgraphs of random graphs*, Random Structures Algorithms **2** (1991), no. 2, 225–231.

[30] B. Bollobás and A. Thomason, *Random graphs of small order*, Random graphs '83 (Poznań, 1983), North-Holland Math. Stud., vol. 118, North-Holland, Amsterdam, 1985, pp. 47–97.

[31] _____, *Threshold functions*, Combinatorica **7** (1987), no. 1, 35–38.

[32] J. Böttcher, Y. Kohayakawa, and A. Taraz, *Almost spanning subgraphs of random graphs after adversarial edge removal*, Combin. Probab. Comput. **22** (2013), no. 5, 639–683.

[33] J. Böttcher, K. P. Pruessmann, A. Taraz, and A. Würfl, *Bandwidth, expansion, treewidth, separators and universality for bounded-degree graphs*, European J. Combin. **31** (2010), no. 5, 1217–1227.

[34] J. Böttcher, M. Schacht, and A. Taraz, *Proof of the bandwidth conjecture of Bollobás and Komlós*, Math. Ann. **343** (2009), no. 1, 175–205.

[35] G. Brightwell, K. Panagiotou, and A. Steger, *Extremal subgraphs of random graphs*, Random Structures Algorithms **41** (2012), no. 2, 147–178.

[36] M. R. Capalbo, *A small universal graph for bounded-degree planar graphs*, Proceedings of the Tenth Annual ACM-SIAM Symposium on Discrete Algorithms (Baltimore, MD, 1999), ACM, New York, 1999, pp. 156–160.

[37] F. R. K. Chung, *Labelings of graphs*, Selected topics in graph theory, 3, Academic Press, San Diego, CA, 1988, pp. 151–168.

[38] D. Conlon, A. Ferber, R. Nenadov, and N. Škorić, *Almost-spanning universality in random graphs*, Random Structures Algorithms, accepted, arXiv:1503.05612.

[39] D. Conlon and W. T. Gowers, *Combinatorial theorems in sparse random sets*, Ann. of Math. (2) **184** (2016), no. 2, 367–454.

[40] D. Conlon, W. T. Gowers, W. Samotij, and M. Schacht, *On the KŁR conjecture in random graphs*, Israel J. Math. **203** (2014), no. 1, 535–580.

[41] K. Corrádi and A. Hajnal, *On the maximal number of independent circuits in a graph*, Acta Math. Acad. Sci. Hungar. **14** (1963), 423–439.

[42] D. Dellamonica, Y. Kohayakawa, M. Marciniszyn, and A. Steger, *On the resilience of long cycles in random graphs*, Electron. J. Combin. **15** (2008), 26 pp., R32.

[43] D. Dellamonica, Y. Kohayakawa, V. Rödl, and A. Ruciński, *Universality of random graphs*, SIAM J. Discrete Math. **26** (2012), no. 1, 353–374.

[44] _____, *An improved upper bound on the density of universal random graphs*, Random Structures Algorithms **46** (2015), no. 2, 274–299.

[45] B. DeMarco and J. Kahn, *Turán's Theorem for random graphs*, arXiv:1501.01340.

[46] _____, *Mantel's theorem for random graphs*, Random Structures Algorithms **47** (2015), no. 1, 59–72.

[47] G. A. Dirac, *Some theorems on abstract graphs*, Proc. London Math. Soc. (3) **2** (1952), 69–81.

[48] M. Drmota, O. Giménez, M. Noy, K. Panagiotou, and A. Steger, *The maximum degree of random planar graphs*, Proc. Lond. Math. Soc. (3) **109** (2014), no. 4, 892–920.

[49] P. Erdős and A. Rényi, *On random graphs. I*, Publ. Math. Debrecen **6** (1959), 290–297.

[50] _____, *On the evolution of random graphs*, Magyar Tud. Akad. Mat. Kutató Int. Közl. **5** (1960), 17–61.

[51] _____, *On random matrices*, Magyar Tud. Akad. Mat. Kutató Int. Közl. **8** (1964), 455–461 (1964).

[52] _____, *On the existence of a factor of degree one of a connected random graph*, Acta Math. Acad. Sci. Hungar. **17** (1966), 359–368.

[53] P. Erdős and A. H. Stone, *On the structure of linear graphs*, Bull. Amer. Math. Soc. **52** (1946), 1087–1091.

[54] A. Ferber, G. Kronenberg, and K. Luh, *Optimal threshold for a random graph to be 2-universal*, arXiv:1612.06026.

[55] A. Ferber, K. Luh, and O. Nguyen, *Embedding large graphs into a random graph*, arXiv:1606.05923.

[56] A. Ferber, R. Nenadov, and U. Peter, *Universality of random graphs and rainbow embedding*, Random Structures Algorithms **48** (2016), no. 3, 546–564.

[57] W. Fernandez de la Vega, *Long paths in random graphs*, Studia Sci. Math. Hungar. **14** (1979), no. 4, 335–340.

[58] ――――, *Trees in sparse random graphs*, J. Combin. Theory Ser. B **45** (1988), no. 1, 77–85.

[59] P. Frankl and V. Rödl, *Large triangle-free subgraphs in graphs without K_4*, Graphs Combin. **2** (1986), no. 2, 135–144.

[60] E. Friedgut, *Sharp thresholds of graph properties, and the k-sat problem*, J. Amer. Math. Soc. **12** (1999), no. 4, 1017–1054, With an appendix by Jean Bourgain.

[61] ――――, *Hunting for sharp thresholds*, Random Structures Algorithms **26** (2005), no. 1-2, 37–51.

[62] J. Friedman and N. Pippenger, *Expanding graphs contain all small trees*, Combinatorica **7** (1987), no. 1, 71–76.

[63] A. Frieze and M. Karoński, *Introduction to random graphs*, Cambridge University Press, 2015.

[64] A. Frieze and M. Krivelevich, *On two Hamilton cycle problems in random graphs*, Israel J. Math. **166** (2008), 221–234.

[65] Z. Füredi, *Random Ramsey graphs for the four-cycle*, Discrete Math. **126** (1994), no. 1-3, 407–410.

[66] S. Gerke, *Random graphs with constraints*, 2005, Habilitationsschrift, Institut für Informatik, TU München.

[67] S. Gerke, Y. Kohayakawa, V. Rödl, and A. Steger, *Small subsets inherit sparse ϵ-regularity*, J. Combin. Theory Ser. B **97** (2007), no. 1, 34–56.

[68] S. Gerke and A. McDowell, *Nonvertex-balanced factors in random graphs*, J. Graph Theory **78** (2015), no. 4, 269–286, (arXiv:1304.3000).

[69] S. Gerke, H. J. Prömel, T. Schickinger, A. Steger, and A. Taraz, K_4-free subgraphs of random graphs revisited, Combinatorica **27** (2007), no. 3, 329–365.

[70] S. Gerke, T. Schickinger, and A. Steger, K_5-free subgraphs of random graphs, Random Structures Algorithms **24** (2004), no. 2, 194–232.

[71] Y. Gurevich and S. Shelah, Expected computation time for Hamiltonian path problem, SIAM J. Comput. **16** (1987), no. 3, 486–502.

[72] A. Hajnal and E. Szemerédi, Proof of a conjecture of P. Erdős, Combinatorial theory and its applications, II (Proc. Colloq., Balatonfüred, 1969), North-Holland, Amsterdam, 1970, pp. 601–623.

[73] P. E. Haxell, Tree embeddings, J. Graph Theory **36** (2001), no. 3, 121–130.

[74] P. E. Haxell, Y. Kohayakawa, and T. Łuczak, Turán's extremal problem in random graphs: forbidding even cycles, J. Combin. Theory Ser. B **64** (1995), no. 2, 273–287.

[75] ――――, Turán's extremal problem in random graphs: forbidding odd cycles, Combinatorica **16** (1996), no. 1, 107–122.

[76] D. Hefetz, M. Krivelevich, and T. Szabó, Hamilton cycles in highly connected and expanding graphs, Combinatorica **29** (2009), no. 5, 547–568.

[77] ――――, Sharp threshold for the appearance of certain spanning trees in random graphs, Random Structures Algorithms **41** (2012), no. 4, 391–412.

[78] H. Huang, C. Lee, and B. Sudakov, Bandwidth theorem for random graphs, J. Combin. Theory Ser. B **102** (2012), no. 1, 14–37.

[79] S. Janson, T. Łuczak, and A. Ruciński, Random graphs, Wiley-Interscience, New York, 2000.

[80] D. Johannsen, M. Krivelevich, and W. Samotij, Expanders are universal for the class of all spanning trees, Combin. Probab. Comput. **22** (2013), no. 2, 253–281.

[81] A. Johansson, J. Kahn, and V. Vu, Factors in random graphs, Random Structures Algorithms **33** (2008), no. 1, 1–28.

[82] J. Kahn and G. Kalai, Thresholds and expectation thresholds, Combin. Probab. Comput. **16** (2007), no. 3, 495–502.

[83] J. Kahn, E. Lubetzky, and N. Wormald, *Cycle factors and renewal theory*, Comm. Pure Appl. Math., accepted, arXiv:1401.2707.

[84] ———, *The threshold for combs in random graphs*, Random Structures Algorithms **48** (2016), no. 4, 794–802.

[85] J. H. Kim and S. J. Lee, *Universality of random graphs for graphs of maximum degree two*, SIAM J. Discrete Math. **28** (2014), no. 3, 1467–1478.

[86] J. H. Kim and V. H. Vu, *Concentration of multivariate polynomials and its applications*, Combinatorica **20** (2000), no. 3, 417–434.

[87] ———, *Sandwiching random graphs: universality between random graph models*, Adv. Math. **188** (2004), no. 2, 444–469.

[88] Y. Kohayakawa, *Szemerédi's regularity lemma for sparse graphs*, Foundations of computational mathematics, Springer, 1997, pp. 216–230.

[89] Y. Kohayakawa, B. Kreuter, and A. Steger, *An extremal problem for random graphs and the number of graphs with large even-girth*, Combinatorica **18** (1998), no. 1, 101–120.

[90] Y. Kohayakawa, T. Łuczak, and V. Rödl, *On K^4-free subgraphs of random graphs*, Combinatorica **17** (1997), no. 2, 173–213.

[91] Y. Kohayakawa and V. Rödl, *Regular pairs in sparse random graphs. I*, Random Structures Algorithms **22** (2003), no. 4, 359–434.

[92] ———, *Szemerédi's regularity lemma and quasi-randomness*, Recent advances in algorithms and combinatorics, Springer, 2003, pp. 289–351.

[93] Y. Kohayakawa, V. Rödl, and M. Schacht, *The Turán theorem for random graphs*, Combin. Probab. Comput. **13** (2004), no. 1, 61–91.

[94] Y. Kohayakawa, V. Rödl, M. Schacht, and E. Szemerédi, *Sparse partition universal graphs for graphs of bounded degree*, Adv. Math. **226** (2011), no. 6, 5041–5065.

[95] J. Komlós, *The blow-up lemma*, Combin. Probab. Comput. **8** (1999), no. 1-2, 161–176, Recent trends in combinatorics (Mátraháza, 1995).

[96] J. Komlós, G. N. Sárközy, and E. Szemerédi, *Proof of a packing conjecture of Bollobás*, Combin. Probab. Comput. **4** (1995), no. 3, 241–255.

[97] _____, *Blow-up lemma*, Combinatorica **17** (1997), no. 1, 109–123.

[98] _____, *An algorithmic version of the blow-up lemma*, Random Structures Algorithms **12** (1998), no. 3, 297–312.

[99] _____, *Spanning trees in dense graphs*, Combin. Probab. Comput. **10** (2001), no. 5, 397–416.

[100] J. Komlós, A. Shokoufandeh, M. Simonovits, and E. Szemerédi, *The regularity lemma and its applications in graph theory*, Theoretical aspects of computer science (Tehran, 2000), Lecture Notes in Comput. Sci., vol. 2292, Springer, Berlin, 2002, pp. 84–112.

[101] J. Komlós and M. Simonovits, *Szemerédi's regularity lemma and its applications in graph theory*, Combinatorics, Paul Erdős is eighty, Vol. 2 (Keszthely, 1993), Bolyai Soc. Math. Stud., vol. 2, János Bolyai Math. Soc., Budapest, 1996, pp. 295–352.

[102] J. Komlós and E. Szemerédi, *Limit distribution for the existence of Hamiltonian cycles in a random graph*, Discrete Math. **43** (1983), no. 1, 55–63.

[103] A. Korshunov, *Solution of a problem of Erdős and Rényi on Hamiltonian cycles in nonoriented graphs.*, Sov. Math., Dokl. **17** (1976), 760–764.

[104] _____, *Solution of a problem of P. Erdős and A. Rényi on Hamiltonian cycles in undirected graphs*, Metody Diskretn. Anal. **31** (1977), 17–56.

[105] M. Krivelevich, *Triangle factors in random graphs*, Combin. Probab. Comput. **6** (1997), no. 3, 337–347.

[106] _____, *Embedding spanning trees in random graphs*, SIAM J. Discrete Math. **24** (2010), no. 4, 1495–1500.

[107] M. Krivelevich, C. Lee, and B. Sudakov, *Resilient pancyclicity of random and pseudorandom graphs*, SIAM J. Discrete Math. **24** (2010), no. 1, 1–16.

[108] D. Kühn and D. Osthus, *On Pósa's conjecture for random graphs*, SIAM J. Discrete Math. **26** (2012), no. 3, 1440–1457.

[109] C. Lee and W. Samotij, *Pancyclic subgraphs of random graphs*, J. Graph Theory **71** (2012), no. 2, 142–158.

[110] C. Lee and B. Sudakov, *Dirac's theorem for random graphs*, Random Structures Algorithms **41** (2012), no. 3, 293–305.

[111] T. Łuczak and A. Ruciński, *Tree-matchings in graph processes*, SIAM J. Discrete Math. **4** (1991), no. 1, 107–120.

[112] C. McDiarmid and B. Reed, *On the maximum degree of a random planar graph*, Combin. Probab. Comput. **17** (2008), no. 4, 591–601.

[113] R. Montgomery, *Embedding bounded degree spanning trees in random graphs*, arXiv:1405.6559.

[114] ———, *Sharp threshold for embedding combs and other spanning trees in random graphs*, arXiv:1405.6560.

[115] J. W. Moon, *On the maximum degree in a random tree*, Michigan Math. J. **15** (1968), 429–432.

[116] R. Nenadov and N. Škorić, *Powers of cycles in random graphs and hypergraphs*, arXiv:1601.04034.

[117] A. Noever and A. Steger, *Local resilience for squares of almost spanning cycles in sparse random graphs*, arXiv:1606.02958.

[118] O. Parczyk and Y. Person, *Spanning structures and universality in sparse hypergraphs*, Random Structures Algorithms, accepted, arXiv:1504.02243.

[119] L. Pósa, *Hamiltonian circuits in random graphs*, Discrete Math. **14** (1976), no. 4, 359–364.

[120] O. Riordan, *Spanning subgraphs of random graphs*, Combin. Probab. Comput. **9** (2000), no. 2, 125–148.

[121] V. Rödl and A. Ruciński, *Perfect matchings in ε-regular graphs and the blow-up lemma*, Combinatorica **19** (1999), no. 3, 437–452.

[122] V. Rödl, A. Ruciński, and A. Taraz, *Hypergraph packing and graph embedding*, Combin. Probab. Comput. **8** (1999), no. 4, 363–376, Random graphs and combinatorial structures (Oberwolfach, 1997).

[123] A. Ruciński, *Matching and covering the vertices of a random graph by copies of a given graph*, Discrete Math. **105** (1992), no. 1-3, 185–197.

[124] D. Saxton and A. Thomason, *Hypergraph containers*, Invent. Math. **201** (2015), no. 3, 925–992.

[125] M. Schacht, *Extremal results for random discrete structures*, Ann. of Math. (2) **184** (2016), no. 2, 333–365.

[126] A. Scott, *Szemerédi's regularity lemma for matrices and sparse graphs*, Combin. Probab. Comput. **20** (2011), no. 3, 455–466.

[127] E. Shamir, *How many random edges make a graph Hamiltonian?*, Combinatorica **3** (1983), no. 1, 123–131.

[128] J. Spencer, *Threshold functions for extension statements*, J. Combin. Theory Ser. A **53** (1990), no. 2, 286–305.

[129] B. Sudakov and V. H. Vu, *Local resilience of graphs*, Random Structures Algorithms **33** (2008), no. 4, 409–433.

[130] T. Szabó and V. H. Vu, *Turán's theorem in sparse random graphs*, Random Structures Algorithms **23** (2003), no. 3, 225–234.

[131] A. Thomason, *A simple linear expected time algorithm for finding a Hamilton path*, Discrete Math. **75** (1989), no. 1-3, 373–379, Graph theory and combinatorics (Cambridge, 1988).

[132] P. Turán, *Eine Extremalaufgabe aus der Graphentheorie*, Mat. Fiz. Lapok **48** (1941), 436–452.

Department of Mathematics, London School of Economics
Houghton St, London WC2A 2AE, UK
j.boettcher@lse.ac.uk

The spt-Function of Andrews

William Y. C. Chen

Abstract

The spt-function $\mathrm{spt}(n)$ was introduced by Andrews as the weighted counting of partitions of n with respect to the number of occurrences of the smallest part. Andrews showed that $\mathrm{spt}(5n + 4) \equiv 0$ (mod 5), $\mathrm{spt}(7n + 5) \equiv 0$ (mod 7) and $\mathrm{spt}(13n + 6) \equiv 0$ (mod 13). Since then, congruences of $\mathrm{spt}(n)$ have been extensively studied. Folsom and Ono obtained congruences of $\mathrm{spt}(n)$ mod 2 and 3. They also showed that the generating function of $\mathrm{spt}(n)$ mod 3 is related to a weight 3/2 Hecke eigenform with Nebentypus. Combinatorial interpretations of congruences of $\mathrm{spt}(n)$ mod 5 and 7 have been found by Andrews, Garvan and Liang by introducing the spt-crank of a vector partition. Chen, Ji and Zang showed that the set of partitions counted by $\mathrm{spt}(5n + 4)$ (or $\mathrm{spt}(7n + 5)$) can be divided into five (or seven) equinumerous classes according to the spt-crank of a doubly marked partition. Let $N_S(m, n)$ denote the net number of S-partitions of n with spt-crank m. Andrews, Dyson and Rhoades conjectured that $\{N_S(m, n)\}_m$ is unimodal for any n. Chen, Ji and Zang gave a constructive proof of this conjecture. In this survey, we summarize developments on congruence properties of $\mathrm{spt}(n)$ established by Andrews, Bringmann, Folsom, Garvan, Lovejoy and Ono et al., as well as their combinatorial interpretations. Generalizations and variations of the spt-function are also discussed. We also give an overview of asymptotic formulas of $\mathrm{spt}(n)$ obtained by Ahlgren, Andersen and Rhoades et al. We conclude with some conjectures on inequalities on $\mathrm{spt}(n)$, which are reminiscent of inequalities on $p(n)$ due to DeSalvo and Pak, and Bessenrodt and Ono. Furthermore, we observe that, beyond the log-concavity, $p(n)$ and $\mathrm{spt}(n)$ satisfy higher order inequalities based on polynomials arising in the invariant theory of binary forms. In particular, we conjecture that the higher order Turán inequality $4(a_n^2 - a_{n-1}a_{n+1})(a_{n+1}^2 - a_n a_{n+2}) - (a_n a_{n+1} - a_{n-1}a_{n+2})^2 > 0$ holds for $p(n)$ when $n \geq 95$ and for $\mathrm{spt}(n)$ when $n \geq 108$.

1 Introduction

Andrews [12] introduced the spt-function $\mathrm{spt}(n)$ as the weighted counting of partitions with respect to the number of occurrences of the smallest part and he discovered that the spt-function bears striking resemblance to the classical partition function $p(n)$. Since then, the spt-function has drawn much attention and has been extensively studied. In this survey, we shall summarize developments on the spt-function including congruence properties derived from q-identities and modular forms, along with their combinatorial interpretations, as well as generalizations, variations and asymptotic properties. For the background on partitions, we refer to [8, 10, 20], and for the background on modular forms, we refer to [26, 59, 98, 111].

The spt-function spt(n), called the smallest part function, is defined to be the total number of smallest parts in all partitions of n. More precisely, for a partition λ of n, we use $n_s(\lambda)$ to denote the number of occurrences of the smallest part in λ. Let $P(n)$ denote the set of partitions of n, then

$$\text{spt}(n) = \sum_{\lambda \in P(n)} n_s(\lambda). \tag{1.1}$$

For example, for $n = 4$, we have spt(4) $= 10$. Partitions in $P(4)$ and the values of $n_s(\lambda)$ are listed below:

$\lambda \in P(4)$	(4)	$(3,1)$	$(2,2)$	$(2,1,1)$	$(1,1,1,1)$
$n_s(\lambda)$	1	1	2	2	4

The spt-function spt(n) can also be interpreted by marked partitions, see Andrews, Dyson and Rhoades [19]. A marked partition of n is meant to be a pair (λ, k), where $\lambda = (\lambda_1, \lambda_2, \ldots, \lambda_l)$ is an ordinary partition of n and k is an integer identifying one of its smallest parts. If λ_k is the identified smallest part of λ, we then use (λ, k) to denote this marked partition. For example, there are ten marked partitions of 4.

$$((4),1), \qquad ((3,1),2), \qquad ((2,2),1), \qquad ((2,2),2), \qquad ((2,1,1),2),$$
$$((2,1,1),3), \quad ((1,1,1,1),1), \quad ((1,1,1,1),2), \quad ((1,1,1,1),3), \quad ((1,1,1,1),4).$$

Using the definition (1.1), it is easy to derive the following generating function, see Andrews [12],

$$\sum_{n=1}^{\infty} \text{spt}(n)q^n = \sum_{n=1}^{\infty} \frac{q^n}{(1 - q^n)^2 (q^{n+1}; q)_\infty}. \tag{1.2}$$

Here we have adopted the common notation [10]:

$$(a; q)_\infty = \prod_{n=0}^{\infty} (1 - aq^n) \quad \text{and} \quad (a; q)_n = \frac{(a; q)_\infty}{(aq^n; q)_\infty}.$$

The spt-function is closely related to the rank and the crank of a partition. Recall that the rank of a partition was introduced by Dyson [63] as the largest part of the partition minus the number of parts. The crank of a partition was defined by Andrews and Garvan [21] as the largest part if the partition contains no ones, otherwise as the number of parts larger than the number of ones minus the number of ones. For $n \geq 1$, let $N(m, n)$ denote the number of partitions of n with rank m, and for $n > 1$, let $M(m, n)$ denote the number of partitions of n with crank m. For $n = 1$, set

$$M(0, 1) = -1, \; M(1, 1) = M(-1, 1) = 1,$$

and for $n = 1$ and $m \neq -1, 0, 1$, set

$$M(m, 1) = 0.$$

Atkin and Garvan [28] defined the k-th moment $N_k(n)$ of ranks as

$$N_k(n) \quad = \quad \sum_{m=-\infty}^{\infty} m^k N(m, n), \tag{1.3}$$

and the k-th moment $M_k(n)$ of cranks as

$$M_k(n) = \sum_{m=-\infty}^{\infty} m^k M(m, n).$$

It is worth mentioning that Atkin and Garvan [28] showed that the generating functions of the moments of cranks are related to quasimodular forms. Bringmann, Garvan and Mahlburg [41] showed that the generating functions of the moments of ranks are related to quasimock theta functions. Asymptotic formulas for the moments of ranks and cranks were derived by Bringmann, Mahlburg and Rhoades [47].

Based on the generating function (1.2) and Watson's q-analog of Whipple's theorem [80, p. 43, eq. (2.5.1)], Andrews [12] showed that the spt-function can be expressed in terms of the second moment $N_2(n)$ of ranks introduced by Atkin and Garvan [28],

$$\mathrm{spt}(n) = np(n) - \frac{1}{2} N_2(n). \tag{1.4}$$

Ji [93] found a combinatorial proof of (1.4) using rooted partitions.

By means of a relation due to Dyson [64], namely,

$$M_2(n) = 2np(n), \tag{1.5}$$

Garvan [72] observed that the expression

$$\mathrm{spt}(n) = \frac{1}{2} M_2(n) - \frac{1}{2} N_2(n) \tag{1.6}$$

implies that $M_2(n) > N_2(n)$ for $n \geq 1$. In general, he conjectured and later proved that $M_{2k}(n) > N_{2k}(n)$ for $k \geq 1$ and $n \geq 1$, see [72, 73].

In view of the relation (1.4) and identities on refinements of $N(m, n)$ established by Atkin and Swinnerton-Dyer [30] and O'Brien [108], Andrews proved that $\mathrm{spt}(n)$ satisfies congruences mod 5, 7 and 13 which are reminiscent of Ramanujan's congruences for $p(n)$. Let ℓ be a prime. A Ramanujan congruence modulo ℓ for the sequence $\{a(n)\}_{n \geq 0}$ means a congruence of the form

$$a(\ell n + \beta) \equiv 0 \pmod{\ell}$$

for all nonnegative integers n and a fixed integer β.

Ramanujan [122] discovered the following congruences for $p(n)$,

$$p(5n + 4) \equiv 0 \pmod 5, \tag{1.7}$$

$$p(7n + 5) \equiv 0 \pmod 7, \tag{1.8}$$

$$p(11n + 6) \equiv 0 \pmod{11}, \tag{1.9}$$

and proclaimed that "it appears that there are no equally simple properties for any moduli involving primes other than these three (i.e. $\ell = 5, 7, 11$)." See also Berndt [34, p. 27].

Elementary proofs of the congruences (1.7) and (1.8) were given by Ramanujan [122] and an elementary proof of the congruence (1.9) was given by Winquist [133]. Alternative proofs of (1.9) were found by Berndt, Chan, Liu and Yesilyurt [36] and Hirschhorn [84]. Recently, Paule and Radu [116] found a recurrence relation of the generating function of $p(11n + 6)$, from which (1.9) is an immediate consequence. Berndt [35] provided simple proofs of (1.7)–(1.9) by using Ramanujan's differential equations for the Eisenstein series. Uniform proofs of (1.7)–(1.9) were found by Hirschhorn [83].

Concerning Ramanujan's conjecture, Kiming and Olsson [97] showed that if there exists a Ramanujan's congruence $p(\ell n + \beta) \equiv 0 \pmod{\ell}$, then $24\beta \equiv 1 \pmod{\ell}$. According to this condition, Ahlgren and Boylan [4] confirmed Ramanujan's conjecture. More precisely, they showed that for a prime ℓ, if there is a Ramanujan's congruence modulo ℓ for $p(n)$, then it must be one of the congruences (1.7), (1.8) and (1.9).

Combinatorial studies of Ramanujan's congruences of $p(n)$ go back to Dyson [63]. He conjectured that the rank of a partition can be used to divide the set of partitions of $5n + 4$ (or $7n + 5$) into five (or seven) equinumerous classes. More precisely, let $N(i, t, n)$ denote the number of partitions of n with rank congruent to i modulo t. Dyson [63] conjectured that

$$N(i, 5, 5n + 4) = \frac{p(5n + 4)}{5} \quad \text{for} \quad 0 \le i \le 4, \qquad (1.10)$$

$$N(i, 7, 7n + 5) = \frac{p(7n + 5)}{7} \quad \text{for} \quad 0 \le i \le 6. \qquad (1.11)$$

These relations were proved by Atkin and Swinnerton-Dyer [30], which imply (1.7) and (1.8). Dyson also pointed out that the rank of a partition cannot be used to interpret (1.9). To give a combinatorial explanation of this congruence modulo 11, Garvan [70] introduced the crank of a vector partition and showed that this statistic leads to interpretations of the above congruences of $p(n)$ mod 5, 7 and 11. Andrews and Garvan [21] found an equivalent definition of the crank in terms of an ordinary partition. For the history of the rank and the crank, see, for example, Andrews and Berndt [14] and Andrews and Ono [24].

Although Dyson's rank fails to explain Ramanujan's congruence (1.9) combinatorially, the generating functions for the rank differences have been extensively studied. For example, the generating functions for the rank differences $N(s, \ell, \ell n + d) - N(t, \ell, \ell n + d)$ for $\ell = 2, 9, 11, 12, 13$ have been determined by Atkin and Hussain [29], O'Brien [108], Lewis [102, 103] and Santa-Gadea [126].

By the relations (1.4), (1.10) and (1.11), Andrews [12] showed that

$$\text{spt}(5n + 4) \equiv 0 \pmod{5}, \qquad (1.12)$$

$$\text{spt}(7n + 5) \equiv 0 \pmod{7}. \qquad (1.13)$$

He also obtained that

$$\text{spt}(13n + 6) \equiv 0 \pmod{13}, \qquad (1.14)$$

by considering the properties of $N(i, 13, 13n + 6)$ due to O'Brien [108]. Let

$$r_{a,b}(d) = \sum_{n=0}^{\infty} (N(a, 13, 13n + d) - N(b, 13, 13n + d))q^{13n},$$

and for $1 \le i \le 5$, and let

$$S_i(d) = r_{(i-1),i}(d) - (7 - i)r_{5,6}(d).$$

O'Brien [108] deduced that

$$S_1(6) + 2S_2(6) - 5S_5(6) \equiv 0 \pmod{13} \qquad (1.15)$$

and

$$S_2(6) + 5S_3(6) + 3S_4(6) + 3S_5(6) \equiv 0 \pmod{13}. \qquad (1.16)$$

Employing (1.4), Andrews derived an expression for $\operatorname{spt}(13n+6)$ in terms of $N(i, 13, 13n+6)$ modulo 13. Then the congruence (1.14) follows from (1.15) and (1.16).

This paper is organized as follows. In Section 2, we recall the spt-crank of an S-partition defined by Andrews, Garvan and Liang, which leads to combinatorial interpretations of the congruences of the spt-function mod 5 and 7. Motivated by a problem of Andrews, Garvan and Liang on constructive proofs of the congruences mod 5 and 7, Chen, Ji and Zang introduced the notion of a doubly marked partition and its spt-crank. Such an spt-crank can be used to divide the set counted by $\operatorname{spt}(5n+4)$ (resp. $\operatorname{spt}(7n+5)$) into five (resp. seven) equinumerous classes. The unimodality of the spt-crank and related topics are also discussed. In Section 3, we begin with Ramanujan-type congruences of $\operatorname{spt}(n)$ mod 11,17, 19, 29, 31 and 37 obtained by Garvan. We then consider Ramanujan-type congruences of $\operatorname{spt}(n)$ modulo any prime $\ell \geq 5$ due to Ono and the ℓ-adic generalization due to Ahlgren, Bringmann and Lovejoy. The congruences of $\operatorname{spt}(n)$ mod powers of 5, 7 and 13 established by Garvan will also be discussed. We finish this section with congruences of $\operatorname{spt}(n)$ mod $2,3$ and powers of 2 due to Folsom and Ono, and Garvan and Jennings-Shaffer. Section 4 is devoted to generalizations and variations of the spt-function. We first recall the higher order spt-function defined by Garvan, as a generalization of the spt-function. We then concentrate on two generalizations of the spt-function based on the j-rank, given by Dixit and Yee. The first variation of the spt-function was defined by Andrews, Chan and Kim as the difference between the first rank and crank moments. At the end of this section, we present three variations of the spt-function, which are restrictions of the spt-function to three classes of partitions. The generating functions, combinatorial interpretations and congruences of these generalizations and variations of the spt-function will also be discussed. In Section 5, we summarize asymptotic formulas of the spt-function and its variations. Section 6 contains some conjectures on inequalities on $\operatorname{spt}(n)$, which are analogous to those on $p(n)$, due to DeSalvo and Pak, and Bessenrodt and Ono. Beyond the log-concavity, we conjecture that $p(n)$ and $\operatorname{spt}(n)$ satisfy higher order inequalities induced from invariants of binary forms. In particular, we conjecture that the higher order Turán inequality holds for both $p(n)$ and $\operatorname{spt}(n)$ when n is large enough.

2   The spt-crank

To give combinatorial interpretations of congruences on spt(n), Andrews, Garvan and Liang [22] introduced the spt-crank of an S-partition, which is analogous to Garvan's crank of a vector partition [70]. They showed that the spt-crank of an S-partition can be used to divide the set of S-partitions with signs counted by spt($5n + 4$) (or spt($7n + 5$)) into five (or seven) equinumerous classes which leads to the congruences (1.12) and (1.13).

Andrews, Dyson and Rhoades [19] proposed the problem of finding an equivalent definition of the spt-crank for a marked partition. Chen, Ji and Zang [53] introduced the structure of a doubly marked partition and established a bijection between marked partitions and doubly marked partitions. Then they defined the spt-crank of a doubly marked partition in order to divide the set of marked partitions counted by spt($5n + 4$) (or spt($7n + 5$)) into five (or seven) equinumerous classes. Hence, in principle, the spt-crank of a doubly marked partition can be considered as a solution to the problem of Andrews, Dyson and Rhoades. It would be interesting to find an spt-crank directly defined on marked partitions.

Let $N_S(m, n)$ denote the net number, or the sum of signs, of S-partitions of n with spt-crank m. Andrews, Dyson and Rhoades [19] conjectured that $\{N_S(m,n)\}_m$ is unimodal for any given n and showed that this conjecture is equivalent to an inequality between the rank and the crank of a partition. Using the notion of the rank-set of a partition introduced by Dyson [64], Chen, Ji and Zang [52] gave a proof of this conjecture by constructing an injection from the set of partitions of n such that m appears in the rank-set to the set of partitions of n with rank not less than $-m$.

2.1   The spt-crank of an S-partition

Based on (1.2), Andrews, Garvan and Liang [22] noticed that the generating function of spt(n) can be expressed as

$$\sum_{n=1}^{\infty} \text{spt}(n)q^n = \sum_{n=1}^{\infty} \frac{q^n(q^{n+1};q)_\infty}{(q^n;q)_\infty^2}, \tag{2.1}$$

and they introduced the structure of S-partitions to interpret the right-hand side of (2.1) as the generating function of the net number of S-partitions of n, that is, the sum of signs of S-partitions of n. More precisely, let $\mathcal{D}$ denote the set of partitions into distinct parts and $\mathcal{P}$ denote the set of partitions. For $\lambda \in \mathcal{P}$, we use $s(\lambda)$ to denote the smallest part of λ with the convention that $s(\emptyset) = +\infty$. The set of S-partitions is defined by

$$S = \{(\pi_1, \pi_2, \pi_3) \in \mathcal{D} \times \mathcal{P} \times \mathcal{P} \mid \pi_1 \neq \emptyset \text{ and } s(\pi_1) \leq \min\{s(\pi_2), s(\pi_3)\}\}. \tag{2.2}$$

For $\pi = (\pi_1, \pi_2, \pi_3) \in S$, Andrews, Garvan and Liang [22] defined the weight of π to be $|\pi_1| + |\pi_2| + |\pi_3|$ and defined the sign of π to be

$$\omega(\pi) = (-1)^{l(\pi_1)-1},$$

where $|\pi|$ denotes the sum of parts of π and $l(\pi)$ denotes the number of parts of π.

They showed that

$$\mathrm{spt}(n) = \sum_{\pi} \omega(\pi),$$

where π ranges over S-partitions of n. To give combinatorial interpretations of the congruences (1.12) and (1.13), Andrews, Garvan and Liang [22] defined the spt-crank of an S-partition, which takes the same form as the crank of a vector partition.

Let π be an S-partition, the spt-crank of π, denoted $r(\pi)$, is defined to be the number of parts of π_2 minus the number of parts of π_3, i.e.,

$$r(\pi) = l(\pi_2) - l(\pi_3).$$

Let $N_S(m, n)$ denote the net number of S-partitions of n with spt-crank m, that is,

$$N_S(m, n) = \sum_{\substack{|\pi|=n \\ r(\pi)=m}} \omega(\pi), \tag{2.3}$$

and let $N_S(k, t, n)$ denote the net number of S-partitions of n with spt-crank congruent to $k \pmod{t}$, namely,

$$N_S(k, t, n) = \sum_{m \equiv k \pmod{t}} N_S(m, n).$$

Andrews, Garvan and Liang [22] obtained the following relations.

Theorem 2.1 (Andrews, Garvan and Liang) *For* $0 \leq k \leq 4$,

$$N_S(k, 5, 5n + 4) = \frac{\mathrm{spt}(5n + 4)}{5},$$

and for $0 \leq k \leq 6$,

$$N_S(k, 7, 7n + 5) = \frac{\mathrm{spt}(7n + 5)}{7}.$$

Andrews, Garvan and Liang [22] defined an involution on the set of S-partitions:

$$\iota(\vec{\pi}) = \iota(\pi_1, \pi_2, \pi_3) = (\pi_1, \pi_3, \pi_2),$$

which leads to the symmetry property of $N_S(m, n)$:

$$N_S(m, n) = N_S(-m, n). \tag{2.4}$$

Using the generating function of $N_S(m, n)$, Andrews, Garvan and Liang [22] proved its positivity.

Theorem 2.2 (Andrews, Garvan and Liang) *For all integers m and positive integers n,*

$$N_S(m, n) \geq 0. \tag{2.5}$$

Dyson [65] gave an alternative proof of this property by establishing the relation:

$$N_S(m, n) = \sum_{k=1}^{\infty} (-1)^{k-1} \sum_{j=0}^{k-1} p(n - k(m + j) - (k(k + 1)/2)).$$

Andrews, Garvan and Liang [22] posed the problem of finding a combinatorial interpretation of $N_S(m, n)$. Chen, Ji and Zang [53] introduced the structure of a doubly marked partition which leads to a combinatorial interpretation of $N_S(m, n)$.

2.2   The spt-crank of a doubly marked partition

In this section, we first give a definition of a doubly marked partition and then define its spt-crank. To this end, we assume that a partition λ of n is represented by its Ferrers diagram, and we use $D(\lambda)$ to denote size of the Durfee square of λ, see [10, p. 28]. For each partition $\lambda = (\lambda_1, \lambda_2, \ldots, \lambda_l)$ of n, the associated Ferrers diagram is the arrangement of n dots in l rows with the dots being left-justified and the i-th row having λ_i dots for $1 \leq i \leq l$. The Durfee square of λ is the largest-size square contained within the Ferrers diagram of λ.

For a partition λ, let λ' denote its conjugate. A doubly marked partition of n is a partition λ of n along with two distinguished columns indexed by s and t, denoted (λ, s, t), where

(1) $1 \leq s \leq D(\lambda)$;

(2) $s \leq t \leq \lambda_1$;

(3) $\lambda'_s = \lambda'_t$.

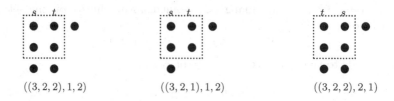

$$((3,2,2),1,2) \qquad\qquad ((3,2,1),1,2) \qquad\qquad ((3,2,2),2,1)$$

Figure 1: An illustration of the conditions for a doubly marked partition

For example, $((3,2,2),1,2)$ is a doubly marked partition, whereas $((3,2,1),1,2)$ and $((3,2,2),2,1)$ are not doubly marked partitions, see Figure 1.

To define the spt-crank of a doubly marked partition (λ, s, t), let

$$g(\lambda, s, t) = \lambda'_s - s + 1, \tag{2.6}$$

As $s \leq D(\lambda)$, we see that $\lambda'_s \geq s$, which implies that $g(\lambda, s, t) \geq 1$.

Let (λ, s, t) be a doubly marked partition, and let $g = g(\lambda, s, t)$. The spt-crank of (λ, s, t) is defined by

$$c(\lambda, s, t) = g - \lambda_g + t - s. \tag{2.7}$$

For example, for the doubly marked partition $((4,4,1,1),2,3)$, we have $g = 2 - 1 = 1$ and the spt-crank equals $1 - \lambda_1 + 3 - 2 = -2$.

The following theorem in [53] gives a combinatorial interpretation of $N_S(m, n)$.

Theorem 2.3 (Chen, Ji and Zang) *For any integer m and any positive integer n, $N_S(m, n)$ equals the number of doubly marked partitions of n with spt-crank m.*

For example, for $n = 4$, the sixteen S-partitions of 4, their spt-cranks and the ten doubly marked partitions of 4 and their spt-cranks are listed in Table 1.

The proof of Theorem 2.3 relies on the generating function of $N_S(m, n)$ given by Andrews, Garvan and Liang [22].

Andrews, Dyson and Rhoades [19] proposed the problem of finding a definition of the spt-crank for a marked partition so that the set of marked partitions of $5n + 4$ (or $7n + 5$) can be divided into five (or seven) equinumerous classes. Chen, Ji and Zang [53] established a bijection Δ between the set of marked partitions of n and the set of doubly marked partitions of n.

S-partition	sign	spt-crank	doubly marked partition	spt-crank
$((1),(1,1,1),\emptyset)$	$+1$	3	$((1,1,1,1),1,1)$	3
$((1),(2,1),\emptyset)$	$+1$	2	$((2,1,1),1,1)$	2
$((1),(1,1),(1))$	$+1$	1	$((3,1),1,1)$	1
$((1),(3),\emptyset)$	$+1$	1	$((2,2),1,2)$	1
$((2,1),(1),\emptyset)$	-1	1		
$((2),(2),\emptyset)$	$+1$	1		
$((1),(2),(1))$	$+1$	0	$((2,2),1,1)$	0
$((1),(1),(2))$	$+1$	0	$((4),1,4)$	0
$((3,1),\emptyset,\emptyset)$	-1	0		
$((4),\emptyset,\emptyset)$	$+1$	0		
$((1),(1),(1,1))$	$+1$	-1	$((2,2),2,2)$	-1
$((1),\emptyset,(3))$	$+1$	-1	$((4),1,3)$	-1
$((2,1),\emptyset,(1))$	-1	-1		
$((2),\emptyset,(2))$	$+1$	-1		
$((1),\emptyset,(2,1))$	$+1$	-2	$((4),1,2)$	-2
$((1),\emptyset,(1,1,1))$	$+1$	-3	$((4),1,1)$	-3

Table 1: S-partitions and doubly marked partitions

Theorem 2.4 (Chen, Ji and Zang) *There is a bijection Δ between the set of marked partitions (μ, k) of n and the set of doubly marked partitions (λ, s, t) of n.*

To prove the above theorem, we adopt the notation (λ, s, t) for a partition λ with two distinguished columns λ'_s and λ'_t in the Ferrers diagram. Let Q_n denote the set of doubly marked partitions of n, and let

$$U_n = \{(\lambda, s, t) \mid |\lambda| = n, \ 1 \le s \le D(\lambda), \ 1 \le t \le \lambda_1\}.$$

Obviously, $Q_n \subseteq U_n$.

Before we give a description of the bijection Δ, we introduce a transformation τ from $U_n \setminus Q_n$ to U_n.

The transformation τ: Assume that $(\lambda, s, t) \in U_n \setminus Q_n$, that is, λ is a partition of n with two distinguished columns indexed by s and t such that $1 \le s \le D(\lambda)$ and either $1 \le t < s$ or $\lambda'_s > \lambda'_t$. We wish to construct a partition μ with two distinguished columns indexed by a and b. Let p

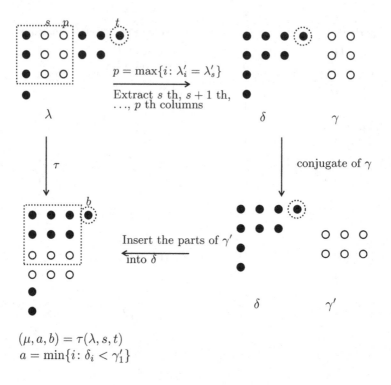

$(\mu, a, b) = \tau(\lambda, s, t)$
$a = \min\{i : \delta_i < \gamma'_1\}$

Figure 2: An illustration of the map τ

be the maximum integer such that $\lambda'_p = \lambda'_s$. Define

$$\delta = (\lambda_1 - p + s - 1, \lambda_2 - p + s - 1, \ldots, \lambda_{\lambda'_s} - p + s - 1, \lambda_{\lambda'_s + 1}, \ldots, \lambda_\ell). \quad (2.8)$$

Set a to be the minimum integer such that $\delta_a < \lambda'_s$ and

$$\mu = (\delta_1, \ldots, \delta_{a-1}, \lambda'_s, \ldots, \lambda'_p, \delta_a, \ldots, \delta_\ell). \quad (2.9)$$

If $t < s$, then set $b = t$ and if $\lambda'_s > \lambda'_t$, then set $b = t - p + s - 1$.
Define $\tau(\lambda, s, t) = (\mu, a, b)$. Figure 2 gives an illustration of the map
$\tau \colon ((6, 5, 3, 1), 2, 6) \mapsto ((4, 3, 3, 3, 1, 1), 3, 4)$.

It was proved in [53] that the map τ is indeed an injection. Using
this property, they described the bijection Δ in Theorem 2.4 based on the
injection τ.

The definition of Δ: Let (μ, k) be a marked partition of n, we proceed to
construct a doubly marked partition (λ, s, t) of n.

$((2,1,1,1,1),5)$ $((2,2,1,1),2,2)$

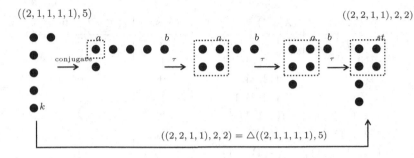

$$((2,2,1,1),2,2) = \Delta((2,1,1,1,1),5)$$

Figure 3: The bijection $\Delta\colon ((2,1,1,1,1),5) \mapsto ((2,2,1,1),2,2)$

We first consider $(\mu',1,k)$. If $(\mu',1,k)$ is already a doubly marked partition, then there is nothing to be done and we just set $(\lambda,s,t) = (\mu',1,k)$. Otherwise, we iteratively apply the map τ to $(\mu',1,k)$ until we get a doubly marked partition (λ,s,t). We then define

$$\Delta(\mu,k) = (\lambda,s,t).$$

It can be shown that this process terminates and it is reversible. Thus Δ is well-defined and is a bijection between the set of marked partitions (μ,k) of n and the set of doubly marked partitions (λ,s,t) of n.

To give an example of the map Δ, let $n = 6$, $\mu = (2,1,1,1,1)$ and $k = 5$. We have $\mu' = (5,1)$. Note that $(\mu',1,k) = ((5,1),1,5)$, which is not a doubly marked partition. It can be checked that $\tau(\mu',1,k) = ((4,2),2,4)$, which is not a doubly marked partition. Repeating this process, we get $\tau((4,2),2,4) = ((3,2,1),2,3)$, and $\tau((3,2,1),2,3) = ((2,2,1,1),2,2)$, which is eventually a doubly marked partition. See Figure 3. Thus, we obtain

$$\Delta((2,1,1,1,1),5) = ((2,2,1,1),2,2).$$

Utilizing the bijection Δ and the spt-crank for a doubly marked partition, one can divide the set of marked partitions of $5n+4$ (or $7n+5$) into five (or seven) equinumerous classes. Hence, in principle, the spt-crank of a doubly marked partition can be considered as a solution to the problem of Andrews, Dyson and Rhoades. It would be interesting to find an spt-crank directly defined on marked partitions.

For example, for $n = 4$, we have $\mathrm{spt}(4) = 10$. The ten marked partitions of 4, the corresponding doubly marked partitions, and the spt-crank modulo 5 are listed in Table 2.

For $n = 5$, we have $\mathrm{spt}(5) = 14$. The fourteen marked partitions of 5,

(μ, k)	$(\lambda, s, t) = \Delta(\mu, k)$	$c(\lambda, s, t)$	$c(\lambda, s, t) \mod 5$
$((4), 1)$	$((1, 1, 1, 1), 1, 1)$	3	3
$((3, 1), 2)$	$((3, 1), 1, 1)$	1	1
$((2, 2), 1)$	$((2, 2), 1, 1)$	0	0
$((2, 2), 2)$	$((2, 2), 1, 2)$	1	1
$((2, 1, 1), 2)$	$((2, 1, 1), 1, 1)$	2	2
$((2, 1, 1), 3)$	$((2, 2), 2, 2)$	-1	4
$((1, 1, 1, 1), 1)$	$((4), 1, 1)$	-3	2
$((1, 1, 1, 1), 2)$	$((4), 1, 2)$	-2	3
$((1, 1, 1, 1), 3)$	$((4), 1, 3)$	-1	4
$((1, 1, 1, 1), 4)$	$((4), 1, 4)$	0	0

Table 2: The case for $n = 4$

the corresponding doubly marked partitions, and the spt-crank modulo 7 are listed in Table 3.

(μ, k)	$(\lambda, s, t) = \Delta(\mu, k)$	$c(\lambda, s, t)$	$c(\lambda, s, t) \mod 7$
$((5), 1)$	$((1, 1, 1, 1, 1), 1, 1)$	4	4
$((4, 1), 2)$	$((4, 1), 1, 1)$	1	1
$((3, 2), 2)$	$((3, 1, 1), 1, 1)$	2	2
$((3, 1, 1), 2)$	$((3, 2), 1, 1)$	0	0
$((3, 1, 1), 3)$	$((3, 2), 1, 2)$	1	1
$((2, 2, 1), 3)$	$((2, 2, 1), 1, 1)$	2	2
$((2, 1, 1, 1), 2)$	$((2, 1, 1, 1), 1, 1)$	3	3
$((2, 1, 1, 1), 3)$	$((3, 2), 2, 2)$	-2	5
$((2, 1, 1, 1), 4)$	$((2, 2, 1), 2, 2)$	-1	6
$((1, 1, 1, 1, 1), 1)$	$((5), 1, 1)$	-4	3
$((1, 1, 1, 1, 1), 2)$	$((5), 1, 2)$	-3	4
$((1, 1, 1, 1, 1), 3)$	$((5), 1, 3)$	-2	5
$((1, 1, 1, 1, 1), 4)$	$((5), 1, 4)$	-1	6
$((1, 1, 1, 1, 1), 5)$	$((5), 1, 5)$	0	0

Table 3: The case for $n = 5$

2.3 The unimodality of the spt-crank

The unimodality of the spt-crank was first studied by Andrews, Dyson and Rhoades [19]. They showed that the unimodality of the spt-crank is equivalent to an inequality between the rank and the crank of a partition. Define

$$N_{\leq m}(n) = \sum_{|r| \leq m} N(r, n), \tag{2.10}$$

$$M_{\leq m}(n) = \sum_{|r| \leq m} M(r, n). \tag{2.11}$$

Andrews, Dyson and Rhoades [19] established the following relation.

Theorem 2.5 (Andrews, Dyson and Rhoades) *For* $m \geq 0$ *and* $n > 1$,

$$N_S(m, n) - N_S(m + 1, n) = \frac{1}{2}\left(N_{\leq m}(n) - M_{\leq m}(n)\right). \tag{2.12}$$

They also posed a conjecture on the spt-crank.

Conjecture 2.6 (Andrews, Dyson and Rhoades) *For* $m, n \geq 0$,

$$N_S(m, n) \geq N_S(m + 1, n). \tag{2.13}$$

By the symmetry (2.4) of $N_S(m, n)$ and the relation (2.13), we see that

$$N_S(-n, n) \leq \cdots \leq N_S(-1, n) \leq N_S(0, n) \geq N_S(1, n) \geq \cdots \geq N_S(n, n).$$

In view of (2.12), Andrews, Dyson and Rhoades pointed out that Conjecture 2.6 is equivalent to the assertion

$$N_{\leq m}(n) \geq M_{\leq m}(n), \tag{2.14}$$

where $m, n \geq 0$. It was remarked in [19] that (2.14) was conjectured by Bringmann and Mahlburg [44]. When $m = 0$, (2.14) was conjectured by Kaavya [95].

Andrews, Dyson and Rhoades [19] obtained an asymptotic formula for $N_{\leq m}(n) - M_{\leq m}(n)$, which implies that Conjecture 2.6 holds for fixed m and sufficiently large n.

Theorem 2.7 (Andrews, Dyson and Rhoades) *For each* $m \geq 0$,

$$\left(N_{\leq m}(n) - M_{\leq m}(n)\right) \sim \frac{(2m + 1)\pi^2}{192\sqrt{3}n^2}e^{\pi\sqrt{\frac{2n}{3}}} \quad as \quad n \to \infty. \tag{2.15}$$

$n \setminus m$	-6	-5	-4	-3	-2	-1	0	1	2	3	4	5	6
0							1						
1							1						
2						1	1	1					
3					1	1	1	1	1				
4				1	1	2	2	2	1	1			
5			1	1	2	2	2	2	2	1	1		
6		1	1	2	3	4	4	4	3	2	1	1	
7	1	1	2	3	4	4	5	4	4	3	2	1	1

Table 4: An illustration of the unimodality of $N_S(m,n)$

Using the rank-set of a partition, Chen, Ji and Zang [52] constructed an injection from the set of partitions of n such that m appears in the rank-set to the set of partitions of n with rank not less than $-m$. This proves the inequality (2.14) for all $m \geq 0$ and $n \geq 1$, and hence Conjecture 2.6 is confirmed.

In fact, the relation (2.14) was stated by Bringmann and Mahlburg [44] in a different notation. For an integer m and a positive integer n, let

$$\overline{\mathcal{M}}(m,n) = \sum_{r \leq m} M(r,n),$$

and

$$\overline{\mathcal{N}}(m,n) = \sum_{r \leq m} N(r,n).$$

By the symmetry properties of the rank and the crank, that is,

$$N(m,n) = N(-m,n) \quad \text{and} \quad M(m,n) = M(-m,n),$$

see [63] and [70], it is not difficult to verify that (2.14) is equivalent to the following inequality for $m < 0$ and $n \geq 1$:

$$\overline{\mathcal{N}}(m,n) \leq \overline{\mathcal{M}}(m,n). \tag{2.16}$$

It turns out that the constructive approach in [52] can be used to prove the other part of the conjecture (2.16) of Bringmann and Mahlburg, that is,

$$\overline{\mathcal{M}}(m,n) \leq \overline{\mathcal{N}}(m+1,n), \tag{2.17}$$

for $m < 0$ and $n \geq 1$. A proof of (2.17) was given in [54].

In the notation $N_{\leq m-1}(n)$ and $M_{\leq m}(n)$, the inequality (2.17) can be expressed as

$$M_{\leq m}(n) \geq N_{\leq m-1}(n), \qquad (2.18)$$

for $m \geq 1$ and $n \geq 1$.

Bringmann and Mahlburg [44] also pointed out that the inequalities (2.16) and (2.17) can be restated as the existence of a re-ordering τ_n on the set of partitions of n such that $|\text{crank}(\lambda)| - |\text{rank}(\tau_n(\lambda))| = 0$ or 1 for all partitions λ of n. Chen, Ji and Zang [54] defined a re-ordering τ_n on the set of partitions of n and showed that this re-ordering τ_n satisfies the relation $|\text{crank}(\lambda)| - |\text{rank}(\tau_n(\lambda))| = 0$ or 1 for any partition λ of n. Appealing to this re-ordering τ_n, they gave a new combinatorial interpretation of the function $\text{ospt}(n)$ defined by Andrews, Chan and Kim [15], which leads to an upper bound for $\text{ospt}(n)$ due to Chan and Mao [49].

Bringmann and Mahlburg [44] also remarked that using the Cauchy-Schwartz inequality, the bijection τ_n leads to an upper bound for $\text{spt}(n)$, namely, for $n \geq 1$,

$$\text{spt}(n) \leq \sqrt{2n}p(n). \qquad (2.19)$$

Chan and Mao [49] posed a conjecture on a sharper upper bound and a lower bound for $\text{spt}(n)$.

Conjecture 2.1 (Chan and Mao) *For $n \geq 3$,*

$$\frac{\sqrt{6n}}{\pi}p(n) \leq \text{spt}(n) \leq \sqrt{n}p(n). \qquad (2.20)$$

The following upper bound and lower bound for $\text{spt}(n)$ were conjectured by Hirschhorn and later proved by Eichhorn and Hirschhorn [66].

Theorem 2.8 (Eichhorn and Hirschhorn) *For $n \geq 2$,*

$$p(0) + p(1) + \cdots + p(n-1) < \text{spt}(n) < p(0) + p(1) + \cdots + p(n). \qquad (2.21)$$

3   More congruences

Garvan [72] obtained Ramanujan-type congruences of $\text{spt}(n)$ mod 11, 17, 19, 29, 31 and 37.

Theorem 3.1 (Garvan) *For $n \geq 0$,*

$$\text{spt}(11 \cdot 19^4 \cdot n + 22006) \equiv 0 \pmod{11}, \qquad (3.1)$$

$$\text{spt}(17 \cdot 7^4 \cdot n + 243) \equiv 0 \pmod{17}, \tag{3.2}$$

$$\text{spt}(19 \cdot 5^4 \cdot n + 99) \equiv 0 \pmod{19}, \tag{3.3}$$

$$\text{spt}(29 \cdot 13^4 \cdot n + 18583) \equiv 0 \pmod{29}, \tag{3.4}$$

$$\text{spt}(31 \cdot 29^4 \cdot n + 409532) \equiv 0 \pmod{31}, \tag{3.5}$$

$$\text{spt}(37 \cdot 5^4 \cdot n + 1349) \equiv 0 \pmod{37}. \tag{3.6}$$

Bringmann [39] showed that $\text{spt}(n)$ possesses a congruence property analogous to the following theorem for $p(n)$, due to Ono [110].

Theorem 3.2 (Ono) *For any prime $\ell \geq 5$, there are infinitely many arithmetic progressions $an + b$ such that*

$$p(an + b) \equiv 0 \pmod{\ell}. \tag{3.7}$$

As for $\text{spt}(n)$, Bringmann [39] proved the following assertion.

Theorem 3.3 (Bringmann) *For any prime $\ell \geq 5$, there are infinitely many arithmetic progressions $an + b$ such that*

$$\text{spt}(an + b) \equiv 0 \pmod{\ell}.$$

The above theorem is a consequence of (1.4), Theorem 3.2 and the following theorem of Bringmann [39].

Theorem 3.4 (Bringmann) *For any prime $\ell \geq 5$, there are infinitely many arithmetic progressions $an + b$ such that*

$$N_2(an + b) \equiv 0 \pmod{\ell}. \tag{3.8}$$

Bringmann [39] constructed a weight $3/2$ harmonic weak Maass form $\mathcal{M}(z)$ on $\Gamma_0(576)$ with Nebentypus $\chi_{12}(\bullet) = \left(\frac{12}{\bullet}\right)$, which is related to the generating function of $\text{spt}(n)$. This implies that the generating function of $\text{spt}(n)$ is essentially a mock theta function with Dedekind eta-function $\eta(q)$ as its shadow just as pointed out by Rhoades [124]. Ono [112] found a weight $(\ell^2 + 3)/2$ holomorphic modular form on $SL_2(\mathbb{Z})$ which contains the holomorphic part of $\mathcal{M}(z)$. Using this modular form, Ono [112] derived Ramanujan-type congruences of $\text{spt}(n)$ modulo ℓ for any prime $\ell \geq 5$.

Theorem 3.5 (Ono) *Let $\ell \geq 5$ be a prime and let $\left(\frac{\bullet}{\circ}\right)$ denote the Legendre symbol.*

(i) *For $n \geq 1$, if $\left(\frac{-n}{\ell}\right) = 1$,*

$$\mathrm{spt}\left((\ell^2 n + 1)/24\right) \equiv 0 \pmod{\ell}.$$

(ii) *For $n \geq 0$,*

$$\mathrm{spt}\left((\ell^3 n + 1)/24\right) \equiv \left(\frac{3}{\ell}\right) \mathrm{spt}\left((\ell n + 1)/24\right) \pmod{\ell}.$$

Ahlgren, Bringmann and Lovejoy [5] extended Theorem 3.5 to any prime power. An analogous congruence for $p(n)$ was found by Ahlgren [1].

Theorem 3.6 (Ahlgren, Bringmann and Lovejoy) *Let $\ell \geq 5$ be a prime and let $m \geq 1$.*

(i) *For $n \geq 1$, if $\left(\frac{-n}{\ell}\right) = 1$,*

$$\mathrm{spt}\left((\ell^{2m} n + 1)/24\right) \equiv 0 \pmod{\ell^m}.$$

(ii) *For $n \geq 0$,*

$$\mathrm{spt}\left((\ell^{2m+1} n + 1)/24\right) \equiv \left(\frac{3}{\ell}\right) \mathrm{spt}\left((\ell^{2m-1} n + 1)/24\right) \pmod{\ell^m}.$$

Recall the following congruences of $p(n)$:

$$p(5^a n + \delta_a) \equiv 0 \pmod{5^a}, \tag{3.9}$$

$$p(7^b n + \lambda_b) \equiv 0 \pmod{7^{\lfloor \frac{b+2}{2} \rfloor}}, \tag{3.10}$$

$$p(11^c n + \varphi_c) \equiv 0 \pmod{11^c}, \tag{3.11}$$

where a, b, c are positive integers and δ_a, λ_b and φ_c are the least nonnegative residues of the reciprocals of 24 mod $5^a, 7^b$ and 11^c, respectively. The congruences (3.9) and (3.10) were proved by Watson [132] and the congruence (3.11) was proved by Atkin [27]. Folsom, Kent and Ono [67] provided alternative proofs of the congruences (3.9)–(3.11) with the aid of the theory of ℓ-adic modular forms. Recently, Paule and Radu [117] found a unified algorithmic approach to (3.9)–(3.11) resorting to elementary modular function tools only.

In the case of the spt-function, although Theorem 3.6 gives congruences for all primes $\ell \geq 5$, the congruences (1.12)–(1.14) do not follow from Theorem 3.6. Congruences for these missing cases have been obtained by Garvan [74], which are analogous to (3.9)–(3.11).

Theorem 3.7 (Garvan) *For $n \geq 0$,*

$$\mathrm{spt}(5^a n + \delta_a) \equiv 0 \pmod{5^{\lfloor \frac{a+1}{2} \rfloor}},$$

$$\mathrm{spt}(7^b n + \lambda_b) \equiv 0 \pmod{7^{\lfloor \frac{b+1}{2} \rfloor}},$$

$$\mathrm{spt}(13^c n + \gamma_c) \equiv 0 \pmod{13^{\lfloor \frac{c+1}{2} \rfloor}},$$

where a, b, c are positive integers, and δ_a, λ_b and γ_c are the least nonnegative residues of the reciprocals of $24 \bmod 5^a$, 7^b and 13^c respectively.

Setting $a = b = c = 1$, Theorem 3.7 reduces to (1.12)–(1.14). Belmont, Lee, Musat and Trebat-Leder [32] provided another proof of the above theorem by generalizing techniques of Folsom, Kent and Ono [67] and by utilizing refinements due to Boylan and Webb [38].

Before we get into the discussions about the parity of $\mathrm{spt}(n)$, let us look back at the parity of $p(n)$. Subbarao [129] conjectured that in every arithmetic progression $r \pmod{t}$, there are infinitely many integers $N \equiv r \pmod{t}$ for which $p(N)$ is even, and infinitely many integers $M \equiv r \pmod{t}$ for which $p(M)$ is odd. This conjecture has been confirmed for $t = 1, 2, 3, 4, 5, 10, 12, 16$ and 40 by Garvan and Stanton [79], Hirschhorn [82], Hirschhorn and Subbarao [85], Kolberg [99] and Subbarao [129]. The even case of Subbarao's conjecture was settled by Ono [109] and the odd case was solved by Radu [121]. Radu [121] also showed that for every arithmetic progression $r \pmod{t}$, there are infinitely many integers $N \equiv r \pmod{t}$ such that $p(N) \not\equiv 0 \pmod{3}$. This confirms a conjecture posed by Ahlgren and Ono [6].

For $n \geq 1$, the parity of $\mathrm{spt}(n)$ is determined by Folsom and Ono [68]. They constructed a pair of harmonic weak Maass forms with equal nonholomorphic parts, whose difference contains the generating function of $\mathrm{spt}(n)$ as a component. Based on the results in [40], Folsom and Ono showed that the difference of such pair of harmonic weak Maass forms can be expressed as the sum of the generating function for $\mathrm{spt}(n)$ and a modular form. This enables us to completely determine the parity of $\mathrm{spt}(n)$.

To be more specific, Folsom and Ono [68] first defined the mock theta functions:

$$D(z) = \frac{q^{-\frac{1}{24}}}{(q;q)_\infty} \left(1 - 24 \sum_{n=1}^{\infty} \frac{nq^n}{1 - q^n} \right) = \frac{q^{-\frac{1}{24}}}{(q;q)_\infty} E_2(z)$$

and

$$L(z) = \frac{(q^6; q)_\infty^2 (q^{24}; q)_\infty^2}{(q^{12}; q)_\infty^5} \left(\sum_{n=-\infty}^{\infty} \frac{(12n-1)q^{6n^2 - \frac{1}{24}}}{1 - q^{12n-1}} - \sum_{n=-\infty}^{\infty} \frac{(12n-5)q^{6n^2 - \frac{25}{24}}}{1 - q^{12n-5}} \right).$$

Then they obtained the following modular form.

Theorem 3.8 (Folsom and Ono) *The function*

$$D(24z) - 12L(24z) - 12q^{-1}S(24z)$$

is a weight $3/2$ weakly holomorphic modular form on $\Gamma_0(576)$ with Nebentypus $\left(\frac{12}{\bullet}\right)$, where

$$S(z) = \sum_{n=0}^{\infty} \operatorname{spt}(n)q^n.$$

By Theorem 3.8, Folsom and Ono [68] obtained a characterization of the parity of $\operatorname{spt}(n)$.

Theorem 3.9 (Folsom and Ono) *The function $\operatorname{spt}(n)$ is odd if and only if $24n - 1 = pm^2$, where m is an integer and $p \equiv 23$ (mod 24) is prime.*

As pointed out by Andrews, Garvan and Liang [23], Theorem 3.9 contains an error. For example, for $n = 507$, it is clear that $507 \times 24 - 1 = 12167 = 23 \times 23^2 = pm^2$, where $p = m = 23$. Obviously, 507 satisfies the condition of Theorem 3.9. But $\operatorname{spt}(507) = 60470327737556285225064$ is even. This error has been corrected by Andrews, Garvan and Liang [23]. By using the notion of S-partitions as defined in (2.2), they noticed that the number of S-partitions of n has the same parity as $\operatorname{spt}(n)$. Then they built an involution ι on the set of S-partitions of n as follows:

$$\iota(\vec{\pi}) = \iota(\pi_1, \pi_2, \pi_3) = (\pi_1, \pi_3, \pi_2).$$

Clearly, an S-partition (π_1, π_2, π_3) is a fixed point of ι if and only if $\pi_2 = \pi_3$. Denote the number of such S-partitions of n by $N_{SC}(n)$. It is not difficult to see that

$$\operatorname{spt}(n) \equiv N_{SC}(n) \pmod 2.$$

By computing the generating function of $N_{SC}(n)$, Andrews, Garvan and Liang [23] established a corrected version of Theorem 3.9.

Theorem 3.10 (Andrews, Garvan and Liang) *The function $\operatorname{spt}(n)$ is odd if and only if $24n - 1 = p^{4a+1}m^2$ for some prime $p \equiv 23$ (mod 24) and some integers a, m with $(p, m) = 1$.*

The spt-function is also related to some combinatorial sequences, see, for example, Andrews, Rhoades and Zwegers [25] and Bryson, Ono, Pitman and Rhoades [48]. Bryson, Ono, Pitman and Rhoades [48] showed that the number of strongly unimodal sequences of size n has the same parity as $\text{spt}(n)$. More specifically, a sequence of integers $\{a_i\}_{i=1}^s$ is said to be a strongly unimodal sequence of size n if $a_1 + \cdots + a_s = n$ and for some k,

$$0 < a_1 < a_2 < \cdots < a_k > a_{k+1} > a_{k+2} > \cdots > a_s > 0.$$

Let $u(n)$ be the number of strongly unimodal sequences of size n. By [13, Theorem 1], Bryson, Ono, Pitman and Rhoades [48] observed that

$$u(n) \equiv \text{spt}(n) \pmod 2.$$

As for congruences of $\text{spt}(n)$ modulo powers of 2, Garvan and Jennings-Shaffer [77] obtained congruences mod $2^3, 2^4$ and 2^5. Let

$$s_\ell = \frac{\ell^2 - 1}{24}.$$

Theorem 3.11 (Garvan and Jennings-Shaffer) *Let $\ell \geq 5$ be a prime, and define*

$$\beta = \begin{cases} 3, & \text{if } \ell \equiv 7, 9 \pmod{24}, \\ 4, & \text{if } \ell \equiv 13, 23 \pmod{24}, \\ 5, & \text{if } \ell \equiv 1, 11, 17, 19 \pmod{24}. \end{cases}$$

Then for $n \geq 1$,

$$\text{spt}(\ell^2 n - s_\ell) + \left(\frac{3 - 72n}{\ell}\right) \text{spt}(n) + \ell \, \text{spt}\left((n + s_\ell)/\ell^2\right)$$

$$\equiv \left(\frac{3}{\ell}\right)(1 + \ell)\, \text{spt}(n) \pmod{2^\beta}.$$

By using the Hecke algebra of a Maass form, Folsom and Ono [68] derived a congruence of $\text{spt}(n)$ modulo 3.

Theorem 3.12 (Folsom and Ono) *Let $\ell \geq 5$ be a prime, then for $n \geq 1$,*

$$\text{spt}(\ell^2 n - s_\ell) + \left(\frac{3 - 72n}{\ell}\right) \text{spt}(n) + \ell \, \text{spt}\left((n + s_\ell)/\ell^2\right)$$

$$\equiv \left(\frac{3}{\ell}\right)(1 + \ell)\, \text{spt}(n) \pmod{3}.$$

Corollary 3.13 (Folsom and Ono) *Let $\ell \geq 5$ be a prime such that $\ell \equiv 2 \pmod 3$. If $0 < k < \ell - 1$, then for $n \geq 1$,*

$$\mathrm{spt}(\ell^4 n + \ell^3 k - (\ell^4 - 1)/24) \equiv 0 \pmod 3.$$

For example, for $\ell = 5$, we have

$$
\begin{aligned}
\mathrm{spt}(625n + 99) &\equiv \mathrm{spt}(625n + 224) \\
&\equiv \mathrm{spt}(625n + 349) \\
&\equiv \mathrm{spt}(625n + 474) \\
&\equiv 0 \pmod 3.
\end{aligned}
$$

Garvan [75] derived congruences mod $5, 7, 13$ and 72.

Theorem 3.14 (Garvan) (i) *If $\ell \geq 5$ is prime, then for $n \geq 1$*

$$\mathrm{spt}(\ell^2 n - s_\ell) + \left(\frac{3 - 72n}{\ell}\right) \mathrm{spt}(n) + \ell \, \mathrm{spt}\left((n + s_\ell)/\ell^2\right)$$

$$\equiv \left(\frac{3}{\ell}\right)(1 + \ell) \mathrm{spt}(n) \pmod{72}. \tag{3.12}$$

(ii) *If $\ell \geq 5$ is prime, $t = 5, 7$ or 13 and $\ell \neq t$, then for $n \geq 1$*

$$\mathrm{spt}(\ell^2 n - s_\ell) + \left(\frac{3 - 72n}{\ell}\right) \mathrm{spt}(n) + \ell \, \mathrm{spt}\left((n + s_\ell)/\ell^2\right)$$

$$\equiv \left(\frac{3}{\ell}\right)(1 + \ell) \mathrm{spt}(n) \pmod{t}. \tag{3.13}$$

Note that Theorem 3.12 can be deduced from (3.12). Moreover, writing $32760 = 2^3 \cdot 3^2 \cdot 5 \cdot 7 \cdot 13$, from (3.12) and (3.13), it is easy to deduce a congruence of $\mathrm{spt}(n)$ modulo 32760.

Corollary 3.15 (Garvan) *If ℓ is prime and $\ell \notin \{2, 3, 5, 7, 13\}$, then for $n \geq 1$*

$$\mathrm{spt}(\ell^2 n - s_\ell) + \left(\frac{3 - 72n}{\ell}\right) \mathrm{spt}(n) + \ell \, \mathrm{spt}\left((n + s_\ell)/\ell^2\right)$$

$$\equiv \left(\frac{3}{\ell}\right)(1 + \ell) \mathrm{spt}(n) \pmod{32760}.$$

Garrett, McEachern, Frederick and Hall-Holt [69] obtained a recurrence relation for spt(n). To compute spt(n), they introduced two integer arrays $A(n, j)$ and $B(n, j)$, where $A(n, j)$ denotes the number of partitions of n with the smallest part at least j and $B(n, j)$ denotes the number of times that j occurs as the smallest part of partitions of n. From the definitions of $A(n, j)$ and $B(n, j)$, it is not difficult to deduce the following recurrence relations:

$$A(n, j) = A(n - j, j) + A(n, j + 1),$$
$$B(n, j) = A(n - j, j) + B(n - j, j),$$

where $A(n, j) = B(n, j) = 0$ whenever $n < j$ and $A(n, n) = B(n, n) = 1$.
    Thus we have

$$\mathrm{spt}(n) = \sum_{j=1}^{n} B(n, j).$$

By the above relation, Garrett, McEachern, Frederick and Hall-Holt computed the first million values of spt(n), and found many conjectures on congruences of spt(n).

$$\mathrm{spt}(1331n + 479) \equiv 0 \pmod{11}, \tag{3.14}$$

$$\mathrm{spt}(1331n + 842) \equiv 0 \pmod{11}, \tag{3.15}$$

$$\mathrm{spt}(1331n + 1084) \equiv 0 \pmod{11}, \tag{3.16}$$

$$\mathrm{spt}(1331n + 1205) \equiv 0 \pmod{11}, \tag{3.17}$$

$$\mathrm{spt}(1331n + 1326) \equiv 0 \pmod{11}, \tag{3.18}$$

$$\mathrm{spt}(4913n + 566) \equiv 0 \pmod{17}, \tag{3.19}$$

$$\mathrm{spt}(4913n + 2300) \equiv 0 \pmod{17}, \tag{3.20}$$

$$\mathrm{spt}(4913n + 2878) \equiv 0 \pmod{17}, \tag{3.21}$$

$$\mathrm{spt}(4913n + 3167) \equiv 0 \pmod{17}, \tag{3.22}$$

$$\mathrm{spt}(4913n + 3456) \equiv 0 \pmod{17}, \tag{3.23}$$

$$\mathrm{spt}(4913n + 4323) \equiv 0 \pmod{17}, \tag{3.24}$$

$$\mathrm{spt}(4913n + 4612) \equiv 0 \pmod{17}, \tag{3.25}$$

$$\mathrm{spt}(4913n + 4901) \equiv 0 \pmod{17}, \tag{3.26}$$

$$\mathrm{spt}(11875n + 99) \equiv 0 \pmod{19}, \tag{3.27}$$

$$\mathrm{spt}(12167n + 9500) \equiv 0 \pmod{23}, \tag{3.28}$$

$$\text{spt}(24389n + 806) \equiv 0 \pmod{29}. \tag{3.29}$$

All the above conjectures have been confirmed. The congruence (3.27) has been proved by Garvan [72], and the rest are consequences of Theorem 3.5. Indeed, Theorem 3.5 (i) implies that if $\left(\frac{-\delta}{\ell}\right) = 1$, then

$$\text{spt}\left(\frac{\ell^2(\ell n + \delta) + 1}{24}\right) \equiv 0 \pmod{\ell}. \tag{3.30}$$

When $\ell = 11, 17, 23, 29$, (3.30) becomes (3.14)–(3.26), (3.28) and (3.29), respectively.

4   Generalizations and variations

In this section, we discuss three generalizations and one variation of the spt-function based on the relation (1.6) and three variations based on the combinatorial definition.

4.1   The higher order spt-function of Garvan

The first generalization of the spt-function was due to Garvan [73]. He defined a higher order spt-function in terms of the k-th symmetrized rank function and the k-th symmetrized crank function.

The k-th symmetrized rank function $\eta_k(n)$ was introduced by Andrews [11], and it is defined by

$$\eta_k(n) = \sum_{m=-n}^{n} \binom{m + \lfloor \frac{k-1}{2} \rfloor}{k} N(m, n). \tag{4.1}$$

By using q-identities, Andrews [11] found a combinatorial interpretation of $\eta_k(n)$ in terms of k-marked Durfee symbol. Ji [94] and Kursungoz [101] found combinatorial derivations of this combinatorial interpretation of $\eta_k(n)$ directly from the definition (4.1). When $k = 2$, it is easy to check that

$$\eta_2(n) = \frac{1}{2} N_2(n),$$

where the second rank moment $N_2(n)$ is defined as in (1.3).

Garvan [73] introduced the k-th symmetrized crank function $\mu_k(n)$ as follows:

$$\mu_k(n) = \sum_{m=-n}^{n} \binom{m + \lfloor \frac{k-1}{2} \rfloor}{k} M(m, n). \tag{4.2}$$

A combinatorial interpretation of $\mu_k(n)$ was given by Chen, Ji and Shen [51]. When $k = 2$, it is not difficult to derive that

$$\mu_2(n) = \frac{1}{2}M_2(n).$$

Garvan [73] introduced the higher order spt-function $\mathrm{spt}_k(n)$.

Definition 4.1 *For $k \geq 1$, define*

$$\mathrm{spt}_k(n) = \mu_{2k}(n) - \eta_{2k}(n). \tag{4.3}$$

In view of (1.6), it is easy to see that $\mathrm{spt}_k(n)$ reduces to $\mathrm{spt}(n)$ when $k = 1$. Making use of Bailey pairs [9], Garvan obtained the generating function of $\mathrm{spt}_k(n)$.

Theorem 4.2 (Garvan) *For $k \geq 1$,*

$$\sum_{n=1}^{\infty} \mathrm{spt}_k(n)q^n$$

$$= \sum_{n_k \geq n_{k-1} \geq \cdots \geq n_1 \geq 1} \frac{q^{n_1+n_2+\cdots+n_k}}{(1-q^{n_k})^2(1-q^{n_{k-1}})^2\cdots(1-q^{n_1})^2(q^{n_1+1};q)_{\infty}}. \tag{4.4}$$

Setting $k = 1$ in (4.4), we get the generating function (1.2) of $\mathrm{spt}(n)$. Furthermore, it can be seen from (4.4) that $\mathrm{spt}_k(n) \geq 0$ for $n, k \geq 1$. Together with (4.3), we find that

$$\mu_{2k}(n) \geq \eta_{2k}(n). \tag{4.5}$$

The inequality (4.5) plays a key role in the proof of an inequality between the rank moments and the crank moments, as conjectured by Garvan [72].

Conjecture 4.3 (Garvan) *For $n, k \geq 1$,*

$$M_{2k}(n) \geq N_{2k}(n). \tag{4.6}$$

Bringmann and Mahlburg [44] showed that the above conjecture is true for $k = 1, 2$ and sufficiently large n. For each fixed k, Garvan's conjecture was proved for sufficiently large n by Bringmann, Mahlburg and Rhoades [46]. Garvan [73] confirmed his conjecture for all k and n. He introduced an analogue of the Stirling numbers of the second kind, denoted by $S^*(k, j)$. It is defined recursively as follows:

(1) $S^*(1,1) = 1$;

(2) $S^*(k,j) = 0$ if $j \leq 0$ or $j > k$;

(3) $S^*(k+1,j) = S^*(k, j-1) + j^2 S^*(k, j)$ for $1 \leq j \leq k+1$.

It is clear from the above recurrence relation that $S^*(k, j) \geq 0$. Garvan established the following relations between the ordinary moments and symmetrized moments in terms of $S^*(k, j)$:

$$M_{2k}(n) = \sum_{j=1}^{k} (2j)! S^*(k, j) \mu_{2j}(n) \tag{4.7}$$

and

$$N_{2k}(n) = \sum_{j=1}^{k} (2j)! S^*(k, j) \eta_{2j}(n). \tag{4.8}$$

It follows from (4.7) and (4.8) that

$$M_{2k}(n) - N_{2k}(n) = \sum_{j=1}^{k} (2j)! S^*(k, j) \left(\mu_{2j}(n) - \eta_{2j}(n) \right). \tag{4.9}$$

Invoking (4.5) we deduce that $M_{2k}(n) - N_{2k}(n) \geq 0$ for $n, k \geq 1$, and hence Conjecture 4.3 is proved.

Garvan [73] gave a combinatorial explanation of the right-hand side of (4.4). Thus Theorem 4.2 leads to a combinatorial interpretation of $\mathrm{spt}_k(n)$.

Theorem 4.4 (Garvan) *Let λ be a partition with m different parts*

$$n_1 < n_2 < \cdots < n_m.$$

Let $k \geq 1$, define the weight $\omega_k(\lambda)$ of λ as follows:

$$\omega_k(\lambda) = \sum_{\substack{m_1 + \cdots + m_r = k \\ 1 \leq r \leq k}} \binom{f_1 + m_1 - 1}{2m_1 - 1}$$

$$\times \sum_{2 \leq j_2 < j_3 < \cdots < j_r} \binom{f_{j_2} + m_2}{2m_2} \binom{f_{j_3} + m_3}{2m_3} \cdots \binom{f_{j_r} + m_r}{2m_r},$$

where $f_j = f_j(\lambda)$ denotes the multiplicity of the part n_j in λ. Then

$$\mathrm{spt}_k(n) = \sum_{\lambda \in P(n)} \omega_k(\lambda).$$

Garvan [73] also obtained congruences of $\mathrm{spt}_2(n)$, $\mathrm{spt}_3(n)$ and $\mathrm{spt}_4(n)$.

Theorem 4.5 (Garvan) *For $n \geq 1$,*

$$\mathrm{spt}_2(n) \equiv 0 \pmod{5}, \qquad if\ n \equiv 0, 1, 4 \pmod{5},$$

$$\mathrm{spt}_2(n) \equiv 0 \pmod{7}, \qquad if\ n \equiv 0, 1, 5 \pmod{7},$$

$$\mathrm{spt}_2(n) \equiv 0 \pmod{11}, \qquad if\ n \equiv 0 \pmod{11},$$

$$\mathrm{spt}_3(n) \equiv 0 \pmod{7}, \qquad if\ n \not\equiv 3, 6 \pmod{7},$$

$$\mathrm{spt}_3(n) \equiv 0 \pmod{2}, \qquad if\ n \equiv 1 \pmod{4},$$

$$\mathrm{spt}_4(n) \equiv 0 \pmod{3}, \qquad if\ n \equiv 0 \pmod{3}.$$

4.2 Generalized higher order spt-functions of Dixit and Yee

Other generalizations of the spt-function have been given by Dixit and Yee [62], which are based on the j-rank introduced by Garvan [71]. The j-rank is a generalization of Dyson's rank. For a partition λ and $j \geq 2$, let $n_j(\lambda)$ denote the size of the j-th successive Durfee square of λ, let $c_j(\lambda)$ denote the number of columns in the Ferrers diagram of λ with length not exceeding $n_j(\lambda)$ and let $r_j(\lambda)$ denote the number of parts of λ that lie below the j-th Durfee square. Then the j-rank of λ is defined to be $c_{j-1}(\lambda) - r_{j-1}(\lambda)$. It should be noted that the 2-rank coincides with Dyson's rank.

For example, the 3-rank of $\lambda = (9, 9, 7, 7, 7, 5, 3, 3, 2, 2, 1)$ is equal to -1, since $n_2(\lambda) = 3$, $c_2(\lambda) = 2$ and $r_2(\lambda) = 3$, see Figure 4.

Let $N_j(m, n)$ denote the number of partitions of n with j-rank m. Garvan [71] showed that for $j \geq 2$,

$$\sum_{n=0}^{\infty} N_j(m, n) q^n = \frac{1}{(q; q)_\infty} \sum_{n=1}^{\infty} (-1)^{n-1} q^{\frac{n((2j-1)n-1)}{2} + |m|n} (1 - q^n). \quad (4.10)$$

Dixit and Yee [62] defined the j-rank moment $_j N_k(n)$ by

$$_j N_k(n) = \sum_{m=-\infty}^{\infty} m^k N_j(m, n). \quad (4.11)$$

In the notation $_j N_k(n)$, they defined $\mathrm{Spt}_j(n)$ as follows.

Definition 4.6 *For $n, j \geq 1$,*

$$\mathrm{Spt}_j(n) = np(n) - \frac{1}{2}\, _{j+1} N_2(n). \quad (4.12)$$

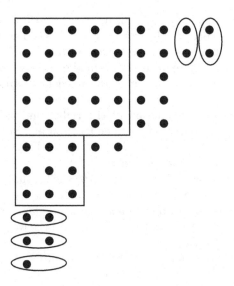

Figure 4: An illustration of 3-rank of $(9, 9, 7, 7, 7, 5, 3, 3, 2, 2, 1)$

In light of (1.4), it is easy to see that $\mathrm{Spt}_j(n)$ reduces to $\mathrm{spt}(n)$ when $j = 1$.

Dixit and Yee [62] derived the generating function of $\mathrm{Spt}_j(n)$.

Theorem 4.7 (Dixit and Yee) *For $j \geq 1$,*

$$\sum_{n=1}^{\infty} \mathrm{Spt}_j(n) q^n$$

$$= \sum_{n_j \geq 1} \sum_{n_{j-1} \geq \cdots \geq n_1 \geq 0} \frac{q^{n_j}}{(1 - q^{n_j})(q^{n_j}; q)_{\infty}} \begin{bmatrix} n_j \\ n_{j-1} \end{bmatrix} \cdots \begin{bmatrix} n_2 \\ n_1 \end{bmatrix} q^{n_1^2 + \cdots + n_{j-1}^2},$$

$$(4.13)$$

where the q-binomial coefficients or the Gaussian coefficients are defined by

$$\begin{bmatrix} n \\ k \end{bmatrix} = \frac{(q; q)_n}{(q; q)_k (q; q)_{n-k}}. \qquad (4.14)$$

Dixit and Yee also found a combinatorial interpretation of $\mathrm{Spt}_j(n)$. To give a combinatorial explanation of the right-hand side of (4.13), they introduced the k-th lower-Durfee square of a partition λ. For a partition

λ, take the largest square that fits inside the Ferrers diagram of λ starting from the lower left corner. This square is called the lower-Durfee square. If there are remaining parts above the lower-Durfee square, then take the second lower-Durfee square in the diagram above the lower-Durfee square. Repeating this process, we are led to the third lower-Durfee square, if it exists, and so on.

The combinatorial explanation of the right-hand side of (4.13) also requires a labeling of a partition, as given by Dixit and Yee. For a partition λ, let f_i denote the multiplicity of i in λ. For the f_i occurrences of i, we label these f_i parts from left to right by $1, 2, \ldots, f_i$. The labels are represented by subscripts. For instance, $(9, 8, 8, 8, 8, 6, 6, 5, 4, 4, 3)$ can be labeled as $(9_1, 8_1, 8_2, 8_3, 8_4, 6_1, 6_2, 5_1, 4_1, 4_2, 3_1)$.

Using the lower-Durfee squares and the above labeling of a partition, for a partition λ and $j \geq 1$, Dixit and Yee defined the weight of λ, denoted $W_j(\lambda)$. There are two cases:

Case 1: λ does not contain the $(j-1)$-th lower-Durfee square. Then $W_j(\lambda)$ is defined to be the sum of the labels of λ.

Case 2: λ contains the $(j-1)$-th lower-Durfee square. Then $W_j(\lambda)$ is defined to be the sum of labels of all the parts that are contained in and below the $(j-1)$-th lower-Durfee square and the label of the part just right above the $(j-1)$-th lower-Durfee square.

For example, for $\lambda = (9, 8, 8, 8, 8, 6, 6, 5, 4, 4, 3)$ and $j = 3$, we have $W_3(\lambda) = 2 + 3 + 4 + 1 + 2 + 1 + 1 + 2 + 1 = 17$, see Figure 5.

We are now ready to state the combinatorial interpretation of $\mathrm{Spt}_j(n)$.

Theorem 4.8 (Dixit and Yee) *For $j \geq 1$,*

$$\mathrm{Spt}_j(n) = \sum_{\lambda \in P(n)} W_j(\lambda).$$

Analogous to the k-th symmetrized rank moments $\eta_k(n)$ and the k-th symmetrized crank moments $\mu_k(n)$, Dixit and Yee [62] defined the k-th symmetrized j-rank function $_j\mu_k(n)$ by

$$_j\mu_k(n) = \sum_{m=-\infty}^{\infty} \binom{m + \lfloor \frac{k-1}{2} \rfloor}{k} N_j(m, n).$$

It can be checked that $_1\mu_k(n) = \mu_k(n)$ and $_2\mu_k(n) = \eta_k(n)$. By the definition (4.3) of the higher order spt-function $\mathrm{spt}_k(n)$, we see that

$$\mathrm{spt}_k(n) = {_1\mu_{2k}(n)} - {_2\mu_{2k}(n)}. \tag{4.15}$$

The generalized higher order spt-function $_j\mathrm{spt}_k$ is defined as follows.

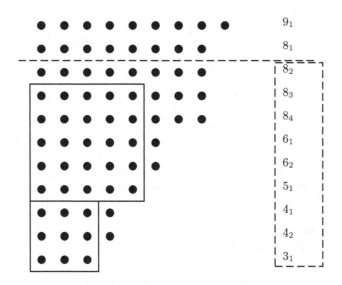

Figure 5: An illustration of weight $W_3(\lambda)$

Definition 4.9 *For $j, k \geq 1$,*

$$_j\operatorname{spt}_k(n) = {}_j\mu_{2k}(n) - {}_{j+1}\mu_{2k}(n).$$

Dixit and Yee [62] derived the generating function of $_j\operatorname{spt}_k(n)$:

Theorem 4.10 (Dixit and Yee) *For $j, k \geq 1$,*

$$\sum_{n=1}^{\infty} {}_j\operatorname{spt}_k(n)q^n = \sum_{\substack{n_k \geq \cdots \geq n_1 \geq \\ m_1 \geq \cdots \geq m_{j-1} \geq 1}} \left(\frac{q^{n_k + \cdots + n_1}(q;q)_{n_1}}{(1 - q^{n_k})^2 \cdots (1 - q^{n_1})^2 (q^{n_1+1};q)_{\infty}} \right.$$

$$\left. \times \frac{q^{m_1^2 + \cdots + m_{j-1}^2}}{(q;q)_{n_1-m_1}(q;q)_{m_1-m_2}\cdots(q;q)_{m_{j-1}}} \right). \quad (4.16)$$

They also gave a combinatorial explanation of the right-hand side of (4.16). Let λ be a partition, and let f_t denote the number of occurrences of t in λ. We shall use the same labeling of λ as given before. For a positive integer k and a part t in λ with label a, define

$$g_k(\lambda, t_a) = \binom{a + k - 1}{2k - 1}$$

$$+\sum_{r=2}^{k} \sum_{\substack{m_1,m_2,\dots,m_r\geq 1 \\ m_1+\cdots+m_r=k \\ t<t_2<\cdots<t_r\leq\lambda_1}} \binom{a+m_1-1}{2m_1-1}\binom{f_{t_2}+m_2}{2m_2}\cdots\binom{f_{t_r}+m_r}{2m_r}.$$

Definition 4.11 *For $j,k\geq 1$, define*

$$_j\omega_k(\lambda) = \sum_{t_a} g_k(\lambda, t_a), \qquad (4.17)$$

where the sum ranges over the parts that are contained in the $(j-1)$-th lower-Durfee square except for the last part, but also contains the part immediately above the $(j-1)$-th lower-Durfee square.

For example, let $\lambda = (5,5,5,3,3,2,2,2)$, $j=3$ and $k=2$. Label λ as $(5_1,5_2,5_3,3_1,3_2,2_1,2_2,2_3)$. Then

$$g_2(\lambda, 3_1) = 0 + 1 \cdot \binom{3+1}{2} = 6$$

and

$$g_2(\lambda, 3_2) = 1 + 2 \cdot \binom{3+1}{2} = 13.$$

Moreover, from (4.17) we find that

$$_3\omega_2(\lambda) = g_2(\lambda, 3_1) + g_2(\lambda, 3_2) = 6 + 13 = 19.$$

Figure 6 gives an illustration of this example.

Dixit and Yee [62] proved that $_j\,\mathrm{spt}_k(n)$ can be expressed in terms of $_j\omega_k(\lambda)$.

Theorem 4.12 (Dixit and Yee) *We have*

$$_j\,\mathrm{spt}_k(n) = \sum_{\lambda\in P(n)} {}_j\omega_k(\lambda).$$

4.3   The ospt-function of Andrews, Chan and Kim

A variation of the spt-function based on relation (1.6) was given by Andrews, Chan and Kim [15]. In view of the symmetry properties $N(-m,n) = N(m,n)$ and $M(-m,n) = M(m,n)$, it is known that

$$N_{2k+1}(n) = M_{2k+1}(n) = 0.$$

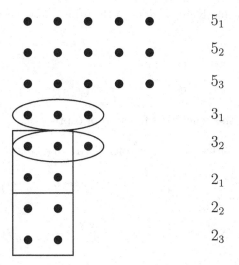

$$5_1$$
$$5_2$$
$$5_3$$
$$3_1$$
$$3_2$$
$$2_1$$
$$2_2$$
$$2_3$$

Figure 6: An illustration of $_j\omega_k(\pi)$

To avoid the trivial odd moments, Andrews, Chan and Kim [15] introduced the modified rank and crank moments $N_j^+(n)$ and $M_j^+(n)$ by considering the unilateral sums:

$$N_j^+(n) = \sum_{m \geq 0} m^j N(m, n)$$

and

$$M_j^+(n) = \sum_{m \geq 0} m^j M(m, n).$$

They proved the following inequality.

Theorem 4.13 (Andrews, Chan and Kim) *For $n, k \geq 1$,*

$$M_k^+(n) > N_k^+(n). \tag{4.18}$$

Bringmann and Mahlburg [45] proved that the above inequality (4.18) holds for any fixed positive integer k and sufficiently large n by deriving an asymptotic formula for $M_k^+(n) - N_k^+(n)$ stated in Theorem 5.2. When k is even, this inequality (4.18) is equivalent to the inequality (4.6) of Garvan between the rank moments and the crank moments. Chen, Ji and Zang [52] showed that the Andrews-Dyson-Rhoades conjecture (2.13) implies the inequality (4.18).

Andrews, Chan and Kim [15] defined the ospt-function ospt(n) as given below:

Definition 4.14 *For $n \geq 1$,*

$$\text{ospt}(n) = M_1^+(n) - N_1^+(n). \tag{4.19}$$

They obtained the generating function of ospt(n).

Theorem 4.15 (Andrews, Chan and Kim) *We have*

$$\sum_{n=0}^{\infty} \text{ospt}(n)q^n = \frac{1}{(q;q)_\infty} \sum_{i=0}^{\infty} \left(\sum_{j=0}^{\infty} q^{6i^2+8ij+2j^2+7i+5j+2}(1-q^{4i+2})(1-q^{4i+2j+3}) \right.$$

$$\left. + \sum_{j=0}^{\infty} q^{6i^2+8ij+2j^2+5i+3j+1}(1-q^{2i+1})(1-q^{4i+2j+2}) \right). \tag{4.20}$$

Andrews, Chan and Kim found a combinatorial interpretation of the right-hand side of (4.20), which leads to a combinatorial interpretation of ospt(n). In doing so, they defined even strings and odd strings of a partition.

Definition 4.16 *Let λ be a partition. A maximal consecutive sequence $(r, r-1, \ldots, s)$ in λ is called an even string of λ if it satisfies the following restrictions:*

(1) $r \geq 2s - 2$;

(2) r and s are even.

Similarly, a consecutive sequence $(r, r-1, \ldots, s)$ in λ, not necessarily maximal, is called an odd string of λ if it satisfies the following restrictions:

(1) $r + 1$ is not a part of λ;

(2) s is odd and it appears only once in λ;

(3) $r \geq 2s - 1$.

For example, the partition $\lambda = (5, 4, 4, 3, 2, 2)$ contains only one odd string $(5, 4, 3)$, and it does not contain any even string. For $\lambda = (6, 4, 4, 3, 2)$, it contains an even string $(4, 3, 2)$, but it does not contain any odd string.

Andrews, Chan and Kim [15] found a combinatorial interpretation of ospt(n).

Theorem 4.17 (Andrews, Chan and Kim) *For a partition* λ, *let* $ST(\lambda)$ *denote the total number of even strings and odd strings in* λ. *For* $n \geq 1$,

$$\text{ospt}(n) = \sum_{\lambda \in P(n)} ST(\lambda).$$

In light of Theorem 4.17, Bringmann and Mahlburg [45] proved a monotone property of $\text{ospt}(n)$ by a combinatorial argument.

Theorem 4.18 (Bringmann and Mahlburg) *For* $n \geq 1$,

$$\text{ospt}(n+1) \geq \text{ospt}(n).$$

They also noticed that $\text{ospt}(n)$ and $\text{spt}(n)$ have the same parity. This fact can be justified as follows: Since

$$M_1^+(n) = \sum_{m \geq 0} mM(m,n) \equiv \sum_{m \geq 0} m^2 M(m,n) = M_2^+(n) \pmod 2$$

and

$$N_1^+(n) = \sum_{m \geq 0} mN(m,n) \equiv \sum_{m \geq 0} m^2 N(m,n) = N_2^+(n) \pmod 2,$$

we see that

$$\text{ospt}(n) = M_1^+(n) - N_1^+(n) \equiv M_2^+(n) - N_2^+(n) = \text{spt}(n) \pmod 2.$$

With the aid of the characterization of the parity of $\text{spt}(n)$, Bringmann and Mahlburg [45] determined the parity of $\text{ospt}(n)$.

Theorem 4.19 (Bringmann and Mahlburg) *The ospt-function* $\text{ospt}(n)$ *is odd if and only if* $24n - 1 = p^{4a+1}m^2$ *for some prime* $p \equiv 23 \pmod{24}$ *and some integers* a, m, *where* $(p, m) = 1$.

Chan and Mao [49] established an upper bound and a lower bound for $\text{ospt}(n)$, leading to an asymptotic estimate of $\text{ospt}(n)$.

Theorem 4.20 (Chan and Mao) *We have*

$$\text{ospt}(n) > \frac{p(n)}{4} + \frac{N(0,n)}{2} - \frac{M(0,n)}{4} \quad \text{for } n \geq 8, \tag{4.21}$$

$$\text{ospt}(n) < \frac{p(n)}{4} + \frac{N(0,n)}{2} - \frac{M(0,n)}{4} + \frac{N(1,n)}{2} \quad \text{for } n \geq 7, \tag{4.22}$$

$$\text{ospt}(n) < \frac{p(n)}{2} \quad \text{for } n \geq 3. \tag{4.23}$$

An asymptotic estimate of $\text{ospt}(n)$ can be deduced from the bounds (4.21) and (4.22), along with an asymptotic property of $M(m,n)$ and $N(m,n)$ due to Mao [106].

Theorem 4.21 (Mao) *For any integer m, as $n \to \infty$*

$$M(m,n) \sim N(m,n) \sim \frac{\pi}{4\sqrt{6n}} p(n). \tag{4.24}$$

By (4.24), we see that as $n \to \infty$,

$$\frac{p(n)}{4} + \frac{N(0,n)}{2} - \frac{M(0,n)}{4} \sim \frac{p(n)}{4} + \frac{N(0,n)}{2} - \frac{M(0,n)}{4} + \frac{N(1,n)}{2} \sim \frac{1}{4} p(n).$$

Combining (4.21) and (4.22), we arrive at the asymptotic estimate (5.2) due to Bringmann and Mahlburg [45] as given in Section 5.

4.4 The first variation of Ahlgren, Bringmann and Lovejoy

We now turn to three variations of the spt-function based on the combinatorial definition. The first variation of the spt-function was given by Ahlgren, Bringmann and Lovejoy [5]. They defined the M2spt-function as follows.

Definition 4.22 *The function M2spt(n) is defined to be the total number of smallest parts in all partitions of n without repeated odd parts and the smallest part is even.*

For example, there are two partitions of 7 without repeated odd parts and the smallest part is even, namely,

$$(5, \mathbf{2}), (3, \mathbf{2}, \mathbf{2}).$$

So we have $\text{M2spt}(7) = 3$.

By [43, Section 7], Ahlgren, Bringmann and Lovejoy [5] derived the generating function of M2spt(n).

Theorem 4.23 (Ahlgren, Bringmann and Lovejoy) *We have*

$$\sum_{n=1}^{\infty} \text{M2spt}(n)q^n = \frac{(-q; q^2)_\infty}{(q^2; q^2)_\infty}$$

$$\times \left(\sum_{n=1}^{\infty} \frac{nq^{2n}}{1 - q^{2n}} + \sum_{\substack{n=-\infty \\ n \neq 0}}^{\infty} \frac{(-1)^n q^{2n^2+n}}{(1 - q^{2n})^2} \right). \tag{4.25}$$

Jennings-Shaffer [88] showed that the function M2spt(n) can be expressed as the difference between the symmetrized M_2-rank moments and the symmetrized residue crank moments of partitions without repeated odd parts. Let us first recall the definitions of the M_2-rank of a partition without repeated odd parts and the residue crank of a partition without repeated odd parts.

Let λ be a partition without repeated odd parts, the M_2-rank of λ was defined by Berkovich and Garvan [33] as stated below:

$$M_2\text{-rank}(\lambda) = \left\lceil \frac{\lambda_1}{2} \right\rceil - l(\lambda). \qquad (4.26)$$

The residue crank of λ was defined by Garvan and Jennings-Shaffer [76] which is related to the crank of an ordinary partition. Let $\lambda = (\lambda_1, \lambda_2, \ldots, \lambda_l)$ be a partition without repeated odd parts, define λ^e to be the ordinary partition obtained from λ by omitting odd parts of λ and dividing each even part by 2. The residue crank of λ is defined to be the crank of λ^e.

For example, let $\lambda = (11, 7, 6, 5, 4, 4, 3, 2, 2)$, then $\lambda_1 = 11$, $l(\lambda) = 9$ and $\lambda^e = (3, 2, 2, 1, 1)$. Hence the M_2-rank of λ is equal to -3 and the residue crank of λ is equal to the crank of λ^e, which equals -1.

Let $N2(m, n)$ denote the number of partitions of n without repeated odd parts such that M_2-rank is equal to m. Let $M2(m, n)$ denote the number of partitions of n without repeated odd parts such that the residue crank is equal to m. The k-th symmetrized M_2-rank moments $\eta 2_k(n)$ and the k-th symmetrized residue crank moments $\mu 2_k(n)$ of partitions without repeated odd parts were defined by Jennings-Shaffer [88] as follows:

$$\eta 2_k(n) = \sum_{m=-\infty}^{\infty} \binom{m + \lfloor \frac{k-1}{2} \rfloor}{k} N2(m, n),$$

$$\mu 2_k(n) = \sum_{m=-\infty}^{\infty} \binom{m + \lfloor \frac{k-1}{2} \rfloor}{k} M2(m, n).$$

Analogue to the relation (1.6) for spt(n), Jennings-Shaffer [88] established the following connection.

Theorem 4.24 (Jennings-Shaffer) *For* $n \geq 1$,

$$\text{M2spt}(n) = \mu 2_2(n) - \eta 2_2(n). \qquad (4.27)$$

The following congruences of M2spt(n) mod 3 and 5 were given by Garvan and Jennings-Shaffer [76].

Theorem 4.25 (Garvan and Jennings-Shaffer) *For $n \geq 0$,*

$$
\begin{aligned}
\text{M2spt}(3n + 1) &\equiv 0 \pmod{3}, \\
\text{M2spt}(5n + 1) &\equiv 0 \pmod{5}, \\
\text{M2spt}(5n + 3) &\equiv 0 \pmod{5}.
\end{aligned}
$$

Jennings-Shaffer [89] provided alternative proofs of the above congruences. Furthermore, he showed that

Theorem 4.26 (Jennings-Shaffer) *For $n \geq 0$,*

$$
\begin{aligned}
\text{M2spt}(27n + 26) &\equiv 0 \pmod{5}, \\
\text{M2spt}(125n + 97) &\equiv 0 \pmod{5}, \\
\text{M2spt}(125n + 122) &\equiv 0 \pmod{5}.
\end{aligned}
$$

Ahlgren, Bringmann and Lovejoy [5] established Ramanujan-type congruences of $\text{M2spt}(n)$ modulo powers of ℓ for any prime $\ell \geq 3$.

Theorem 4.27 (Ahlgren, Bringmann and Lovejoy) *Let $\ell \geq 3$ be a prime, and let $m, n \geq 1$.*

(i) *If $\left(\frac{-n}{\ell}\right) = 1$, then*

$$
\text{M2spt}\left((\ell^{2m}n + 1)/8\right) \equiv 0 \pmod{\ell^m}.
$$

(ii)

$$
\text{M2spt}\left((\ell^{2m+1}n + 1)/8\right) \equiv \left(\frac{2}{\ell}\right) \text{M2spt}\left((\ell^{2m-1}n + 1)/8\right) \pmod{\ell^m}.
$$

Hecke-type congruences of $\text{M2spt}(n) \bmod 2, 2^2, 2^3, 3$ and 5 have been found by Andersen [7].

Theorem 4.28 (Andersen) *Let $\ell \geq 3$ be a prime. Define $s_\ell = (\ell^2 - 1)/8$ and*

$$
\beta = \begin{cases}
1, & \text{if } \ell \equiv 3 \pmod{8}, \\
2, & \text{if } \ell \equiv 5 \pmod{8}, \\
3, & \text{if } \ell \equiv 1, 7 \pmod{8}.
\end{cases}
$$

For $t \in \{2^\beta, 3, 5\}, \ell \neq t$ and $n \geq 1$,

$$
\text{M2spt}(\ell^2 n - s_\ell) + \left(\frac{2}{\ell}\right)\left(\frac{1 - 8n}{\ell}\right) \text{M2spt}(n) + \ell \, \text{M2spt}\left((n + s_\ell)/\ell^2\right)
$$

$$\equiv \left(\frac{2}{\ell}\right)(1+\ell)\,\mathrm{M2spt}(n) \pmod{t}.$$

In analogy with the higher order spt-function $\mathrm{spt}_k(n)$, Jennings-Shaffer [89] defined the higher order function $\mathrm{M2spt}_k(n)$ in terms of the k-th symmetrized M_2-rank moments $\eta 2_k(n)$ and the k-th symmetrized residue crank moments $\mu 2_k(n)$ for partitions without repeated odd parts.

Definition 4.29 *For $k \geq 1$, define*

$$\mathrm{M2spt}_k(n) = \mu 2_{2k}(n) - \eta 2_{2k}(n).$$

Using (4.27), it is clear to see that $\mathrm{M2spt}_k(n)$ reduces to $\mathrm{M2spt}(n)$ when $k = 1$. Jennings-Shaffer [88] also obtained the generating function of $\mathrm{M2spt}_k(n)$.

Theorem 4.30 (Jennings-Shaffer) *We have*

$$\sum_{n=1}^{\infty} \mathrm{M2spt}_k(n)q^n$$

$$= \sum_{n_k \geq n_{k-1} \geq \cdots \geq n_1 \geq 1} \frac{(-q^{2n_1+1};q^2)_\infty\, q^{2n_1+2n_2+\cdots+2n_k}}{(1-q^{2n_k})^2(1-q^{2n_{k-1}})^2\cdots(1-q^{2n_1})^2(q^{2n_1+2};q^2)_\infty}.$$
$$(4.28)$$

By interpreting the right-hand side of (4.28) combinatorially, Jennings-Shaffer [88] found a combinatorial interpretation of $\mathrm{M2spt}_k(n)$. Let $P_o(n)$ denote the set of partitions of n without repeated odd parts and the s-mallest part is even. For a partition $\lambda \in P_o(n)$, assume that there are r different even parts in λ, namely,

$$2t_1 < 2t_2 < \cdots < 2t_r.$$

Let $f_j = f_j(\lambda)$ denote the frequency of the part $2t_j$ in λ. For a fixed integer $k \geq 1$, Jennings-Shaffer [88] defined $\omega_k(\lambda)$ as follows:

$$\omega_k(\lambda) = \sum_{\substack{m_1+m_2+\cdots+m_s=k \\ 1\leq s\leq k}} \binom{f_1+m_1-1}{2m_1-1} \times \sum_{2\leq j_2<j_3<\cdots<j_s} \prod_{i=2}^{s}\binom{f_{j_i}+m_i}{2m_i}.$$
$$(4.29)$$

For example, let $k = 2$ and $\lambda = (10, 10, 9, 5, 4, 3, 2, 2, 2)$ be a partition in $P_o(47)$, there are three distinct even parts in λ. Thus $r = 3$, $f_1 = 3$, $f_2 = 1$ and $f_3 = 2$. By the definition (4.29) of $\omega_k(\lambda)$, we have $\omega_2(\lambda) = 16$.

With the above notation, Jennings-Shaffer [88] found a combinatorial interpretation of $\mathrm{M2spt}_k(n)$.

Theorem 4.31 (Jennings-Shaffer) *For $n \geq 1$,*

$$M2spt_k(n) = \sum_{\lambda \in P_o(n)} \omega_k(\lambda). \qquad (4.30)$$

Jennings-Shaffer [89] also obtained the following congruences of $M2spt_2(n)$.

Theorem 4.32 (Jennings-Shaffer) *For $n \geq 1$,*

$$M2spt_2(n) \equiv 0 \pmod{3}, \; if \; n \equiv 0 \pmod{9},$$

$$M2spt_2(n) \equiv 0 \pmod{5}, \; if \; n \equiv 0 \pmod{5},$$

$$M2spt_2(n) \equiv 0 \pmod{5}, \; if \; n \equiv 1 \pmod{5},$$

$$M2spt_2(n) \equiv 0 \pmod{5}, \; if \; n \equiv 3 \pmod{5}.$$

4.5 The second variation of Bringmann, Lovejoy and Osburn

The second variation of the spt-function was due to Bringmann, Lovejoy and Osburn [42], which is defined on overpartitions. Recall that Corteel and Lovejoy [56] defined an overpartition of n as a partition of n in which the first occurrence of a part may be overlined. Bringmann, Lovejoy and Osburn [42] introduced three spt-type functions.

Definition 4.33 (Bringmann, Lovejoy and Osburn)

(1) *The function* $\overline{spt}(n)$ *is defined to be the total number of smallest parts in all overpartitions of n.*

(2) *The function* $\overline{spt\,1}(n)$ *is defined to be the total number of smallest parts in all overpartitions of n with the smallest part being odd.*

(3) *The function* $\overline{spt\,2}(n)$ *is defined to be the total number of smallest parts in all overpartitions of n with the smallest part being even.*

For example, there are 14 overpartitions of 4:

$$(4) \quad (\bar{4}) \quad (3,1) \quad (\bar{3},1) \quad (3,\bar{1}) \quad (\bar{3},\bar{1}) \quad (2,2),$$
$$(\bar{2},2) \quad (2,1,1) \quad (\bar{2},1,1) \quad (2,\bar{1},1) \quad (\bar{2},\bar{1},1) \quad (1,1,1,1) \quad (\bar{1},1,1,1).$$

We have $\overline{spt}(4) = 26$, $\overline{spt\,1}(4) = 20$ and $\overline{spt\,2}(4) = 6$.

Analogous to the relation (1.6) for the spt-function, the functions $\overline{spt}(n)$ and $\overline{spt\,2}(n)$ can also be expressed as the differences of the rank and the crank moments of overpartitions. The definitions of the rank and the crank moments of overpartitions are based on the two definitions of the rank of an overpartition and the two definitions of the crank of an overpartition. Although there are four possibilities, only two of them have been studied.

For an overpartition λ, there are two kinds of ranks. One is called the D-rank introduced by Lovejoy [104] and the other is called the M_2-rank introduced by Lovejoy [105]. The D-rank of λ is defined as the largest part minus the number of parts. To define the M_2-rank, let λ_o denote the partition consisting of non-overlined odd parts of λ. Then $M_2\text{-rank}(\lambda)$ can be defined as follows:

$$M_2\text{-rank}(\lambda) = \left\lceil \frac{\lambda_1}{2} \right\rceil - l(\lambda) + l(\lambda_o) - \chi(\lambda),$$

where $\chi(\lambda) = 1$ if the largest part of λ is odd and non-overlined and $\chi(\lambda) = 0$ otherwise.

For example, for an overpartition $\lambda = (\bar{9}, 9, 7, \bar{6}, 5, 5, \bar{4}, 3, 2, \bar{1})$, we see that D-rank$(\lambda) = 9 - 10 = -1$. Moreover, since $\lambda_o = (9, 7, 5, 5, 3)$ and $\chi(\lambda) = 0$, we have $M_2\text{-rank}(\lambda) = 0$.

Bringmann, Lovejoy and Osburn [42] defined the first and second residue crank of an overpartition. The first residue crank of an overpartition is defined as the crank of the partition consisting of non-overlined parts. The second residue crank is defined as the crank of the subpartition consisting of all the even non-overlined parts divided by two.

For example, for $\lambda = (\bar{9}, 9, 7, \bar{6}, 5, 5, \bar{4}, 4, 3, 2, \bar{1})$, the partition consisting of non-overlined parts of λ is $(9, 7, 5, 5, 4, 3, 2)$. The first residue crank of λ is 9. The partition formed by even non-overlined parts of λ is $(4, 2)$. So the second residue crank of λ is equal to the crank of $(2, 1)$, which is equal to 0.

We are now in a position to present the definitions of the rank and the crank moments of overpartitions. Let $\overline{N}(m, n)$ denote the number of overpartitions of n with the D-rank m, and let $\overline{N2}(m, n)$ denote the number of overpartitions of n with the M_2-rank m. Notice that there are two kinds of ranks of overpartitions. Consequently, there are two possibilities to define the rank moments of overpartitions. The two rank moments are defined as follows:

$$\overline{N}_k(n) = \sum_{m=-\infty}^{\infty} m^k \overline{N}(m, n), \qquad (4.31)$$

$$\overline{N2}_k(n) = \sum_{m=-\infty}^{\infty} m^k \overline{N2}(m, n). \qquad (4.32)$$

Similarly, let $\overline{M}(m,n)$ denote the number of overpartitions of n with the first residue crank m and let $\overline{M2}(m,n)$ denote the number of overpartitions of n with the second residue crank m. The two crank moments are defined by

$$\overline{M}_k(n) \;=\; \sum_{m=-\infty}^{\infty} m^k \overline{M}(m,n), \qquad (4.33)$$

$$\overline{M2}_k(n) \;=\; \sum_{m=-\infty}^{\infty} m^k \overline{M2}(m,n). \qquad (4.34)$$

Bringmann, Lovejoy and Osburn [42] deduced the following relations on $\overline{\mathrm{spt}}(n)$ and $\overline{\mathrm{spt}\,2}(n)$.

Theorem 4.34 (Bringmann, Lovejoy and Osburn) *For $n \geq 1$,*

$$\overline{\mathrm{spt}}(n) \;=\; \overline{M}_2(n) - \overline{N}_2(n), \qquad (4.35)$$

$$\overline{\mathrm{spt}\,2}(n) \;=\; \overline{M2}_2(n) - \overline{N2}_2(n). \qquad (4.36)$$

In light of Theorem 4.34, Bringmann, Lovejoy and Osburn [42] proved the following congruences:

Theorem 4.35 (Bringmann, Lovejoy and Osburn) *For $n \geq 1$,*

$$\overline{\mathrm{spt}\,2}(n) \equiv \overline{\mathrm{spt}\,2}(n) \;\equiv\; 0 \pmod 3, \; \textit{if } n \equiv 0,1 \pmod 3, \quad (4.37)$$

$$\overline{\mathrm{spt}}(n) \;\equiv\; 0 \pmod 3, \; \textit{if } n \equiv 0 \pmod 3, \qquad (4.38)$$

$$\overline{\mathrm{spt}\,2}(n) \;\equiv\; 0 \pmod 5, \; \textit{if } n \equiv 3 \pmod 5, \qquad (4.39)$$

$$\overline{\mathrm{spt}\,1}(n) \;\equiv\; 0 \pmod 5, \; \textit{if } n \equiv 0 \pmod 5. \qquad (4.40)$$

Moreover, if $\ell \geq 5$ is a prime, then the following congruence holds:

$$\overline{\mathrm{spt}\,1}(\ell^2 n) + \left(\frac{-n}{\ell}\right) \overline{\mathrm{spt}\,1}(n) + \ell \overline{\mathrm{spt}\,1}\left(n/\ell^2\right) \equiv (\ell+1)\overline{\mathrm{spt}\,1}(n) \pmod 3.$$

An alternative proof of the congruence (4.37) was given by Jennings-Shaffer [87]. The combinatorial interpretations of the congruences (4.37)–(4.40) were given by Garvan and Jennings-Shaffer [76]. Ahlgren, Bringmann and Lovejoy [5] derived Ramanujan-type congruences of $\overline{\mathrm{spt}\,1}(n)$ modulo powers of a prime ℓ, which are similar to the Ramanujan-type congruences of $\mathrm{spt}(n)$ modulo powers of a prime ℓ.

Theorem 4.36 (Ahlgren, Bringmann and Lovejoy) *Let $\ell \geq 3$ be a prime, and let $m, n \geq 1$.*

(1) *If $\left(\frac{-n}{\ell}\right) = 1$, then*

$$\overline{\mathrm{spt}\,1}(\ell^{2m}n) \equiv 0 \pmod{\ell^m}.$$

(2)

$$\overline{\mathrm{spt}\,1}(\ell^{2m+1}n) \equiv \overline{\mathrm{spt}\,1}(\ell^{2m-1}n) \pmod{\ell^m}.$$

Andersen [7] obtained Hecke-type congruences of $\overline{\mathrm{spt}\,1}(n)$ mod 2^6, 2^7, 2^8, 3 and 5.

Theorem 4.37 (Andersen) *Let $\ell \geq 3$ be a prime, and define*

$$\beta = \begin{cases} 6, & \text{if } \ell \equiv 3 \pmod 8, \\ 7, & \text{if } \ell \equiv 5, 7 \pmod 8, \\ 8, & \text{if } \ell \equiv 1 \pmod 8. \end{cases}$$

For $t \in \{2^\beta, 3, 5\}$, $\ell \neq t$ and $n \geq 1$,

$$\overline{\mathrm{spt}\,1}(\ell^2 n) + \left(\frac{-n}{\ell}\right)\overline{\mathrm{spt}\,1}(n) + \ell\overline{\mathrm{spt}\,1}(n/\ell^2) \equiv (1+\ell)\overline{\mathrm{spt}\,1}(n) \pmod t.$$

It is readily seen that $\overline{\mathrm{spt}\,1}(n)$, $\overline{\mathrm{spt}\,2}(n)$ and $\overline{\mathrm{spt}}(n)$ are all even. Congruences of these functions modulo 4 were investigated by Garvan and Jennings-Shaffer [76].

Theorem 4.38 (Garvan and Jennings-Shaffer) *For $n \geq 1$,*

(1) $\overline{\mathrm{spt}}(n) \equiv 2 \pmod 4$ *if and only if n is a square or twice a square;*

(2) $\overline{\mathrm{spt}\,1}(n) \equiv 2 \pmod 4$ *if and only if n is an odd square;*

(3) $\overline{\mathrm{spt}\,2}(n) \equiv 2 \pmod 4$ *if and only if n is an even square or twice a square.*

Moreover, they introduced a statistic $\overline{\mathrm{sptcrank}}$ defined on a marked overpartition, which leads to combinatorial interpretations of the above congruences.

The following recurrence relation of $\overline{\mathrm{spt}\,1}(n)$ was given by Ahlgren and Andersen [2].

Theorem 4.39 (Ahlgren and Andersen) *Let*

$$s(n) = \sum_{d|n} \min\left(d, \frac{n}{d}\right).$$

For $n > 0$,

$$\sum_k (-1)^k \overline{\mathrm{spt}\,1}(n - k^2) = b(n),$$

where

$$b(n) = \begin{cases} 2s(n), & \text{if } n \text{ is odd,} \\ -4s(n/4), & \text{if } n \equiv 0 \pmod 4, \\ 0, & \text{if } n \equiv 2 \pmod 4. \end{cases}$$

In view of the symmetry properties $\overline{N}(-m, n) = \overline{N}(m, n)$ and $\overline{M}(-m, n) = \overline{M}(m, n)$, we see that

$$\overline{N}_{2k+1}(n) = \overline{M}_{2k+1}(n) = 0.$$

Similarly, to avoid the trivial odd moments, Andrews, Chan, Kim and Osburn [16] introduced the modified rank and crank moments $\overline{N}_k^+(n)$ and $\overline{M}_k^+(n)$ for overpartitions:

$$\overline{N}_k^+(n) = \sum_{m \geq 1} m^k \overline{N}(m, n) \tag{4.41}$$

and

$$\overline{M}_k^+(n) = \sum_{m \geq 1} m^k \overline{M}(m, n). \tag{4.42}$$

They defined the ospt-function $\overline{\mathrm{ospt}}(n)$ for overpartitions which is in the spirit of the ospt-function $\mathrm{ospt}(n)$ for ordinary partitions.

Definition 4.40 *For $n \geq 1$,*

$$\overline{\mathrm{ospt}}(n) = \overline{M}_1^+(n) - \overline{N}_1^+(n). \tag{4.43}$$

Andrews, Chan, Kim and Osburn [16] defined even strings and odd strings of an overpartition, and provided a combinatorial interpretation of $\overline{\mathrm{ospt}}(n)$.

Jennings-Shaffer [88] defined the higher order spt-functions for overpartitions by using the k-th symmetrized rank and crank moments for overpartitions. There are two symmetrized rank moments for overpartitions:

$$\overline{\eta}_k(n) = \sum_{m=-n}^{n} \binom{m + \lfloor \frac{k-1}{2} \rfloor}{k} \overline{N}(m, n) \tag{4.44}$$

and

$$\overline{\eta 2}_k(n) = \sum_{m=-n}^{n} \binom{m + \lfloor \frac{k-1}{2} \rfloor}{k} \overline{N2}(m, n). \tag{4.45}$$

There are also two symmetrized crank moments for overpartitions:

$$\overline{\mu}_k(n) = \sum_{m=-n}^{n} \binom{m + \lfloor \frac{k-1}{2} \rfloor}{k} \overline{M}(m, n) \tag{4.46}$$

and

$$\overline{\mu 2}_k(n) = \sum_{m=-n}^{n} \binom{m + \lfloor \frac{k-1}{2} \rfloor}{k} \overline{M2}(m, n). \tag{4.47}$$

The two higher order spt-functions for overpartitions are defined as follows.

Definition 4.41 *For $k \geq 1$,*

$$\overline{\text{spt}}_k(n) = \overline{\mu}_{2k}(n) - \overline{\eta}_{2k}(n), \tag{4.48}$$

$$\overline{\text{spt} 2}_k(n) = \overline{\mu 2}_{2k}(n) - \overline{\eta 2}_{2k}(n). \tag{4.49}$$

Using Bailey pairs, Jennings-Shaffer [88] obtained the generating functions of $\overline{\text{spt}}_k(n)$ and $\overline{\text{spt} 2}_k(n)$.

Theorem 4.42 (Jennings-Shaffer) *For $k \geq 1$,*

$$\sum_{n=1}^{\infty} \overline{\text{spt}}_k(n) q^n$$

$$= \sum_{n_k \geq n_{k-1} \geq \cdots \geq n_1 \geq 1} \frac{q^{n_1 + n_2 + \cdots + n_k}(-q^{n_1+1}; q)_\infty}{(1 - q^{n_k})^2(1 - q^{n_{k-1}})^2 \cdots (1 - q^{n_1})^2(q^{n_1+1}; q)_\infty}, \tag{4.50}$$

$$\sum_{n=1}^{\infty} \overline{\text{spt} 2}_k(n) q^n$$

$$= \sum_{n_k \geq n_{k-1} \geq \cdots \geq n_1 \geq 1} \frac{q^{2n_1 + 2n_2 + \cdots + 2n_k}(-q^{2n_1+1}; q)_\infty}{(1 - q^{2n_k})^2(1 - q^{2n_{k-1}})^2 \cdots (1 - q^{2n_1})^2(q^{2n_1+1}; q)_\infty}. \tag{4.51}$$

By interpreting the right-hand sides of (4.50) and (4.51) based on vector partitions, Jennings-Shaffer found combinatorial explanations of $\overline{\text{spt}}_k(n)$ and $\overline{\text{spt}\,2}_k(n)$.

4.6   The third variation of Andrews, Dixit and Yee

The third variation of the spt-function was introduced by Andrews, Dixit and Yee [18]. Let $p_\omega(n)$ denote the number of partitions of n in which each odd part is less than twice the smallest part. They defined $\text{spt}_\omega(n)$ as follows.

Definition 4.43 *The function* $\text{spt}_\omega(n)$ *is defined to be the number of s-mallest parts in the partitions enumerated by* $p_\omega(n)$.

For example, for $n = 4$, there are four partitions counted by $p_\omega(4)$, namely,

$$(4) \quad (2,2) \quad (2,1,1) \quad (1,1,1,1).$$

We have $p_\omega(4) = 4$ and $\text{spt}_\omega(4) = 9$.

They derived the generating function of $\text{spt}_\omega(n)$.

Theorem 4.44 (Andrews, Dixit and Yee) *We have*

$$\sum_{n=1}^{\infty} \text{spt}_\omega(n)q^n$$

$$= \frac{1}{(q^2;q^2)_\infty} \sum_{n=1}^{\infty} \frac{nq^n}{1-q^n} + \frac{1}{(q^2;q^2)_\infty} \sum_{n=1}^{\infty} \frac{(-1)^n(1+q^{2n})q^{n(3n+1)}}{(1-q^{2n})^2}. \quad (4.52)$$

Using the above generating function, Andrews, Dixit and Yee [18] proved the following congruences of $\text{spt}_\omega(n)$.

Theorem 4.45 (Andrews, Dixit and Yee) *For* $n \geq 0$,

$$\text{spt}_\omega(5n+3) \quad \equiv \quad 0 \pmod{5}, \quad (4.53)$$

$$\text{spt}_\omega(10n+7) \quad \equiv \quad 0 \pmod{5}, \quad (4.54)$$

$$\text{spt}_\omega(10n+9) \quad \equiv \quad 0 \pmod{5}. \quad (4.55)$$

Employing the generating function (4.52), Wang [131] derived the generating function of $\text{spt}_\omega(2n+1)$.

Theorem 4.46 (Wang) *We have*

$$\sum_{n=0}^{\infty} \operatorname{spt}_\omega(2n+1)q^n = \frac{(q^2;q^2)_\infty^8}{(q;q)_\infty^5}. \tag{4.56}$$

Wang [131] also posed two conjectures on congruences of $\operatorname{spt}_\omega(n)$ modulo arbitrary powers of 5.

Conjecture 4.47 (Wang) *For $k \geq 1$ and $n \geq 0$,*

$$\operatorname{spt}_\omega\left(2\cdot 5^{2k-1}n + \frac{7\cdot 5^{2k-1}+1}{12}\right) \equiv 0 \pmod{5^{2k-1}}.$$

Conjecture 4.48 (Wang) *For $k \geq 1$ and $n \geq 0$,*

$$\operatorname{spt}_\omega\left(2\cdot 5^{2k}n + \frac{11\cdot 5^{2k}+1}{12}\right) \equiv 0 \pmod{5^{2k}}.$$

Jang and Kim [86] obtained a congruence of $\operatorname{spt}_\omega(n)$ via the mock modularity of its generating function.

Theorem 4.49 (Jang and Kim) *Let $\ell \geq 5$ be a prime, and let j, m and n be positive integers with $\left(\frac{n}{\ell}\right) = -1$. If m is sufficiently large, then there are infinitely many primes $Q \equiv -1 \pmod{576\ell^j}$ satisfying*

$$\operatorname{spt}_\omega\left(\frac{Q^3\ell^m n + 1}{12}\right) \equiv 0 \pmod{\ell^j}. \tag{4.57}$$

An overpartition analogue of the function $\operatorname{spt}_\omega(n)$ was defined by Andrews, Dixit, Schultz and Yee [17].

Definition 4.50 *The function $\overline{\operatorname{spt}}_\omega(n)$ is defined to be the number of smallest parts in the overpartitions of n in which the smallest part is always overlined and all odd parts are less than twice the smallest part.*

They obtained the generating function of $\overline{\operatorname{spt}}_\omega(n)$.

Theorem 4.51 (Andrews, Dixit, Schultz and Yee) *We have*

$$\sum_{n=1}^{\infty} \overline{\operatorname{spt}}_\omega(n)q^n = \frac{(-q^2;q^2)_\infty}{(q^2;q^2)_\infty}\sum_{n=1}^{\infty}\frac{nq^n}{(1-q^n)} + 2\frac{(-q^2;q^2)_\infty}{(q^2;q^2)_\infty}\sum_{n=1}^{\infty}\frac{(-1)^n q^{2n(n+1)}}{(1-q^{2n})^2}. \tag{4.58}$$

Based on the generating function (4.58), they derived the following congruences of $\overline{\mathrm{spt}}_\omega(n)$ mod 3, 5 and 6.

Theorem 4.52 (Andrews, Dixit, Schultz and Yee) *For $n \geq 0$,*

$$\overline{\mathrm{spt}}_\omega(3n) \equiv 0 \pmod{3},$$
$$\overline{\mathrm{spt}}_\omega(3n+2) \equiv 0 \pmod{3},$$
$$\overline{\mathrm{spt}}_\omega(10n+6) \equiv 0 \pmod{5},$$
$$\overline{\mathrm{spt}}_\omega(6n+5) \equiv 0 \pmod{6}.$$

They also characterized the parity of $\overline{\mathrm{spt}}_\omega(n)$.

Theorem 4.53 (Andrews, Dixit, Schultz and Yee) *For $n \geq 1$, $\overline{\mathrm{spt}}_\omega(n)$ is odd if and only if $n = k^2$ or $2k^2$ for some $k \geq 1$.*

Moreover, they found a congruence of $\overline{\mathrm{spt}}_\omega(n)$ modulo 4.

Theorem 4.54 (Andrews, Dixit, Schultz and Yee) *For $n \geq 1$,*

$$\overline{\mathrm{spt}}_\omega(7n) \equiv \overline{\mathrm{spt}}_\omega(n/7) \pmod{4},$$

where we adopt the convention that $\overline{\mathrm{spt}}_\omega(x) = 0$ if x is not a positive integer.

By (4.52), Wang [131] obtained the generating function of $\overline{\mathrm{spt}}_\omega(2n+1)$.

Theorem 4.55 (Wang) *We have*

$$\sum_{n=0}^{\infty} \overline{\mathrm{spt}}_\omega(2n+1)q^n = \frac{(q^2;q^2)_\infty^9}{(q;q)_\infty^6}. \tag{4.59}$$

In light of (4.59), Wang derived the following congruences of $\overline{\mathrm{spt}}_\omega(n)$.

Theorem 4.56 (Wang) *For $n \geq 0$,*

$$\overline{\mathrm{spt}}_\omega(8n+7) \equiv 0 \pmod{4},$$
$$\overline{\mathrm{spt}}_\omega(6n+5) \equiv 0 \pmod{9},$$
$$\overline{\mathrm{spt}}_\omega(18n+r) \equiv 0 \pmod{9}, \qquad \text{for } r = 9 \text{ or } 15,$$
$$\overline{\mathrm{spt}}_\omega(22n+r) \equiv 0 \pmod{11}, \qquad \text{for } r = 7, 11, 13, 17, 19, \text{ or } 21,$$
$$\overline{\mathrm{spt}}_\omega(162n+r) \equiv 0 \pmod{27}, \qquad \text{for } r = 81 \text{ or } 135.$$

There are other variations of the spt-function, and we just mention the main ideas of these variations. Jennings-Shaffer [90–92] introduced several spt-type functions arising from Bailey pairs and derived several Ramanujan-type congruences. Garvan and Jennings-Shaffer [78] discovered more spt-type functions and found some congruences of these spt-type functions. Patkowski [113–115] also defined several spt-type functions based on Bailey pairs. Furthermore, Patkowski obtained generating functions and congruences of these functions. Sarma, Reddy, Gunakala and Comissiong [127] defined a more general function, in the notation $\text{spt}_i(n)$, as the total number of the i-th smallest part in all partitions of n.

5 Asymptotic properties

In this section, we present asymptotic formulas for the spt-function and its variations. By applying the circle method to the second symmetrized rank moment $\eta_2(n)$, Bringmann [39] obtained an asymptotic expression of the spt-function $\text{spt}(n)$.

Theorem 5.1 (Bringmann) *As $n \to \infty$,*

$$\text{spt}(n) \sim \frac{\sqrt{6}}{\pi}\sqrt{n}p(n) \sim \frac{1}{2\sqrt{2}\pi\sqrt{n}}e^{\pi\sqrt{\frac{2n}{3}}}. \tag{5.1}$$

The above formula also follows from an asymptotic estimate of the difference of the positive rank moments and the positive crank moments, due to Bringmann and Mahlburg [45].

Theorem 5.2 (Bringmann and Mahlburg) *For $r \geq 1$, as $n \to \infty$,*

$$M_r^+(n) - N_r^+(n) \sim \delta_r n^{\frac{r}{2}-\frac{3}{2}}e^{\pi\sqrt{\frac{2n}{3}}},$$

where

$$\delta_r = r!\zeta(r-2)\left(1 - 2^{3-r}\right)\frac{6^{\frac{r-1}{2}}}{4\sqrt{3}\pi^{r-1}}.$$

Using Hardy and Ramanujan's asymptotic formula

$$p(n) \sim \frac{1}{4\sqrt{3}n}e^{\pi\sqrt{\frac{2n}{3}}}, \quad \text{as} \quad n \to \infty,$$

the $r = 2$ case of Theorem 5.2 implies Theorem 5.1, since

$$\text{spt}(n) = M_2^+(n) - N_2^+(n).$$

Bringmann and Mahlburg [45] pointed out that for $r = 1$, Theorem 5.2 leads to an asymptotic formula for $\text{ospt}(n)$, as defined in (4.19).

Theorem 5.3 (Bringmann and Mahlburg) *As* $n \to \infty$,

$$\mathrm{ospt}(n) \sim \frac{p(n)}{4} \sim \frac{1}{16\sqrt{3}n} e^{\pi \sqrt{\frac{2n}{3}}}. \tag{5.2}$$

Eichhorn and Hirschhorn [66] provided an alternative proof of Theorem 5.1. In fact, they showed that

$$\frac{\mathrm{spt}(n)}{p(n)} \sim \frac{\sqrt{6}}{\pi}\sqrt{n}, \quad \text{as} \quad n \to \infty. \tag{5.3}$$

Let λ be a partition of n, define $n_s(\lambda)$ to be the number of smallest parts of λ. It is clear that the left-hand side of (5.3) can be viewed as the mean of the statistic $n_s(\lambda)$ over all partitions of n. Eichhorn and Hirschhorn [66] obtained formulas for the mean and the standard deviation of $n_s(\lambda)$.

Theorem 5.4 (Eichhorn and Hirschhorn) *As* $n \to \infty$, *the statistic* $n_s(\lambda)$ *is distributed roughly as a negative exponential, with mean*

$$\mu = \frac{\sqrt{6}}{\pi}\sqrt{n} + \frac{3}{\pi^2} + O\left(\frac{1}{\sqrt{n}}\right) \tag{5.4}$$

and standard derivation

$$\sigma = \frac{\sqrt{6}}{\pi}\sqrt{n} - \frac{1}{4} + O\left(\frac{1}{\sqrt{n}}\right). \tag{5.5}$$

An asymptotic formula with a power saving error term for $\mathrm{spt}(n)$ has been obtained by Banks, Barquero-Sanchez, Masri, Sheng [31] based on an asymptotic formula for $p(n)$ due to Masri [107].

In analogy with the explicit formula for $p(n)$ due to Rademacher [118–120], Ahlgren and Andersen [3] obtained an exact expression for the spt-function.

Theorem 5.5 (Ahlgren and Andersen) *For* $n \geq 1$,

$$\mathrm{spt}(n) = \frac{\pi}{6}(24n-1)^{\frac{1}{4}} \sum_{c=1}^{\infty} \frac{A_c(n)}{c} (I_{1/2} - I_{3/2})\left(\frac{\pi\sqrt{24n-1}}{6c}\right),$$

where I_ν *is the I-Bessel function,* $A_c(n)$ *is the Kloosterman sum*

$$A_c(n) = \sum_{\substack{d \mod c \\ (d,c)=1}} e^{\pi i s(d,c) - 2i\pi \frac{dn}{c}},$$

and $s(d,c)$ is the Dedekind sum

$$s(d,c) = \sum_{r=1}^{c-1} \frac{r}{c}\left(\frac{dr}{c} - \left\lfloor \frac{dr}{c} \right\rfloor - \frac{1}{2}\right).$$

Asymptotic properties of generalizations and variations of the spt-function have also been well-studied. Recall that the higher order spt-function $\text{spt}_k(n)$ introduced by Garvan is defined in (4.3). Its asymptotic property was first conjectured by Bringmann and Mahlburg [44], and then confirmed by Bringmann, Mahlburg and Rhoades [46].

Theorem 5.6 (Bringmann, Mahlburg and Rhoades) *As* $n \to \infty$,

$$\text{spt}_k(n) \sim \beta_{2k} n^{k-\frac{1}{2}} p(n),$$

where $\beta_{2k} \in \frac{\sqrt{6}}{\pi}\mathbb{Q}$ *is positive.*

The following asymptotic formula for $\text{Spt}_j(n)$, as defined in (4.12), is due to Rhoades [123].

Theorem 5.7 (Rhoades) *As* $n \to \infty$,

$$\text{Spt}_j(n) = \frac{j}{2\pi\sqrt{2n}} e^{\pi\sqrt{\frac{2n}{3}}}(1 + o_j(1)).$$

Waldherr [130] obtained an asymptotic property of the j-rank moment $_jN_k(n)$ defined in (4.11).

Theorem 5.8 (Waldherr) *For* $1 \le j \le 12$, *as* $n \to \infty$,

$$_jN_{2k}(n) \sim 2\sqrt{3}(-1)^k B_{2k}\left(\frac{1}{2}\right)(24n)^{k-1}e^{\pi\sqrt{\frac{2n}{3}}}, \tag{5.6}$$

where $B_r(\cdot)$ *is a Bernoulli polynomial. Furthermore,*

$$_{j-1}N_{2k}(n) - {}_jN_{2k}(n) \sim \sqrt{3}\frac{(2k)!}{(2k-2)!}(-1)^{k+1}B_{2k-2}(24n)^{k-\frac{3}{2}}e^{\pi\sqrt{\frac{2n}{3}}}. \tag{5.7}$$

In particular, $_{j-1}N_{2k}(n) > {}_jN_{2k}(n)$ *for all sufficiently large* n.

Kim, Kim and Seo [96] derived an asymptotic expression for $\overline{\text{ospt}}(n)$, as defined in (4.43).

Theorem 5.9 (Kim, Kim and Seo) *As $n \to \infty$,*

$$\overline{\text{ospt}}(n) \sim \frac{1}{64n} e^{\pi\sqrt{n}} \sim \frac{\bar{p}(n)}{8}, \tag{5.8}$$

where $\bar{p}(n)$ denotes the number of overpartitions of n.

The above theorem is a consequence of an asymptotic formula for the difference of the modified rank and crank moments for overpartitions due to Rolon [125].

Theorem 5.10 (Rolon) *As $n \to \infty$,*

$$\overline{M}_r^+(n) - \overline{N}_r^+(n) \sim \delta_r n^{\frac{r}{2} - \frac{3}{2}} e^{\pi\sqrt{n}}, \tag{5.9}$$

where

$$\delta_r = r! \pi^{-r+1} 2^{r-5} \zeta(r-2) \left(1 - 2^{3-r}\right).$$

Combining (4.43) and (5.9) with $r = 1$, we arrive at (5.8).

6   Conjectures on inequalities

In this section, we pose some conjectures on inequalities on the spt-function, which are reminiscent of inequalities on $p(n)$. We first state some results and conjectures on $p(n)$. Then we present corresponding conjectures on $\text{spt}(n)$.

Recall that a sequence $\{a_n\}_{n \geq 0}$ is called log-concave if for $n \geq 1$,

$$a_n^2 - a_{n-1}a_{n+1} \geq 0. \tag{6.1}$$

It was conjectured in [50] that the partition function $p(n)$ is log-concave for $n \geq 26$, that is, (6.1) is true for $p(n)$ when $n \geq 26$. DeSalvo and Pak [58] confirmed this conjecture by using the Hardy-Ramanujan-Rademacher formula for $p(n)$ and Lehmer's error bound.

Theorem 6.1 (DeSalvo and Pak) *For $n \geq 26$,*

$$p(n)^2 > p(n-1)p(n+1). \tag{6.2}$$

They also proved the following inequalities conjectured in [50].

Theorem 6.2 (DeSalvo and Pak) *For $n \geq 2$,*

$$\frac{p(n-1)}{p(n)} \left(1 + \frac{1}{n}\right) > \frac{p(n)}{p(n+1)}. \tag{6.3}$$

Theorem 6.3 (DeSalvo and Pak) *For $n > m > 1$,*

$$p(n)^2 \geq p(n-m)p(n+m). \tag{6.4}$$

DeSalvo and Pak further proved that the term $(1 + 1/n)$ in (6.3) can be improved to $(1 + O(n^{-3/2}))$.

Theorem 6.4 (DeSalvo and Pak) *For $n \geq 7$,*

$$\frac{p(n-1)}{p(n)}\left(1 + \frac{240}{(24n)^{3/2}}\right) > \frac{p(n)}{p(n+1)}. \tag{6.5}$$

DeSalvo and Pak [58] conjectured that the coefficient of $1/n^{3/2}$ in the inequality (6.5) can be improved to $\pi/\sqrt{24}$, which was proved by Chen, Wang and Xie [55].

Theorem 6.5 (Chen, Wang and Xie) *For $n \geq 45$,*

$$\frac{p(n-1)}{p(n)}\left(1 + \frac{\pi}{\sqrt{24}n^{3/2}}\right) > \frac{p(n)}{p(n+1)}. \tag{6.6}$$

Bessenrodt and Ono [37] obtained an inequality on $p(n)$.

Theorem 6.6 (Bessenrodt and Ono) *If a, b are integers with $a, b > 1$ and $a + b > 8$, then*

$$p(a)p(b) \geq p(a+b), \tag{6.7}$$

where the equality can occur only if $\{a, b\} = \{2, 7\}$.

We now turn to conjectures on spt(n).

Conjecture 6.7 *For $n \geq 36$,*

$$\text{spt}(n)^2 > \text{spt}(n-1)\,\text{spt}(n+1). \tag{6.8}$$

Conjecture 6.8 *For $n \geq 13$,*

$$\frac{\text{spt}(n-1)}{\text{spt}(n)}\left(1 + \frac{1}{n}\right) > \frac{\text{spt}(n)}{\text{spt}(n+1)}. \tag{6.9}$$

Like the case for $p(n)$, the term $(1 + 1/n)$ in Conjecture 6.8 can be sharpened to $(1 + O(n^{-3/2}))$.

Conjecture 6.9 *For $n \geq 73$,*

$$\frac{\text{spt}(n-1)}{\text{spt}(n)} \left(1 + \frac{\pi}{\sqrt{24}n^{3/2}}\right) > \frac{\text{spt}(n)}{\text{spt}(n+1)}. \tag{6.10}$$

The following conjectures are analogous to (6.4) and (6.7).

Conjecture 6.10 *For $n > m > 1$,*

$$\text{spt}(n)^2 > \text{spt}(n-m)\,\text{spt}(n+m). \tag{6.11}$$

Conjecture 6.11 *If a, b are integers with $a, b > 1$ and $(a, b) \neq (2, 2)$ or $(3, 3)$, then*

$$\text{spt}(a)\,\text{spt}(b) > \text{spt}(a + b). \tag{6.12}$$

Beyond quadratic inequalities, we observe that many combinatorial sequences including $\{p(n)\}_{n \geq 1}$ and $\{\text{spt}(n)\}_{n \geq 1}$ seem to satisfy higher order inequalities except for a few terms at the beginning. Notice that $I(a_0, a_1, a_2) = a_1^2 - a_0 a_2$ is an invariant of the quadratic binary form

$$a_2 x^2 + 2a_1 xy + a_0 y^2.$$

For a sequence $a_0, a_1, a_2, \ldots$ of indeterminates, let

$$I_{n-1}(a_0, a_1, a_2) = I(a_{n-1}, a_n, a_{n+1}) = a_n^2 - a_{n-1}a_{n+1}.$$

Then Conjecture 6.7 says that for $a_n = \text{spt}(n)$, $I_{n-1}(a_0, a_1, a_2) > 0$ holds when $n \geq 36$.

This phenomenon occurs for other invariants as well. For the background on the theory of invariants, see, for example, Hilbert [81], Kung and Rota [100] and Sturmfels [128]. A binary form $f(x, y)$ of degree n is a homogeneous polynomial of degree n in two variables x and y:

$$f(x, y) = \sum_{i=0}^{n} \binom{n}{i} a_i x^i y^{n-i},$$

where the coefficients a_i are complex numbers.

Let

$$C = \begin{pmatrix} c_{11} & c_{12} \\ c_{21} & c_{22} \end{pmatrix}$$

be an invertible complex matrix. Under the linear transformation

$$x = c_{11}\overline{x} + c_{12}\overline{y},$$

$$y = c_{21}\overline{x} + c_{22}\overline{y},$$

the binary form $f(x, y)$ is transformed into another binary form

$$\overline{f}(\overline{x}, \overline{y}) = \sum_{i=0}^{n} \binom{n}{i} \overline{a}_i \, \overline{x}^i \, \overline{y}^{n-i},$$

where the coefficients $\overline{a}_i$ are polynomials in a_i and c_{ij}. Let g be a nonnegative integer. A polynomial $I(a_0, a_1, \ldots, a_n)$ in the coefficients $a_0, a_1, \ldots, a_n$ is an invariant of index g of the binary form $f(x, y)$ if for any invertible matrix C,

$$I(\overline{a}_0, \overline{a}_1, \ldots, \overline{a}_n) = (c_{11}c_{22} - c_{12}c_{21})^g I(a_0, a_1, \ldots, a_n).$$

For example,

$$I(a_0, a_1, a_2, a_3) = 3a_1^2 a_2^2 - 4a_1^3 a_3 - 4a_0 a_2^3 - a_0^2 a_3^2 + 6a_0 a_1 a_2 a_3 \qquad (6.13)$$

is an invariant of the cubic binary form

$$f(x, y) = a_3 x^3 + 3a_2 x^2 y + 3a_1 x y^2 + a_0 y^3. \qquad (6.14)$$

Note that $27I(a_0, a_1, a_2, a_3)$ is called the discriminant of (6.14). The polynomial $I(a_{n-1}, a_n, a_{n+1}, a_{n+2})$ is related to the higher order Turán inequality. Recall that a sequence $\{a_n\}_{n \geq 0}$ satisfies the higher order Turán inequality if for $n \geq 1$,

$$4(a_n^2 - a_{n-1}a_{n+1})(a_{n+1}^2 - a_n a_{n+2}) - (a_n a_{n+1} - a_{n-1}a_{n+2})^2 > 0, \qquad (6.15)$$

and we say that $\{a_n\}_{n \geq 0}$ satisfies the Turán inequality if it is log-concave.

A simple calculation shows that for $n = 1$, the polynomial in (6.15) reduces to the invariant $I(a_0, a_1, a_2, a_3)$ in (6.13), namely,

$$3a_1^2 a_2^2 - 4a_1^3 a_3 - 4a_0 a_2^3 - a_0^2 a_3^2 + 6a_0 a_1 a_2 a_3$$
$$= 4(a_1^2 - a_0 a_2)(a_2^2 - a_1 a_3) - (a_1 a_2 - a_0 a_3)^2.$$

Csordas, Norfolk and Varga [57] proved that the coefficients of the Riemann ξ-function satisfy the Turán inequality. This settles a conjecture of Pólya. Dimitrov [60] showed under the Riemann hypothesis, the coefficients of the Riemann ξ-function satisfy the higher order Turán inequality. Dimitrov and Lucas [61] proved this assertion without the Riemann hypothesis.

Numerical evidence indicates that both $p(n)$ and $\text{spt}(n)$ satisfy the high order Turán inequality.

Conjecture 6.12 *For $n \geq 95$, $p(n)$ satisfies the higher order Turán inequality (6.15), whereas* spt(n) *satisfies (6.15) for $n \geq 108$.*

We next consider the invariant of the quartic binary form

$$f(x, y) = a_4 x^4 + 4a_3 x^3 y + 6a_2 x^2 y^2 + 4a_1 xy^3 + a_0 y^4. \tag{6.16}$$

It appears that for large n, both $p(n)$ and spt(n) satisfy the inequalities derived from the following invariants of (6.16):

$$A(a_0, a_1, a_2, a_3, a_4) = a_0 a_4 - 4a_1 a_3 + 3a_2^2,$$

$$B(a_0, a_1, a_2, a_3, a_4) = -a_0 a_2 a_4 + a_2^3 + a_0 a_3^2 + a_1^2 a_4 - 2a_1 a_2 a_3,$$

$$I(a_0, a_1, a_2, a_3, a_4) = A(a_0, a_1, a_2, a_3, a_4)^3 - 27 B(a_0, a_1, a_2, a_3, a_4)^2.$$

Notice that $256 I(a_0, a_1, a_2, a_3, a_4)$ is the discriminant of $f(x, y)$ in (6.16). To be more specific, we have the following conjectures: Setting $a_n = p(n)$,

$$A(a_{n-1}, a_n, a_{n+1}, a_{n+2}, a_{n+3}) > 0, \qquad \text{for } n \geq 185,$$

$$B(a_{n-1}, a_n, a_{n+1}, a_{n+2}, a_{n+3}) > 0, \qquad \text{for } n \geq 221,$$

$$I(a_{n-1}, a_n, a_{n+1}, a_{n+2}, a_{n+3}) > 0, \qquad \text{for } n \geq 207.$$

Setting $a_n = $ spt(n),

$$A(a_{n-1}, a_n, a_{n+1}, a_{n+2}, a_{n+3}) > 0, \qquad \text{for } n \geq 205,$$

$$B(a_{n-1}, a_n, a_{n+1}, a_{n+2}, a_{n+3}) > 0, \qquad \text{for } n \geq 241,$$

$$I(a_{n-1}, a_n, a_{n+1}, a_{n+2}, a_{n+3}) > 0, \qquad \text{for } n \geq 227.$$

In general, it would be interesting to further study higher order inequalities on $p(n)$ and spt(n) based on polynomials arising in the invariant theory of binary forms.

Acknowledgments. I wish to thank George E. Andrews, Kathrin Bringmann, Mike Hirschhorn, Joseph Kung, Karl Mahlburg, Ken Ono and Peter Paule for valuable comments and suggestions. This work was supported by the National Science Foundation of China.

References

[1] S. Ahlgren, Distribution of the partition function modulo composite integers M, Math. Ann. 318 (4) (2000) 795–803.

[2] S. Ahlgren and N. Andersen, Euler-like recurrences for smallest parts functions, Ramanujan J. 36 (1-2) (2015) 237–248.

[3] S. Ahlgren and N. Andersen, Algebraic and transcendental formulas for the smallest parts function, Adv. Math. 289 (2016) 411–437.

[4] S. Ahlgren and M. Boylan, Arithmetic properties of the partition function, Invent. Math. 153 (3) (2003) 487–502.

[5] S. Ahlgren, K. Bringmann and J. Lovejoy, ℓ-adic properties of smallest parts functions, Adv. Math. 228 (1) (2011) 629–645.

[6] S. Ahlgren and K. Ono, Congruences and conjectures for the partition function, In q-Series with Applications to Combinatorics, Number Theory, and Physics, 1–10, Contemp. Math., 291, American Mathematical Society, Providence, RI, 2001.

[7] N. Andersen, Hecke-type congruences for two smallest parts functions, Int. J. Number Theory 9 (3) (2013) 713–728.

[8] G.E. Andrews, Partitions: Yesterday and Today, New Zealand Mathematical Society, Wellington, 1979.

[9] G.E. Andrews, q-Series: Their Development and Application in Analysis, Number Theory, Combinatorics, Physics and Computer Algebra, CBMS Regional Conference Series in Mathematics, 66, American Mathematical Society, Providence, RI, 1986.

[10] G.E. Andrews, The Theory of Partitions, Cambridge University Press, Cambridge, 1998.

[11] G.E. Andrews, Partitions, Durfee symbols, and the Atkin-Garvan moments of ranks, Invent. Math. 169 (1) (2007) 37–73.

[12] G.E. Andrews, The number of smallest parts in the partitions of n, J. Reine Angew. Math. 624 (2008) 133–142.

[13] G.E. Andrews, Concave and convex compositions, Ramanujan J. 31 (1-2) (2013) 67–82.

[14] G.E. Andrews and B.C. Berndt, Ramanujan's Lost Notebook. Part III, Springer, New York, 2012.

[15] G.E. Andrews, S.H. Chan and B. Kim, The odd moments of ranks and cranks, J. Combin. Theory Ser. A 120 (1) (2013) 77–91.

[16] G.E. Andrews, S.H. Chan, B. Kim and R. Osburn, The first positive rank and crank moments for overpartitions, Ann. Combin. 20 (2) (2016) 193–207.

[17] G.E. Andrews, A. Dixit, D. Schultz and A.J. Yee, Overpartitions related to the mock theta function $\omega(q)$, arXiv:1603.04352.

[18] G.E. Andrews, A. Dixit and A.J. Yee, Partitions associated with the Ramanujan/Watson mock theta functions $\omega(q)$, $\nu(q)$ and $\phi(q)$, Res. Number Theory 1 (2015) Art. 19.

[19] G.E. Andrews, F.J. Dyson and R.C. Rhoades, On the distribution of the spt-crank, Mathematics 1 (3) (2013) 76–88.

[20] G.E. Andrews and K. Eriksson, Integer Partitions, Cambridge University Press, Cambridge, 2004.

[21] G.E. Andrews and F.G. Garvan, Dyson's crank of a partition, Bull. Amer. Math. Soc. 18 (2) (1988) 167–171.

[22] G.E. Andrews, F.G. Garvan and J. Liang, Combinatorial interpretations of congruences for the spt-function, Ramanujan J. 29 (1-3) (2012) 321–338.

[23] G.E. Andrews, F.G. Garvan and J. Liang, Self-conjugate vector partitions and the parity of the spt-function, Acta Arith. 158 (3) (2013) 199–218.

[24] G.E. Andrews and K. Ono, Ramanujan's congruences and Dyson's crank, Proc. Natl. Acad. Sci. USA 102 (43) (2005) 15277.

[25] G.E. Andrews, R.C. Rhoades and S.P. Zwegers, Modularity of the concave composition generating function, Algebra Number Theory 7 (9) (2013) 2103–2139.

[26] T.M. Apostol, Modular Functions and Dirichlet Series in Number Theory, Graduate Texts in Mathematics, 41, Springer-Verlag, New York, 1990.

[27] A.O.L. Atkin, Proof of a conjecture of Ramanujan, Glasgow Math. J. 8 (1967) 14–32.

[28] A.O.L. Atkin and F.G. Garvan, Relations between the ranks and cranks of partitions, Ramanujan J. 7 (1-3) (2003) 343–366.

[29] A.O.L. Atkin and S.M. Hussain, Some properties of partitions. II, Trans. Amer. Math. Soc. 89 (1958) 184–200.

[30] A.O.L. Atkin and P. Swinnerton-Dyer, Some properties of partitions, Proc. London Math. Soc. 4 (3) (1954) 84–106.

[31] J. Banks, A. Barquero-Sanchez, R. Masri and Y. Sheng, The asymptotic distribution of Andrews' smallest parts function, Arch. Math. 105 (6) (2015) 539–555.

[32] E. Belmont, H. Lee, A. Musat and S. Trebat-Leder, ℓ-adic properties of partition functions, Monatsh. Math. 173 (1) (2014) 1–34.

[33] A. Berkovich and F.G. Garvan, Some observations on Dyson's new symmetries of partitions, J. Combin. Theory Ser. A 100 (1) (2002) 61–93.

[34] B.C. Berndt, Number Theory in the Spirit of Ramanujan, Student Mathematical Library, 34, American Mathematical Society, Providence, RI, 2006.

[35] B.C. Berndt, Ramanujan's congruences for the partition function modulo 5, 7, and 11, Int. J. Number Theory 3 (3) (2007) 349–354.

[36] B.C. Berndt, S.H. Chan, Z.G. Liu and H. Yesilyurt, A new identity for $(q;q)_\infty^{10}$ with an application to Ramanujan's partition congruence modulo 11, Q. J. Math. 55 (1) (2004) 13–30.

[37] C. Bessenrodt and K. Ono, Maximal multiplicative properties of partitions, Ann. Combin. 20 (1) (2016) 59–64.

[38] M. Boylan and J.J. Webb, The partition function modulo prime powers, Trans. Amer. Math. Soc. 365 (4) (2013) 2169–2206.

[39] K. Bringmann, On the explicit construction of higher deformations of partition statistics, Duke Math. J. 144 (2) (2008) 195–233.

[40] K. Bringmann, A. Folsom and K. Ono, q-series and weight 3/2 Maass forms, Compos. Math. 145 (3) (2009) 541–552.

[41] K. Bringmann, F.G. Garvan and K. Mahlburg, Partition statistics and quasiharmonic Maass forms, Int. Math. Res. Not. IMRN (1) (2009) 63–97.

[42] K. Bringmann, J. Lovejoy and R. Osburn, Rank and crank moments for overpartitions, J. Number Theory 129 (7) (2009) 1758–1772.

[43] K. Bringmann, J. Lovejoy and R. Osburn, Automorphic properties of generating functions for generalized rank moments and Durfee symbols, Int. Math. Res. Not. IMRN (2) (2010) 238–260.

[44] K. Bringmann and K. Mahlburg, Inequalities between ranks and cranks, Proc. Amer. Math. Soc. 137 (8) (2009) 2567–2574.

[45] K. Bringmann and K. Mahlburg, Asymptotic inequalities for positive crank and rank moments, Trans. Amer. Math. Soc. 366 (2) (2014) 1073–1094.

[46] K. Bringmann, K. Mahlburg and R.C. Rhoades, Asymptotics for rank and crank moments, Bull. London Math. Soc. 43 (4) (2011) 661–672.

[47] K. Bringmann, K. Mahlburg and R.C. Rhoades, Taylor coefficients of mock-Jacobi forms and moments of partition statistics, Math. Proc. Cambridge Philos. Soc. 157 (2) (2014) 231–251.

[48] J. Bryson, K. Ono, S. Pitman and R.C. Rhoades, Unimodal sequences and quantum and mock modular forms, Proc. Natl. Acad. Sci. USA 109 (40) (2012) 16063–16067.

[49] S.H. Chan and R. Mao, Inequalities for ranks of partitions and the first moment of ranks and cranks of partitions, Adv. Math. 258 (2014) 414–437.

[50] W.Y.C. Chen, Recent developments on log-concavity and q-log-concavity of combinatorial polynomials, In: FPSAC 2010 Conference Talk Slides, http://www.billchen.org/talks/2010-FPSAC.pdf (2010).

[51] W.Y.C. Chen, K.Q. Ji and E.Y.Y. Shen, k-marked Dyson symbols and congruences for moments of cranks, arXiv:1312.2080.

[52] W.Y.C. Chen, K.Q. Ji and W.J.T. Zang, Proof of the Andrews-Dyson-Rhoades conjecture on the spt-crank, Adv. Math. 270 (2015) 60–96.

[53] W.Y.C. Chen, K.Q. Ji and W.J.T. Zang, The spt-crank for ordinary partitions, J. Reine Angew. Math. 711 (2016) 231–249.

[54] W.Y.C. Chen, K.Q. Ji and W.J.T. Zang, Nearly equal distributions of the rank and the crank of partitions, arXiv:1704.00882.

[55] W.Y.C. Chen, L.X.W. Wang and G.Y.B. Xie, Finite differences of the logarithm of the partition function, Math. Comp. 85 (298) (2016) 825–847.

[56] S. Corteel and J. Lovejoy, Overpartitions, Trans. Amer. Math. Soc. 356 (4) (2004) 1623–1635.

[57] G. Csordas, T.S. Norfolk and R.S. Varga, The Riemann hypothesis and the Turán inequalities, Trans. Amer. Math. Soc. 296 (2) (1986) 521–541.

[58] S. DeSalvo and I. Pak, Log-concavity of the partition function, Ramanujan J. 38 (1) (2015) 61–73.

[59] F. Diamond and J. Shurman, A First Course in Modular Forms, Graduate Texts in Mathematics, 228, Springer-Verlag, New York, 2005.

[60] D.K. Dimitrov, Higher order Turán inequalities, Proc. Amer. Math. Soc. 126 (7) (1998) 2033–2037.

[61] D.K. Dimitrov and F.R. Lucas, Higher order Turán inequalities for the Riemann ξ-function, Proc. Amer. Math. Soc. 139 (3) (2011) 1013–1022.

[62] A. Dixit and A.J. Yee, Generalized higher order spt-functions, Ramanujan J. 31 (1-2) (2013) 191–212.

[63] F.J. Dyson, Some guesses in the theory of partitions, Eureka (Cambridge) 8 (1944) 10–15.

[64] F.J. Dyson, Mappings and symmetries of partitions, J. Combin. Theory Ser. A 51 (2) (1989) 169–180.

[65] F.J. Dyson, Partitions and the grand canonical ensemble, Ramanujan J. 29 (1-3) (2012) 423–429.

[66] D.A. Eichhorn and D.M. Hirschhorn, Notes on the spt function of George E. Andrews, Ramanujan J. 38 (1) (2015) 17–34.

[67] A. Folsom, Z.A. Kent and K. Ono, ℓ-adic properties of the partition function, Adv. Math. 229 (3) (2012) 1586–1609.

[68] A. Folsom and K. Ono, The spt-function of Andrews, Proc. Natl. Acad. Sci. USA 105 (51) (2008) 20152–20156.

[69] K.C. Garrett, C. McEachern, T. Frederick and O. Hall-Holt, Fast computation of Andrews' smallest part statistic and conjectured congruences, Discrete Appl. Math. 159 (13) (2011) 1377–1380.

[70] F.G. Garvan, New combinatorial interpretations of Ramanujan's partition congruences mod 5, 7 and 11, Trans. Amer. Math. Soc. 305 (1) (1988) 47–77.

[71] F.G. Garvan, Generalizations of Dyson's rank and non-Rogers-Ramanujan partitions, Manuscripta Math. 84 (3-4) (1994) 343–359.

[72] F.G. Garvan, Congruences for Andrews' smallest parts partition function and new congruences for Dyson's rank, Int. J. Number Theory 6 (2) (2010) 281–309.

[73] F.G. Garvan, Higher order spt-functions, Adv. Math. 228 (1) (2011) 241–265.

[74] F.G. Garvan, Congruences for Andrews' spt-function modulo powers of 5, 7 and 13, Trans. Amer. Math. Soc. 364 (9) (2012) 4847–4873.

[75] F.G. Garvan, Congruences for Andrews' spt-function modulo 32760 and extension of Atkin's Hecke-type partition congruences, Number Theory and Related Fields, Springer Proc. Math. Stat. 43, Springer, New York, (2013) 165–185.

[76] F.G. Garvan and C. Jennings-Shaffer, The spt-crank for overpartitions, Acta Arith. 166 (2) (2014) 141–188.

[77] F.G. Garvan and C. Jennings-Shaffer, Hecke-type congruences for Andrews' SPT-function modulo 16 and 32, Int. J. Number Theory 10 (2) (2014) 375–390.

[78] F.G. Garvan and C. Jennings-Shaffer, Exotic Bailey-Slater SPT-functions II: Hecke-Rogers-type double sums and Bailey pairs from groups A, C, E, Adv. Math. 299 (2016) 605–639.

[79] F.G. Garvan and D. Stanton, Sieved partition functions and q-binomial coefficients, Math. Comp. 55 (191) (1990) 299–311.

[80] G. Gasper and M. Rahman, Basic Hypergeometric Series, 2nd ed., Encyclopedia of Mathematics and Its Applications, Vol. 96, Cambridge University Press, Cambridge, 2004.

[81] D. Hilbert, Theory of Algebraic Invariants, Cambridge University Press, Cambridge, 1993.

[82] M.D. Hirschhorn, On the parity of $p(n)$. II, J. Combin. Theory Ser. A 62 (1) (1993) 128–138.

[83] M.D. Hirschhorn, Ramanujan's partition congruences, Discrete Math. 131 (1-3) (1994) 351–355.

[84] M.D. Hirschhorn, A short and simple proof of Ramanujan's mod 11 partition congruence, J. Number Theory 139 (2014) 205–209.

[85] M.D. Hirschhorn and M.V. Subbarao, On the parity of $p(n)$, Acta Arith. 50 (4) (1988) 355–356.

[86] M.J. Jang and B. Kim, On spt-crank-type functions, Ramanujan J. (2016) doi:10.1007/s11139-016-9838-5.

[87] C. Jennings-Shaffer, Another SPT crank for the number of smallest parts in overpartitions with even smallest part, J. Number Theory 148 (2015) 196–203.

[88] C. Jennings-Shaffer, Higher order SPT-functions for overpartitions, overpartitions with smallest part even, and partitions with smallest part even and without repeated odd parts, J. Number Theory 149 (2015) 285–312.

[89] C. Jennings-Shaffer, Rank and crank moments for partitions without repeated odd parts, Int. J. Number Theory 11 (3) (2015) 683–703.

[90] C. Jennings-Shaffer, Exotic Bailey-Slater spt-functions III: Bailey pairs from groups B, F, G, and J, Acta Arith. 173 (4) (2016) 317–364.

[91] C. Jennings-Shaffer, Exotic Bailey-Slater spt-functions I: Group A, Adv. Math. 305 (2017) 479–514.

[92] C. Jennings-Shaffer, Some smallest parts functions from variations of Bailey's lemma, arXiv:1506.05344.

[93] K.Q. Ji, A combinatorial proof of Andrews' smallest parts partition function, Electron. J. Combin. 15 (1) (2008) #N12.

[94] K.Q. Ji, The combinatorics of k-marked Durfee symbols, Trans. Amer. Math. Soc. 363 (2) (2011) 987–1005.

[95] S.J. Kaavya, Crank 0 partitions and the parity of the partition function, Int. J. Number Theory 7 (3) (2011) 793–801.

[96] B. Kim, E. Kim and J. Seo, On the number of even and odd strings along the overpartitions of n, Arch. Math. (Basel) 102 (4) (2014) 357–368.

[97] I. Kiming and J.B. Olsson, Congruences like Ramanujan's for powers of the partition function, Arch. Math. (Basel) 59 (4) (1992) 348–360.

[98] N. Koblitz, Introduction to Elliptic Curves and Modular Forms, Graduate Texts in Mathematics, 97, Springer-Verlag, New York, 1993.

[99] O. Kolberg, Note on the parity of the partition function, Math. Scand. 7 (1959) 377–378.

[100] J.P.S. Kung and G.C. Rota, The invariant theory of binary forms, Bull. Amer. Math. Soc. 10 (1) (1984) 27–85.

[101] K. Kursungöz, Counting k-marked Durfee symbols, Electron. J. Combin. 18 (1) (2011) #P41.

[102] R.P. Lewis, On the ranks of partitions modulo 9, Bull. London Math. Soc. 23 (5) (1991) 417–421.

[103] R.P. Lewis, The ranks of partitions modulo 2, Discrete Math. 167/168 (1997) 445–449.

[104] J. Lovejoy, Rank and conjugation for the Frobenius representation of an overpartition, Ann. Combin. 9 (3) (2005) 321–334.

[105] J. Lovejoy, Rank and conjugation for a second Frobenius representation of an overpartition, Ann. Combin. 12 (1) (2008) 101–113.

[106] R. Mao, Asymptotic inequalities for k-ranks and their cumulation functions, J. Math. Anal. Appl. 409 (2) (2014) 729–741.

[107] R. Masri, Fourier coefficients of harmonic weak Maass forms and the partition function, Amer. J. Math. 137 (4) (2015) 1061–1097.

[108] J.N. O'Brien, Some properties of partitions, with special reference to primes other than 5, 7 and 11, Ph. D. thesis, Durham University, 1965.

[109] K. Ono, Parity of the partition function in arithmetic progressions, J. Reine Angew. Math. 472 (1996) 1–15.

[110] K. Ono, Distribution of the partition function modulo m, Ann. of Math. (2) 151 (1) (2000) 293–307.

[111] K. Ono, The Web of Modularity: Arithmetic of the Coefficients of Modular Forms and q-Series, CBMS Regional Conference Series in Mathematics, 102, American Mathematical Society, Providence, RI, 2004.

[112] K. Ono, Congruences for the Andrews spt function, Proc. Natl. Acad. Sci. USA 108 (2) (2011) 473–476.

[113] A.E. Patkowski, A strange partition theorem related to the second Atkin–Garvan moment, Int. J. Number Theory 11 (7) (2015) 2191–2197.

[114] A.E. Patkowski, Another smallest part function related to Andrews' spt function, Acta Arith. 168 (2) (2015) 101–105.

[115] A.E. Patkowski, An interesting q-series related to the 4-th symmetrized rank function, arXiv:1310.5282.

[116] P. Paule and C.-S. Radu, A new witness identity for $11|p(11n+6)$, Preprint, 2017.

[117] P. Paule and C.-S. Radu, A unified algorithmic framework for Ramanujan's congruences modulo powers of 5, 7, and 11, Preprint, 2017.

[118] H. Rademacher, On the partition function $p(n)$, Proc. London Math. Soc. S2-43 (4) 241.

[119] H. Rademacher, Fourier expansions of modular forms and problems of partition, Bull. Amer. Math. Soc. 46 (1940) 59–73.

[120] H. Rademacher, On the expansion of the partition function in a series, Ann. of Math. (2) 44 (1943) 416–422.

[121] C.-S. Radu, A proof of Subbarao's conjecture, J. Reine Angew. Math. 672 (2012) 161–175.

[122] S. Ramanujan, Some properties of $p(n)$, the number of partitions of n, Proc. Cambridge Philos. Soc. 19 (1919) 207–210.

[123] R.C. Rhoades, Soft asymptotics for generalized spt-functions, J. Combin. Theory Ser. A 120 (3) (2013) 637–643.

[124] R.C. Rhoades, On Ramanujan's definition of mock theta function, Proc. Natl. Acad. Sci. USA 110 (19) (2013) 7592–7594.

[125] J.M.Z. Rolon, Asymptotics of higher order ospt-functions for overpartitions, Ann. Combin. 20 (1) (2016) 177–191.

[126] N. Santa-Gadea, On some relations for the rank moduli 9 and 12, J. Number Theory 40 (2) (1992) 130–145.

[127] I.R. Sarma, K.H. Reddy, S.R. Gunakala and D.M.G. Comissiong, Relation between the smallest and the greatest parts of the partitions of n, J. Math. Research 3 (4) (2011) 133–140.

[128] B. Sturmfels, Algorithms in Invariant Theory, Texts and Monographs in Symbolic Computation, Springer-Verlag, Vienna, 1993.

[129] M.V. Subbarao, Some remarks on the partition function, Amer. Math. Monthly 73 (1966) 851–854.

[130] M. Waldherr, Asymptotic for moments of higher ranks, Int. J. Number Theory 9 (3) (2013) 675–712.

[131] L. Wang, New congruences for partitions related to mock theta functions, J. Number Theory 175 (2017) 51–65.

[132] G.N. Watson, Ramanujans Vermutung über Zerfällungszahlen, J. Reine Angew. Math. 179 (1938) 97–128.

[133] L. Winquist, An elementary proof of $p(11m + 6) \equiv 0 \bmod 11$, J. Combin. Theory 6 (1969) 56–59.

Center for Applied Mathematics
Tianjin University
Tianjin 300072
P. R. China

Center for Combinatorics
Nankai University
Tianjin 300071
P. R. China
chen@nankai.edu.cn

Combinatorial structures in finite classical polar spaces

Antonio Cossidente

Abstract

Some recent results on regular systems and intriguing sets of finite classical polar spaces are surveyed.

1 Introduction

The finite classical polar spaces are the geometries associated with non–degenerate reflexive sesquilinear and non–singular quadratic forms on vector spaces of finite dimension over a finite field $GF(q)$. Let $PG(n, q)$ denote the n–dimensional projective space over $GF(q)$. A polar space in $PG(n, q)$ consists of its projective subspaces that are totally isotropic with respect to a given non–degenerate reflexive sesquilinear form or that are totally singular with respect to a given non–singular quadratic form. The *rank* of a polar space is the vector space dimension of a maximal totally isotropic or totally singular subspace, called here *maximal*. In this paper, the term *polar space* always refers to a finite classical polar space. A polar space of rank two is a *generalised quadrangle*. In the last decades, intensive investigations on combinatorial structures in finite polar spaces, such as spreads, ovoids, blocking sets, covers, have been carried out. More recently, other structures, such as m–systems, m–ovoids, i–tight sets (*intriguing sets*) have been studied. In this paper, some recent results on regular systems and intriguing sets of finite polar spaces are surveyed, with special emphasis on Hermitian polar spaces, that is, polar spaces arising from a non–degenerate unitary form.

Throughout the paper the following notation is adopted for finite polar spaces:

(1) $\mathcal{H}(n, q^2)$ is the space associated with a non–degenerate hermitian form on a vector space of dimension $n + 1$ over $GF(q^2)$;

(2) $\mathcal{Q}^-(n, q)$ is the space associated with a non –singular quadratic form of non–maximal Witt index on a vector space of dimension $n + 1$ even over $GF(q)$;

(3) $\mathcal{Q}^+(n, q)$ is the space associated with a non–singular quadratic form of maximal Witt index on a vector space of even dimension $n + 1$ over $GF(q)$;

(4) $\mathcal{Q}(n, q)$ is the space associated with a non–singular quadratic form on a vector space of odd dimension $n + 1$ over $\mathrm{GF}(q)$;

(5) $\mathcal{W}(n, q)$ is the space associated with a non–degenerate symplectic form on a vector space of even dimension $n + 1$ over $\mathrm{GF}(q)$.

2 Regular Systems

2.1 Hemisystems and Relative Hemisystems

The notion of regular system was introduced for the first time by Beniamino Segre in his monumental paper [59] in the context of Hermitian polar spaces.

Definition 2.1 Let $\mathcal{P} = (P, M, I)$ be an incidence structure with two types, called here points and maximals. A *regular system* of order m of $\mathcal{P}$ (m–regular system) is a set $\mathcal{R}$ of maximals in M with the property that every point lies on exactly m maximals of $\mathcal{R}$, with $0 < m < a$, where a denotes the number of maximals of $\mathcal{P}$ passing through a point in P. A regular system having the same order as its complement in M is said to be a *hemisystem*; here, a must be even.

In [59], Segre proved that a regular system of order m of $\mathcal{H}(3, q^2)$ must be a *hemisystem* of $\mathcal{H}(3, q^2)$, and hence $m = (q+1)/2$. He also constructed a hemisystem of $\mathcal{H}(3, 9)$ admitting the linear group $PSL(3, 4)$ and proved that this hemisystem is unique. See also [36] for an alternative construction. A simple proof that a regular system of $\mathcal{H}(3, q^2)$ is a hemisystem (and so q is odd) was given by Thas in [63], by showing that the concurrency graph of the lines of a regular system on $\mathcal{H}(3, q^2)$ of order m is a strongly regular graph $srg(v, k, \lambda, \mu)$, with

$$v = (q^3 + 1)(q + 1) - m, \quad k = (q^2 + 1)(q - m),$$
$$\lambda = q - m - 1, \quad \mu = q^2 + 1 - m(q + 1),$$

where

$$(v - k - 1)\mu = k(k - \lambda - 1).$$

In [14], the nonexistence of regular systems of $\mathcal{H}(3, q^2)$ for q even was established.

Even more general was the work of Cameron–Goethals–Seidel [18], who defined a hemisystem of a generalised quadrangle of order (s, s^2), s odd, to be a set of points meeting every line in $(s + 1)/2$ points and showed that the collinearity graph of such a set is strongly regular.

Much of the motivation for the early study of hemisystems comes from their connection to *partial quadrangles* introduced by Cameron [16], as each hemisystem of $\mathcal{H}(3, q^2)$ gives rise to a partial quadrangle.

A partial quadrangle $PQ(s, t, \mu)$ is an incidence structure of points and lines such that any two points are incident with at most one line, every point is incident with $t + 1$ lines, every line is incident with $s + 1$ points, two non–collinear points are jointly collinear with exactly μ points, and for every line ℓ and point P not on ℓ, there is at most one point Q on ℓ collinear with P. It follows from these conditions that the point graph of a partial quadrangle is a strongly regular graph.

In particular, a generalised quadrangle is a partial quadrangle with $\mu = t + 1$. Partial quadrangles that are not generalised quadrangles are rare - most arise from deleting a point P, all lines through P, and all points collinear with P, from a generalised quadrangle of order (s, s^2). This construction gives rise to a partial quadrangle of order $(s - 1, s^2, s(s - 1))$. Examples of partial quadrangles that do not come from this construction either arise from one of the seven known triangle free strongly regular graphs, or from caps of projective spaces, or from a hemisystem of a generalised quadrangle of order (s, s^2): the points of the partial quadrangle are the points of the hemisystem, and the lines of the partial quadrangle are the lines of the generalised quadrangle. The resulting partial quadrangle has order $((s - 1)/2, s^2, (s - 1)^2/2)$, and therefore does not arise from deleting points and lines from a generalised quadrangle as explained above. Since the complement of a hemisystem is also a hemisystem, and partial quadrangles give strongly regular graphs, any hemisystem of a $GQ(s, s^2)$ leads to two strongly regular graphs, which may not be isomorphic.

Another combinatorial structure involved in the context of hemisystems is that of an association scheme.

Definition 2.2 A *d–class association scheme* is a set X, together with $d+1$ symmetric relations R_i on X such that $\{R_0, \ldots, R_d\}$ partitions $X \times X$, the identity relation on $X \times X$ is R_0, and for any $(x, y) \in R_k$, the numbers

$$p_{ij}^k = |\{z \in X : (x, z) \in R_i \text{ and } (z, y) \in R_j\}|$$

depend only on i, j, k and not on x and y.

Association schemes were first defined in 1952 by Bose and Shimamoto [12]. A strongly regular graph gives a 2-class association scheme by defining $(x, y) \in R_1$ if x is adjacent to y, and $(x, y) \in R_2$ if x and y are distinct and non–adjacent. Any 2-class association scheme can be viewed as coming from a strongly regular graph in this way. Similarly, a distance regular graph with diameter i gives an i–class association scheme by defining $(x, y) \in R_i$ if $d(x, y) = i$.

A hemisystem H of a hermitian generalised quadrangle $\mathbf{X} = \mathcal{H}(3, q^2)$ gives a 4–class association scheme via the following construction.

The members $x, y \in \mathbf{X}$ are in the same *half* of $\mathbf{X}$ if either x and y are both in H or both in the complement H^c of H. Otherwise, the members x and y are in *opposite halves*. Now define R_1, R_2, R_3, R_4 as follows:

(a) $(x, y) \in R_1$ if x and y are incident and in the same half;

(b) $(x, y) \in R_2$ if x and y are incident and in opposite halves;

(c) $(x, y) \in R_3$ if x and y are not incident and in the same half;

(d) $(x, y) \in R_4$ if x and y are not incident and in opposite halves.

This construction is of some interest because the resulting association scheme is cometric but not metric; equivalently, Q–polynomial but not P–polynomial. Such association schemes are quite rare in the literature. See [54] for definitions of these terms and a survey of cometric but not metric association schemes.

Thirty years after Segre's hemisystem of $\mathcal{H}(3, 9)$, no new hemisystems had been found, and J.A. Thas stated the following conjecture.

Conjecture 2.3 *(1995) [64] There are no hemisystems of $\mathcal{H}(3, q^2)$ for $q > 3$.*

This conjecture was shown to be false ten years later: Cossidente and Penttila [34] constructed an infinite family of hemisystems of $\mathcal{H}(3, q^2)$, q odd, and a sporadic example in $\mathcal{H}(3, 25)$.

Theorem 2.4 *There exists a hemisystem of $\mathcal{H}(3, q^2)$, for all odd prime powers q, admitting $P\Omega^-(4, q)$ as an automorphism group. There exists a hemisystem of $\mathcal{H}(3, 25)$, admitting $3 \cdot A_7$ as an automorphism group.*

The construction of such hemisystems relies on the geometry of commuting polarities.

Consider $\mathcal{H}(3, q^2)$ in $\mathrm{PG}(3, q^2)$ and let $\mathcal{U}$ be the associated Hermitian polarity Let $\mathcal{B}$ be an orthogonal polarity *commuting* with the Hermitian polarity $\mathcal{U}$: $\mathcal{B}\mathcal{U} = \mathcal{U}\mathcal{B} := \mathcal{V}$. Then $\mathcal{V}$ is a non–linear collineation and from [59], the fixed points of $\mathcal{V}$ on $\mathcal{H}(3, q^2)$ form a non–degenerate quadric $\mathcal{Q}$ which is either elliptic or hyperbolic. In particular, $\mathcal{Q} = \mathcal{H}(3, q^2) \cap \Sigma$, where Σ is a suitable subgeometry of $\mathrm{PG}(3, q^2)$ isomorphic to $\mathrm{PG}(3, q)$, [59]. Assume that $\mathcal{Q}$ is elliptic.

Let $\bar{G}$ denote the stabiliser of $\mathcal{Q}$ in $PSU_4(q^2)$. From [34],

$$\bar{G} = PGO_4^-(q) \cap PSU_4(q^2) = PSO_4^-(q) \cdot 2.$$

The following results have been proved in [34].

Proposition 2.5 *The group $\bar{G}$ has three orbits on points of $\mathcal{H}(3, q^2)$.*

The $\bar{G}$–orbits have sizes $q^2 + 1$, $q^2(q^2 + 1)(q + 1)/2$, $q^2(q^2 + 1)(q - 1)/2$: they are points on $\mathcal{Q}$, points on generators tangent to $\mathcal{Q}$ but not on $\mathcal{Q}$, and the complements of these in $\mathcal{H}(3, q^2)$.

Proposition 2.6 *$\bar{G}$ has two orbits on lines of $\mathcal{H}(3, q^2)$.*

The quadric $\mathcal{Q}$ is a partial ovoid of $\mathcal{H}(3, q^2)$ and so each generator (line) of $\mathcal{H}(3, q^2)$ is either disjoint from $\mathcal{Q}$ or meets $\mathcal{Q}$ in exactly one point. The quadric $\mathcal{Q}$ is a *special set* of $\mathcal{H}(3, q^2)$, that is, it is a subset of $q^2 + 1$ points of $\mathcal{H}(3, q^2)$ such that each point of $\mathcal{H}(3, q^2) \setminus \mathcal{Q}$ is conjugate to 0 or 2 points of $\mathcal{Q}$, or equivalently, any three points of $\mathcal{Q}$ generate a non–tangent plane to $\mathcal{H}(3, q^2)$; see [34] and references therein. Through each point P on $\mathcal{Q}$ there are $q + 1$ generators of $\mathcal{H}(3, q^2)$, which are permuted by the stabiliser of P in $\bar{G}$. Since $\mathcal{Q}$ has $q^2 + 1$ points and $\bar{G}$ acts transitively on $\mathcal{Q}$, there are $(q + 1)(q^2 + 1)$ generators of $\mathcal{H}(3, q^2)$ permuted in a single orbit under the action of $\bar{G}$. The second orbit consists of all generators of $\mathcal{H}(3, q^2)$ disjoint from $\mathcal{Q}$.

The group $G = P\Omega^-(4, q)$ has the same orbits on points of $\mathcal{H}(3, q^2)$ as $\bar{G}$. Under the action of G the two $\bar{G}$–line–orbits given in Proposition 2.6 split into four orbits, two of size $(q^2 + 1)(q + 1)/2$, say O_1 and O_2, and two of size $q^2(q^2 - 1)/2$, say O_3 and O_4 . Since G acts transitively on $\mathcal{Q}$, each orbit of size $(q^2 + 1)(q + 1)/2$ represents a partial hemisystem of size $q^2 + 1$. The block-tactical decomposition matrix for this orbit decomposition is

$$\begin{bmatrix} 1 & 1 & 0 & 0 \\ 0 & 0 & \frac{q^2+1}{2} & \frac{q^2+1}{2} \\ q^2 & q^2 & \frac{q^2+1}{2} & \frac{q^2+1}{2} \end{bmatrix},$$

and hence the point-tactical decomposition matrix is

$$\begin{bmatrix} \frac{q+1}{2} & \frac{q+1}{2} & 0 & 0 \\ 0 & 0 & \frac{q+1}{2} & \frac{q+1}{2} \\ 1 & 1 & \frac{q-1}{2} & \frac{q-1}{2} \end{bmatrix}.$$

It follows that amalgamation of an orbit of size $(q^2 + 1)(q + 1)/2$ and an orbit of size $q^2(q^2 - 1)/2$ yields a G-invariant hemisystem of $\mathcal{H}(3, q^2)$.

The resulting partial quadrangles and strongly regular graphs arising from these hemisystems are new for $q > 3$.

These constructions led again to renewed interest in hemisystems of polar space.

Indeed, four years later, Cossidente and Penttila [35] used a construction similar to that proposed for $\mathcal{H}(3, q^2)$ to get three new families of hemisystems of $\mathcal{H}(5, q^2)$, q odd.

Theorem 2.7 *There exists a hemisystem of* $\mathcal{H}(5, q^2)$ *admitting* $P\Omega^\epsilon(6, q)$, $\epsilon = \pm$, *for all odd prime powers* q. *There exists a hemisystem of* $\mathcal{H}(5, q^2)$ *admitting* $P\Omega(5, q)$, *for all odd prime powers* q.

In 2013, Luke Bayens [10] generalised the Cossidente–Penttila construction and proved the following result.

Theorem 2.8 *For* $r > 2$, q *odd, there exists a hemisystem of* $\mathcal{H}(2r-1, q^2)$ *admitting* $PSO^-(2r, q)$ *as an automorphism group.*

Bayen's approach relies on a group–theoretic argument valid in any polar space. There are actually two groups involved in the construction of each hemisystem.

Lemma 2.9 *(The AB-Lemma) Let* $\mathcal{P} = (P, M, I)$ *be an incidence structure with two types called points and maximals. Let* A *and* B *be two subgroups of* $Aut(\mathcal{P})$ *such that*

(1) B *is a normal subgroup of* A,

(2) A *and* B *have the same orbits on* P,

(3) *each* A-*orbit on* M *splits into two* B-*orbits.*

Then there are 2^n *hemisystems admitting* B *as an automorphism group, where* n *is the number of* A-*orbits on maximals.*

The existence of hemisystems of hermitian polar spaces led to the renewed interest in hemisystems of any generalised quadrangle.

In 2010, Bamberg, Giudici and Royle [4] proved the following beautiful theorem, valid for a special class of generalised quadrangles of order (s^2, s), including the Hermitian generalised quadrangle $\mathcal{H}(3, q^2)$, known as *flock generalised quadrangles*. Such generalised quadrangles arise from BLT–sets of lines of $\mathcal{W}(3, q)$, q odd [50].

Theorem 2.10 *Every flock generalised quadrangle of order* (s^2, s), s *odd, has a hemisystem.*

For further results on hemisystems of (flock) generalised quadrangles, see [5].

Very recently, a new family of hemisystems of $\mathcal{H}(3, q^2)$, $q \geq 7$ has been found [28] admitting a subgroup of the stabiliser of a point $P \in \mathcal{Q}^-(3, q)$ in the group $P\Omega^-(4, q)$ as an automorphism group. Below is a detailed description.

From the point-tactical decomposition used to construct the hemisystems of Cossidente–Penttila, it also follows that through any point on a generator tangent to $\mathcal{Q}$ there passes a unique line of $\mathcal{O}_1$ and a unique line of $\mathcal{O}_2$. Let $\mathcal{R}$ be the G–invariant hemisystem of $\mathcal{H}(3, q^2)$ obtained by glueing together the orbits O_1 and O_3.

Let a point $P \in \mathcal{Q}$ be fixed. The $q + 1$ generators on P lie on the tangent plane π to $\mathcal{H}(3, q^2)$ at P and are such that $(q + 1)/2$ of them are in O_1 and so belong to $\mathcal{R}$, and $(q + 1)/2$ are in O_2. Denote by π_i the points of π on the generators of O_i on P, $i = 1, 2$. The stabiliser of P in G contains a subgroup K of order $q^2(q + 1)$ fixing P and acting transitively on $\mathcal{Q} \setminus \{P\}$. Also, it fixes other $q - 1$ permutable Baer elliptic quadrics, say $\mathcal{Q}_1, \dots \mathcal{Q}_{q-1}$, pairwise intersecting just in P, all embedded in $\mathcal{H}(3, q^2)$ and obtained by intersecting $\mathcal{H}(3, q^2)$ with suitable Baer subgeometries.

The union $\mathcal{O} = \mathcal{Q} \cup \mathcal{Q}_1 \cup \dots \cup \mathcal{Q}_{q-1}$ is an ovoid of $\mathcal{H}(3, q^2)$, the *semiclassical ovoid*, [25]. Since K fixes $\mathcal{R}$, the set of generators tangent to an elliptic quadric $\mathcal{Q}_i$ not on P splits into two orbits, say $\mathcal{L}_{i1}$ and $\mathcal{L}_{i2}$ of size $q^2(q + 1)/2$ according as a generator meets π_1 or π_2, $i = 1, \dots, q - 1$. Analogously, the generators tangent to $\mathcal{Q}$ not at P split into two orbits, say $\mathcal{L}_1$ and $\mathcal{L}_2$, of size $q^2(q + 1)/2$, according as a generator meets π_1 or π_2. In particular, $\mathcal{L}_2$ belongs to $\mathcal{R}$. A generator of $\mathcal{R}$ not tangent to $\mathcal{Q}$ is necessarily tangent to a unique quadric $\mathcal{Q}_i$ due to the fact that O is an ovoid. Also, it is always possible to assume that $\mathcal{L}_{i1}$ belongs to $\mathcal{R}$ for $i = 1, \dots, (q - 1)/2$. Hence $\mathcal{L}_{i2}$ belongs to $\mathcal{R}$, for $i = (q + 1)/2, \dots, q - 1$.

Let $X := \{\mathcal{Q}_{(q+1)/2}, \dots \mathcal{Q}_{(q-1)}, \mathcal{Q}\}$ and let $Y := \{\mathcal{Q}_1, \dots \mathcal{Q}_{(q-1)/2}\}$. Consider a set Y' consisting of $(q-1)/2$ quadrics arbitrarily chosen among those in $\mathcal{O}$. It follows that, if $|X \cap Y'| = x$, then $|Y \setminus Y'| = x$. Let us denote by $\mathcal{Q}_j$, $j = 1, \dots, x$, the quadrics in $X \cap Y'$ and by $\mathcal{Q}_k$, $k = 1, \dots, x$, the quadrics in $Y \setminus Y'$. The set

$$\mathcal{R}' := \bigcup_{j=1}^{x} \left((\mathcal{R} \setminus \mathcal{L}_{j2}) \cup \mathcal{L}_{j1} \right) \cup \bigcup_{k=1}^{x} \left((\mathcal{R} \setminus \mathcal{L}_{k1}) \cup \mathcal{L}_{k2} \right),$$

is a hemisystem of $\mathcal{H}(3, q^2)$.

Let $x = 1$ and begin from the hemisystem $\mathcal{R}$. Then

$$\mathcal{R}' := (\mathcal{R} \setminus \mathcal{L}_{j2}) \cup \mathcal{L}_{j1} \cup (\mathcal{R} \setminus \mathcal{L}_{k1}) \cup \mathcal{L}_{k2}.$$

Let T be a point of $\mathcal{H}(3, q^2)$ in π_1. There is a unique generator of O_1 passing through T and $(q-1)/2$ generators of $\mathcal{R} \setminus \{O_1\}$ passing through T of which exactly one belongs to $\mathcal{L}_{k1}$. This is due to the fact that $\mathcal{Q}_k$ is a special set of $\mathcal{H}(3, q^2)$. Substituting for the unique line of $\mathcal{R}$ in $\mathcal{L}_{k1}$ the unique line of $\mathcal{L}_{j1}$ passing through T, it follows that there exist $(q+1)/2$ generators of $\mathcal{R}'$ on T.

Now, let $T \in \pi_2$. Through T there pass a unique line ℓ of $\mathcal{L}_{j2}$ contained in $\mathcal{R}$ and a unique line m of $\mathcal{L}_{k1}$. Substituting for ℓ the unique line of $\mathcal{L}_{j1}$ on T and for m the unique line of $\mathcal{L}_{k2}$ through T, it follows that there are $(q+1)/2$ generators of $\mathcal{R}'$ passing through T. Assume that T is a point on a generator tangent to $\mathcal{Q}_j$ not at P. Through T there passes a unique generator of $\mathcal{L}_{j2}$ belonging to $\mathcal{R}$. Substituting for such a generator the unique generator of $\mathcal{L}_{j1}$ on T, there are again $(q+1)/2$ generators of $\mathcal{R}'$ through T. A similar situation occurs if T is a point on a generator tangent to $\mathcal{Q}_k$ not at P. Such a procedure is called *derivation* of $\mathcal{R}$ with respect to the quadrics $\mathcal{Q}_j$, $\mathcal{Q}_k$. This procedure can be iterated and at each step a hemisystem arises. In the general case, $\mathcal{R}'$ is obtained from $\mathcal{R}$ by applying the derivation procedure x times; this is *multiple derivation*. It turns out that the multiple procedure described above produces $\binom{q}{(q-1)/2}$ hemisystems of $\mathcal{H}(3, q^2)$.

Theorem 2.11 *There exists a hemisystem of $\mathcal{H}(3, q^2)$ admitting a group K of order $q^2(q+1)$ for all odd prime powers $q \geq 7$. Dually, there exists a K–invariant $(q+1)/2$–ovoid of $\mathcal{Q}^-(5, q)$, $q \geq 7$.*

Proof From the discussion above, the group G produces exactly four G–invariant hemisystems. Since in this construction a G–orbit of generators tangent to $\mathcal{Q}$ is fixed, the number of G–invariant hemisystems is $2q$. It follows that, for $q \geq 7$, there exist hemisystems of $\mathcal{H}(3, q^2)$ that are K–invariant but not G–invariant. $\qquad\square$

Remark 1 Let $(\mathcal{P}, \mathcal{L})$ be the point-line incidence structure of a generalised quadrangle of order (t^2, t) with t odd. Let Γ_0 be the line graph: its vertex set is $X = \mathcal{L}$ with two vertices adjacent if the lines have a point in common. This is a strongly regular graph with parameters

$$((t^3 + 1)(t + 1), t(t^2 + 1), t - 1, t^2 + 1)$$

and with eigenvalues $k_0 = t(t^2+1)$, $r_0 = t-1$, and $s_0 = -1-t^2$. A hemisystem in $(\mathcal{P}, \mathcal{L})$ is a subset $U_1 \subset \mathcal{L}$ with the property that every point in $\mathcal{P}$ lies on exactly $(t + 1)/2$ lines in U_1 and $(t + 1)/2$ lines in $U_2 = X \setminus U_1$. Cameron, Delsarte, and Goethals [17] showed that any hemisystem in a generalised quadrangle of order (t^2, t) corresponds to a strongly regular decomposition of the line graph of the corresponding generalised quadrangle. Since the complementary set U_2 of lines of a hemisystem is also a hemisystem, this decomposition $X = U_1 \cup U_2$ has equally sized parts. Moreover, the parameters of the parts are the same: each U_i induces a subgraph Γ_i which is strongly regular with parameters

$$(n, k, \lambda, \mu) = ((t^3 + 1)(t + 1)/2, (t^2 + 1)(t - 1)/2, (t - 3)/2, (t - 1)^2/2)$$

and eigenvalues $k = (t^2 + 1)(t - 1)/2$, $r = t - 1$, $s = (t^2 - t + 2)/2$. To our hemisystem a probably new cometric Q–antipodal association scheme is associated.

Remark 2 Another infinite family of hemisystems of $\mathcal{H}(3, q^2)$, for q odd, $q \equiv 3 \pmod 4$, admitting the group $C_{(q^3+1)/4} : C_3$, has been constructed in [9].

Now relative hemisystems of $\mathcal{H}(3, q^2)$, q even, are considered. In 2011, Penttila and Williford [57] introduced the notion of *relative hemisystem*, an analogous concept to hemisystems that exists on $\mathcal{H}(3, q^2)$, q even. They were motivated by the desire to construct an example of an infinite family of primitive cometric association scheme that do not arise from distance regular graphs. Prior to their paper, only sporadic examples of such association schemes were known.

Let Q be a generalised quadrangle of order (q^2, q), containing a generalised quadrangle Q' of order (q, q), where q is a power of two. Each of the lines in Q meets Q' in either $q + 1$ points or is disjoint from it. A subset $\mathcal{R}$ of the lines in $Q \setminus Q'$ is a *relative hemisystem* of Q with respect to Q' if for every point $P \in Q \setminus Q'$ exactly half of the lines on P disjoint from Q' lie in $\mathcal{R}$. Note that the lines through P disjoint from Q' is q.

It can be proved that the generalised quadrangle $\mathcal{H}(3, q^2)$, that has order (q^2, q), has the symplectic generalised quadrangle $\mathcal{W}(3, q)$, that has order (q, q), as a subquadrangle.

Penttila and Williford proved the following result on $\mathcal{H}(3, q^2)$ [57, Theorem 4]

Theorem 2.12 *If $\mathcal{H}(3, q^2)$, $q > 2$, has a relative hemisystem with respect to $\mathcal{W}(3, q)$, then a primitive Q–polynomial 3–class scheme can be constructed on the lines of the relative hemisystem for the following relations R_1, R_2, R_3 :*

(1) $(l, m) \in R_1$ *if and only if l and m are not concurrent and $|O_l \cap O_m| = 1$;*

(2) $(l, m) \in R_2$ if and only if l and m are not concurrent and, in addition, $|O_l \cap O_m| = q + 1$;

(3) $(l, m) \in R_3$ if and only if l and m are concurrent.

Here, for a fixed line l which does not meet $\mathcal{W}(3, q)$, the set O_l consists of the lines meeting both l and $\mathcal{W}(3, q)$.

Also, they showed that $\mathcal{H}(3, q^2)$, q even, has a relative hemisystem with respect to an embedded $\mathcal{W}(3, q)$, admitting the orthogonal group $P\Omega^-(4, q)$ [57, Theorem 5]. They concluded their paper with an open question on the existence of non–isomorphic relative hemisystems on $\mathcal{H}(3, q^2)$, q even.

Cossidente [23] resolved this question two years later by constructing an infinite family of relative hemisystems of $\mathcal{H}(3, q^2)$, each admitting the group $PSL(2, q)$ as an automorphism group, and later on another infinite family admitting a group of order $q^2(q + 1)$, for each $q \geq 8$, q a power of two [24]. Also, in 2014, Cossidente and Pavese constructed a relative hemisystem arising from a Suzuki–Tits ovoid on $\mathcal{H}(3, 64)$ and conjectured that this relative hemisystem is sporadic.

Below is a description of the known relative hemisystems of $\mathcal{H}(3, q^2)$.

2.2   The Penttila–Williford relative hemisystems

The Penttila–Williford construction considers the action of the normaliser of a Singer cyclic group of $P\Omega^-(4, q)$ on $\mathcal{H}(3, q^2)$. They proved that $P\Omega^-(4, q)$, q even, $q \geq 2$, has two orbits on lines of $\mathcal{H}(3, q^2)$ disjoint from $\mathcal{W}(3, q)$. These orbits form two relative hemisystems, H_1 and H_2, They further showed that there exists an involution t, which fixes the points of $\mathcal{W}(3, q)$ and switches H_1 and H_2.

2.3   The Cossidente relative hemisystems

Apart from the Penttila–Williford family of relative hemisystems, the only known infinite families of relative hemisystems are the two discovered by Cossidente in 2013 and 2015 [23], [24]. Both of these families are *perturbations* of the Penttila–Williford relative hemisystems.

The first family on $\mathcal{H}(3, q^2)$, q even and $q \geq 4$, admits the linear group $PSL(2, q)$ as an automorphism group. It was constructed by taking the two Penttila–Williford relative hemisystems H_1 and H_2 and considering the stabiliser of a conic section of the elliptic quadric $\mathcal{Q}^-(3, q)$ in $\mathcal{W}(3, q)$ fixed by $P\Omega^-(4, q)$. This stabiliser is isomorphic to $PSL(2, q)$ and does not act transitively on H_1 and H_2. Cossidente then uses the involution t, switching H_1 and H_2, to delete some orbits on H_1 under $PSL(2, q)$ and replace them with their images under t. Since the number of ways this can be done outnumbers the number of Penttila–Williford hemisystems, Cossidente constructed a new infinite family.

The second infinite family of relative hemisystems discovered by Cossidente admits a group of order $q^2(q+1)$ for each q even and $q > 4$. The construction of this infinite family is very similar to the previous one.

Choose a point P of an elliptic quadric $\mathcal{Q}^-(3,q)$, which is an ovoid of $\mathcal{W}(3,q)$. Let M be the subgroup of the stabiliser of P in $P\Omega^-(4,q)$ of order $q^2(q+1)$. Instead of orbits under $\mathrm{PSL}(2,q)$, Cossidente considers orbits on H_1 and H_2 under M, deleting orbits of H_1 and replacing them by their image under the involution t. Since the number of relative hemisystems invariant under M exceeds that of the Penttila–Williford relative hemisystems, so there exists another infinite family of relative hemisystems.

2.4　The sporadic Cossidente–Pavese relative hemisystem

Let $F = GF(64)$. Let ω be a primitive element of F over $K = GF(2)$, with $\omega^6 + \omega^4 + \omega^3 + \omega + 1 = 0$. Let $\mathcal{H}(3,64)$ be the Hermitian surface of $\mathrm{PG}(3,64)$ with equation $X_1^8 X_4 + X_1 X_4^8 + X_2^8 X_3 + X_2 X_3^8 = 0$ and let $\mathcal{W}(3,8)$ be the canonical symplectic subgeometry of $\mathcal{H}(3,64)$. Let

$$\mathcal{O} := \{(1,x,y,x^4+xy+y^6) : x,y \in K\} \cup \{(0,0,0,1)\}$$

be a Suzuki–Tits ovoid of $\mathcal{W}(3,8)$ with semilinear group $G = 2.Sz(8).3$. Let $P := (0,0,0,1)$. The stabiliser of P in G contains a subgroup K of order 168 fixing a plane section of $\mathcal{O}$. The generators of K are

$$
\begin{bmatrix} 1 & 0 & 0 & 0 \\ 0 & \omega^{27} & 0 & 0 \\ 0 & 0 & \omega^{18} & 0 \\ 0 & 0 & 0 & \omega^{45} \end{bmatrix},
\begin{bmatrix} 1 & \omega^{27} & \omega^9 & \omega^{18} \\ 0 & \omega^{36} & 0 & \omega^{45} \\ 0 & \omega^{18} & \omega^{45} & 1 \\ 0 & 0 & 0 & \omega^{18} \end{bmatrix},
\begin{bmatrix} 1 & 1 & 0 & 1 \\ 0 & 1 & 0 & 0 \\ 0 & 0 & 1 & 1 \\ 0 & 0 & 0 & 1 \end{bmatrix}.
$$

With the aid of MAGMA [20], Cossidente and Pavese were able to construct a relative hemisystem of $\mathcal{H}(3,64)$ as union of line K-orbits: it is formed by considering two orbits of size 28, two orbits of size 56, two orbits of size 84 and ten orbits of size 168, for which representatives are as follows:

$L_1 : \langle (0,1,0,\omega^{49}), (1,0,\omega^{14},\omega^{45}) \rangle;$ $L_2 : \langle (0,1,0,\omega14), (1,\omega^{40},\omega^{49},0) \rangle;$

$L_3 : \langle (0,0,1\omega^{22}), (1,\omega^{50},\omega^{14},0) \rangle;$ $L_4 : \langle (1,\omega,\omega,0), (1,\omega^{55},0,1) \rangle;$

$L_5 : \langle (0,1,0,\omega^{6}), (1,\omega^{48},\omega^{48},0) \rangle;$ $L_6 : \langle (0,1,0,\omega^{21}), (1,\omega^{42},\omega^{42},0) \rangle;$

$L_7 : \langle (0,1,\omega^{54},\omega^{7}), (1,0,\omega^{56},\omega^{36}) \rangle;$ $L_8 : \langle (1,\omega^{59},0,\omega^{36}), (0,1,\omega^{27},\omega^{58}) \rangle;$

$L_9 : \langle (1,0,\omega^{21},\omega^{45}), (0,1,\omega^{18},\omega^{42}) \rangle;$ $L_{10} : \langle (1,0,\omega^{4},1), (0,1,\omega^{27},\omega^{32}) \rangle;$

$L_{11} : \langle (1,0,\omega^{56},\omega^{36}), (0,1,\omega^{9},\omega^{7}) \rangle;$ $L_{12} : \langle (0,0,1,\omega^{43}), (1,\omega^{29},\omega^{47},0) \rangle;$

$L_{13} : \langle (1,0,\omega^{44},0), (0,1,\omega^{54},\omega^{37}) \rangle;$ $L_{14} : \langle (0,1,\omega^{36},\omega^{12}), (1,0,\omega^{33},\omega^{36}) \rangle;$

$L_{15} : \langle (1,\omega^{22},0,0), (1,0,\omega^{4},\omega^{54}) \rangle;$ $L_{16} : \langle (0,1,\omega^{27},\omega^{28}), (1,\omega^{8},0,\omega^{54}) \rangle.$

Finally, in 2015 Bamberg, Lee and Swartz [8] have proposed a group–theoretic approach that unifies all known infinite families of relative hemisystems of $\mathcal{H}(3,q^2)$.

2.5   An alternative approach to the relative hemisystems of Penttila–Williford

Here a new description of the Penttila–Williford relative hemisystem is provided, suggesting an alternative construction of relative hemisystems on the Hermitian surface [26].

Coordinatise $PG(3,q^2)$ in such a way that $\mathcal{H}(3,q^2)$ has equation

$$X_1^q X_4 + X_1 X_4^2 - X_2^{q+1} - X_3^{q+1} = 0$$

and $\mathcal{Q}^-(3,q)$ is the rational curve

$$C := \{(1,t,t^q,t^{q+1}) : t \in \mathrm{GF}(q^2)\} \cup \{(0,0,0,1)\},$$

[34]. Since q is even, the orthogonal polarity associated to $\mathcal{Q}^-(3,q)$ is also symplectic and hence there exists a symplectic Baer subgeometry $\mathcal{W}(3,q)$ such that $\mathcal{Q}^-(3,q) \subset \mathcal{W}(3,q) \subset \mathcal{H}(3,q^2)$. The symplectic Baer subgeometry is given by

$$\{(a,b,b^q,c) : b \in \mathrm{GF}(q^2), a,c \in GF(q)\}.$$

Let τ denote the Baer involution of $\mathrm{P\Gamma L}(4, q^2)$ fixing $\mathcal{W}(3, q)$ pointwise. Then t has matrix

$$\begin{bmatrix} 1 & 0 & 0 & 0 \\ 0 & 0 & 1 & 0 \\ 0 & 1 & 0 & 0 \\ 0 & 0 & 0 & 1 \end{bmatrix}.$$

Now, embed $\mathrm{PG}(3, q^2)$ into $\mathrm{PG}(3, q^4)$ and let t_1 denote the Baer involution of $\mathrm{P\Gamma L}(4, q^4)$ fixing $\mathrm{PG}(3, q^2)$ pointwise. Regarding C as a rational curve in $\mathrm{PG}(3, q^4)$ it turns out that C has $q^4 + 1$ points and hence there is a set, say X, of $q^4 - q^2$ points of C in $\mathrm{PG}(3, q^4) \setminus \mathrm{PG}(3, q^2)$. Note that the Hermitian surface $\mathcal{H}(3, q^2)$, once embedded in $\mathrm{PG}(3, q^4)$ (or more generally in $\mathrm{PG}(3, K)$, where K is the algebraic closure of $\mathrm{PG}(3, q^2)$), remains a surface, say H of degree $q + 1$. Also, C is a curve of degree $q + 1$ embedded in the hyperbolic quadric $\mathcal{Q}^+(3, q^4)$ with equation $X_1 X_4 + X_2 X_3 = 0$.

Lemma 2.13 C is an ovoid of $\mathcal{Q}^+(3, q^4)$.

Proof Represent $\mathcal{Q}^+(3, q^4)$ as the set of all 2×2 singular matrices

$$\begin{bmatrix} X_1 & X_2 \\ X_3 & X_4 \end{bmatrix}.$$

Then the two reguli of $\mathcal{Q}^+(3, q^4)$, say R_1, R_2 are given by

$$X_1 + \lambda X_2 = 0, \quad X_3 + \lambda X_4 = 0;$$
$$X_1 + \lambda X_3 = 0, \quad X_2 + \lambda X_4 = 0,$$

with $\lambda \in \mathrm{GF}(q^4)$. A simple calculation shows that each of these lines meets C exactly once. $\qquad \Box$

Note that the Baer involution τ switches R_1 and R_2.

Lemma 2.14 C is a cap of $\mathrm{PG}(3, q^4)$.

Proof Assume that there exist three collinear points on C on the line r. Since $C \subset \mathcal{Q}^+(3, q^2)$, so r is a line of R_1 or R_2, a contradiction. $\qquad \Box$

Lemma 2.15 The non–linear collineation t_1 fixes X setwise and the pairs $(P, t_1(P))$ define a set $\mathcal{R}$ of $(q^4 - q^2)/2$ lines of $\mathrm{PG}(3, q^2)$ that are generators of $\mathcal{H}(3, q^2)$ and are disjoint from $\mathcal{W}(3, q)$.

Proof First, t_1 fixes X setwise and, for $P \in X$, the line $\ell = \langle P, t(P) \rangle$ is a line of $\mathrm{PG}(3, q^2)$. Since a tangent line to $\mathcal{H}(3, q^2)$ remains tangent to H and H contains X, so ℓ cannot be a tangent to H. Assume that ℓ is secant to $\mathcal{H}(3, q^2)$ Since H has degree $q + 1$ and ℓ has more than $q + 1$ points in common with H, it turns out that ℓ is necessarily a generator of $\mathcal{H}(3, q^2)$. Also, ℓ is disjoint from $\mathcal{W}(3, q)$. Indeed, assume that ℓ meets $\mathcal{W}(3, q)$. Then ℓ is necessarily a line of $\mathcal{W}(3, q)$ (a generator of $\mathcal{H}(3, q^2)$ is either disjoint from $\mathcal{W}(3, q)$ or meets $\mathcal{W}(3, q)$ in a line). Since $C \cap \mathcal{W}(3, q)$ is an ovoid of $\mathcal{W}(3, q)$, so ℓ is tangent to $C \cap \mathcal{W}(3, q)$. But C is a cap of $\mathrm{PG}(3, q^4)$, a contradiction. $\qquad\square$

The size of $\mathcal{R}$ is exactly the half of the set of generators of $\mathcal{H}(3, q^2)$ that are disjoint from $\mathcal{W}(3, q)$ and hence a good candidate to be a relative hemisystem of $\mathcal{H}(3, q^2)$ with respect to $\mathcal{W}(3, q)$. Also, C is left invariant by the subgroup $\mathrm{P\Omega}^-(4, q) \leq \mathrm{PGU}(4, q^2)$ fixing $\mathcal{Q}^-(3, q)$. From [57] the group $\mathrm{P\Omega}^-(4, q)$ acts transitively on $\mathcal{H}(3, q^2) \setminus \mathcal{W}(3, q)$. This means that one can choose P in such a way that $P^\perp$ has equation $X_1 + \alpha X_2 + \alpha X_3 + X_4 = 0$, $\alpha \in \mathrm{GF}(q^2) \setminus \mathrm{GF}(q)$, where $\perp$ denotes the unitary polarity associated with $\mathcal{H}(3, q^2)$. Also P lies on a unique extended line of $\mathcal{W}(3, q)$. The plane $P^\perp$ is secant to $\mathcal{Q}^+(3, q^4)$ and meets C at the points satisfying the equation $t^{q+1} + \alpha t^q + \alpha t + 1 = 0$. From [58] and [11], this equation admits $q + 1$ solutions one of which correspond to a point of C in $\mathcal{W}(3, q)$. Hence $P^\perp$ meets X at exactly q points. Let g be a generator of $\mathcal{H}(3, q^2)$ through P. Assume that P meets X at a unique point T. Since g is fixed by t_1, $t_1(T) \in g$ and then g is a line of $\mathcal{R}$. By construction the number N of lines of $\mathcal{R}$ through P is equal to $q/2$ and $\mathcal{R}$ is a relative hemisystem with respect to $\mathcal{W}(3, q)$, as desired. Clearly, R admits $\mathrm{P\Omega}^-(4, q)$ as an automorphism group and hence it is the Penttila–Williford relative hemisystem.

Remark 3 The complement R^c of R arises from the curve

$$D := \{(1, t^q, t, t^{q+1}) : t \in \mathrm{GF}(q^4)\} \cup \{(0, 0, 0, 1)\}$$

as the image of C under τ.

The above construction suggests the following general construction.

Let $\mathcal{Y}$ be a subset of $q^4 + 1$ points of $\mathrm{PG}(3, q^4)$ that form a cap, having a subset $\mathcal{Z}$ of $q^2 + 1$ points in common with $\mathcal{H}(3, q^2)$ and forming an ovoid in a symplectic Baer subgeometry $\mathcal{W}(3, q) \subset \mathcal{H}(3, q^2)$. Assume that the Baer involution τ fixing $\mathrm{PG}(3, q^2)$ fixes $\mathcal{Y} \setminus \mathcal{Z}$. Assume also that, for a point $P \in \mathcal{H}(3, q^2) \setminus \mathcal{W}(3, q)$, the set $P^\perp$ meets $\mathcal{Y}$ in a number of points less or equal to q. Then the set $\mathcal{R}$ of lines joining $\langle P, \tau(P) \rangle$ is a relative hemisystem of $\mathcal{H}(3, q^2)$ with respect to $\mathcal{W}(3, q)$.

Remark 4 Recently the notion of relative hemisystem has been generalized in [33]. See also [1]. Let $\mathcal{Q}^-(2n+1,q)$ be an elliptic quadric of $\mathrm{PG}(2n+1,q)$. A *relative m–ovoid* of $\mathcal{Q}^-(2n+1,q)$ (with respect to a parablic section $\mathcal{Q} := \mathcal{Q}(2n,q) \subset \mathcal{Q}^-(2n+1,q)$) is a subset $\mathcal{R}$ of points of $\mathcal{Q}^-(2n+1,q) \setminus \mathcal{Q}$ such that every generator of $\mathcal{Q}^-(2n+1,q)$ not contained in $\mathcal{Q}$ meets $\mathcal{R}$ in precisely m points. A relative m–ovoid having the same size as its complement (in $\mathcal{Q}^-(2n+1,q) \setminus \mathcal{Q}$) is called a *relative hemisystem*. It has been proved that a nontrivial relative m–ovoid of $\mathcal{Q}^-(2n+1,q)$ is necessarily a relative hemisystem, forcing q to be even. Also, an infinite family of relative hemisystems of $\mathcal{Q}^-(4n+1,q)$, $n \geq 2$, admitting $\mathrm{PSp}(2n,q^2)$ as an automorphism group has been constructed.

2.6   Hemisystems of $\mathcal{Q}(6,q)$

In [30] Cossidente and Pavese constructed the first infinite family of hemisystems of the parabolic quadric $\mathcal{Q}(6,q)$, q odd, admitting the linear group $\mathrm{PSL}(2,q^2)$ as an automorphism group. The construction relies on the geometry of the classical BLT–set of $\mathcal{Q}^-(5,q)$ and on the classical 1–system of $\mathcal{Q}^-(7,q)$.

Let $\mathcal{Q}(6,q)$ be a parabolic quadric of $\mathrm{PG}(6,q)$, q odd, and also let $Q := \mathcal{Q}^-(5,q)$ be an elliptic quadric obtained as hyperplane section of $\mathcal{Q}(6,q)$. Then $Q = R^\perp \cap \mathcal{Q}(6,q)$ for some point $R \in \mathrm{PG}(6,q) \setminus \mathcal{Q}(6,q)$; here $\perp$ is the polarity induced by $\mathcal{Q}(6,q)$ in $\mathrm{PG}(6,q)$.

Let $\mathcal{H}(3,q^2)$ be the dual of Q, that is, a Hermitian surface of $\mathrm{PG}(3,q^2)$ [55]. A special set of $\mathcal{H}(3,q^2)$ gives, by duality, a BLT–set of lines of Q, that is a set of mutually skew lines of $\mathcal{Q}^-(5,q)$ such that no three of them have a common transversal on Q.

First, recall the construction of the classical BLT–set of Q. Let $V(3,q^2)$ be a three–dimensional vector space over the field $\mathrm{GF}(q^2)$, and let $\mathbf{Q}$ be a non–degenerate quadratic form on $V(3,q^2)$. Using the *trace trick*, [51], starting from $\mathbf{Q}$, a non–degenerate quadratic form $\bar{\mathbf{Q}}$ can be defined in a vector space $V(6,q)$. In this setting, isotropic one-dimensional subspaces of $V(3,q^2)$ correspond to totally isotropic two–dimensional subspaces of $V(6,q)$, and non–isotropic one-dimensional subspaces of $V(3,q^2)$ correspond to non–isotropic two–dimensional subspaces of $V(6,q)$. By multiplying $\mathbf{Q}$ by an appropriate non–zero scalar in $\mathrm{GF}(q^2)$, it may be assumed that $\bar{\mathbf{Q}}$ is elliptic. Let Q be the elliptic quadric of $\mathrm{PG}(5,q)$ associated with $\bar{\mathbf{Q}}$. Let $\mathcal{B}$ be the set of q^2+1 lines of Q obtained by considering the q^2+1 isotropic one–dimensional vector subspaces of $V(3,q^2)$ of $\mathbf{Q}$. Then the lines of $\mathcal{B}$ have the BLT property: any line of $Q \setminus \mathcal{B}$ is either disjoint from $\mathcal{B}$ or meets exactly two lines of $\mathcal{B}$. Any three lines of $\mathcal{B}$ generate $\mathrm{PG}(5,q)$.

Remark 5 Any linear map on $V(3, q^2)$ preserving $\mathbf{Q}$ gives rise to a linear map on $V(6, q)$ preserving $\bar{\mathbf{Q}}$.

A more geometric way to describe $\mathcal{B}$ is as follows. There exist two disjoint planes Γ and $\bar{\Gamma}$ in the extension $\mathrm{PG}(5, q^2)$ of $\mathrm{PG}(5, q)$, that are conjugate with respect to $\mathrm{GF}(q^2)$ and polar with respect to the polarity $\perp'$ of the extension $\mathcal{Q}^+(5, q^2)$ of $\mathcal{Q}^-(5, q)$. Also, $\Gamma \cap \mathcal{Q}^+(5, q^2)$ is a conic $\mathcal{Q}(2, q^2)$ and $\mathcal{B}$ consists of all lines $P\bar{P} \cap \mathrm{PG}(5, q)$, where P varies on $\mathcal{Q}(2, q^2)$ and $\bar{P}$ is its conjugate and hence it is a point of $\bar{\Gamma} \cap \mathcal{Q}^+(5, q^2)$.

If $\mathcal{X}$ is a hyperbolic quadric of $\mathrm{PG}(3, q^2)$, q odd, commuting with $\mathcal{H}(3, q^2)$ such that $\mathcal{X} \cap \mathcal{H}(3, q^2)$ is a Baer elliptic quadric, then $\mathcal{X} \cap \mathcal{H}(3, q^2)$ is a *special–set* of $\mathcal{H}(3, q^2)$. Such a special–set corresponds, under the Klein map, to the classical BLT–set of Q [27, Theorem 2.1].

From Remark 5, the stabiliser of $\mathcal{B}$ in $\mathrm{PGO}^-(6, q)$ contains a subgroup, say G, isomorphic to $\mathrm{PSL}(2, q^2)$.

The following results have been proved in [30].

Lemma 2.16 *The group G has four orbits on points of Q; these are two orbits of size $(q^2 + 1)(q + 1)/2$ and two orbits of size $q^2(q^2 - 1)/2$.*

Lemma 2.17 *The group G has q orbits on 4–dimensional parabolic quadrics embedded in Q; these are $q + 1$ orbits of size $q^2(q^2 + 1)/2$ and $q - 1$ orbits of size $q^2(q^2 - 1)/2$.*

Proposition 2.18 *The group G has q orbits on points of $\mathcal{Q}(6, q) \backslash Q$; these are $(q - 1)/2$ orbits of size $q^2(q^2 - 1)$ and $(q + 1)/2$ orbits of size $q^2(q^2 + 1)$.*

Below are some important properties of the classical 1–system of the elliptic quadric $\mathcal{Q}^-(7, q)$.

A 1–*system* $\mathcal{M}$ of $\mathcal{Q}^-(7, q)$ is a set of $q^4 + 1$ lines $\ell_0, \ldots, \ell_{q^4}$ of $\mathcal{Q}^-(7, q)$ such that every plane of $\mathcal{Q}^-(7, q)$ containing a line $\ell_i \in \mathcal{M}$ has an empty intersection with $(\ell_0, \ell_1, \ldots, \ell_{q^4}) \setminus \ell_i$; see [62] for more details. There is just one 1–system of $\mathcal{Q}^-(7, q)$, [53], [52]. This is the *classical* 1–system, which arises from the trace trick applied to the elliptic quadric $\mathcal{Q}^-(3, q^2)$ considered as an ovoid of itself. From [49, Table 4.3.A, p. 112], the group $O^-(4, q^2)$ embeds in $O^-(8, q)$ and the trace trick applied to the quadratic form associated to $\mathcal{Q}^-(3, q^2)$ produces a pencil $\mathcal{P}$ of elliptic quadrics of $\mathrm{PG}(7, q)$ having as base locus the 1–system $\mathcal{M}$. The stabiliser of $\mathcal{M}$ in $\mathrm{PGO}^-(8, q)$ contains a group isomorphic to $J = \mathrm{PSL}(2, q^4)$.

A more geometric way to describe the classical 1–system of $\mathcal{Q}^-(7, q)$ is as follows. There exist two disjoint 3–subspaces Σ and $\bar{\Sigma}$ in the extension $\mathrm{PG}(7, q^2)$ of $\mathrm{PG}(7, q)$, that are conjugate with respect to $GF(q^2)$ and polar with respect to the polarity $\perp'$ of the extension $\mathcal{Q}^+(7, q^2)$ of $\mathcal{Q}^-(7, q)$. Also, $\Sigma \cap \mathcal{Q}^+(7, q^2)$ is an elliptic quadric $\mathcal{Q}^-(3, q^2)$ and $\mathcal{M}$ consists of all lines $X\bar{X} \cap \mathrm{PG}(7, q)$, where X varies on $\mathcal{Q}^-(3, q^2)$ and $\bar{X}$ is its conjugate and hence it is a point of $\bar{\Sigma} \cap \mathcal{Q}^+(7, q^2)$.

Lemma 2.19 *The group J has three orbits on points of $\mathcal{Q}^-(7, q)$.*

Proposition 2.20 *The group G has $3(q+1)$ orbits on planes of $\mathcal{Q}(6, q)$. These consist of $q+1$ orbits of each size q^2+1, $q^2(q^2+1)(q+1)/2$ and $q^2(q^2+1)(q-1)/2$.*

Now, consider the involution $\alpha \in \mathrm{PGO}(7, q)$ having as an axis the hyperplane Π containing Q and as centre $R = \Pi^\perp$. Such an involution normalises G. Consider the group $K = \langle G, \alpha \rangle$. The groups K and G have the same orbits on points of $\mathcal{Q}(6, q)$. Indeed, let X be a parabolic quadric section of Q such that $X^\perp$ is secant to $\mathcal{Q}(6, q)$ at the points R_1 and R_2. The involution α interchanges R_1 and R_2. On the other hand, the group K has $3(q+1)/2$ orbits on planes of $\mathcal{Q}(6, q)$. Choosing $B = G$ and $A = K$ the hypotheses of the AB–Lemma are satisfied.

Theorem 2.21 *There exist $2^{3(q+1)/2}$ hemisystems of $\mathcal{Q}(6, q)$, q odd, admitting G as an automorphism group.*

Other regular systems of $\mathcal{Q}(6, q)$ are described in the following remarks.

Remark 6 Let $\mathcal{F}$ be any line–spread of $\mathcal{Q}^-(5, q) \subset \mathcal{Q}(6, q)$, that is, a partition of the point set of $\mathcal{Q}^-(5, q)$ into totally singular lines. Then $\mathcal{F}$ is a 1–system of $\mathcal{Q}(6, q)$. Through any line of $\mathcal{F}$ there pass $q+1$ planes of $\mathcal{Q}(6, q)$. From [62, Theorem 5] the number x of planes of $\mathcal{Q}(6, q)$ containing an element of $\mathcal{F}$ and a given point $P \in \mathcal{Q}(6, q)$ not in an element of $\mathcal{F}$ is independent of the choice of P and $x = q+1$. As a consequence, every 1–system of $\mathcal{Q}(6, q)$ gives rise to a regular $(q+1)$–system of $\mathcal{Q}(6, q)$.

Remark 7 Let $\mathrm{PGO}(7, q)$ be the projective orthogonal group of $\mathcal{Q}(6, q)$. The Cartan–Dickson–Chevalley exceptional group $G_2(q)$ is a subgroup of $\mathrm{PGO}(7, q)$. It occurs as the stabiliser in $\mathrm{PGO}(7, q)$ of a configuration F of points, lines and planes on $\mathcal{Q}(6, q)$. The group $G_2(q)$ is also contained in the automorphism group of a classical generalised hexagon, the *split Cayley hexagon*, and denoted by $H(q)$. The points of $H(q)$ are all the points of $\mathcal{Q}(6, q)$ and the lines of $H(q)$ are certain lines of $\mathcal{Q}(6, q)$. The number of points of $\mathcal{Q}(6, q)$ is $(q^6-1)/(q-1)$, and this is also the number of lines of $\mathcal{Q}(6, q)$ involved in $H(q)$. The generalised hexagon $H(q)$ has order (q, q), that is, through any point of $H(q)$ there pass $q+1$ lines of

$H(q)$, and any line of $H(q)$ contains $q+1$ points of $H(q)$. Under the action of $G_2(q)$ the set of planes of $\mathcal{Q}(6,q)$ splits into two orbits. An orbit S of size $(q^6-1)/(q-1)$ and its complement. The orbit S turns out to be a (q^2+q+1)–regular system of $\mathcal{Q}(6,q)$.

Remark 8 With the aid of MAGMA [20], a hemisystem of $\mathcal{Q}(6,3)$ admitting the alternating group A_5 as an automorphism group was found. There is a well-defined geometric setting for the construction of such a hemisystem. Consider the subgroup $K := \mathrm{PGO}(3,3) \times \mathrm{PGO}^-(4,3)$ of $\mathrm{PGO}(7,3)$ stabilising a conic section C of $\mathcal{Q}(6,3)$ whose conjugate space $C^\perp$ is of elliptic type. The group K contains a subgroup isomorphic to A_5 fixing C pointwise and permuting the points of $C^\perp \cap \mathcal{Q}(6,3)$ in a single orbit. A Magma computation shows that K has 28 orbits on points of $\mathcal{Q}(6,3)$ and 32 orbits on generators of $\mathcal{Q}(6,3)$ of which 20 have size 20 and 12 have size 60. The normaliser of K in $\mathrm{PGO}(7,3)$ contains a subgroup N with structure $2.A_5$ that contains K as a normal subgroup. The group N has the same structure of point orbits as K and each N–orbit on generators of $\mathcal{Q}(6,3)$ splits into two K–orbits. From Lemma 2.9 the desired hemisystem is obtained. It is not known whether or not such a construction can be generalised to any odd q.

Remark 9 Considering the planes of $\mathcal{Q}(6,q)$ as vertices of a graph $\mathcal{G}$ with two vertices being adjacent when they are not disjoint, it turns out that $\mathcal{G}$ is strongly regular with

$$v = (q+1)(q^2+1)(q^3+1), \quad k = (q^2+1)(q^2+q+1)q,$$
$$\lambda = (q^3+q^2)(q+1) + (q-1), \quad \mu = (q^2+1)(q^2+q+1).$$

This is induced by the strongly regular graph $\bar{\mathcal{G}}$ arising from one family of solids of the triality quadric $\mathcal{Q}^+(7,q)$ on one of its non–degenerate hyperplane sections. In $\bar{\mathcal{G}}$ two solids are adjacent whenever they meet non–trivially. Following [44], a m–regular system $\mathcal{M}$ of $\mathcal{Q}(6,q)$ is an example of cocliquoid of $\mathcal{G}$. On the other hand, a family $\mathcal{P}$ of planes on a hyperbolic hyperplane section K of $\mathcal{Q}(6,q)$ is an example of cliquoid of $\mathcal{G}$. From [44, Corollary 2.3] it follows that $\mathcal{M}$ and $\mathcal{P}$ always share m planes. Note that the set of all planes of K is again a cliquoid and hence shares $2m$ planes with $\mathcal{M}$. This depends on the fact that the union of two disjoint cliquoids is again a cliquoid.

3 Intriguing sets

Tight sets and m-ovoids of finite classical polar spaces are well studied objects, [7], [6], [42], [62], [65], as well as their connections with two–

character sets and strongly regular graphs.

Let $\mathbf{X}$ be a finite polar space. An m-*ovoid* of $\mathbf{X}$ is a set of points which meets every maximal of $\mathbf{X}$ in precisely m points. A *tight set* of $\mathbf{X}$ is a set T of points of $\mathbf{X}$ with the property that the average number of points of T collinear with a given point of T attains a maximal possible value. A set X of points of $\mathbf{X}$ is called *intriguing* if the number of points of X collinear with a point x of $\mathbf{X}$ is equal to either h_1 or h_2 depending on whether $x \in X$ or $x \notin X$. From [6] an intriguing set of $\mathbf{X}$ is either an m-ovoid or an i-tight set. More precisely, an m-ovoid of a polar space $\mathbf{X}$ of rank r is an intriguing set with intersection numbers $h_1 = m\theta_{r-1} - \theta_{r-1} + 1$ and $h_2 = m\theta_{r-1}$, where θ_r is the *ovoid number* of $\mathbf{X}$. An i-tight set of $\mathbf{X}$ is an intriguing set with intersection numbers $h_1 = i(q^{r-1} - 1)/(q - 1) + q^{r-1}$ and $h_2 = i(q^{r-1} - 1)/(q - 1)$.

In the context of intriguing sets, an important and very useful result is the so called mi-lemma [6, Theorem 4, Corollary 5]. According to this lemma, an m-ovoid and an i-tight set of $\mathbf{X}$ share mi points.

3.1 Intriguing sets of symplectic polar spaces: $(q + 1)$-ovoids of $\mathcal{W}(5, q)$ from relative hemisystems

Many of the results in this section are contained in [31].

Let $\mathcal{R}$ be a relative hemisystem of $\mathcal{H}(3, q^2)$ with respect to a $\mathcal{W}(3, q)$ embedded in $\mathcal{H}(3, q^2)$. In the dual setting $\mathcal{H}(3, q^2)$ is an elliptic quadric $\mathcal{Q}^-(5, q)$ in $\mathrm{PG}(5, q)$ and $\mathcal{W}(3, q)$ is represented by a parabolic section $\mathcal{Q}(4, q)$ of $\mathcal{Q}^-(5, q)$. The ambient space $\mathrm{PG}(3, q)$ of $\mathcal{W}(3, q)$ is a Klein quadric $\mathcal{Q}^+(5, q)$ embedded in the same $\mathrm{PG}(5, q)$ containing $\mathcal{Q}^-(5, q)$ and of course they share $\mathcal{Q}(4, q)$. Actually the above 5-dimensional quadrics generate a pencil $\mathcal{P}$ of quadrics comprising $q/2$ elliptic quadrics, $q/2$ hyperbolic quadrics and a cone $\mathcal{C}$ with basis the parabolic quadric $\mathcal{Q}(4, q)$ and as vertex its nucleus N. Since q is even, an orthogonal polarity is also a symplectic polarity and the orthogonal polarities induced by the non-degenerate quadrics of $\mathcal{P}$ in $\mathrm{PG}(5, q)$ all give the same symplectic polarity $\perp$ of $\mathrm{PG}(5, q)$ giving rise to a symplectic polar space $\mathcal{W}(5, q)$. There exists a group $F \leq \mathrm{PSp}(6, q)$ of non-trivial elations with axis the 4-space containing $\mathcal{C}$ and centre N permuting in a single orbit the $q/2$ non-degenerate quadrics of $\mathcal{P}$.

Each elliptic quadric Q_i^- of $\mathcal{P}$ is partitioned as $Q_i^- = \mathcal{R}_i \cup \mathcal{R}_i^c \cup \mathcal{Q}(4, q)$, $i = 1 \ldots q/2$, where $\mathcal{R}_i$ and $\mathcal{R}_i^c$ we have denote the points on Q_i^- corresponding to $\mathcal{R}$ and its complement in $\mathcal{H}(3, q^2)$, respectively, and their images under f.

Let Q_i^- and Q_j^- be two elliptic quadrics of $\mathcal{P}$. Let A be one of the two subsets $\mathcal{R}_i$ and $\mathcal{R}_i^c$ on Q_i^- and let B be one of the two subsets $\mathcal{R}_j$ and $\mathcal{R}_j^c$

on Q_j^-, $i \neq j$. Then $X = A \cup B \cup Q(4, q)$ is a $(q + 1)$-ovoid of $W(5, q)$.

Proposition 3.1 *A relative hemisystem of $\mathcal{H}(3, q^2)$ gives rise to $2^{q/2}$ non-classical $(q + 1)$-ovoids of $W(5, q)$.*

This gives the following result.

Corollary 3.2 *The set X is an elliptic quasi-quadric.*

Proof The set X has the same size of an elliptic quadric $Q^-(5, q)$ of $PG(5, q)$ and from [6, Lemma 1] is a two-character set with intersection numbers $h_1 = q^3 + q + 1$ and $h_2 = (q + 1)(q^2 + 1)$. $\square$

Remark 10 This construction of non-classical $(q + 1)$-ovoids of $W(5, q)$ can be realised by taking on the two elliptic quadrics Q_i^- and Q_j^- two subsets A and B corresponding to two non-isomorphic relative hemisystems. As a consequence it is possible to construct many infinite families of $(q + 1)$-ovoids of $W(5, q)$ combining the three known infinite families of relative hemisystems of $\mathcal{H}(3, q^2)$.

3.2 $(q + 1)$-ovoids of $W(5, q)$ admitting the Suzuki group

Let $W(3, q) \subset \mathcal{H}(3, q^2)$, $q = 2^h$, h odd. Let $\mathcal{O}$ be a Suzuki-Tits ovoid of $W(3, q)$. The stabiliser of $\mathcal{O}$ in the symplectic group $PSp(4, q)$ of $W(3, q)$ is the simple Suzuki group $Sz(q)$.

Consider the action of $Sz(q)$ on points and generators of $\mathcal{H}(3, q^2)$.

Proposition 3.3 *The Suzuki group $Sz(q)$ has four orbits on both the points and the generators of $\mathcal{H}(3, q^2)$.*

(a) *For the points, they are the following:*

    (i) *the Suzuki-Tits ovoid $O_1 = \mathcal{O}$;*

    (ii) *its complement O_2 in $W(3, q)$;*

    (iii) *an orbit O_3 of size $(q^2+1)(q^2-q)$ consisting of points on $GF(q^2)$-extended lines of the Lüneburg spread $\mathcal{L}$;*

    (iv) *an orbit O_4 of size $(q^3 - q^2)(q^2 + 1)$ consisting of points on $GF(q^2)$-extended lines of $W(3, q)$ not in $\mathcal{L}$.*

(b) *For the generators, they are the following:*

    (i) *the Lüneburg spread $L_1 = \mathcal{L}$;*

    (ii) *its complement L_2 in the set of totally isotropic lines of $W(3, q)$;*

(iii) *an orbit L_3 of size $(q^3-q^2)(q-\sqrt{2q}+1)/2$ consisting of generators meeting O_3 in $q + \sqrt{2q} + 1$ points;*

(iv) *an orbit L_4 of size $(q^3-q^2)(q+\sqrt{2q}+1)/2$ consisting of generators meeting O_3 in $q - \sqrt{2q} + 1$ points.*

Proof The action of the group $Sz(q)$ on points and lines of $\mathcal{W}(3,q)$ is classical, [47]. The stabiliser of a line ℓ of $\mathcal{W}(3,q)$ in $Sz(q)$ is transitive on the imaginary points of ℓ. Any generator of $\mathcal{H}(3,q^2)$ not arising from a line of $\mathcal{W}(3,q)$ is disjoint from $\mathcal{W}(3,q)$ and defines a regular spread $\mathcal{S}$ of $\mathcal{W}(3,q)$. The second part of the proposition follows from [12, Theorem 1a)] and the fact that $Sz(q)$ has exactly two orbits on regular spreads of $\mathcal{W}(3,q)$. □

The block-tactical decomposition matrix for this orbit decomposition is

$$\begin{bmatrix} 1 & q & q^2-q & 0 \\ 1 & q & 0 & q^2-q \\ 0 & 0 & q+\sqrt{2q}+1 & q^2-q-\sqrt{2q} \\ 0 & 0 & q-\sqrt{2q}+1 & q^2-q+\sqrt{2q} \end{bmatrix},$$

and hence the point-tactical decomposition matrix is

$$\begin{bmatrix} 1 & q & 0 & 0 \\ 1 & q & 0 & 0 \\ 1 & 0 & q/2 & q/2 \\ 0 & 1 & (q-\sqrt{2q})/2 & (q+\sqrt{2q})/2 \end{bmatrix}.$$

As above, consider the geometric setting described in Section 3.1. It turns out that any elliptic quadric Q_i^- of $\mathcal{P}$ is partitioned into the parabolic quadric $\mathcal{Q}(4,q)$, an orbit A_i of size $(q^3 - q^2)(q - \sqrt{2q} + 1)/2$ and an orbit B_i of size $(q^3 - q^2)(q + \sqrt{2q} + 1)/2$, $i = 1, \ldots, q/2$. Let π be a totally isotropic plane of $\mathcal{W}(5,q)$. Then π meets any elliptic quadric of $\mathcal{P}$ into a line. By using the above point-tactical decomposition matrix, gluing together $\mathcal{Q}(4,q)$, A_i and B_j, $i \neq j$, a $(q+1)$-ovoid of $\mathcal{W}(5,q)$ is obtained.

Proposition 3.4 *There exist non-classical $(q + 1)$-ovoids of $\mathcal{W}(5,q)$ admitting the group $Sz(q)$ as an automorphism group.*

Proposition 3.5 *A Suzuki–Tits ovoid of $\mathcal{W}(3,q) \subset \mathcal{H}(3,q^2)$ gives rise to $(q^2/2 - q)$ non-classical $(q + 1)$-ovoids of $\mathcal{W}(5,q)$.*

Again as a by-product, we note the following fact.

Corollary 3.6 *The set $\mathcal{Q}(4,q) \cup A_i \cup B_i$ is an elliptic quasi-quadric.*

Proof The set X has the same size as an elliptic quadric $\mathcal{Q}^-(5,q)$ of $PG(5,q)$ and from [6, Lemma 1] is a two-character set with intersection numbers $h_1 = q^3 + q + 1$ and $h_2 = (q+1)(q^2+1)$.   □

Remark 11 Let Q^+ be a hyperbolic quadric of $PG(5,q)$ that gives the the symplectic polarity of $\mathcal{W}(5,q)$ and not belonging to $\mathcal{P}$. Since Q^+ is a (q^2+1)-tight set of $\mathcal{W}(5,q)$ it follows that Q^+ meets X in $(q+1)(q^2+1)$ points and forming a $(q+1)$-ovoid, say Y of Q^+. Equivalently, Y represents a $(q+1)$-cover of $PG(3,q)$. A $(q+1)$-ovoid as constructed above gives rise to a $(q+1)$-cover of $PG(3,q)$.

3.3 (q^2+1)-**tight sets of** $\mathcal{W}(5,q)$ **from ovoids of** $\mathcal{W}(3,q)$

Let $\mathcal{H}(3,q^2)$ be a Hermitian surface of $PG(3,q^2)$, q even, and let $\mathcal{W}(3,q)$ a symplectic subgeometry embedded in $\mathcal{H}(3,q^2)$. Let $PG(3,q)$ be the ambient space of $\mathcal{W}(3,q)$ and let $\mathcal{O}$ be an ovoid of $\mathcal{W}(3,q)$. There are $(q+1)(q^2+1)$ lines of $PG(3,q)$ that are tangent to $\mathcal{O}$ and are totally isotropic with respect to the symplectic polarity of $\mathcal{W}(3,q)$. The non-isotropic lines of $PG(3,q)$ are either secant to or external to $\mathcal{O}$. There are $q^2(q^2+1)/2$ external lines to $\mathcal{O}$ and the same number of secant lines to $\mathcal{O}$.

As above, consider the geometric setting described in Section 3.1. Each hyperbolic quadric Q_i^+ of $\mathcal{P}$ is partitioned as follows: $Q_i^- = S_i \cup E_i \cup Q(4,q)$, $i = 1, \ldots, q/2$, where S_i is the point set corresponding to secant lines to $\mathcal{O}$ and E_i is the point set corresponding to external lines of $\mathcal{O}$. Let A be one of the two subsets S_i and E_i on Q_i^+ and let B be one of the two subsets S_j and E_j on Q_j^+, $i \neq j$.

Proposition 3.7 The set $X = A \cup B \cup Q(4,q)$ is a (q^2+1)-tight set of $\mathcal{W}(5,q)$.

Proposition 3.8 An ovoid of $\mathcal{W}(3,q)$ gives $2^{q/2}$ non-classical (q^2+1)-tight sets of $\mathcal{W}(5,q)$.

Corollary 3.9 The set X is a hyperbolic quasi-quadric.

Remark 12 Let Q^- be an elliptic quadric of $PG(5,q)$ that gives the symplectic polarity of $\mathcal{W}(5,q)$ and not belonging to $\mathcal{P}$. Since Q^- is a $(q+1)$-ovoid of $\mathcal{W}(5,q)$ it follows that Q^- meets X in $(q+1)(q^2+1)$ points and forming a (q^2+1)-tight set of Q^- corresponding to a (q^2+1)-tight set of lines on $\mathcal{H}(3,q^2)$.

3.4   Small tight sets of $\mathcal{W}(4n \pm 1, q^2)$ from disjoint Baer subgeometries

In a recent paper, [32], small tight sets of $\mathcal{W}(4n \pm 1, q^2)$ have been constructed. Here, the construction is shown in the case of $\mathcal{W}(4n + 1, q^2)$.

Consider the projective line $\ell := \mathrm{PG}(1, q^{2(2n+1)})$. Each point P of ℓ defines a projective subspace $X(P)$ of dimension $2n$ of the projective space $\mathrm{PG}(4n + 1, q^2)$ and the set $\mathcal{D} = \{X(P) : P \in \ell\}$ is a *Desarguesian spread* of $\mathrm{PG}(4n+1, q^2)$ ([60, Section 25]). The set $\mathcal{D}$ is isomorphic to ℓ and is the $\mathrm{GF}(q^2)$-*linear representation* of ℓ, with respect to the Desarguesian spread $\mathcal{D}$. Also, there always exists a symplectic polarity $\mathcal{A}$ of $\mathrm{PG}(4n + 1, q^2)$ with respect to which the members of $\mathcal{D}$ are totally isotropic. The linear representation of ℓ with respect to $\mathcal{D}$ gives rise to the embedding of the group $\mathrm{PSL}(2, q^{2(2n+1)})$ into the group $\mathrm{PSp}(4n+2, q^2)$ as a stabiliser of the spread $\mathcal{D}$. In particular, the full stabiliser of $\mathcal{D}$ in $\mathrm{PSp}(4n + 2, q^2)$ is the group $\mathrm{PSL}(2, q^{2(2n+1)}) \cdot C_{2n+1}$, and the stabiliser of $\mathcal{D}$ in $\mathrm{PGL}(4n+2, q^2)$ is the group $\mathrm{PGL}(2, q^{2(2n+1)}) \cdot C_{2n+1} \cdot C_{(q^{4n+2}-1)/(q^2-1)}$ [43]. As shown in [43], there exists a net $\mathcal{N}$ of non–singular linear complexes in $\mathrm{PG}(4n + 1, q^2)$, having $\mathcal{D}$ as base locus which are permuted in a single orbit by the group $C_{(q^{4n+2}-1)/(q^2-1)}$. In particular, $C_{(q^{4n+2}-1)/(q^2-1)}$ fixes each member of $\mathcal{D}$ setwise. In other words, the group $C_{(q^{4n+2}-1)/(q^2-1)}$ lies in the kernel of the representation of $\mathrm{PGL}(2, q^{2n+1}) \cdot C_{2n+1} \cdot C_{(q^{4n+2}-1)/(q^2-1)}$ as stabiliser of $\mathcal{D}$.

A Baer subline of ℓ, regarded as a Hermitian variety $\mathcal{H}(1, q^{2(2n+1)})$ of ℓ, determines a set $\mathcal{D}'$ consisting of $q^{2n+1} + 1$ members of $\mathcal{D}$ and lying on a Hermitian variety $\mathcal{H}(4n + 1, q^2)$ of $\mathrm{PG}(4n + 1, q^2)$, [49]. The stabiliser S of the partial spread $\mathcal{D}'$ in the unitary group $\mathrm{PGU}(4n + 2, q^2)$ certainly contains a central extension of $\mathrm{PSL}(2, q^{2n+1})$ by a cyclic group K of order $(q^{2n+1} + 1)/(q + 1)$, that is actually a subgroup of the cyclic subgroup $C_{(q^{4n+2}-1)/(q^2-1)}$ stabilising $\mathcal{D}$, as observed above. For the structure of the full stabiliser of $\mathcal{D}'$, see [13, Table 2.6]. It turns out that K lies in the kernel of the representation of S as a stabiliser of $\mathcal{D}'$ and hence stabilises $\mathcal{D}'$ component-wise.

Here, $\mathcal{D}'$ arises from a Desarguesian spread of the Baer symplectic subgeometry Σ of $\mathcal{H}(4n + 1, q^2)$, where $\Sigma = \mathrm{Fix}(\mathcal{V})$, with $\mathcal{V} = \mathcal{AU} = \mathcal{UA}$ and $\mathcal{U}$ the non–degenerate unitary polarity of $\mathrm{PG}(4n+1, q^2)$ associated to $\mathcal{H}(4n + 1, q^2)$ [59]. It turns out that each member of $\mathcal{D}'$, say X_i, meets Σ in a subspace of Σ isomorphic to $\mathrm{PG}(2n, q)$. Since K fixes $\mathcal{D}'$ component-wise, K induces in X_i a partition into Baer subspaces of dimension $2n$, see [48]. This means that Σ^K is a set consisting of mutually disjoint Baer subgeometries of $\mathcal{H}(4n + 1, q^2)$, say $\Sigma := \Sigma_1, \ldots, \Sigma_{q^{2n+1}+1)/(q+1)}$, $\Sigma_i = \mathrm{Fix}(\mathcal{UA}_i)$, where $\mathcal{A}_i$ is the symplectic polarity associated to a certain member of $\mathcal{N}$ and $\mathcal{A} = \mathcal{A}_1$.

Lemma 3.10 *The symplectic polarity $\mathcal{A}$ does not induce a symplectic polarity of Σ_i with $\Sigma_i \neq \Sigma$.*

Proof From [59], the symplectic polar space $\mathcal{W}(4n+1, q^2)$ arising from $\mathcal{A}$ and the Hermitian variety $\mathcal{H}(4n+1, q^2)$ have the number of maximal totally isotropic subspaces as the symplectic polar space $\mathcal{W}(4n+1, q)$ of Σ arising from the symplectic polarity induced by $\mathcal{A}$ on Σ.

Assume that $\mathcal{A}$ induces a symplectic polarity of $\Sigma_i \neq \Sigma$. In this case, the symplectic polar space $\mathcal{W}(4n+1, q^2)$ arising from $\mathcal{A}$ should have in common with $\mathcal{H}(4n+1, q^2)$ the number of maximal totally isotropic subspaces of two symplectic polar spaces $\mathcal{W}(4n+1, q)$ sharing $\mathcal{D}'$. As observed above, this cannot happen. $\qquad\square$

Definition 3.11 Two Baer subgeometries Σ_1, Σ_2 of $\mathrm{PG}(2r+1, q^2)$ are *symplectically paired* with respect to a symplectic polarity $\mathcal{A}$ if, for all $P \in \Sigma_1$, the intersection $P^{\mathcal{A}} \cap \Sigma_2 = \mathrm{PG}(2r, q)$.

Lemma 3.12 *The Baer subgeometries Σ_i, Σ_j, $i, j \neq 1, i \neq j$, are symplectically paired with respect to the symplectic polarity $\mathcal{A}$.*

Proof Let P be a point of Σ_i. Then $P^{\mathcal{A}} = P^{\mathcal{VU}} = P^{\mathcal{UV}} = P^{\mathcal{U}^{\mathcal{V}}}$, and $P^{\mathcal{U}}$ is a tangent hyperplane to $\mathcal{H}(4n+1, q^2)$ inducing a hyperplane, say Π, of Σ_i. Let $\mathcal{V}_i$ be the Baer involution fixing Σ_i pointwise. Then, the product $g = \mathcal{VV}_i$, $i \neq 1$, lies in K. Then $\Pi^{\mathcal{V}} = \Pi^{\mathcal{V}_i g} = \Pi^g$. Since $\langle g \rangle$ is fixed point free and $\langle g \rangle \leq K$ the Baer subgeometry Σ_i is sent to another Baer subgeometry Σ_j, $j \neq i$ and hence $P^{\mathcal{A}}$ is a hyperplane of Σ_j, as required. $\qquad\square$

Theorem 3.13 *Any two symplectically paired Baer subgeometries Σ_i, Σ_j give rise to a $2(q^{2n}+1)$–tight set of $\mathcal{W}(4n+1, q^2)$.*

3.5 Tight sets of $\mathcal{W}(5, q^2)$ and the simple group $G_2(q)$

When q is even, the simple exceptional group $G_2(q)$ is a subgroup of $\mathrm{PSp}_6(q)$ and it is the automorphism group of the *perfect symplectic hexagon* $\mathrm{H}(q)$, [66, p.74]: the points of $\mathrm{H}(q)$ are all the points of $\mathrm{PG}(5, q)$ and the lines of $\mathrm{H}(q)$ are certain totally isotropic lines of $\mathrm{PG}(5, q)$ with respect to a symplectic polarity ρ.

Now, embed $\mathrm{PG}(5, q)$ in $\mathrm{PG}(5, q^2)$, thereby extending $\mathrm{H}(q)$ to $\mathrm{H}(q^2)$ and ρ extends (in a unique way, actually) to $\mathrm{PG}(5, q^2)$ in such a way that all lines of $H(q)$ are totally isotropic. The points of $H(q^2)$ are all points of $PG(5, q^2)$. Every hyperplane H in $PG(5, q^2)$ is the image of a unique point P of $H(q^2)$ under ρ. The following results have been proved in [38].

Proposition 3.14 *Let* $\mathcal{P}_i$ *be the set of points of* $H(q^2)$ *at distance* i *from* $H(q)$, $i = 0, 1, 2, 3$. *Then*

(a) $\mathcal{P}_1$ *is precisely the set of points of* $PG(5, q^2) \setminus PG(5, q)$ *incident to lines of* $H(q)$;

(b) $\mathcal{P}_2$ *is the set of points of* $PG(5, q^2) \setminus PG(5, q)$ *incident to a totally isotropic line of* $PG(5, q)$ *with respect to* ρ;

(c) $\mathcal{P}_3$ *is the set of points of* $PG(5, q^2) \setminus PG(5, q)$ *incident to a non-isotropic line of* $PG(5, q)$;

(d)

$$|\mathcal{P}_0| = q^5 + q^4 + q^3 + q^2 + q + 1, \qquad |\mathcal{P}_1| = q^7 - q,$$
$$|\mathcal{P}_2| = q^9 - q^3, \qquad |\mathcal{P}_3| = q^{10} - q^9 + q^8 - q^7 + q^6 - q^5.$$

Theorem 3.15 *Let* $\mathcal{P}_i$ *be the set of points of* $H(q^2)$ *at distance* i *from* $H(q)$, $i = 0, 1, 2, 3$. *Then each* $\mathcal{P}_i$, $i \in \{0, 1, 2, 3\}$ *is a two-character set of* $PG(5, q^2)$, *and so is each* $\mathcal{P}_i \cup \mathcal{P}_j$, $i, j \in \{0, 1, 2, 3\}$, $i \neq j$, *and also each* $\mathcal{P}_i \cup \mathcal{P}_j \cup \mathcal{P}_k$, $|\{0, 1, 2, 3\} \setminus \{i, j, k\}| = 1$. *Omitting the Baer subgeometry* $\mathcal{P}_0$ *and its complement* $\mathcal{P}_1 \cup \mathcal{P}_2 \cup \mathcal{P}_3$, *the corresponding weights* w_1 *and* w_2 *are as follows.*

	w_1	w_2
$\mathcal{P}_1$	$q^7 - q^5$	$q^7 - q^5 - q^4$
$\mathcal{P}_2$	$q^9 - q^7$	$q^9 - q^7 - q^4$
$\mathcal{P}_3$	$q^{10} - q^9$	$q^{10} - q^9 - q^4$
$\mathcal{P}_0 \cup \mathcal{P}_1$	q^7	$q^7 + q^4$
$\mathcal{P}_0 \cup \mathcal{P}_2$	$q^9 - q^7 + q^5$	$q^9 - q^7 + q^5 + q^4$
$\mathcal{P}_0 \cup \mathcal{P}_3$	$q^{10} - q^9 + q^5$	$q^{10} - q^9 + q^5 + q^4$
$\mathcal{P}_1 \cup \mathcal{P}_2$	$q^9 - q^7 + q^6 - q^5$	$q^9 - q^7 + q^6 - q^5 - q^4$
$\mathcal{P}_1 \cup \mathcal{P}_3$	$q^{10} - q^9 + q^7 - q^5$	$q^{10} - q^9 + q^7 - q^5 - q^4$
$\mathcal{P}_2 \cup \mathcal{P}_3$	$q^{10} - q^7$	$q^{10} - q^7 - q^4$
$\mathcal{P}_0 \cup \mathcal{P}_1 \cup \mathcal{P}_2$	q^9	$q^9 + q^4$
$\mathcal{P}_0 \cup \mathcal{P}_1 \cup \mathcal{P}_3$	$q^{10} - q^9 + 2q^7$	$q^{10} - q^9 + 2q^7 + q^4$
$\mathcal{P}_0 \cup \mathcal{P}_2 \cup \mathcal{P}_3$	$q^{10} - q^7 + q^5$	$q^{10} - q^7 + q^5 + q^4$

The Baer subgeometry $W(5, q)$ is $(q + 1)$-tight and its complement in $W(5, q^2)$ is $(q^6 - q)$-tight [6, Theorem 8]. Each of the sets $\mathcal{P}_i$, $i = 1, 2, 3$, in the table above turns out to be an i-tight set of $\mathcal{W}(5, q^2)$ for i as in the following table:

	i
$\mathcal{P}_1$	$q^3 - q$
$\mathcal{P}_2$	$q^5 - q^3$
$\mathcal{P}_3$	$q^6 - q^5$

The tightness of the other sets is a consequence of [6, p. 1296].

3.6   Higher dimensions

The technique for constructing symplectic intriguing sets sometimes can be generalised to higher dimensions by using suitable collineation groups. This is the case of the Second Janko group J_2 which is a subgroup of $PSp(6, 4) \simeq P\Omega(7, 4)$. Starting from a parabolic quadric $\mathcal{Q}(6, 4)$ of $\mathcal{W}(7, 4)$, the group J_2 stabilises $\mathcal{Q}(6, 4)$ and the pencil $\mathcal{P}$ of quadrics on $\mathcal{Q}(6, 4)$. With the aid of Magma, J_2 has 11 orbits on points of $\mathcal{W}(7, 4)$. The two hyperbolic quadrics Q_1^+ and Q_2^+ in $\mathcal{P}$ are partitioned into the $\mathcal{Q}(6, 4)$, an orbit of size 560 and an orbit of size 3600. Gluing together $\mathcal{Q}(6, 4)$, the orbit of size 560 on Q_i^+ and an orbit of size 3600 on Q_j^+, $i \neq j$, a set X is obtained which is a hyperbolic quasi-quadric and a 65-tight set of $\mathcal{W}(7, 4)$ whose automorphism group is $J_2 : 2$.

Intersecting X with an elliptic quadric $\mathcal{Q}^-(7, 4)$ of $\mathcal{W}(7, 4)$ polarising to the symplectic polarity and meeting $\mathcal{Q}(6, 4)$ into a Klein quadric we get a point set of $\mathcal{Q}^-(7, 4)$ of size 1365 forming a 65-tight set of $\mathcal{Q}^-(7, 4)$ admitting the group $PSL(2, 7) : 2$. When the elliptic quadric $\mathcal{Q}^-(7, 4)$ of $\mathcal{W}(7, 4)$ polarising to the symplectic polarity meets $\mathcal{Q}(6, 4)$ in a quadric cone we get a point set of $\mathcal{Q}^-(7, 4)$ of size 1365 forming a 65-tight set of $\mathcal{Q}^-(7, 4)$ admitting the group $A_5 : 2^3$. If the intersection between $\mathcal{Q}^-(7, 4)$ and $\mathcal{Q}(6, 4)$ is an elliptic quadric we get a point set of $\mathcal{Q}^-(7, 4)$ of size 1365 forming a 65-tight set of $\mathcal{Q}^-(7, 4)$ admitting a soluble group of order $2^2.3.5^2$.

4   Intriguing sets of quadrics

There exist two very nice and important objects in finite geometry admitting the same automorphism group and apparently not related to each other. They are the hemisystem of the Hermitian generalised quadrangle $\mathcal{H}(3, q^2)$, q odd [34], presented in Section 2.1, and a Cameron–Liebler line class in $PG(3, q)$, q odd, with parameter $(q^2 + 1)/2$, [15], both admitting the classical group $G := P\Omega^-(4, q)$ stabilising an elliptic quadric $\mathcal{Q}^-(3, q)$ of $PG(3, q)$. Under duality [55], they are both examples of intriguing sets on quadrics of $PG(5, q)$.

Although a possible connection between these objects has not been established, the following question arises. Does there exist a suitable chosen subgroup of G producing new hemisystems of the Hermitian generalised quadrangle and new Cameron–Liebler line classes of $PG(3,q)$? The answer is affirmative. By considering a suitable subgroup of the stabiliser of a point $P \in \mathcal{Q}^-(3,q)$ we were able to construct, for $q \geq 5$, a new infinite family of hemisystems of $\mathcal{H}(3,q^2)$ (see Section 2.1). Here the existence is shown of a new infinite family of Cameron–Liebler line classes with parameter $(q^2 + 1)/2$ for $q \geq 5$, and admitting the same automorphism group.

A *Cameron–Liebler line class* $\mathcal{L}$ is a set of lines in $PG(3,q)$ such that for any line ℓ of $PG(3,q)$,

$$|\{m \in \mathcal{L} : |m \cap \ell| = 1, m \neq \ell\}| = \begin{cases} (q+1)x + (q^2 - 1) & \text{if } \ell \in \mathcal{L}, \\ (q+1)x & \text{if } \ell \notin \mathcal{L}. \end{cases}$$

for some fixed integer x, the *parameter* of $\mathcal{L}$.

There are many equivalent characterisations of Cameron–Liebler line classes; see [56]. Under the Klein correspondence between the lines of $PG(3,q)$ and points of a Klein quadric $\mathcal{Q}^+(5,q)$, a Cameron–Liebler line class produces a *tight* set of $\mathcal{Q}^+(5,q)$. There exist classical examples of Cameron-Liebler line classes $\mathcal{L}$ with parameters $x = 1, 2$ and $x = q^2, q^2 - 1$. These are the set of lines through a point P and, analogously, the set of lines in a plane π form a Cameron–Liebler line class with parameter $x = 1$. The union of these two sets for $P \notin \pi$ forms a Cameron-Liebler line class with parameter $x = 2$.

In general, the complement of a Cameron-Liebler line class with parameter x is a Cameron–Liebler line class with parameter $q^2 + 1 - x$. It was conjectured that no other examples of Cameron-Liebler line classes exist, [19], but Bruen and Drudge [15] found an infinite class of Cameron–Liebler line classes in $PG(3,q)$, q odd, with $x = (q^2 + 1)/2$. Govaerts and Penttila [46] found a Cameron–Liebler line class in $PG(3,4)$ with parameter $x = 7$, and so also with parameter $x = 10$. Very recently, a new infinite Cameron–Liebler line class was found for $q \equiv 9 \mod 12$ in [45].

4.1   The new family of Cameron–Liebler line classes

Let q be odd, and let $\mathcal{Q}^-(3,q) \subset PG(3,q)$ be an elliptic quadric with quadratic form Q. Each point $P \in \mathcal{Q}^-(3,q)$ lies on q^2 secants to $\mathcal{Q}^-(3,q)$, and so lies on $q + 1$ tangent lines. Let $\mathcal{L}_P$ be a set of $(q+1)/2$ of the tangent lines to $\mathcal{Q}^-(3,q)$ through P, and let S be the set of secant lines to $Q^-(3,q)$; then the set

$$\mathcal{L} = \cup_{P \in Q^-(3,q)} \mathcal{L}_P \cup S$$

has size $(q^2+1)(q^2+q+1)/2$, which is the number of lines of $PG(3,q)$ in a Cameron–Liebler line class with parameter $(q^2+1)/2$. A suitably choice of the set $\mathcal{L}_P$ makes $\mathcal{L}$ a Cameron–Liebler line class, [15].

Let $G = P\Omega_4^-(q)$ be the commutator subgroup of the full stabiliser of $Q^-(3,q) \subset PG(3,q)$ in $PGL(4,q)$. The group G has three orbits on the points of $PG(3,q)$, that is, the points of $Q^-(3,q)$ and other two orbits O_s and O_n of size $q^2(q^2+1)/2$. The two orbits O_s, O_n correspond to points of $PG(3,q)$ such that the evaluation of the quadratic form associated to $Q^-(3,q)$ is a square or a non–square in $GF(q)$. In its action on lines of $PG(3,q)$, the group G has four orbits: two orbits, say $\mathcal{L}_1$ and $\mathcal{L}_2$, both of size $(q+1)(q^2+1)/2$, consisting of lines tangent to $Q^-(3,q)$ and two orbits, say $\mathcal{L}_3$ and $\mathcal{L}_4$, both of size $q^2(q^2+1)/2$ consisting of lines secant or external to $Q^-(3,q)$.

The block-tactical decomposition matrix for this orbit decomposition is

$$\begin{bmatrix} 1 & 1 & 2 & 0 \\ q & 0 & \frac{q-1}{2} & \frac{q+1}{2} \\ 0 & q & \frac{q-1}{2} & \frac{q+1}{2} \end{bmatrix},$$

and hence the point-tactical decomposition matrix is

$$\begin{bmatrix} \frac{q+1}{2} & \frac{q+1}{2} & q^2 & 0 \\ q+1 & 0 & \frac{q(q-1)}{2} & \frac{q(q+1)}{2} \\ 0 & q+1 & \frac{q(q-1)}{2} & \frac{q(q+1)}{2} \end{bmatrix}.$$

Simple group–theoretic arguments show that a line of $\mathcal{L}_1$ contains q points of O_s, that a line of $\mathcal{L}_2$ contains q points of O_n, that a secant line to $Q^-(3,q)$ always contains $(q-1)/2$ points of O_s and $(q-1)/2$ points of O_n, and that a line external to $Q^-(3,q)$ contains $(q+1)/2$ points of O_s and $(q+1)/2$ points of O_n.

Let $\mathcal{L} = \mathcal{L}_1 \cup \mathcal{L}_3$. From the orbit–decompositions above it follows that $\mathcal{L}$ is the Cameron–Liebler line class in [15] . In particular, $\mathcal{L}$ has (a) the characters $q^2+(q+1)/2$, $(q^2-q)/2$, $(q^2+q+2)/2$ with respect to line–stars of $PG(3,q)$, and (b) the characters $(q+1)/2$, $(q^2+q)/2$, $(q^2+3q+2)/2$, with respect to the line sets in planes of $PG(3,q)$.

Let $P \in Q^-(3,q)$ and let π be the tangent plane to $Q^-(3,q)$ at P. The following sets of lines of $PG(3,q)$ can be distinguished:

(1) $t_1 :=$ the $(q+1)/2$ lines in $\mathcal{L}_1$ through P;

(2) $t_2 :=$ the $(q+1)/2$ lines in $\mathcal{L}_2$ through P;

(3) $s_1 :=$ the q^2 secants on P;

(4) $s_2 := \mathcal{L}_3 \setminus s_1$;

(5) $u_1 :=$ lines in $\mathcal{L}_1 \setminus t_1$;

(6) $u_2 :=$ lines in $\mathcal{L}_2 \setminus t_2$;

(7) $e_1 := q^2$ external line lying on π;

(8) $e_2 := q^2(q^2 - 1)/2$ external lines not in π.

Let $\mathcal{L}' := t_1 \cup s_1 \cup u_1 \cup e_2$. Then $\mathcal{L}'$ is a Cameron–Liebler line class with parameter $(q^2 + 1)/2$. Note that $\mathcal{L}'$ has size

$$(q+1)/2 + q^2 + q^2(q+1)/2 + q^2(q^2 - 1)/2 = (q^2 + 1)(q^2 + q + 1)/2.$$

Also, $\mathcal{L}'$ has the following five characters with respect to line–stars of $PG(3, q)$:

$$(q+3)/2, \quad (q^2 - q)/2, \quad (q^2 + q + 2)/2,$$
$$(q^2 + 3q + 4)/2, q^2 + (q+1)/2.$$

It has the following characters with respect to the line sets in planes of $PG(3, q)$:

$$(q+1)/2, \quad (q^2 - q - 2)/2, \quad (q^2 + q)/2,$$
$$(q^2 + 3q + 2)/2, \quad q^2 + (q-1)/2.$$

It turns out that, for $q > 3$, these characters are distinct from those of a Bruen–Drudge and Cameron–Liebler line class.

Theorem 4.1 *If $q \geq 5$, there exists a Cameron–Liebler line class with parameter $(q^2 + 1)/2$ not equivalent to the Bruen–Drudge example.*

References

[1] J. Bamberg, M. Lee, A relative m–cover of a Hermitian surface is a relative hemisystem, *arXiv 1608.03055v1*.

[2] Laura Bader, Guglielmo Lunardon, and Joseph A. Thas, *Derivation of flocks of quadratic cones*, Forum Math. **2** (1990), 163–174.

[3] Simeon Ball, Aart Blokhuis, and Francesco Mazzocca, *Maximal arcs in Desarguesian planes of odd order do not exist*, Combinatorica **17** (1997), 31–41.

[4] John Bamberg, Michael Giudici, and Gordon F. Royle, *Every flock generalized quadrangle has a hemisystem*, Bull. London Math. Soc. **42** (2010), 795–810.

[5] John Bamberg, Michael Giudici, and Gordon F. Royle, *Hemisystems of small flock generalized quadrangles*, Des. Codes Cryptogr. **67** (2013), 137–157.

[6] John Bamberg, Shane Kelly, Maska Law, and Tim Penttila, *Tight sets and m-ovoids of finite polar spaces*, J. Combin. Theory Ser. A **114** (2007), 1293–1314.

[7] John Bamberg, Mashka Law, and Tim Penttila, *Tight sets and m-ovoids of finite generalised quadrangles*, Combinatorica **29** (2009), 1–17.

[8] John Bamberg, Melissa Lee, and Eric Swartz, *A note on relative hemisystems of Hermitian generalized quadrangles*, Des. Codes Cryptogr. (2015), to appear.

[9] John Bamberg, Melissa Lee, Koji Momihara, and Qing Xiang, *A new infinite family of hemisystems of the Hermitian surface*, preprint.

[10] Luke Bayens, *Hyperovals, Laguerre planes and hemisystems – An approach via symmetry*, Ph.D. Thesis, Colorado State University, Fort Collins, Colorado, 2013.

[11] Antonia W. Bluher, *On $x^{q+1} + ax + b$*, Finite Fields Appl. **10** (2004), 285–305.

[12] Ray C. Bose and T. Shimamoto, *Classification and analysis of partially balanced incomplete block designs with two associate classes*, J. Amer. Statist. Assoc. **47** (1952), 151–184.

[13] John Bray, Derek Holt, and Colva Roney-Dougal, *The maximal subgroups of the low-dimensional finite classical groups*, London Mathematical Society, LNS 407, Cambridge University Press, Cambridge, 2013.

[14] Aiden A. Bruen and James W.P. Hirschfeld, *Applications of line geometry over finite fields. II. The Hermitian surface*, Geom. Dedicata **7** (1978), 333–353.

[15] Aiden A. Bruen and Keldon Drudge, *The construction of Cameron-Liebler line classes in* PG$(3, q)$, Finite Fields Appl. **5** (1999), 35–45.

[16] Peter J. Cameron, *Partial quadrangles*, Quart. J. Math. Oxford Ser. **26** (1975), 61–73.

[17] Peter J. Cameron, Philippe Delsarte, and Jean-Marie Goethals, *Hemisystems, orthogonal configurations and dissipative conference matrices*, Philips J. Res. **34** (1979), 147–162.

[18] Peter J. Cameron, Jean-Marie Goethals, and Johan J. Seidel, *Strongly regular graphs having strongly regular subconstituents*, J. Algebra **55** (1978), 257–280.

[19] Peter J. Cameron and Robert A. Liebler, *Tactical decompositions and orbits of projective groups*, Linear Algebra Appl. **46** (1982), 91–102.

[20] John Cannon and Catherine Playoust, *An introduction to MAGMA*, University of Sydney, Sydney, Australia, 1993.

[21] Miroslava Cimráková and Veerle Fack, *Searching for maximal partial ovoids and spreads in generalized quadrangles*, Bull. Belg. Math. Soc. Simon Stevin **12** (2005), 697–705.

[22] Miroslava Cimráková and Veerle Fack, *Clique algorithms for finding substructures in generalized quadrangles*, J. Combin. Math. Combin. Comput. **63** (2007), 129–143.

[23] Antonio Cossidente, *Relative hemisystems on the Hermitian surface*, J. Algebraic Combin. **38** (2013), 275–284.

[24] Antonio Cossidente, *A new family of relative hemisystems on the Hermitian surface*, Des. Codes Cryptogr. **75** (2005), 213–221.

[25] Antonio Cossidente and Gary L. Ebert, *Permutable polarities and a class of ovoids of the Hermitian surface*, European J. Combin. **25** (2004), 1059–1066.

[26] Antonio Cossidente, Gabor Korchmáros, and Francesco Pavese, *A new approach to relative hemisystems*, preprint (2016).

[27] Antonio Cossidente, Giuseppe Marino, and Olga Polverino, *Special sets of the Hermitian surface and indicator sets*, J. Combin. Des. **16** (2008), 18–24.

[28] Antonio Cossidente and Francesco Pavese, *Intriguing sets of quadrics in* PG$(5, q)$, Adv. Geom., to appear.

[29] Antonio Cossidente and Francesco Pavese, *On the geometry of unitary involutions*, Finite Fields Appl. **36** (2015), 14–28.

[30] Antonio Cossidente and Francesco Pavese, *Hemisystems of $\mathcal{Q}(6,q)$, q odd*, J. Combin. Theory Ser. A, to appear.

[31] Antonio Cossidente and Francesco Pavese, *Intriguing sets of $\mathcal{W}(5,q)$, q even*, J. Combin Theory Ser. A **127** (2014), 303–313.

[32] Antonio Cossidente and Francesco Pavese, *Small tight sets of $\mathcal{W}(4n\pm 1,q^2)$*, preprint.

[33] Antonio Cossidente and Francesco Pavese, *Relative m–ovoids of elliptic quadrics*, submitted.

[34] Antonio Cossidente and Tim Penttila, *Hemisystems on the Hermitian surface*, J. London Math. Soc. **72** (2005), 731–741.

[35] Antonio Cossidente and Tim Penttila, *On m-regular systems on $\mathcal{H}(5,q^2)$*, J. Algebraic Combin. **29** (2009), 437–445.

[36] Antonio Cossidente and Tim Penttila, *Segre's hemisystem and McLaughlin's graph*, J. Combin. Theory Ser. A **115** (2008), 686–692.

[37] Antonio Cossidente and Tim Penttila, *Subquadrangle m-regular systems on generalized quadrangles*, J. Combin. Des. **19** (2011), 28–41.

[38] Antonio Cossidente and Hendrik Van Maldeghem, *The simple exceptional group $G_2(q)$, q even, and two-character sets*, J. Combin. Theory Ser. A **114** (2007), 964–969.

[39] Antonio Cossidente and Sam K.J. Vereecke, *Some geometry of the isomorphism $Sp(4,q) \simeq O(5,q)$, q even*, J. Geom. **70** (2001), 28–37.

[40] Jan De Beule, Patrick Govaerts, Anja Hallez, and Leo Storme, *Tight sets, weighted m–covers, weighted m–ovoids, and minihypers*, Des. Codes Cryptogr. **50** (2009), 187–201.

[41] Franck De Clerck, Nikias De Feyter, Nicola Durante, *Two-intersection sets with respect to lines on the Klein quadric*, Bull. Belg. Math. Soc. Simon Stevin **12** (2005), 743–750.

[42] Keldon Drudge, *Extremal sets in projective and polar spaces*, Ph. D. Thesis, University of Western Ontario, Canada, 1998.

[43] Roger H. Dye, *Spreads and classes of maximal subgroups of* $GL_n(q), SL_n(q), PGL_n(q)$ *and* $PSL_n(q)$, Ann. Mat. Pura Appl. **158** (1991), 33–50.

[44] Joergen Eisfeld, *On the common nature of spreads and pencils in* $PG(d, q)$, Discrete Math. **189** (1998), 95–104.

[45] Tao Feng, Koji Momihara, and Qing Xiang, *Cameron–Liebler line classes with parameter* $x = (q^2 - 1)/2$, J. Combin. Theory Ser. A **133** (2015), 307–338.

[46] Patrick Govaerts and Tim Penttila, *Cameron-Liebler line classes in* $PG(3, 4)$, Bull. Belg. Math. Soc. Simon Stevin **12** (2005), 793–804.

[47] James W.P. Hirschfeld, *Finite projective spaces of three dimensions*, Oxford Mathematical Monographs, Oxford Science Publications,The Clarendon Press, Oxford University Press, Oxford, 1985.

[48] Barbu C. Kestenband, *Projective geometries that are disjoint unions of caps*, Canad. J. Math., **32** (1980), 1299–1305.

[49] Peter Kleidman and Martin Liebeck, *The Subgroup Structure of the Finite Classical Groups*, Cambridge University Press, Cambridge (1990).

[50] Norbert Knarr, *A geometric construction of generalized quadrangles from polar spaces of rank three*, Results Math. **21** (1992), 332–344.

[51] Michel Lavrauw and Gertrude Van de Voorde, *Field reduction and linear sets in finite geometry*, Topics in finite fields, 271–293, *Contemp. Math.*, **632**, Amer. Math. Soc., Providence, RI, 2015.

[52] Deirdre Luyckx, *m–systems of finite classical polar spaces*, Ph.D. Thesis, Ghent University (2002).

[53] Deirdre Luyckx and Joseph A. Thas, *The uniqueness of the 1-system of* $Q^-(7, q)$, *q odd*, J. Combin. Theory Ser. A **98** (2002), 253–267.

[54] William J. Martin, Mikhail Muzychuk, and Jason Williford, *Imprimitive cometric association schemes: constructions and analysis*, J. Algebraic Combin. **25** (2007), 399–415.

[55] Stanley E. Payne and Joseph A. Thas, *Finite Generalized Quadrangles,* Res. Notes Math. 110, Pitman, 1984.

[56] Tim Penttila, *Cameron–Liebler line classes in* $PG(3, q)$, Geom. Dedicata **37** (1991), 245–252.

[57] Tim Penttila and J. Williford, *New families of Q-polynomial associa-tion schemes*, J. Combin. Theory Ser. A **118** (2011), 502–509.

[58] Joergen Rathmann, *The uniform principle for curves in characteristic p*, Math Ann. **276** (1987), 565–579.

[59] Beniamino Segre, *Forme e geometrie hermitiane, con particolare riguardo al caso finito*, Ann. Mat. Pura Appl. **70** (1965), 1–201.

[60] Beniamino Segre, *Teoria di Galois, fibrazioni proiettive e geometrie non desarguesiane*, Ann. Mat. Pura Appl. **64** (1964), 1–76.

[61] Ernest E. Shult and Joseph A. Thas, *Construction of polygons from buildings*, Proc. London Math. Soc. **71** (1995), 397–440.

[62] Ernest E., Shult and Joseph A. Thas, *m-systems of polar spaces*, J. Combin. Theory Ser. A **68** (1994), 184–204.

[63] Joseph A. Thas, *Ovoids and spreads of finite classical polar spaces*, Geom. Dedicata **10** (1981), 135–143.

[64] Joseph A. Thas, *Projective geometry over a finite field*, Handbook of Incidence Geometry, (Buekenhout, F., ed.), 295–347, North-Holland, Amsterdam, 1995.

[65] Joseph A. Thas, *Interesting pointsets in generalized quadrangles and partial geometries*, Linear Algebra Appl. **114/115** (1989), 103–131.

[66] Hendrik Van Maldeghem, *Generalized polygons*, Monographs in Math-ematics, vol. 93, Birkhäuser, Basel, (1998).

Dipartimento di Matematica Informatica ed Economia
Università della Basilicata
Contrada Macchia Romana
I-85100 Potenza
Italy
antonio.cossidente@unibas.it

Switching techniques for
edge decompositions of graphs

Daniel Horsley

Abstract

This article concerns a class of techniques, herein referred to as
edge switching techniques, that enable a new edge decomposition to
be obtained from an existing one by interchanging edges between the
subgraphs in the decomposition. These techniques can be viewed as
generalisations of classical path switching methods for proper edge
colourings. Their use in other edge decomposition settings dates
back at least to 1980, but the last ten years have seen them rapidly
developed and employed to resolve Lindner's conjecture on embedding partial Steiner triple systems, Alspach's cycle decomposition
problem, and numerous other questions. Here we aim to give the
reader a gentle introduction to these techniques and to some of their
most significant applications beyond edge colouring.

1 Overview

An *edge decomposition* (hereafter simply a *decomposition*) of a graph G
is a set of subgraphs of G such that each edge of G occurs in exactly one of
the subgraphs. Questions concerning edge decompositions of graphs form
a major theme in graph theory, combinatorial design theory, and finite
geometry. Significantly for our purposes here, proper edge colourings of
graphs are simply edge decompositions into matchings.

The *edge switching techniques*[1] that we discuss here are methods that
enable a new decomposition to be obtained from an existing one by interchanging edges between the graphs in the decomposition. They can be seen
as an extension of the classical switching methods in graph colouring used
by, for example, Kempe [66] and Vizing [92]. In 1980 Andersen, Hilton
and Mendelsohn [4] employed a seminal example of edge switching in the
setting of triangle decompositions. This result was eventually extended to
4- and 5-cycle decompositions by Raines and Szaniszló [81] in 1999 and
again to decompositions into cycles of arbitrary lengths in 2005 [27]. Since
then, edge switching techniques have been rapidly developed and applied

---

[1] These techniques are distinct from the switching methods used in enumerative and
probabilistic combinatorics (see [47]), although there are some similarities. They are,
however, closely related to the so called *method of amalgamations* (see [5]).

to new settings and problems. This article aims to survey these techniques and some of their major applications beyond edge colouring.

The next section of this article gives an elementary introduction to edge switching techniques. Sections 3, 4 and 5 then discuss three of the most significant applications of edge switching techniques outside of edge colouring to date: to cycle decomposition of graphs, to embedding partial Steiner triple systems, and to almost regular decomposition of graphs.

2   An introduction to edge switching

This section aims to introduce the reader to edge switching techniques, assuming no prior familiarity. We detail how the fundamental idea can be generalised from the classical setting of proper edge colourings to decompositions into triangles and finally to decompositions into cycles of arbitrary lengths.

2.1   Switching for proper edge colourings

The idea of taking a proper (vertex) colouring of a graph and interchanging the colours on a maximal connected 2-coloured subgraph in order to produce a new proper colouring dates back to Kempe [66] in 1879. He famously employed the idea to prove that a minimal counterexample to the four colour theorem must have minimum degree at least five, and the idea has since been used extensively in results on vertex coloring.

A *proper edge colouring* (hereafter simply an *edge colouring*) of a graph G is usually defined as an assignment of colours to the edges of G such that no vertex has two edges of the same colour incident on it. This condition is the same as requiring that the edges assigned any given colour form a matching, and hence we can equivalently define an edge colouring as a decomposition into matchings. Each matching is referred to as a *colour class* of the colouring.

When applied in the setting of edge colourings, Kempe's concept reduces to interchanging the colours of edges along a maximal 2-coloured path. Again, this idea has been used in the proofs of a great many edge colouring results, Vizing's theorem [92] being an archetypical example. Let $\Delta(G)$ denote the maximum degree of a graph G.

Theorem 2.1 ([92]) *Any graph G has an edge colouring with $\Delta(G) + 1$ colours.*

Later, in Section 4.2, we will see how this theorem emerges as a corollary of an edge switching result concerning triangle decompositions.

The union $G_1 \cup G_2$ of graphs G_1 and G_2 is the graph with vertex set $V(G_1) \cup V(G_2)$ and edge set $E(G_1) \cup E(G_2)$. Let $S_1 \triangle S_2$ denote the symmetric difference of sets S_1 and S_2. We extend the usual definition of $\deg_H(x)$ so that $\deg_H(x) = 0$ for any vertex x not in $V(H)$. The following lemma formalises the concept of switching along a maximal 2-coloured path in an edge colouring.

Lemma 2.2 *Let $\{M_1, \ldots, M_t\}$ be an edge colouring of a graph G, and let α, β be distinct elements of $\{1, \ldots, t\}$. There is a partition Π of the set $V(M_\alpha) \triangle V(M_\beta)$ into pairs such that, for each $\{u, v\} \in \Pi$, there is an edge colouring $\mathcal{P}'_{\{u,v\}} = \{M'_1, \ldots, M'_t\}$ of G satisfying*

- $V(M'_\alpha) = V(M_\alpha) \triangle \{u, v\}$ *and* $V(M'_\beta) = V(M_\beta) \triangle \{u, v\}$;

- $M'_i = M_i$ *for* $i \in \{1, \ldots, t\} \setminus \{\alpha, \beta\}$ *and* $M'_\alpha \cup M'_\beta = M_\alpha \cup M_\beta$.

Proof Because $M_\alpha \cup M_\beta$ is a union of two matchings, it is a vertex-disjoint union of paths and even length cycles. The set $V(M_\alpha) \triangle V(M_\beta)$ is the set of end vertices of maximal paths in $M_\alpha \cup M_\beta$. These maximal paths induce a natural partition into pairs, which we take to be Π.

Let $\{u, v\} \in \Pi$ and let Q be the path in $M_\alpha \cup M_\beta$ from u to v. Form the edge colouring $\mathcal{P}'_{\{u,v\}} = \{M'_1, \ldots, M'_t\}$ from $\{M_1, \ldots, M_t\}$ by reassigning[2] the edges in $E(M_\alpha) \cap E(Q)$ to M'_β and the edges in $E(M_\beta) \cap E(Q)$ to M'_α. Then $\mathcal{P}'_{\{u,v\}}$ has the required properties. $\square$

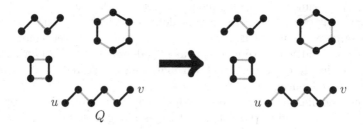

Figure 1: Illustration of Lemma 2.2.

As a simple example of the use of Lemma 2.2 we prove an elementary result on edge colourings from [77].

---

[2]Throughout, when we say that a decomposition $\{G'_1, \ldots, G'_t\}$ is formed from another $\{G_1, \ldots, G_t\}$ by reassigning certain edges, we imply that $e \in E(G_i)$ if and only if $e \in E(G'_i)$ for any edge e that is not reassigned and $i \in \{1, \ldots, t\}$.

Lemma 2.3 *If a graph G has an edge colouring with t colours, then it has an edge colouring with t colours such that the sizes of any two colour classes differ by at most one.*

Proof Let $\{M_1, \ldots, M_t\}$ be an edge colouring of G with t colours. Suppose there are two colour classes whose sizes differ by more than one. Then we can apply Lemma 2.2 with α and β chosen so that $|V(M_\alpha)| > |V(M_\beta)| + 2$. Because $|V(M_\alpha) \setminus V(M_\beta)| > |V(M_\beta) \setminus V(M_\alpha)|$, there is a pair $\{u, v\}$ in the partition Π given by Lemma 2.2 such that $u, v \in V(M_\alpha) \setminus V(M_\beta)$. The corresponding edge colouring $\mathcal{P}'_{\{u,v\}} = \{M'_1, \ldots, M'_t\}$ of G has $V(M'_\alpha) = V(M_\alpha) \setminus \{u, v\}$, $V(M'_\beta) = V(M_\beta) \cup \{u, v\}$, and $M'_i = M_i$ for $i \in \{1, \ldots, t\} \setminus \{\alpha, \beta\}$. It is clear that by repeating this procedure we can obtain an edge colouring of G with t colours such that the sizes of any two colour classes differ by at most one. $\square$

Techniques of the kind formalised by Lemma 2.2 have been used extensively in work concerning edge colouring. Our focus here, however, is on the application of this switching idea beyond the setting of edge colouring. In the next section we see how the idea can be extended to decompositions into triangles.

2.2 Switching for triangle decompositions

A *packing* of a graph G is a set of subgraphs of G such that each edge of G occurs in at most one of the subgraphs. The *leave* of a packing is the spanning subgraph of G that contains exactly those edges that do not occur in a subgraph in the packing. (A packing of G may be equivalently considered as a decomposition of a subgraph H of G.) If each subgraph in a packing is a triangle, we say it is a *triangle packing*.

We denote by $(x_1, \ldots, x_k)$ the cycle with vertices $x_1, \ldots, x_k$ and edges $x_1 x_2, \ldots, x_{k-1} x_k, x_k x_1$. We denote by $[x_1, \ldots, x_k]$ the path with vertices $x_1, \ldots, x_k$ and edges $x_1 x_2, \ldots, x_{k-1} x_k$.

Note that there is a natural correspondence between an edge colouring $\{M_1, \ldots, M_t\}$ of a graph G and a triangle packing $\{(i, y, z) : yz \in E(M_i), i \in \{1, \ldots, t\}\}$ of the graph obtained from G by adjoining new vertices $\{1, \ldots, t\}$ each with neighbourhood $V(G)$. The condition that a vertex x in G is incident with at most one edge in each colour class corresponds to the condition that the edge ix is in at most one triangle of the packing for each $i \in \{1, \ldots, t\}$. This correspondence suggests an analogue of Lemma 2.2, stated below as Lemma 2.4. For a graph G and a permutation σ on a superset of $V(G)$, we define $\sigma(G)$ to be the graph with vertex set $\{\sigma(x) : x \in V(G)\}$ and edge set $\{\sigma(x)\sigma(y) : xy \in E(G)\}$.

Two vertices α and β in a graph G are *twin* if the transposition $(\alpha\ \beta)$ is an automorphism of G.

Lemma 2.4 *Let* $\mathcal{P} = \{G_1, \ldots, G_t\}$ *be a triangle packing of a graph* G *with leave* L, *and let* α *and* β *be twin vertices in* G. *There is a partition* Π *of* $(\mathrm{Nbd}_L(\alpha) \triangle \mathrm{Nbd}_L(\beta)) \setminus \{\alpha, \beta\}$ *into pairs such that, for each* $\{u, v\} \in \Pi$, *there is a triangle packing* $\mathcal{P}'_{\{u,v\}} = \{G_1', \ldots, G_t'\}$ *of* G *satisfying*

- $E(L') = E(L) \triangle \{\alpha u, \alpha v, \beta u, \beta v\}$, *where* L' *is the leave of* $\mathcal{P}'_{\{u,v\}}$; *and*

- $G_i' \in \{G_i, \sigma(G_i)\}$ *for* $i \in \{1, \ldots, t\}$, *where* σ *is the transposition* $(\alpha\ \beta)$.

Proof We form an auxiliary graph as the union of two matchings, M_α and M_β, that are defined as follows. For each $i \in \{1, \ldots, t\}$,

- if $G_i = (\alpha, x_1, x_2)$ with $x_1, x_2 \neq \beta$, then add the edge $x_1 x_2$ to M_α;

- if $G_i = (\beta, y_1, y_2)$ with $y_1, y_2 \neq \alpha$, then add the edge $y_1 y_2$ to M_β.

Observe that the set $(\mathrm{Nbd}_L(\alpha) \triangle \mathrm{Nbd}_L(\beta)) \setminus \{\alpha, \beta\}$ is exactly the set of end vertices of maximal paths in $M_\alpha \cup M_\beta$. These maximal paths induce a natural partition into pairs, which we take to be Π.

Let $\{u, v\} \in \Pi$ and let Q be the path in $M_\alpha \cup M_\beta$ from u to v. Because α and β are twin in G, $\{\alpha u, \alpha v, \beta u, \beta v\} \subseteq E(G)$. Because Q is maximal, L contains exactly one of the edges $\{\alpha u, \alpha v\}$ and one of the edges $\{\beta u, \beta v\}$. Form the packing $\mathcal{P}'_{\{u,v\}} = \{G_1', \ldots, G_t'\}$ from $\{G_1, \ldots, G_t\}$ by replacing G_i with $\sigma(G_i)$ for each $G_i \in \{(\gamma, x, y) : \gamma \in \{\alpha, \beta\}, xy \in E(Q)\}$. Then $\mathcal{P}'_{\{u,v\}}$ has the required properties. $\qquad\qquad\square$

This edge switching idea for triangle decompositions was first used in 1980 by Andersen, Hilton and Mendelsohn [4] who proved the following result, analogous to Lemma 2.3, which allows a triangle packing to be "regularised".

Lemma 2.5 ([4]) *If there is a packing of* K_n *with* t *triangles, then there is a packing of* K_n *with* t *triangles whose leave* L' *is such that* $|\deg_{L'}(x) - \deg_{L'}(y)| \leq 2$ *for all* $x, y \in V(K_n)$.

Proof Let $\{G_1, \ldots, G_t\}$ be a triangle packing of K_n and let L be its leave. If $|\deg_L(x) - \deg_L(y)| \leq 2$ for all $x, y \in V(K_n)$, then we are finished immediately. Otherwise, we can apply Lemma 2.4 with α and β chosen to be vertices such that $\deg_L(\alpha) + 2 < \deg_L(\beta)$. Then there is a pair $\{u, v\}$ in the partition Π given by Lemma 2.4 such that $\beta u, \beta v \in E(L)$.

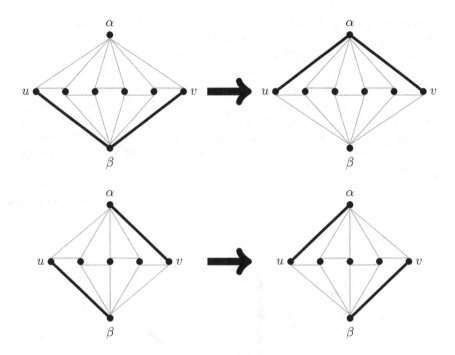

Figure 2: Two triangle switches (in grey) and their effect on the leave of the packing (in black).

The corresponding triangle packing $\mathcal{P}^*_{\{u,v\}}$ has a leave L^* with edge set $(E(L) \setminus \{\beta u, \beta v\}) \cup \{\alpha u, \alpha v\}$. So $\deg_{L^*}(\alpha) = \deg_L(\alpha) + 2$, $\deg_{L^*}(\beta) = \deg_L(\beta) - 2$, and $\deg_{L^*}(x) = \deg_L(x)$ for $x \in V(G) \setminus \{\alpha, \beta\}$. It is clear that by repeating this procedure we can obtain a triangle packing of G with the required properties. $\qquad\square$

Lemma 2.5 represents a very coarse application of Lemma 2.4. The following lemma shows that edge switching can be applied in a much more targeted way.

Lemma 2.6 *If there is a packing of K_n with t triangles whose leave contains a cycle C and an edge incident with exactly one vertex of C, then there is a packing of K_n with $t + 1$ triangles.*

Proof We prove the result by induction on the length of C. Let $\mathcal{P}$ be a packing of K_n that obeys the conditions of the lemma, and let L be the leave of $\mathcal{P}$. If C is a triangle, then $\mathcal{P} \cup \{C\}$ is a packing with the required properties. So suppose that $C = (x_1, \ldots, x_k)$ and $ax_1 \in E(L)$, where $k \geqslant 4$ and $a \notin V(C)$.

Applying Lemma 2.4 with $\alpha = a$ and $\beta = x_3$, there is a unique pair $\{x_1, y\}$ in Π that contains x_1 and a corresponding packing $\mathcal{P}^*_{\{x_1, y\}}$ of K_n with t triangles whose leave L^* has edge set $E(L) \triangle \{ax_1, ay, x_3x_1, x_3y\}$. If $y \neq x_2$, then L^* contains the triangle (x_1, x_2, x_3) and $\mathcal{P} \cup \{(x_1, x_2, x_3)\}$ is a packing with the required properties. If $y = x_2$, then L^* contains the $(k-1)$-cycle $(x_1, x_3, x_4, \ldots, x_k)$ and the edge x_1x_2, and we are finished by our inductive hypothesis. $\square$

In Section 4.2, we will see an application of a result very similar to Lemma 2.6.

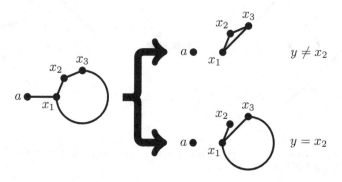

Figure 3: Illustration of Lemma 2.6. The diagrams on the right show only the leave edges needed to complete the proof.

2.3 Switching for cycle decompositions

If each subgraph in a packing is a cycle, we say it is a *cycle packing*, and if each subgraph is a cycle of length m we say it is an *m-cycle packing*. One can attempt to generalise Lemma 2.4 to cycle packings by pretending each cycle $(\ldots, x_{-1}, \alpha, x_1, \ldots)$ is a triangle (x_{-1}, α, x_1) and each cycle $(\ldots, y_{-1}, \beta, y_1, \ldots)$ is a triangle (y_{-1}, β, y_1), and modifying the edges of cycles incident with α or β accordingly. However, this naïve approach fails to deal adequately with cycles containing both α and β. Raines and Szaniszló [81] extended Lemma 2.5 to the case of 4- and 5-cycle packings but were unable to go further because of this issue. The problem can be sidestepped by treating each cycle that contains both α and β as the union of two paths from α to β. This allows a generalisation of Lemma 2.4, stated below as

Lemma 2.7, to decompositions into cycles of (possibly varied) arbitrary lengths.

Lemma 2.7 forms the basis of the results we discuss in the next two sections. The statement below is tailored to our purposes here, but the ideas necessary for its proof were first presented in [27].

Lemma 2.7 *Let* $\mathcal{P} = \{G_1, \ldots, G_t\}$ *be a cycle packing of a graph* G. *Let* L *be the leave of* $\mathcal{P}$ *and let* α *and* β *be twin vertices in* G. *There is a partition* Π *of* $(\mathrm{Nbd}_L(\alpha) \triangle \mathrm{Nbd}_L(\beta)) \setminus \{\alpha, \beta\}$ *into pairs such that, for each* $\{u, v\} \in \Pi$, *there is a cycle packing* $\mathcal{P}'_{\{u,v\}} = \{G'_1, \ldots, G'_t\}$ *of* G *satisfying*

- $E(L') = E(L) \triangle \{\alpha u, \alpha v, \beta u, \beta v\}$ *where* L' *is the leave of* $\mathcal{P}'_{\{u,v\}}$;

- G'_i *is a cycle of the same length as* G_i.

Furthermore, if we let $\sigma = (\alpha\ \beta)$, *then for each* $i \in \{1, \ldots, t\}$,

- $G'_i \in \{G_i, \sigma(G_i)\}$ *if* $\{\alpha, \beta\} \nsubseteq V(G_i)$; *and*

- $G'_i \in \{G_i, \sigma(Y_i) \cup Z_i, Y_i \cup \sigma(Z_i), \sigma(G_i)\}$ *if* $\{\alpha, \beta\} \subseteq V(G_i)$, *where* Y_i *and* Z_i *are the two paths from* α *to* β *whose union is* G_i.

Proof Let $\{H_1, \ldots, H_s\}$ be the packing obtained from $\mathcal{P}$ by, for each $i \in \{1, \ldots, t\}$ such that $\{\alpha, \beta\} \subseteq V(G_i)$, replacing G_i with the two paths from α to β whose union is G_i. Obviously $\{H_1, \ldots, H_s\}$ also has leave L. We form an auxiliary graph as the union of two matchings, M_α and M_β, that are defined as follows. For each $i \in \{1, \ldots, t\}$,

- if H_i is a cycle containing α but not β, then add the edge $x_1 x_2$ to M_α, where $\mathrm{Nbd}_{H_i}(\alpha) = \{x_1, x_2\}$;

- if H_i is a cycle containing β but not α, then add the edge $y_1 y_2$ to M_β, where $\mathrm{Nbd}_{H_i}(\beta) = \{y_1, y_2\}$;

- if H_i is a path from α to β of length at least 3, then add the edge $x z_i$ to M_α and add the edge $y z_i$ to M_β, where z_i is a new vertex, $\mathrm{Nbd}_{H_i}(\alpha) = \{x\}$ and $\mathrm{Nbd}_{H_i}(\beta) = \{y\}$.

(The addition to the auxiliary graph when H_i is a path from α to β can be considered informally as a half red/half green edge.)

Note that, for each $\gamma \in \{\alpha, \beta\}$ and vertex $x \in V(G) \setminus \{\alpha, \beta\}$, $\deg_{M_\gamma}(x) = 1$ if the edge γx is in some graph in $\mathcal{P}$ and $\deg_{M_\gamma}(x) = 0$ otherwise. From this and the fact that α and β are twin in G, it is easy to establish that M_α and M_β are indeed matchings and that $(\mathrm{Nbd}_L(\alpha) \triangle \mathrm{Nbd}_L(\beta)) \setminus \{\alpha, \beta\}$ is exactly the set of end vertices of maximal paths in $M_\alpha \cup M_\beta$. These maximal paths induce a natural partition into pairs, which we take to be Π.

Let $\{u, v\} \in \Pi$ and let Q be the path from u to v in $M_\alpha \cup M_\beta$. Form $\{H_1', \ldots, H_s'\}$ from $\{H_1, \ldots, H_s\}$ by replacing H_i with $\sigma(H_i)$ for each $i \in \{1, \ldots, s\}$ such that H_i corresponds to one or two edges of Q. For $i \in \{1, \ldots, s\}$, H_i' is a cycle if and only if H_i is, and H_i' is a path from α to β if and only if H_i is. Finally, form $\mathcal{P}_{\{u,v\}}' = \{G_1', \ldots, G_t'\}$ from $\{H_1', \ldots, H_s'\}$ by replacing the paths H_i' and H_j' with $H_i' \cup H_j'$, for each pair $\{i, j\} \subseteq \{1, \ldots, s\}$ such that $H_i \cup H_j$ was a cycle of $\mathcal{P}$. Then $\mathcal{P}_{\{u,v\}}'$ satisfies the conditions of the lemma. $\qquad\square$

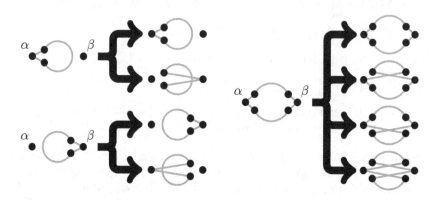

Figure 4: The possible effects of a (α, β)-switch on a cycle in the packing. Note that a path of length 1 or 2 from α to β in a cycle will never be altered.

Given a cycle packing $\mathcal{P}$ of a graph G with a leave L, we usually apply Lemma 2.7 by first selecting vertices α, β and u of G such that u is adjacent to exactly one of α and β in L. Lemma 2.7 then yields a cycle packing $\mathcal{P}_{\{u,v\}}'$, where v is the unique vertex of G such that $\{u, v\} \in \Pi$. We say that $\mathcal{P}_{\{u,v\}}'$ is obtained from $\mathcal{P}$ by *performing the (α, β)-switch from u*, and refer to v as the *terminus* of this switch.

3 Cycle decomposition of complete graphs

3.1 Background

Problems concerning when a complete graph can be decomposed into cycles of specified lengths date back to Kirkman and Walecki in the 19th century. Because we will only be concerned with decompositions into cycles in this section, we adopt the notation $(m_1, \ldots, m_t)$-*decomposition of* G to refer to a decomposition of a graph G into cycles of lengths $m_1, \ldots, m_t$. If there is a $(m_1, \ldots, m_t)$-decomposition of K_n, then obviously n must be odd, $m_1 + \cdots + m_t = \binom{n}{2}$, and $m_i \in \{3, \ldots, n\}$ for $i \in \{1, \ldots, t\}$. For a given odd n, we shall say that a list $m_1, \ldots, m_t$ satisfying these conditions is n-*admissible*. In 1981 Alspach [1] posed the problem of showing that these obvious necessary conditions for a cycle decomposition of a complete graph were also sufficient.

Problem 3.1 ([1]) *Show that there exists an* $(m_1, \ldots, m_t)$-*decomposition of* K_n *for each* n-*admissible list* $m_1, \ldots, m_t$.

Kirkman's result in [67] solves this problem in the case of a $(3, \ldots, 3)$-decomposition and Walecki's in [74] does so in the case of a $(n, \ldots, n)$-decomposition. Many more special cases of this problem were solved from the 1960s onward (see [16] for a survey). Most notably, in 2001 Alspach, Gavlas and Sajna [2, 85] (using earlier results of Bermond, Huang and Sotteau [7] and Hoffman, Lindner and Rodger [55]) gave a complete solution in the case of an $(m, \ldots, m)$-decomposition for any $m \in \{3, \ldots, n\}$.

One of the major difficulties in attacking Alspach's problem in full generality is that, for large n, there is a vast array of n-admissible lists and the existence of a cycle decomposition corresponding to each of them must be established. In [25, 26] edge switching techniques were used to create tools that allow new cycle decompositions to be obtained from existing ones and hence permit the problem to be reduced to a much more restricted class of lists. This reduction was used to solve the problem in [30][3].

Theorem 3.1 ([30]) *There is an* $(m_1, \ldots, m_t)$-*decomposition of* K_n *if and only if* n *is odd,* $m_1 + \cdots + m_t = \binom{n}{2}$, *and* $m_i \in \{3, \ldots, n\}$ *for* $i \in \{1, \ldots, t\}$.

We give an overview of this reduction and the solution to the problem in the next section.

---

[3]In fact, Alspach also posed the problem of showing, when n is even, that the obvious necessary conditions for the decomposition of K_n into a single perfect matching and cycles of arbitrary specified lengths are also sufficient. The tools we discuss adapt easily to this case, and both the n odd and n even cases of the problem are solved in [30]. To keep our discussion here as clean as possible, however, we consider only the n odd case.

More generally, throughout this section we discuss only results concerning cycle decompositions, even when corresponding results for decompositions into a perfect matching and cycles are also known.

3.2   The solution of Alspach's problem

The reduction of Alspach's problem in [26] relies on two key lemmas, presented here as Lemmas 3.4 and 3.6. Given an existing cycle decomposition satisfying appropriate conditions, Lemma 3.4 yields a new cycle decomposition in which two of the original cycle lengths have been merged into one, and Lemma 3.6 yields a new cycle decomposition in which two of the original cycle lengths have been equalised.

We now give two simple examples that suggest how Lemma 2.7 can be applied to merge two cycle lengths. We extend our notation so that an $(m_1^{a_1}, \ldots, m_t^{a_t})$-decomposition refers to a decomposition into a_1 cycles of length m_1, a_2 cycles of length m_2, and so on to a_t cycles of length m_t. The *reduced leave* of a packing is the graph obtained from the leave by removing any isolated vertices.

Example 3.2 Suppose we have a (6^{45})-packing of K_{24} whose reduced leave is a disjoint union of two triangles. Using Lemma 2.7 with α chosen to be a vertex of one triangle and β chosen to be a vertex of the other we can obtain either a (6^{46})-decomposition of K_{24} or a (6^{45})-packing of K_{24} whose reduced leave is a bowtie as shown in Figure 5.

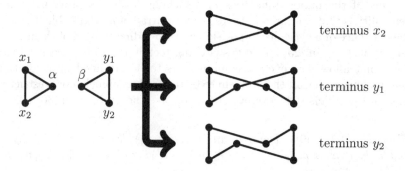

Figure 5: An (α, β)-switch from x_1 here results in a reduced leave that is either a 6-cycle or a bowtie depending on the switch's terminus.

Example 3.3 Suppose that we wish to construct a (6^{46})-decomposition of K_{24} from a $(3^2, 6^{45})$-decomposition of K_{24}. Remove the two 3-cycles from this latter decomposition to obtain a (6^{45})-packing of K_{24} whose reduced leave is either a bowtie or a disjoint union of two triangles. If the reduced leave is a disjoint union of two triangles, we saw in Example 3.2 that we can obtain either the desired (6^{46})-decomposition or a (6^{45})-packing whose reduced leave is a bowtie. So we may assume that we have the latter. We can remove a 6-cycle from this packing to produce a (6^{44})-packing with a reduced leave as pictured in Figure 6. If we then apply Lemma 2.7 as shown in the figure, we can obtain a (6^{44})-packing whose leave has a decomposition into two 6-cycles and we are done.

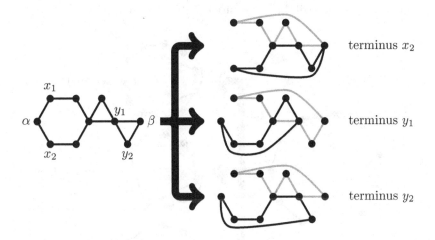

Figure 6: An (α, β)-switch from x_1 here results in a leave that has a decomposition into two 6-cycles (one shown in grey and the other in black).

Using lengthier, more general arguments of this nature the following merging lemma can be proved.

Lemma 3.4 *Suppose there exists an $(m_1, \ldots, m_t, h, x, y)$-decomposition of K_n where $h \geqslant \frac{1}{2}(x + y)$ and $x + y + h \leqslant n + 1$. Then there exists an $(m_1, \ldots, m_t, h, x + y)$-decomposition of K_n.*

Next we give an example that suggests how Lemma 2.7 can be applied to equalise two cycle lengths.

Example 3.5 A result of Thomason [90] implies that if a graph has a decomposition into two cycles that share at least two vertices then it also has another decomposition into two cycles which is distinct from the first. The reduced leave pictured in Figure 7 has a $(4, 11)$-decomposition (shown as the inside cycle and outside cycle) and also has a $(7, 8)$-decomposition (shown as the grey cycle and black cycle). To equalise the cycle lengths by one step we desire a leave with a $(5, 10)$-decomposition. By applying Lemma 2.7 as shown in the figure we can obtain either a leave with the desired $(5, 10)$-decomposition (shown as the inside cycle and outside cycle), or a leave that retains a $(4, 11)$-decomposition (shown as the inside cycle and outside cycle) and also has a $(6, 9)$-decomposition (shown as the grey cycle and black cycle). By continuing to pursue similar switches, keeping track of both decompositions, we can eventually guarantee that one of the two decompositions will be the desired $(5, 10)$-decomposition.

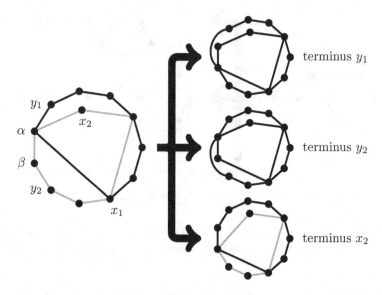

Figure 7: An (α, β)-switch from x_1 here results in either a leave with the desired $(5, 10)$-decomposition (shown as the inside cycle and outside cycle), or a leave that retains a $(4, 11)$-decomposition (shown as the inside cycle and outside cycle) and also has a $(6, 9)$-decomposition (shown as the grey cycle and black cycle).

More intricate arguments of this nature yield the following equalising lemma.

Lemma 3.6 *Suppose there exists an* $(m_1, \ldots, m_t, x, y)$-*decomposition of* K_n *where* $x < y$ *and suppose that* $x + y \geqslant n + 2$. *Then there exists an* $(m_1, m_2, \ldots, m_t, x + 1, y - 1)$-*decomposition of* K_n.

Using Lemmas 3.4 and 3.6 it is not too difficult to show that to completely solve Alspach's problem for complete graphs of some order n, it suffices to find an $(m_1, \ldots, m_t)$-decomposition of K_n for each n-*ancestor list* $(m_1, \ldots, m_t)$. An n-ancestor list is an n-admissible list of the form

$$3, 3, \ldots, 3, 4, 4, \ldots, 4, 5, 5, \ldots, 5, c, n, n, \ldots, n \quad \text{where } 3 \leqslant c \leqslant n$$

that also obeys some further technical conditions (see [26]). A decomposition corresponding to any n-admissible list can be obtained from a decomposition corresponding to some n-ancestor list by first repeatedly applying Lemma 3.4 and then repeatedly applying Lemma 3.6.

In [30] this reduction is used to completely solve Alspach's problem, although a substantial amount of work is required to construct the decompositions corresponding to ancestor lists.

3.3 Related results

Theorem 3.1 was recently extended to complete multigraphs [29]. See [7, 8, 28, 49, 63, 84, 87] for previous results on this problem. We denote by λK_n the complete multigraph with n vertices and λ edges between each pair of vertices.

Theorem 3.7 *There is an* $(m_1, \ldots, m_t)$-*decomposition of* λK_n *if and only if*

(i) $\lambda(n-1)$ *is even;*

(ii) $2 \leqslant m_1, \ldots, m_t \leqslant n;$

(iii) $m_1 + \cdots + m_t = \lambda \binom{n}{2};$

(iv) $\max(m_1, \ldots, m_t) + t - 2 \leqslant \frac{\lambda}{2} \binom{n}{2}$ *if λ is even;*

(v) $\sum_{m_i = 2} m_i \leqslant (\lambda - 1) \binom{n}{2}$ *if λ is odd.*

The major difference from the $\lambda = 1$ case is that the decomposition may contain 2-cycles (pairs of parallel edges), but some decompositions with many 2-cycles are impossible. This serves to complicate the conditions in the result. The conditions can be shown to be necessary reasonably succinctly, although in the case of condition (iv) this is not trivial. Lemma 2.7 can be generalised to deal with multigraphs (this was first done in [28]), and many of the methods used in the proof of Alspach's problem can be adapted to the multigraph case. The proof of sufficiency employs these methods, Theorem 3.1, and induction on both λ and n.

In [18] Bryant uses edge switching techniques to prove the analogue of Theorem 3.7 for decompositions into paths, thus proving a conjecture of Tarsi [89]. For other results on this problem see [46, 51, 52, 62, 69, 70] and the survey [51].

Theorem 3.8 ([18]) *There is a decomposition of λK_n into paths of lengths $m_1, \ldots, m_t$ if and only if $m_1 + \cdots + m_t = \lambda\binom{n}{2}$, and $m_i \in \{1, \ldots, n-1\}$ for $i \in \{1, \ldots, t\}$.*

The proof uses a result similar to Lemma 2.7 that deals with decompositions into paths. One interesting disanalogy is that, because paths P in the decomposition may have $\deg_P(\alpha) + \deg_P(\beta)$ odd, it is sometimes the case that an (α, β)-switch from a vertex u does not have a terminus, and the switch simply results in a leave with edge set $E(L) \triangle \{\alpha u, \beta u\}$.

Edge switching techniques have also been employed to prove a number of results on cycle decompositions of non-complete graphs. Thus far we have only seen Lemma 2.7 applied to decompositions of complete graphs, and so the restriction that α and β be twin vertices when we perform an (α, β)-switch has been trivial. Because of this restriction, however, the other classes of graphs to which edge switching has been successfully applied all have large sets of pairwise twin vertices. In particular, significant progress has been made on cycle decomposition of complete multipartite graphs and of complete graphs with holes. We conclude this section with three examples of such results.

Cycle decompositions of complete bipartite graphs have also been well studied [6, 38, 39, 75]. Most famously, Sotteau [88] completely solved the existence problem for decompositions into cycles of fixed length. In [56], edge switching is used and Lemma 3.4 is successfully adapted to complete bipartite graphs. Thus far, however, this has not been accomplished for Lemma 3.6. Using the adapted joining lemma and a result of Chou, Fu and Huang [39] concerning decompositions of complete bipartite graphs into cycles of lengths 4 and 6, the following partial result is obtained.

Theorem 3.9 ([56]) *Suppose a, b and $m_1, \ldots, m_t$ are integers such that $m_1, \ldots, m_t$ is nondecreasing and $m_t \leqslant \min(a, b, 3m_{t-1})$. There is an $(m_1, \ldots, m_t)$-decomposition of $K_{a,b}$ if and only if*

- *a and b are even;*

- *$m_i \geqslant 4$ is even for $i \in \{1, \ldots, t\}$; and*

- *$m_1 + \cdots + m_t = ab$.*

With some more effort, Theorem 3.9 can be used to establish a result on decompositions of complete multipartite graphs into cycles of uniform even length. See [37, 68] and the survey [15] for previous work on this topic.

Theorem 3.10 ([56]) *Let G be a complete multipartite graph with parts of even size. If $m \geqslant 4$ is an even integer such that m divides $|E(G)|$ and each part of G has at least $m + 2$ vertices, then there is an m-cycle decomposition of G.*

The graph obtained from a complete graph of order v by removing the edges of a complete subgraph of order u is denoted $K_v - K_u$ and is sometimes referred to as a complete graph with a hole. Beginning with the famous result of Doyen and Wilson [45], cycle decompositions of such graphs have received substantial attention [22, 34, 35, 36, 78]. In [56], techniques similar to those used for Theorem 3.9 are employed to construct m-cycle decompositions of $K_v - K_u$ for even $m \leqslant \min(u, v - u)$. In [59], more advanced versions of the techniques are used to obtain a corresponding result for odd m.

Theorem 3.11 ([56, 59]) *Let u, v and m be integers with $3 \leqslant m \leqslant \min(u, v - u)$. There exists an m-cycle decomposition of $K_v - K_u$ if and only if*

- *u and v are odd;*

- *$\binom{v}{2} - \binom{u}{2} \equiv 0 \pmod{m}$; and*

- *$v \geqslant \frac{u(m+1)}{m-1} + 1$ if m is odd.*

Theorem 3.11 has recently been extended to decompositions into cycles of mixed length [60].

4 Embedding partial Steiner triple systems

Edge switching techniques have also proven efficacious in embedding partial Steiner triple systems. For our purposes, a *Steiner triple system of order v* can be defined as a triangle decomposition of a complete graph of order v, and a *partial Steiner triple system of order u* can be defined as a triangle packing of a complete graph of order u. It is easy to show that if a Steiner triple system of order v exists then $v \equiv 1, 3 \pmod{6}$, and Kirkman [67] established that there exists a Steiner triple system of each such positive order. An *embedding* of a partial Steiner triple system $\mathcal{P}$ of order u is a Steiner triple system $\mathcal{D}$ of order $v \geqslant u$ such that $\mathcal{P} \subseteq \mathcal{D}$. We are generally interested in finding for which integers v certain partial Steiner triple systems have embeddings of order v.

Let K_S denote the complete graph with vertex set S and $K_{S,T}$ denote the complete bipartite graph with parts S and T. For vertex disjoint graphs G_1 and G_2, $G_1 \vee G_2$ denotes the graph $G_1 \cup G_2 \cup K_{V(G_1),V(G_2)}$. Our use of this notation implies that G_1 and G_2 are vertex disjoint. Finding an embedding of order v for a partial Steiner triple system $\mathcal{P}$ with a leave L is equivalent to finding a triangle decomposition of $L \vee K_{v-u}$.

4.1 Background

In 1971 Treash [91] proved that every partial Steiner triple system has an embedding. The embeddings she produced had order exponential in the order of the partial systems embedded. In [71], Lindner established the existence of dramatically smaller embeddings when he proved that each partial Steiner triple system of order u has an embedding of order $6u + 3$.

On the other hand, it is not difficult to show that there are partial Steiner triple systems of order u that admit no embeddings of order $v < 2u + 1$. A partial Steiner triple system is *maximal* if its leave is triangle free.

Lemma 4.1 *If a partial Steiner triple system of order u with a leave H has an embedding of order v, then $|E(H)| \geqslant \frac{1}{2}(v - u)(2u + 1 - v)$.*

Proof Suppose that $H \vee K_W$ has a triangle decomposition $\mathcal{D}$ for a set W such that $|W| = v - u$. Then $|E(K_{V(H),W})| \leqslant 2|E(H) \cup E(K_W)|$ because each triangle in $\mathcal{D}$ uses at least one edge of $H \cup K_W$ and at most two edges of $K_{V(H),W}$. Using $|E(K_{V(H),W})| = u(v - u)$ and $|E(K_W)| = \binom{v-u}{2}$, the result follows. $\square$

Corollary 4.2 *For each $u \geqslant 9$ there exists a partial Steiner triple system of order u that has no embedding of order less than $2u + 1$.*

Proof It is known that for each $u \geqslant 9$, there is a maximal partial Steiner triple system with a leave H such that $0 < |E(H)| < u - 1$ and H is not a perfect matching (see [42]). If such a partial Steiner triple system has an embedding of order v, then Lemma 4.1 implies that $v < u + 2$ or $v > 2u - 1$. However $v \neq 2u$ since $v \equiv 1, 3 \pmod 6$, and $v \notin \{u, u + 1\}$ since H is triangle free, nonempty and not a perfect matching. $\square$

In 1977 Lindner [72] made the following conjecture, for which Corollary 4.2 shows that the bound of $2u + 1$ cannot be improved in general.

Conjecture 4.3 ([72]) *Any partial Steiner triple system of order u has an embedding of order v for each $v \geqslant 2u + 1$ such that $v \equiv 1, 3 \pmod 6$.*

Progress toward this conjecture was made in 1980, when Andersen, Hilton and Mendelsohn [4] proved that any partial Steiner triple system of order u has an embedding of order v for all $v \geqslant 4u+1$ with $v \equiv 1,3 \pmod 6$. This bound was reduced to $3u-2$ by Bryant [20] in 2004. Finally, Lindner's conjecture was completely resolved in 2009 [24].

Theorem 4.4 ([24]) *Every partial Steiner triple system of order u has an embedding of order v for each $v \geqslant 2u + 1$ such that $v \equiv 1,3 \pmod 6$*

Edge switching methods are used throughout the proof of Theorem 4.4 in a number of different ways. In the next section, we prove a crucial lemma from [24], presented here as Lemma 4.6, and summarise the main line of the proof.

4.2   The proof of Lindner's conjecture

When finding triangle packings of $H \vee K_W$, where H is the leave of a partial Steiner triple system to be embedded, we perform (α, β)-switches only with $\alpha, \beta \in W$ (note that vertices in W are pairwise twin in $H \vee K_W$). This ensures that the resulting triangle packing is, in a certain sense, unchanged on $V(H)$. We formalise this with the following definition which will also be used in Section 5.   Packings $\mathcal{P}$ and $\mathcal{P}'$ of a graph G are *equivalent on S* for $S \subseteq V(G)$ if we can write $\mathcal{P} = \{G_1, \dots, G_t\}$ and $\mathcal{P}' = \{G_1', \dots, G_t'\}$ such that for $i \in \{1, \dots, t\}$,

- $|E(G_i')| = |E(G_i)|$;

- $\deg_{G_i'}(x) = \deg_{G_i}(x)$ for $x \in S$; and

- $E(G_i') \cap E(K_S) = E(G_i) \cap E(K_S)$.

It follows from Lemma 2.7 that if $\mathcal{P}'$ is the result of performing an (α, β)-switch on a cycle packing $\mathcal{P}$ of a graph G, then $\mathcal{P}$ and $\mathcal{P}'$ are equivalent on $V(G) \setminus \{\alpha, \beta\}$.

To aid in the proof of Lemma 4.6, we first prove a simple preliminary result. The result is reminiscent of the *augmenting path* arguments used, for example, in some proofs of Hall's marriage theorem such as the one given in [44].

Lemma 4.5 *Let $\mathcal{P}$ be a triangle packing of $H \vee K_W$ for a graph H and a set W, let L be the leave of $\mathcal{P}$, and let $\gamma \in V(H) \cup W$. Suppose that there is a sequence of distinct vertices $x_0, \dots, x_{2t+1}$ in $V(H) \cup W$ such that*

- $\gamma x_0, \gamma x_{2t+1} \in E(L)$;

- $x_{2i} \in V(H)$ and $x_{2i+1} \in W$ for $i \in \{0,\dots,t\}$; and

- $x_{2i}x_{2i+1} \in E(L)$ for $i \in \{0,\dots,t\}$ and $(\gamma, x_{2i+1}, x_{2i+2}) \in \mathcal{P}$ for $i \in \{0,\dots,t-1\}$.

Then there exists a triangle packing $\mathcal{P}'$ of $H \vee K_W$ such that $(\gamma, x_0, x_1) \in \mathcal{P}'$ and $\mathcal{P}' \setminus \{(\gamma, x_0, x_1)\}$ is equivalent to $\mathcal{P}$ on $V(H)$.

Proof Let

$$\mathcal{P}' = (\mathcal{P} \setminus \{(\gamma, x_{2i+1}, x_{2i+2}) : i \in \{0,\dots,t-1\}\}) \cup$$
$$\{(\gamma, x_{2i}, x_{2i+1}) : i \in \{0,\dots,t\}\}.$$

It is routine to check that $\mathcal{P}'$ has the required properties. $\square$

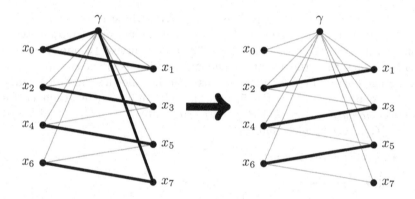

Figure 8: Illustration of Lemma 4.5 when $t = 3$. Vertices in $V(H)$ are drawn on the left and those in W are drawn on the right. The vertex γ may be in $V(H)$ or W.

We now use Lemma 4.5 to prove the critical lemma from [24].

Lemma 4.6 Let $\mathcal{P}$ be a triangle packing of $H \vee K_W$ for a graph H and a set W, let L be the leave of $\mathcal{P}$, and let $\gamma \in V(H) \cup W$. Suppose that

(i) there is a path $[\gamma, x_0, x_1]$ in L such that $x_0 \in V(H)$ and $x_1 \in W$;

(ii) $(\gamma, x, y) \notin \mathcal{P}$ for all $x, y \in W$; and

(iii) $|\mathrm{Nbd}_L(x) \cap W| \geqslant 1$ for each $x \in V(H) \cup \{\gamma\}$.

Then there exists a triangle packing $\mathcal{P}'$ of $H \vee K_W$ such that $(\gamma, x_0, x_1) \in \mathcal{P}'$ and $\mathcal{P}' \setminus \{(\gamma, x_0, x_1)\}$ is equivalent to $\mathcal{P}$ on $V(H)$.

Proof Form a sequence of vertices $x_0, x_1, \ldots, x_{2t+1}$ recursively as follows. Begin with x_0, x_1 as given in the lemma. For $i \in \{0, 1, 2 \ldots, \}$, if $x_{2i+1} \in \text{Nbd}_L(\gamma)$ or if $x_{2i+1} = x_{2j+1}$ for some $j \in \{1, \ldots, i-1\}$, then let $t = i$ and terminate the process. Otherwise, let x_{2i+2} be the vertex such that $(\gamma, x_{2i+1}, x_{2i+2}) \in \mathcal{P}$ ($x_{2i+2} \in V(H)$ by (ii)) and let x_{2i+3} be a vertex in $\text{Nbd}_L(x_{2i+2}) \cap W$ (such a vertex exists by (iii) and is not in $\{\gamma, x_{2i+1}\}$ because $(\gamma, x_{2i+1}, x_{2i+2}) \in \mathcal{P}$). Since W is finite, this process eventually terminates. Note that $x_1, \ldots, x_{2t-1}$ are distinct vertices in W by definition and hence $x_0, x_2, \ldots, x_{2t}$ are distinct vertices in $V(H)$. Furthermore $\gamma \notin \{x_0, \ldots, x_{2t+1}\}$.

If $x_{2t+1} \in \text{Nbd}_L(\gamma)$, then we can apply Lemma 4.5 with sequence $x_0, \ldots, x_{2t+1}$ to complete the proof, so we may assume that $x_{2t+1} = x_{2j+1}$ for some $j \in \{1, \ldots, t-2\}$.

Let $a \in \text{Nbd}_L(\gamma) \cap W$. Note that $a \notin \{x_1, x_3, \ldots, x_{2t-1}\}$ and that a and x_{2j+1} are twin vertices in $H \vee K_W$. Let y be the terminus of the (x_{2j+1}, a)-switch in $\mathcal{P}$ with origin x_{2j}. If $y \neq \gamma$, then let $\mathcal{P}^*$ be the result of performing this switch. If $y = \gamma$, then we instead let $\mathcal{P}^*$ be the result of performing the (x_{2j+1}, a)-switch in $\mathcal{P}$ with origin x_{2t} and let z be the terminus of this switch (note that $z \notin \{x_{2j}, \gamma\}$ because $y = \gamma$). Let L^* be the leave of $\mathcal{P}^*$.

Case 1. Suppose that $y \neq \gamma$. We claim that $\mathcal{P}^*$ satisfies the hypotheses of Lemma 4.5 with sequence $x_0, \ldots, x_{2j}, a$. If this is the case, then we can apply Lemma 4.5 to complete the proof, so it suffices to prove the claim.

Note that $E(L^*) = E(L) \triangle \{a x_{2j}, a y, x_{2j+1} x_{2j}, x_{2j+1} y\}$. In particular,

$$\{x_{2i} x_{2i+1} : i \in \{0, \ldots, j-1\}\} \cup \{\gamma x_0, \gamma a, a x_{2j}\} \subseteq E(L^*)$$

because $a, x_1, x_3, \ldots, x_{2j+1}$ are distinct vertices in W and $y \neq \gamma$. Note also that

$$\{(\gamma, x_{2i+1}, x_{2i+2}) : i \in \{0, \ldots, j-1\}\} \subseteq \mathcal{P}^*$$

because each triangle in $\mathcal{P}$ not containing a or x_{2j+1} is also in $\mathcal{P}^*$.

Case 2. Suppose that $y = \gamma$. We claim that $\mathcal{P}^*$ satisfies the hypotheses of Lemma 4.5 with sequence $x_0, \ldots, x_{2t}, a$. If this is the case, then we can apply Lemma 4.5 to complete the proof, so it suffices to prove the claim.

Note that $E(L^*) = E(L) \triangle \{a x_{2t}, a z, x_{2j+1} x_{2t}, x_{2j+1} z\}$. In particular,

$$\{x_{2i} x_{2i+1} : i \in \{0, \ldots, t-1\}\} \cup \{\gamma x_0, \gamma a, a x_{2t}\} \subseteq E(L^*)$$

because $a, x_1, x_3, \ldots, x_{2t-1}$ are distinct vertices in W and $z \notin \{x_{2j-1}, \gamma\}$. Note also that

$$\{(\gamma, x_{2i+1}, x_{2i+2}) : i \in \{0, \ldots, t-1\}\} \subseteq \mathcal{P}^*$$

because the only triangle in $\mathcal{P}$ that contains both γ and a vertex in $\{a, x_{2j+1}\}$ is $(\gamma, x_{2j+1}, x_{2j+2})$ and the fact that $z \neq \gamma$ implies that the edge γx_{2j+2} was not in the switching path Q in Lemma 2.7 and hence $(\gamma, x_{2j+1}, x_{2j+2}) \in \mathcal{P}_\gamma^*$. $\square$

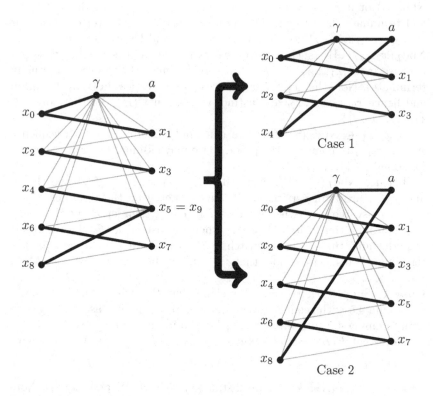

Figure 9: Illustration of Lemma 4.6 when $t = 4$, $j = 2$ and an (x_5, a)-switch is performed. The diagrams on the right show only the triangles and leave edges needed to apply Lemma 4.5.

Lemma 4.6 can be used to prove Vizing's theorem. Let H be a graph and let W be a set of $\Delta(H) + 1$ vertices that is disjoint from $V(H)$. Let $\mathcal{P}$ be a triangle packing of $H \vee K_W$ such that each triangle in $\mathcal{P}$ has exactly one vertex in W. Further, suppose that $\mathcal{P}$ is a largest such packing. If all the edges of H are in triangles in $\mathcal{P}$, then $\mathcal{P}$ induces an edge colouring of H with colour set W. Otherwise, there is an edge of H which is not in a triangle in $\mathcal{P}$ and we can apply Lemma 4.6, with $\gamma, x_0 \in V(H)$ being the endpoints of this edge. The result is a triangle packing of $H \vee K_W$ which contradicts the definition of $\mathcal{P}$ as the largest such packing, and we have Vizing's Theorem. In fact, Lemma 4.6 was arrived at through attempting to apply Vizing's "fan" arguments to triangle packings.

The main line of the proof of Theorem 4.4 in [24] finds an embedding of order $2u + 1$ for a partial Steiner triple system of order $u \geqslant 9$ with a leave containing at least $3u - 18$ edges and having maximum degree at most $u - 5$. Roughly speaking, the argument proceeds as follows. For such a partial Steiner triple system with a leave H, we construct a triangle decomposition of $H \vee K_{W^* \cup \{a,b,c\} \cup \{d,e\}}$ where $W^* = \{1, \ldots, u - 4\}$. Say that a triangle with exactly i vertices in $V(H)$ is a *type i triangle*.

- We first create a triangle packing

$$\mathcal{P}_1 = \{(i, x, y) : xy \in E(M_i), i \in W^*\}$$

of $H \vee K_{W^*}$ from an edge coloring $\{M_1, \ldots, M_{u-4}\}$ of H (such a colouring exists by Vizing's theorem). Note that the leave of $\mathcal{P}_1$ contains no edges of H. This will be true of all the packings created in the proof.

- We then repeatedly apply Lemma 4.6 to add type 1 triangles until we obtain a triangle packing $\mathcal{P}_2$ of $H \vee K_{W^*}$ with a leave L_2 such that each vertex in $V(H)$ has degree 1 in L_2. In each application of Lemma 4.6 we choose $x_0 \in V(H)$ as a vertex whose degree in the leave is greater than 1, and $\gamma, x_1 \in W$ as neighbours of x_0 in the leave such that γ has a neighbour in W^*. An easy edge switching procedure may be required to ensure such vertices exist.

- Next we obtain a triangle packing $\mathcal{P}_3$ of $H \vee K_{W^*}$ with a leave L_3 such that one vertex in $V(H)$ has degree 3 in L_3, every other vertex in $V(H)$ has degree 1 in L_3, $|E(L_3) \cap E(K_{W^*})| = u - 7$, and L_3 has a 3-edge colouring $\{M_a, M_b, M_c\}$. We do this by repeatedly applying a result very similar to Lemma 2.6 (each time choosing the "lasso" to be a subgraph of K_{W^*}) to add type 0 triangles to $\mathcal{P}_3$. Before the final application of the lemma, we remove a judiciously selected type 1 triangle from the packing. Finally, by employing more edge switching techniques, we ensure that the leave has a 3-edge colouring.

- We now obtain a triangle packing $\mathcal{P}_4$ of $H \vee K_{W^* \cup \{a,b,c\}}$ with a leave L_4 in which each vertex has degree 2 except for one isolate in $V(H)$. We do this by applying a technique similar to that used in Lemma 2.5 to the packing

$$\mathcal{P}_3 \cup \{(i, x, y) : xy \in E(M_i), i \in \{a, b, c\}\} \cup \{(a, b, c)\}$$

of $H \vee K_{W^* \cup \{a,b,c\}}$ in order to regularise its leave on $W^* \cup \{a,b,c\}$. It is routine to show that this results in a leave with the required properties.

- The properties of L_4 ensure that it is a union of bipartite cycles in $K_{V(H),W^* \cup \{a,b,c\}}$ and one isolated vertex z. So there is a 2-edge colouring $\{M_d, M_e\}$ of L_4. Then

$$\mathcal{P}_4 \cup \{(i,x,y) : xy \in E(M_i), i \in \{d,e\}\} \cup \{(d,e,z)\}$$

is the required triangle decomposition of $H \vee K_{W^* \cup \{a,b,c\} \cup \{d,e\}}$.

The other major case of the proof in [24], where the leave of the partial Steiner triple system to be embedded has fewer than $3u - 18$ edges, also takes considerable effort. Lemma 4.6 is again employed but in a different context where γ is chosen in $V(H)$. The remaining cases then succumb to more routine arguments.

4.3 Related results

We saw in Section 4.1 that the bound of $2u + 1$ in Theorem 4.4 cannot be improved in general. Of course, many partial Steiner triple systems do have embeddings of order less than $2u + 1$. Much less is known about when these embeddings exist, although there are some results concerning them [17, 23, 32, 41]. Colbourn [40] has shown that it is NP-complete to decide whether a partial Steiner triple system has an embedding of order less than $2u + 1$.

In [57, 58] the edge switching arguments used in [24] are extended to obtain stronger results than those previously achieved on the existence of embeddings of orders less than $2u+1$. In [57] it is established that a partial Steiner triple system of order u whose leave has small size and maximum degree has an embedding of order v for at least half (or nearly half) of the orders $v < 2u + 1$ for which an embedding could feasibly exist.

Theorem 4.7 ([57]) *Any partial Steiner triple system of order u with a leave H such that $\Delta(H) \leqslant \frac{1}{4}(u - 9)$ and $|E(H)| < \frac{1}{32}(u - 5)(u - 11) + 2$ has an embedding of order v for each $v \geqslant \frac{3u+1}{2} + \sqrt{(u+1)^2 - 8|E(H)|}$ such that $v \equiv 1,3 \pmod 6$.*

Note that it follows from Lemma 4.1 that if there exists an embedding of order v for a partial Steiner triple system of order u with a leave H such that $|E(H)| \leqslant \frac{1}{8}(u+1)^2$, then $v \leqslant \frac{3u+1}{2} - \sqrt{(u+1)^2 - 8|E(H)|}$ or $v \geqslant \frac{3u+1}{2} + \sqrt{(u+1)^2 - 8|E(H)|}$. In some cases Theorem 4.7 allows the complete determination of the set of all orders for which a partial Steiner triple system has an embedding.

In [58] it is established that a partial Steiner triple system of order u with few triples has an embedding of order v for each admissible order $v \geqslant \frac{8u+17}{5}$.

Theorem 4.8 ([58]) *Any partial Steiner triple system of order $u \geqslant 62$ with at most $\frac{u^2}{50} - \frac{11u}{100} - \frac{116}{75}$ triples has an embedding of order v for each $v \geqslant \frac{8u+17}{5}$ such that $v \equiv 1,3 \pmod 6$.*

In [21, 33, 76] techniques from [24] are adapted and used to prove various results concerning embeddings of objects related to partial Steiner triple systems. In each case the strongest results to date are obtained on an already well studied problem.

A Steiner triple system is equivalent to an idempotent totally symmetric quasigroup. Embeddings of partial totally symmetric quasigroups, without the restriction to idempotence, have also been studied. In [21] a best-possible result is achieved by showing that every partial totally symmetric quasigroup of order u has an embedding of order v for each even $v \geqslant 2u + 4$. This improved on a previous bound of $4u + 4$ [79, 80].

A generalisation of Conjecture 4.3 states that a partial triple system of index λ (that is, a triangle packing of λK_v) has an embedding of order v for each feasible $v \geqslant 2u + 1$. In [33] this conjecture is completely resolved with the exception of partial systems of order less than 28. The conjecture was previously known to hold for even λ (see [48, 54, 64]), and a bound of $4u + 1$ had been obtained for odd λ (see [82, 83]).

Strong results on embedding partial 5-cycle systems are obtained in [76]. See [61, 73] for previous results on this problem.

5 Almost regular decompositions

5.1 Decompositions into almost regular graphs

Thus far we have seen edge switching applied to decompositions into restricted classes of graphs, namely matchings, cycles, and paths. The idea can also be used on decompositions into arbitrary graphs, but the subgraphs in the resulting decompositions may not be isomorphic to those in the original decompositions. Bryant and Maenhaut [31] have shown, however, that we can guarantee that the subgraphs will be "regularised" on a prescribed set of vertices while leaving the decompositions equivalent on the remaining vertices (recall the definition of equivalence from Section 4.2). A graph H is *almost regular on T* for a set of vertices T if

$|\deg_H(x) - \deg_H(y)| \leqslant 1$ for all $x, y \in T$. The following lemma (altered only slightly from [31]) formalises the fundamental technique.

Lemma 5.1 *Let* $\{G_1, \ldots, G_t\}$ *be a decomposition of a graph* G *and let* T *be a set of pairwise twin vertices in* G. *There exists a decomposition* $\{G'_1, \ldots, G'_t\}$ *of* G *such that*

- $\{G_1, \ldots, G_t\}$ *and* $\{G'_1, \ldots, G'_t\}$ *are equivalent on* $V(G) \setminus T$;

- G'_i *is almost regular on* T *for* $i \in \{1, \ldots, t\}$.

Proof If G_i is almost regular on T for $i \in \{1, \ldots, t\}$, then we are finished immediately, so assume otherwise. Let k be the smallest element of $\{1, \ldots, t\}$ such that G_k is not almost regular on T. Then there are $\alpha, \beta \in T$ such that $\deg_{G_k}(\alpha) > \deg_{G_k}(\beta) + 1$. For a subgraph H of G and $x, y \in V(G)$ let $E_{x \setminus y}(H)$ be the set of edges of H that are incident with x but not with y. Because α and β are twin in G, the transposition $(\alpha \ \beta)$ induces a bijection θ from $E_{\alpha \setminus \beta}(G)$ to $E_{\beta \setminus \alpha}(G)$.

We now construct an auxiliary multigraph A, possibly containing loops, with vertex set $\{1, \ldots, t\}$ and one edge corresponding to each edge in $E_{\alpha \setminus \beta}(G)$. We also define an orientation $\mathcal{O}$ of A. For each $e \in E_{\alpha \setminus \beta}(G)$, let the edge of A corresponding to e be ij, where $e \in E(G_i)$ and $\theta(e) \in E(G_j)$, and let it be oriented from i to j under $\mathcal{O}$ (possibly $i = j$ and this edge is a loop).

It is well known that any multigraph can be oriented so that the in-degree of each vertex differs from its out-degree by at most 1. So we can find an orientation $\mathcal{O}^*$ of A with this property. Form $\{G_1^*, \ldots, G_t^*\}$ from $\{G_1, \ldots, G_t\}$ by reassigning e to G_j^* and $\theta(e)$ to G_i^* for each edge $e \in E_{\alpha \setminus \beta}(G)$ such that the corresponding edge ij of A has opposite orientation under $\mathcal{O}$ and $\mathcal{O}^*$.

Then $\{G_1, \ldots, G_t\}$ and $\{G_1^*, \ldots, G_t^*\}$ are equivalent on $V(G) \setminus T$ because only edges incident with α or β were reassigned. For $i \in \{1, \ldots, t\}$, $\deg_{G_i^*}(x) = \deg_{G_i}(x)$ for $x \in T \setminus \{\alpha, \beta\}$ and $\deg_{G_i^*}(\alpha) + \deg_{G_i^*}(\beta) = \deg_{G_i}(\alpha) + \deg_{G_i}(\beta)$ by the reassignment. Furthermore, for $i \in \{1, \ldots, t\}$, $|\deg_{G_i^*}(\alpha) - \deg_{G_i^*}(\beta)| \leqslant 1$ because $|E_{\alpha \setminus \beta}(G_i^*)|$ and $|E_{\beta \setminus \alpha}(G_i^*)|$ are, respectively, the out-degree and in-degree of the vertex i of A under $\mathcal{O}^*$. In particular, for $i \in \{1, \ldots, k-1\}$, G_i^* is almost regular on T because G_i is almost regular on T.

Iterating this process yields a decomposition equivalent to $\{G_1, \ldots, G_t\}$ on $V(G) \setminus T$ in which the subgraphs indexed $1, \ldots, k$ are almost regular on T. Hence, eventually, it yields a decomposition with the required properties. $\square$

The style of edge switching encapsulated by Lemma 5.1 has so far resulted in the proof of fewer original results than that discussed in the previous sections. However, it does provide a uniform treatment of, and elegant proofs for, a number of existing results. We now give some examples of this. The most obvious application of Lemma 5.1 is to prove the following special case of the main result of [11].

Theorem 5.2 ([11]) *If $m_1, \ldots, m_t$ are nonnegative integers such that $m_1 + \cdots + m_t = \binom{n}{2}$, then there is a decomposition $\{G_1, \ldots, G_t\}$ of K_n such that, for $i \in \{1, \ldots, t\}$, $|E(G_i)| = m_i$ and G_i is almost regular on $V(K_n)$.*

Proof It is easy to decompose K_n into graphs $G_1^*, \ldots, G_t^*$ such that $|E(G_i^*)| = m_i$ for $i \in \{1, \ldots, t\}$. Apply Lemma 5.1 to $\{G_1^*, \ldots, G_t^*\}$ with $T = V(K_n)$ to produce a decomposition $\{G_1, \ldots, G_t\}$ of K_n with the required properties. □

Theorem 5.2 has the following corollary concerning edge colourings.

Corollary 5.3 *There is a t-edge colouring of K_n with colour classes of sizes $m_1, \ldots, m_t$ if and only if $0 \leqslant m_1, \ldots, m_t \leqslant \lfloor \frac{n}{2} \rfloor$ and $m_1 + \cdots + m_t = \binom{n}{2}$.*

A k-*factor* in a graph G is a subgraph F of G such that $\deg_F(x) = k$ for each $x \in V(G)$. A k-*factorisation* of a graph G is a decomposition $\{F_1, \ldots, F_t\}$ of G such that F_i is a k-factor in G for $i \in \{1, \ldots, t\}$. A result of Andersen and Hilton in [3] concerning the extension of a decomposition to a k-factorisation can also be easily proved from Lemma 5.1.

Theorem 5.4 ([3]) *Let U and V be sets such that $U \subseteq V$, let k be a positive integer such that $k|V|$ is even and k divides $|V| - 1$. Let $t = \frac{1}{k}(|V| - 1)$ and let $\{H_1, \ldots, H_t\}$ be a decomposition of K_U (some graphs in this decomposition may be empty). There is a k-factorisation $\{F_1, \ldots, F_t\}$ of K_V such that H_i is a subgraph of F_i for $i \in \{1, \ldots, t\}$ if and only if $\Delta(H_i) \leqslant k$ and $|E(H_i)| \geqslant \frac{k}{2}(2|U| - |V|)$ for $i \in \{1, \ldots, t\}$.*

Proof Suppose there is a k-factorisation $\{F_1, \ldots, F_t\}$ of K_V such that H_i is a subgraph of F_i for $i \in \{1, \ldots, t\}$. Then $\deg_{H_i}(x) \leqslant \deg_{F_i}(x) = k$ for each $x \in U$ and $i \in \{1, \ldots, t\}$. Also, for $i \in \{1, \ldots, t\}$, $|E(H_i)| \geqslant \frac{k}{2}(2|U| - |V|)$ because $|E(F_i)| = \frac{k}{2}|V|$ and $|E(F_i) \setminus E(H_i)| \leqslant k(|V| - |U|)$.

Now suppose $\Delta(H_i) \leqslant k$ and $|E(H_i)| \geqslant \frac{k}{2}(2|U| - |V|)$ for $i \in \{1, \ldots, t\}$. Because $t = \frac{1}{k}(|V| - 1)$ and $\Delta(H_i) \leqslant k$ for $i \in \{1, \ldots, t\}$, it is easy to create a decomposition $\{H_1', \ldots, H_t'\}$ of $K_U \cup K_{U, V \setminus U}$ such that, for each

$i \in \{1, \ldots, t\}$, H_i is a subgraph of H_i' and $\deg_{H_i'}(x) = k$ for $x \in U$. Note that, for $i \in \{1, \ldots, t\}$, $|E(H_i')| = k|U| - |E(H_i)|$ and hence, because $|E(H_i)| \geqslant \frac{k}{2}(2|U| - |V|)$, $|E(H_i')| \leqslant \frac{k}{2}|V|$. Thus, it is easy to create a decomposition $\{H_1'', \ldots, H_t''\}$ of K_V such that, for each $i \in \{1, \ldots, t\}$, H_i' is a subgraph of H_i'' and $|E(H_i'')| = \frac{k}{2}|V|$. Applying Lemma 5.1 to $\{H_1'', \ldots, H_t''\}$ with $T = V \setminus U$ now produces a k-factorisation with the required properties. $\square$

The special case of $k = 1$ in Theorem 5.4 is equivalent, via the well-known correspondence between edge colourings of complete graphs and unipotent symmetric latin squares, to the unipotent case of Cruse's theorem [43] on embeddings of partial symmetric latin squares.

5.2 Decompositions into almost regular hypergraphs

Very recently, Bryant [19] generalised Lemma 2.3 to hypergraph decompositions, marking the first time that edge switching has been applied in this new setting. This generalisation provides elegant and uniform proofs for Baranyai's theorem on factorisations of hypergraphs [11] and a number of its extensions.

Most of our definitions extend naturally to hypergraphs but we must generalise our notion of what it means for two decompositions to be equivalent on a set of vertices. Decompositions $\mathcal{D}$ and $\mathcal{D}'$ of a hypergraph G are *equivalent on* S for $S \subseteq V(G)$ if we can write $\mathcal{D} = \{G_1, \ldots, G_t\}$ and $\mathcal{D}' = \{G_1', \ldots, G_t'\}$ such that

- $|E_A^r(G_i')| = |E_A^r(G_i)|$ for all $A \subseteq S$, $r \geqslant 1$ and $i \in \{1, \ldots, t\}$,

where $E_A^r(H) = \{e \in E(H) : |e| = r, e \cap S = A\}$. In the case of graphs this definition reduces to our previous definition of equivalence.

Lemma 5.5 ([19]) *Let* $\{G_1, \ldots, G_t\}$ *be a decomposition of a hypergraph* G *and let* T *be a set of pairwise twin vertices of* G. *There exists a decomposition* $\{G_1', \ldots, G_t'\}$ *of* G *such that*

- $\{G_1, \ldots, G_t\}$ *and* $\{G_1', \ldots, G_t'\}$ *are equivalent on* $V(G) \setminus T$;

- G_i' *is almost regular on* T *for* $i \in \{1, \ldots, t\}$.

Proof The proof of Lemma 5.1 has been phrased so that it also suffices to prove this lemma. $\square$

The main theorem of Baranyai in [11] concerning factorisations of hypergraphs is an easy corollary of Lemma 5.5. Let K_n^r denote the complete r-uniform hypergraph of order n.

Theorem 5.6 ([11]) *If $m_1, \ldots, m_t$ are nonnegative integers such that $m_1 + \cdots + m_t = \binom{n}{r}$, then there is a decomposition $\{G_1, \ldots, G_t\}$ of K_n^r such that, for $i \in \{1, \ldots, t\}$, $|E(G_i)| = m_i$ and G_i is almost regular on $V(K_n^r)$.*

Proof It is easy to decompose K_n^r into hypergraphs $G_1^*, \ldots, G_t^*$ such that $|E(G_i^*)| = m_i$ for $i \in \{1, \ldots, t\}$. Apply Lemma 5.1 to $\{G_1^*, \ldots, G_t^*\}$ with $T = V(K_n)$ to produce a decomposition of K_n^r with the required properties. □

Note that the result most often referred to as "Baranyai's theorem", that the complete r-uniform hypergraph of order n has a 1-factorisation if r divides n, is recovered by applying Theorem 5.6 with $t = \binom{n-1}{r-1}$ and $m_1 = \cdots = m_t = \frac{n}{r}$.

In [19] Lemma 5.5 is used to easily prove a number of extensions of Baranyai's result that are either known results or closely related to known results. These include a partial generalisation of Theorem 5.4 that characterises when a k-factorisation of K_m^r can be extended to an ℓ-factorisation of K_n^r (see [10, 13, 50] for similar results) and an analogue of Theorem 5.6 for factorisations of complete r-uniform equipartite hypergraphs (see [9, 12, 50] for similar results).

6 Conclusion

Above we have attempted to present an introduction to edge switching techniques and some of their most significant applications outside of edge colouring. Of course, a great many details have been omitted and many results have been mentioned only in passing, but it is hoped that the interested reader can pursue these further through the references. Edge switching has met with a great deal of success over the last ten years, and it continues to be developed and applied in new settings. The recent extension to hypergraphs discussed in Section 5.2 is emblematic of this.

On the other hand, edge switching techniques have so far been usefully applied only to decompositions of highly structured graphs and hypergraphs. In the last few years many of the most dramatic advances concerning edge decomposition have been made using the *randomised algebraic constructions* of Keevash [65] and the *iterative absorption* methods pioneered by Barber et al. [14]. These methods have enabled strong results on edge decomposition of arbitrary dense graphs and hypergraphs to be obtained. Whether edge switching can also be useful in the setting of general dense graphs remains an open question. Results that allow switches of some kind on pairs of non-twin vertices would likely be required.

References

[1] B. Alspach, Research Problem 3, Discrete Math. **36** (1981), 333.

[2] B. Alspach and H. Gavlas, Cycle decompositions of K_n and $K_n - I$, J. Combin. Theory Ser. B **81** (2001), 77–99.

[3] L.D. Andersen and A.J.W. Hilton, Generalized latin rectangles II: embedding, Discrete Math. **31** (1980), 235–260.

[4] L.D. Andersen, A.J.W. Hilton and E. Mendelsohn, Embedding partial Steiner triple systems, Proc. London Math. Soc. **41** (1980), 557–576.

[5] L.D. Andersen and C.A. Rodger, Decompositions of complete graphs: Embedding partial edge-colourings and the method of amalgamations, Surveys in Combinatorics, Lond. Math. Soc. Lect. Note Ser. **307** (2003), 7–41.

[6] D. Archdeacon, M. Debowsky, J. Dinitz and H. Gavlas, Cycle systems in the complete bipartite graph minus a one-factor, Discrete Math. **284** (2004), 37–43.

[7] J-C. Bermond, C. Huang and D. Sotteau, Balanced cycle and circuit designs: even cases, Ars Combin. **5** (1978), 293–318.

[8] J.C. Bermond and D. Sotteau, Cycle and circuit designs odd case, *Contributions to graph theory and its applications* (Internat. Colloq., Oberhof, 1977) (German), pp. 11–32, Tech. Hochschule Ilmenau, Ilmenau, 1977.

[9] M.A. Bahmanian, Detachments of Hypergraphs I: The Berge-Johnson Problem, Combin. Probab. Comput. **21** (2012), 483–495.

[10] A. Bahmanian and C. Rodger, Embedding factorizations for 3-uniform hypergraphs, J. Graph Theory **73** (2013), 216–224.

[11] Z. Baranyai, On the factorization of the complete uniform hypergraph, Infinite and finite sets, Colloquia Math. Soc. János Bolyai **10** (1973), 91–107.

[12] Z. Baranyai, The edge-coloring of complete hypergraphs I, J. Combin. Theory Ser. B **26** (1979),276–294.

[13] Z. Baranyai and A.E. Brouwer, Extension of colorings of the edges of a complete (uniform hyper)graph, Math. Centre Report ZW91 (Mathematisch Centrum Amsterdam). Zbl. 362.05059 (1977).

[14] B. Barber, D. Kühn, A. Lo and D. Osthus, Edge decompositions of graphs with high minimum degree, Advances in Mathematics **288** (2016), 337–385.

[15] E.J. Billington, Multipartite graph decomposition: cycles and closed trails, Matematiche (Catania) **59** (2004), 53–72.

[16] D. Bryant, Cycle decompositions of complete graphs, Surveys in Combinatorics, Lond. Math. Soc. Lect. Note Ser. **346** (2007), 67–97.

[17] D. Bryant, A conjecture on small embeddings of partial Steiner triple systems, J. Combin. Des. **10** (2002), 313–321.

[18] D. Bryant, Packing paths in complete graphs, J. Combin. Theory Ser. B, **100** (2010), 206–215.

[19] D. Bryant, On almost-regular edge colourings of hypergraphs (preprint).

[20] D. Bryant, Embeddings of partial Steiner triple systems, J. Combin. Theory Ser. A **106** (2004), 77–108.

[21] D. Bryant and M. Buchanan, Embedding partial totally symmetric quasigroups, J. Combin. Theory Ser. A **114** (2007), 1046–1088.

[22] D.E. Bryant, D.G. Hoffman and C.A. Rodger, 5-cycle systems with holes, Des. Codes Cryptogr. **8** (1996), 103–108.

[23] D. Bryant and D. Horsley, Steiner triple systems with two disjoint subsystems, J. Combin. Des. **14** (2006), 14–24.

[24] D. Bryant and D. Horsley, A proof of Lindner's conjecture on embeddings of partial Steiner triple systems, J. Combin. Des. **17** (2009), 63–89.

[25] D. Bryant and D. Horsley, Decompositions of complete graphs into long cycles, Bull. London Math. Soc. **41** (2009), 927–934.

[26] D. Bryant and D. Horsley, An asymptotic solution to the cycle decomposition problem for complete graphs, J. Combin. Theory Ser. A **117** (2010), 1258–1284.

[27] D. Bryant, D. Horsley and B. Maenhaut, Decompositions into 2-regular subgraphs and equitable partial cycle decompositions, J. Combin. Theory Ser. B **93** (2005), 67–72.

[28] D. Bryant, D. Horsley, B. Maenhaut and B.R. Smith, Cycle decompositions of complete multigraphs, J. Combin. Des. **19** (2011), 42–69.

[29] D. Bryant, D. Horsley, B. Maenhaut and B.R. Smith, Decompositions of complete multigraphs into cycles of varying lengths, arXiv preprint, arXiv:1508.00645.

[30] D. Bryant, D. Horsley and W. Pettersson, Cycle decompositions V: Complete graphs into cycles of arbitrary lengths, Proc. London Math. Soc. **108** (2014), 1153–1192.

[31] D. Bryant and B. Maenhaut, Almost Regular Edge Colorings and Regular Decompositions of Complete Graphs, J. Combin. Des. **16** (2013), 499–506.

[32] D. Bryant, B. Maenhaut, K. Quinn and B.S. Webb, Existence and embeddings of partial Steiner triple systems of order ten with cubic leaves, Discrete Math. **284** (2004), 83–95.

[33] D. Bryant and G. Martin, Small embeddings for partial triple systems of odd index, J. Combin. Theory Ser. A **119** (2012), 283–309.

[34] D.E. Bryant and C.A. Rodger, On the Doyen-Wilson theorem for m-cycle systems, J. Combin. Des. **2** (1994), 253–271.

[35] D.E. Bryant and C.A. Rodger, The Doyen-Wilson theorem extended to 5-cycles, J. Combin. Theory Ser. A **68** (1994), 218–225.

[36] D.E. Bryant, C.A. Rodger and E.R. Spicer, Embeddings of m-cycle systems and incomplete m-cycle systems: $m \leqslant 14$, Discrete Math. **171** (1997), 55–75.

[37] N.J. Cavenagh and E.J. Billington, Decomposition of complete multipartite graphs into cycles of even length, Graphs Combin. **16** (2000), 49–65.

[38] C-C. Chou and C-M. Fu, Decomposition of $K_{m,n}$ into 4-cycles and $2t$-cycles, J. Comb. Optim. **14** (2007), 205–218.

[39] C-C. Chou, C-M. Fu and W-C. Huang, Decomposition of $K_{m,n}$ into short cycles, Discrete Math. **197/198** (1999), 195–203.

[40] C.J. Colbourn, Embedding partial Steiner triple systems is NP-complete, J. Combin. Theory Ser. A **35** (1983), 100–105.

[41] C.J. Colbourn, M.J. Colbourn and A. Rosa, Completing small partial triple systems, Discrete Math. **45** (1983), 165–179.

[42] C.J. Colbourn and A. Rosa, Triple Systems, Clarendon Press, Oxford (1999).

[43] A. Cruse, On embedding incomplete symmetric latin squares, J. Combin. Theory Ser. A **16** (1974), 18–27.

[44] R. Diestel, Graph Theory (4th edition), Springer, Heidelberg (2010).

[45] J. Doyen and R.M. Wilson, Embeddings of Steiner triple systems, Discrete Math. **5** (1973), 229–239.

[46] H.E. Dudeney, Amusements in Mathematics, Nelson, Edinburgh (1917), reprinted by Dover Publications, New York (1959).

[47] V. Fack and B.D. McKay, A generalized switching method for combinatorial estimation, Australas. J. Combin. **39** (2007), 141–154.

[48] M.N. Ferencak and A.J.W. Hilton, Outline and amalgamated triple systems of even index, Proc. London Math. Soc. **84** (2002), 1–34.

[49] H. Hanani, The existence and construction of balanced incomplete block designs, Ann. Math. Statist. **32** (1961), 361–386.

[50] R. Häggkvist and T. Hellgren, Extensions of edge-colourings in hypergraphs I, Combinatorics, Paul Erdős is eighty, Bolyai Soc. Math. Stud. (1993), 215–238.

[51] K. Heinrich, Path-decompositions, Matematiche (Catania) **47** (1993), 241–258.

[52] P. Hell and A. Rosa, Graph decompositions, handcuffed prisoners and balanced P-designs, Discrete Math. **2** (1972), 229–252.

[53] A.J.W. Hilton, Hamiltonian Decompositions of Complete Graphs, J. Combin. Theory Ser. B **36** (1984), 125–134.

[54] A.J.W. Hilton and C.A. Rodger, The embedding of partial triple systems when 4 divides λ, J. Combin. Theory Ser. A **56** (1991), 109–137.

[55] D.G. Hoffman, C.C. Lindner and C.A. Rodger, On the construction of odd cycle systems, J. Graph Theory **13** (1989), 417–426.

[56] D. Horsley, Decomposing various graphs into short even-length cycles, Ann. Comb. **16** (2012), 571–589.

[57] D. Horsley, Small Embeddings of Partial Steiner Triple Systems, J. Combin. Des. **22** (2014), 343–365.

[58] D. Horsley, Embedding Partial Steiner Triple Systems with Few Triples, SIAM J. Discrete Math. **28** (2014), 1199–1213.

[59] D. Horsley and R.A. Hoyte, Doyen-Wilson Results for Odd Length Cycle Systems, J. Combin. Des. **24** (2016), 308–335.

[60] D. Horsley and R.A. Hoyte, Decomposing $K_{u+w} - K_u$ into cycles of various lengths, arXiv preprint, arXiv:1603.03908.

[61] D. Horsley and D.A. Pike, Embedding partial odd-cycle systems in systems with orders in all admissible congruence classes, J. Combin. Des. **18** (2010), 202–208.

[62] S.H.Y. Hung and N.S. Mendelsohn, Handcuffed designs, Aequationes Math. **11** (1974), 256–266.

[63] C. Huang and A. Rosa, On the existence of balanced bipartite designs, Utilitas Math. **4** (1973), 55–75.

[64] A. Johansson, A note on extending partial triple systems, University of Umea, Sweden, preprint (1997).

[65] P. Keevash, The existence of designs, arXiv preprint, arXiv:1401.3665.

[66] A.B. Kempe, On the geographical problem of the four colours, Amer. J. Math. **2** (1879), 193-200.

[67] T.P. Kirkman, On a problem in combinations, Cambridge and Dublin Math. J. **2** (1847), 191–204.

[68] R. Laskar and B. Auerbach, On decomposition of r-partite graphs into edge-disjoint Hamilton circuits, Discrete Math. **14** (1976), 265–268.

[69] J.F. Lawless, On the construction of handcuffed designs, J. Combin. Theory Ser. A **16** (1974) 76–86.

[70] J.F. Lawless, Further results concerning the existence of handcuffed designs, Aequationes Math. **11** (1974) 97–106.

[71] C.C. Lindner, A partial Steiner triple system of order n can be embedded in a Steiner triple system of order $6n+3$, J. Combin. Theory Ser. A **18** (1975), 349–351.

[72] C.C. Lindner and T. Evans, Finite embedding theorems for partial designs and algebras, SMS **56**, Les Presses de l'Université de Montréal, (1977).

[73] C.C. Lindner and C.A. Rodger, A partial $m = (2k+1)$-cycle system of order n can be embedded in an m-cycle system of order $(2n+1)m$, Discrete Math. **117** (1993), 151–159.

[74] E. Lucas, "Récreations Mathématiqués," Vol II, Gauthier-Villars, Paris (1892).

[75] J. Ma, L. Pu and H. Shen, Cycle decompositions of $K_{n,n} - I$, SIAM J. Discrete Math. **20** (2006), 603–609.

[76] G. Martin and T.A. McCourt, Small embeddings for partial 5-cycle systems, J. Combin. Des. **20** (2012), 199–226.

[77] C.J.H. McDiarmid, The solution of a timetabling problem, J. Inst. Math. Appl. **9** (1972), 23–34.

[78] E. Mendelsohn and A. Rosa, Embedding maximal packings of triples, Congr. Numer. **40** (1983), 235–247.

[79] M.E. Raines, More on embedding partial totally symmetric quasigroups, Australas. J. Combin. **14** (1996), 297–309.

[80] M.E. Raines and C.A. Rodger, Embedding partial extended triple systems and totally symmetric quasigroups, Discrete Math. **176** (1997), 211–222.

[81] M. Raines and Z. Szaniszló, Equitable partial cycle systems, Australas. J. Combin. **19** (1999), 149–156.

[82] C.A. Rodger and S.J. Stubbs, Embedding partial triple systems, J. Combin. Theory Ser. A **44** (1987), 241–252.

[83] C.A. Rodger and S.J. Stubbs, Embedding partial triple systems (Erratum), J. Combin. Theory Ser. A **66** (1994), 182–183.

[84] A. Rosa and C. Huang, Another class of balanced graph designs: balanced circuit designs, Discrete Math. **12** (1975), 269–293.

[85] M. Šajna, Cycle decompositions III: complete graphs and fixed length cycles, J. Combin. Des. **10** (2002), 27–78.

[86] J. Schönheim, On maximal systems of k-tuples, Studia Sci. Math. Hungar. **1** (1966), 363–368.

[87] B.R. Smith, Cycle decompositions of complete multigraphs, J. Combin. Des. **18** (2010), 85–93.

[88] D. Sotteau, Decomposition of $K_{m,n}$ ($K^*_{m,n}$) into cycles (circuits) of length $2k$, J. Combin. Theory Ser. B **30** (1981), 75–81.

[89] M. Tarsi, Decomposition of a complete multigraph into simple paths: Nonbalanced handcuffed designs, J. Combin. Theory Ser. A **34** (1983), 60–70

[90] A.G. Thomason, Hamiltonian cycles and uniquely edge colourable graphs, Ann. Discrete Math. **3** (1978), 259–268.

[91] C. Treash, The completion of finite incomplete Steiner triple systems with applications to loop theory, J. Combin. Theory Ser. A **10** (1971), 259–265.

[92] V.G. Vizing, On an estimate of the chromatic class of a p-graph, Diskret Analiz **3** (1964), 25–30.

School of Mathematical Sciences
Monash University
Vic 3800, Australia
danhorsley@gmail.com

Ramsey-Type and Amalgamation-Type Properties of Permutations

Vít Jelínek

Abstract

This survey deals with some aspects of combinatorics of permutations which are inspired by notions from structural Ramsey theory. Its first main focus is the overview of known results on Ramsey-type and Fraïssé-type properties of hereditary permutation classes, with particular emphasis on the concept of splittability. Secondly, we look at known estimates for Ramsey numbers of permutation matrices, and their relationship to Ramsey numbers of ordered graphs.

1 Introduction

1.1 About this survey

Combinatorics of permutations is an old and well-established field of discrete mathematics. So is Ramsey theory. For a long time these two fields have followed their own separate ways without affecting each other much. However, in the first decade of this century, the situation started to slowly change, as concepts originating from the research on relational structures and on Ramsey classes became adopted (or, occasionally, reinvented) in the study of hereditary permutation classes.

The purpose of this paper is to give an introductory overview of the Ramsey-theoretic and relation-theoretic aspects of combinatorics of permutations, with particular focus on hereditary permutation classes. I do not assume any familiarity with either structural Ramsey theory or permutation combinatorics. In the rest of this first chapter, the reader will find a condensed introduction to the relevant notions from these fields.

Chapter 2 will then present a survey of the known results related to amalgamation, Ramseyness and other related properties of hereditary permutation classes.

The remaining two chapters contain a more detailed treatment of two specific topics related to amalgamation and Ramsey properties of permutations.

Chapter 3 focuses on the notion of unsplittability, a weak form of Ramsey property that has recently found applications in enumerative combinatorics of permutations, and appears to be a promising research direction.

Consequently, splittability and unsplittability play a prominent role in this survey; indeed, Chapter 3 is the longest of the four chapters.

Chapter 4 deals with estimates on Ramsey numbers of permutations. This topic, which is closely connected to graph theory, has received interest only very recently, with only a few nontrivial results known, and many basic questions still open.

This paper does not include any important new results. Most of the results included here are presented without proof, or with only a brief sketch of proof. I do include the proofs of simple basic results, where finding an accurate original reference would be impractical. I also occasionally present proofs which I consider particularly elegant or illuminating.

Open problems are included throughout the text. Some of them, to my knowledge, have not been raised before.

1.2   Amalgamation and Ramseyness in relational structures

The starting point of Ramsey theory is the following classical result of Ramsey [61].

Theorem 1.1 *For every $c, k, n \in \mathbf{N}$ there is an $N \in \mathbf{N}$ such that whenever the k-element subsets of $\{1, \ldots, N\}$ are colored by c colors, there is an n-element subset $A \subseteq \{1, \ldots, N\}$ such that all the k-element subsets of A have the same color.*

Since the publication of this result in 1930, many generalizations and extensions were obtained, often proving analogous Ramsey-type properties in the context of graphs, posets, permutations or other discrete structures. A convenient unifying formalism in which to approach the study of such discrete structures is the concept of relational structure, which we now introduce.

Relational structures.   Let I be an index set, and let $S = (s_i, i \in I)$ be a collection of positive integers. A *relational structure* with signature S is an ordered pair $\alpha = (X, (R_i, i \in I))$ where X is a set, which we call the *domain* or the *vertex set* of α, and each R_i is a relation of arity s_i over X, or in other words, R_i is a set of ordered s_i-tuples of elements of X. The *size* of a relational structure is its number of vertices.

Let $\alpha = (X, (R_i, i \in I))$ be a relational structure. A *substructure* of α is a relational structure $\beta = (Y, (\overline{R_i}, i \in I))$ such that Y is a subset of X and each $\overline{R_i}$ is the restriction of R_i to the set Y, i.e., $\overline{R_i} = R_i \cap Y^{s_i}$. In this situation, we also say that the substructure β is *induced* by the set Y, and we let $\alpha[Y]$ denote the substructure of α induced by Y.

Let $\alpha = (X, (R_i, i \in I))$ and $\alpha' = (X', (R'_i, i \in I))$ be relational structures of the same signature. An *embedding* of α into α' is an injective function $f: X \to X'$ such that, for every $i \in I$ and every s_i-tuple $(x_1, \ldots, x_{s_i}) \in X^{s_i}$, we have

$$(x_1, \ldots, x_{s_i}) \in R_i \iff (f(x_1), \ldots, f(x_{s_i})) \in R'_i.$$

If an embedding f is a bijection between X and X', then it is an *isomorphism*. We say that a relational structure γ is *contained* in α (and α *contains* γ) if γ is isomorphic to a substructure of α. Otherwise we say that α *avoids* γ.

In this paper, we will mostly deal with finite relational structures. To avoid repeating the same assumptions all the time, we will assume, from now on, that each relational structure has a finite and nonempty domain, unless explicitly stated otherwise.

Let $\mathcal{C}$ be a class of relational structures, all sharing the same signature. We say that $\mathcal{C}$ is *hereditary*, if for every $\alpha \in \mathcal{C}$, all the relational structures contained in α belong to $\mathcal{C}$ as well. Notice that if $\mathcal{C}$ is hereditary, it implies that $\mathcal{C}$ is isomorphism-closed.

A class $\mathcal{C}$ has *joint embedding property* if for every $\alpha, \beta \in \mathcal{C}$ there is a $\gamma \in \mathcal{C}$ which contains both α and β.

Let Γ be a possibly infinite relational structure. The *age* of Γ, denoted by $\mathsf{Age}(\Gamma)$, is the class of all the finite structures contained in Γ. We will say that a class $\mathcal{C}$ is an *age* if it is the age of a finite or countable relational structure. The following characterization of ages is due to Fraïssé [29].

Proposition 1.2 ([29]) *A class $\mathcal{C}$ of relational structures is the age of a finite or countable relational structure Γ if and only if $\mathcal{C}$ is hereditary, has the joint embedding property, and has at most countably many members up to isomorphism.*

Note that the assumption that $\mathcal{C}$ has at most countably many members up to isomorphism is automatically satisfied for any class of structures of finite signature, i.e., structures with finitely many relations.

Hereditary classes with joint embedding property are also known as *atomic* classes. This is motivated by the fact that a hereditary class $\mathcal{C}$ has the joint embedding property if and only if it cannot be obtained as a union of two proper hereditary subclasses.

Amalgamation and homogeneity. For a class of structures $\mathcal{C}$ and a structure $\alpha \in \mathcal{C}$, we say that $\mathcal{C}$ has the α-*amalgamation property* if for every β and β' in $\mathcal{C}$ the following holds: if f is an embedding of α into β, and f' is an embedding of α into β', then there exists a structure $\gamma \in C$ and a pair of embeddings $g\colon \beta \to \gamma$ and $g'\colon \beta' \to \gamma$ such that $g \circ f = g' \circ f'$. We say that C has the *amalgamation property*, if it has the α-amalgamation property for every $\alpha \in C$.

Intuitively speaking, α-amalgamation means that any two structures in C can be 'glued together' by identifying a pair of substructures isomorphic to α, with the resulting structure belonging to C.

A possibly infinite relational structure Γ is *homogeneous* if for any two finite isomorphic substructures α and α' of Γ, and any isomorphism $f\colon \alpha \to \alpha'$, there is an automorphism F of Γ whose restriction to α is f. Fraïssé showed that homogeneity can be characterized by the amalgamation property of the age.

Theorem 1.3 ([29]) *Let C be a class of relational structures. Then C is the age of a finite or countable homogeneous structure Γ if and only if C is an age and has the amalgamation property. Moreover, if C has these properties, then the homogeneous structure Γ is determined uniquely up to isomorphism.*

We say that C is a *Fraïssé class* if it is the age of a finite or countable homogeneous structure Γ. The structure Γ itself is called the *Fraïssé limit* of C.

Homogeneous structures have been studied from the point of view of model theory, topological group theory, as well as combinatorics. Significant effort was devoted to the characterization of the homogeneous structures arising as Fraïssé limits of particular types of finite structures. Thus, Lachlan and Woodrow [49] have characterized the infinite homogeneous ordered graphs, Schmerl [67] described the homogeneous partially ordered sets, and more generally, Cherlin [20] classified the infinite homogeneous digraphs. Cameron [19] obtained a classification of infinite homogeneous permutations, to which we shall return in Chapter 2.

The study of homogeneous structures goes well beyond these classification results. A survey of the more recent developments in this area was given by Macpherson [51].

Ramsey property. For relational structures α and β, let $\binom{\beta}{\alpha}$ denote the set of all embeddings of α into β. Let C be a class of relational structures which is an age, and let α be a structure in C. We say that C has the α-*Ramsey property*, if for every $c \in \mathbf{N}$ and every $\beta \in C$ there is a $\gamma \in C$ such that in every coloring of $\binom{\gamma}{\alpha}$ by c colors there is an embedding $g \in \binom{\gamma}{\beta}$ such that all the embeddings in the set $\{g \circ f; \ f \in \binom{\beta}{\alpha}\}$ have the same color. We say that $\mathcal{C}$ has the *Ramsey property* (or $\mathcal{C}$ is a *Ramsey class*) if it has the α-Ramsey property for every $\alpha \in \mathcal{C}$.

In particular, Ramsey's theorem is equivalent to the statement that the class of all finite linearly ordered sets has the Ramsey property.

Actually, to prove that a class $\mathcal{C}$ is a Ramsey class, it is enough to prove the above property for the case $c = 2$, i.e., for coloring with two colors (see, e.g., [11, Lemma 2.5] for the easy proof). It is also not hard to see that α-Ramsey property implies the α-amalgamation property, and in particular, every Ramsey class is also a Fraïssé class (see, e.g., [58]). Moreover, for $\mathcal{C}$ to have the α-Ramsey property, it is necessary that the structure α is *rigid*, i.e., has no non-identical automorphism. However, a Fraïssé class of rigid structures need not be Ramsey (see e.g. [11, Example 2.15]).

A systematic study of Ramsey classes was initiated by the works of Neetil and Rödl in the 1970's. A more recent motivation for the study of Ramsey classes originates in the work of Kechris, Pestov and Todorević [40], who demonstrated a close connection between the Ramsey property and the notion of extreme amenability from topological dynamics.

We remark that there are two different 'flavors' of the notion of Ramsey class present in the literature. Some authors define $\binom{\beta}{\alpha}$ to be the set of substructures of β which are isomorphic to α, and base their notion of Ramseyness upon colorings of substructures rather than colorings of embeddings. Zucker's paper [75] includes an overview of the issues related to the distinction between the two notions of Ramseyness. If the relational structures in question are rigid, as will be mostly the case in this paper, the two notions of Ramseyness coincide.

For an overview of the research concerned with Ramsey classes, see the paper of Neetil [58] or the recent survey by Bodirsky [11].

Restricted amalgamation and Ramseyness. Amalgamation property is quite restrictive, and Ramsey property even more so. Many natural combinatorial structures, for instance permutations, admit only very few Fraïssé or Ramsey classes. When dealing with such structures, it is natural to look at classes satisfying a restricted form of amalgamation or Ramsey property, such as α-amalgamation or α-Ramsey property, for a fixed small α.

Let us say that an atomic class $\mathcal{C}$ of relational structures is *unsplittable*, if for every $c \in \mathbf{N}$ and every $\beta \in \mathcal{C}$ there is a $\gamma \in \mathcal{C}$ such that for every coloring of the vertices of γ with c colors there is a monochromatic copy of β in γ, i.e., a substructure β^* of γ isomorphic to β and with all vertices of the same color. If this property does not hold, we say that $\mathcal{C}$ is *splittable*.

As with the general Ramsey property, to show that a class $\mathcal{C}$ is unsplittable, it is enough to verify the property from the definition in the case $c = 2$.

Notice that an atomic class $\mathcal{A}$ is splittable if and only if it has a proper subclass $\mathcal{B}$ such that for every structure $\alpha \in \mathcal{A}$ the vertices of α can be colored by two colors in such a way that each color induces in α a substructure belonging to $\mathcal{B}$.

If a class $\mathcal{C}$ has only a single element α of size 1, then $\mathcal{C}$ is unsplittable if and only if it has the α-Ramsey property.

Unsplittability has been considered independently by several authors in various contexts. Unfortunately, as a result, the terminology is not consistent, with unsplittability being variously referred to as vertex-partition property [56], vertex Ramsey property [42, 41, 66], Ramsey property [62, 63, 65], or indivisibility [26]. In this survey, we use the term 'unsplittable', which seems to be the most common choice in the context of permutations.

The reader should also beware that some authors, when dealing with unsplittability, do not explicitly require that the class in question is atomic.

Suppose that Γ is a countable relational structure and $\mathcal{C}$ is the age of Γ. El-Zahar and Sauer [26] have shown that in such case, $\mathcal{C}$ is unsplittable if and only if for every partition of the vertex set of Γ into two sets X and Y, one of the two induced substructures $\Gamma[X]$ and $\Gamma[Y]$ has the same age as Γ.

There are several more equivalent ways to define (un)splittability beyond those we have already mentioned, as we will show in Chapter 3.

Historically, the first (implicit) appearance of the splittability concept can be traced to the work of Folkman [27], whose Theorem 2 can be rephrased in our terminology as stating that for any k, the class of all graphs avoiding the complete graph K_k is unsplittable.

Subsequent results on splittability have again dealt with graph classes, with the main focus being to determine which hereditary graph classes are splittable.

Most of the papers dealing with splittability of graph classes have focused on identifying which classes of graphs determined by finitely many forbidden subgraphs are splittable. Let $\mathcal{G}(F_1, \ldots, F_k)$ be the class of graphs not containing any of $F_1, \ldots, F_k$ as induced subgraphs. As we mentioned above, Folkman [27] has shown that $\mathcal{G}(K_k)$ is unsplittable, where K_k is the complete graph on k vertices. Neetil and Rödl [55] generalized this to any class $\mathcal{G}(F_1, \ldots, F_k)$ where all the F_i are 2-connected. Further results in this area were obtained by Rödl and Sauer [62], Rödl, Sauer and Zhu [63] and Sauer [65]. It seems, however, that a full characterization of splittability of graph classes is out of reach.

A particular interest was devoted to the splittability of graph classes of the form $\mathcal{G}(T, K_k)$ where T is a tree. This is motivated by a relationship with a conjecture on χ-boundedness, made independently by Gyárfás [33] and Sumner [71].

A class $\mathcal{G}$ of graphs is said to be χ-bounded if there is a function $f \colon \mathbf{N} \to \mathbf{N}$ such that every graph $G \in \mathcal{G}$ that does not contain the complete graph on k vertices as a subgraph has chromatic number at most $f(k)$. Gyárfás and Sumner conjectured the following.

Conjecture 1.4 *For any tree T, the class $\mathcal{G}(T)$ is χ-bounded.*

Clearly, if $\mathcal{G}(T)$ is χ-bounded, it means the graphs in the class $\mathcal{G}(T, K_k)$ have chromatic number at most $f(k)$ for some function f, and in particular, each graph in $\mathcal{G}(T, K_k)$ can be partitioned into at most $f(k)$ edgeless graphs. This would mean, of course, that $\mathcal{G}(T, K_k)$ is splittable, unless $\mathcal{G}(T, K_k)$ itself contains only edgeless graphs. Motivated by this observation, Sauer [64] proposed the following weakening of Conjecture 1.4.

Conjecture 1.5 *For every tree T on at least three vertices and for every $k \geq 3$ the class $\mathcal{G}(T, K_k)$ is splittable.*

Both of the conjectures have been versified for specific cases of T [41, 42, 43], but in their generality are still open.

1.3 Permutations and their classes

Permutation basics. A *permutation* of size n is a bijection $\pi \colon [n] \to [n]$, where $[n]$ denotes the set $\{1, \dots, n\}$. We typically identify a permutation π with the sequence $\pi(1), \pi(2), \dots, \pi(n)$. When writing out short permutations explicitly, we omit all punctuation an write, e.g., 3124 for the permutation π satisfying $\pi(1) = 3$, $\pi(2) = 1$, $\pi(3) = 2$ and $\pi(4) = 4$. We let $\mathcal{S}_n$ denote the set of permutations of size n, and $\mathcal{S}$ be the set $\bigcup_{n=1}^{\infty} \mathcal{S}_n$.

There is a natural partial order relation on the set of permutations. For two permutations $\sigma = \sigma(1), \dots, \sigma(k) \in \mathcal{S}_k$ and $\pi = \pi(1), \dots, \pi(n) \in \mathcal{S}_n$, we say that π *contains* σ if the sequence $\pi(1), \dots, \pi(n)$ contains a subsequence order-isomorphic to $\sigma(1), \dots, \sigma(k)$, or in other words, if there are indices $1 \leq i_1 < i_2 < \cdots < i_k \leq n$ such that for every $a, b \in [k]$ we have $\pi(i_a) < \pi(i_b)$ if and only if $\sigma(a) < \sigma(b)$.

An alternative way to represent a permutation π is to identify it with a relational structure $(X, (<_H, <_V))$, where $X = \{x_1, x_2, \dots, x_n\}$ is an n-element vertex set in which the vertex x_i is associated with the element $\pi(i)$ of the permutation, the relation $<_H$ is the 'horizontal' linear order $x_1 <_H x_2 <_H \cdots <_H x_n$, while the relation $<_V$ is the 'vertical' linear order defined by $x_i <_V x_j \iff \pi(i) < \pi(j)$. In this way, we obtain a bijective relationship between permutations and isomorphism types of relational structures consisting of a pair of linear orders on a common vertex set.

In this correspondence between permutations and relational structures, the containment relation of permutations defined above corresponds precisely to the containment of relational structures defined in the previous section. If there is no risk of ambiguity, we will make no explicit distinction between permutations represented as bijections, permutations represented as sequences, or permutations represented as relational structures with a pair of linear orders.

As with general relational structures, we always assume that permutations are finite and nonempty, unless noted otherwise. The term *infinite permutation* refers to an infinite relational structure with two linear orders, and is therefore a natural generalization of finite permutations in their relational representation.

Permutation classes. In permutation combinatorics, the term *permutation class* refers to a hereditary class of permutations, and we will use it in this sense as well. A typical example of a permutation class is the class of all permutations that avoid a given forbidden pattern $\pi \in \mathcal{S}$; this class is denoted by $\mathrm{Av}(\pi)$. More generally, if F is a set of permutations, then $\mathrm{Av}(F)$ denotes the class of permutations that avoid all members of F.

For every permutation class $\mathcal{C}$ there is a unique (possibly infinite) antichain $F \subseteq \mathcal{S}$, such that $\mathcal{C} = \mathrm{Av}(F)$. The set F is known as the *basis* of $\mathcal{C}$. Permutation classes whose basis contains a single permutation are known as *principal classes*.

Let $\mathfrak{PC}$ be the set of all permutation classes.

Operations with permutations (and with their classes). Let $\pi = \pi(1) \cdots \pi(m)$ and $\sigma = \sigma(1) \cdots \sigma(n)$ be two permutations. Their *direct sum*, denoted by $\pi \oplus \sigma$ is the permutation

$$\pi(1)\pi(2) \cdots \pi(m)(\sigma(1) + m)(\sigma(2) + m) \cdots (\sigma(n) + m)$$

of size $m + n$. Symmetrically, their *skew sum*, denoted by $\pi \ominus \sigma$ is the permutation

$$(\pi(1) + n)(\pi(2) + n) \cdots (\pi(m) + n)\sigma(1)\sigma(2) \cdots \sigma(n).$$

See Figure 1. A permutation is *sum-decomposable* if it can be expressed as a direct sum of two nonempty permutations. *Skew-decomposable* permutations are defined analogously.

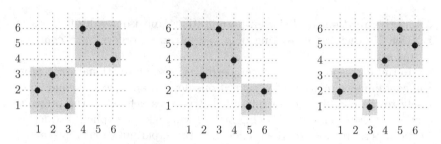

Figure 1: Left: the sum $231 \oplus 321 = 231654$. Center: the skew-sum $3142 \ominus 12 = 536412$. Right: the inflation $213[12, 1, 132] = 231465$.

A more general operation is the inflation, also called wreath product by some authors. Let π be a permutation of size n and let $\sigma_1, \sigma_2, \ldots, \sigma_n$ be an n-tuple of permutations, with σ_i having size k_i. The *inflation* of π by $\sigma_1, \ldots, \sigma_n$, denoted by $\pi[\sigma_1, \ldots, \sigma_n]$, is the permutation obtained by replacing each element $\pi(i)$ in π by a block of k_i consecutive elements which are order-isomorphic to σ_i; see Figure 1. In the special case when all the $\sigma_1, \ldots, \sigma_n$ are equal to the same permutation σ, we write $\pi[\sigma]$ instead of $\pi[\sigma, \sigma, \ldots, \sigma]$.

We say that a permutation is *simple* if it cannot be obtained from smaller permutations by inflation.

The last operation we need is the merge. For two permutations $\pi \in \mathcal{S}_m$ and $\sigma \in \mathcal{S}_n$, a *merge* of π and σ is any permutation $\tau \in \mathcal{S}_{m+n}$ with the property that the elements of τ can be colored by red a blue in a such way that the red elements form a sequence isomorphic to π and the blue ones are isomorphic to σ.

To each of the above operations, we may associate a corresponding operation with permutation classes. Let $\mathcal{A}$ and $\mathcal{B}$ be two sets of permutations. Their direct sum, skew sum and inflation are defined, respectively, as follows:

$$\mathcal{A} \oplus \mathcal{B} = \{\sigma \oplus \pi; \ \sigma \in \mathcal{A}, \ \pi \in \mathcal{B}\} \cup \mathcal{A} \cup \mathcal{B},$$
$$\mathcal{A} \ominus \mathcal{B} = \{\sigma \ominus \pi; \ \sigma \in \mathcal{A}, \ \pi \in \mathcal{B}\} \cup \mathcal{A} \cup \mathcal{B}, \text{ and}$$
$$\mathcal{A}[\mathcal{B}] = \{\pi[\sigma_1, \sigma_2, \ldots, \sigma_{|\pi|}]; \ \pi \in \mathcal{A}, \ \sigma_1, \ldots, \sigma_{|\pi|} \in \mathcal{B}\}.$$

Additionally, we define the merge of $\mathcal{A}$ and $\mathcal{B}$, denoted by $\mathcal{A} \odot \mathcal{B}$, to be the set of all the permutations that belong to $\mathcal{A} \cup \mathcal{B}$ or can be obtained by merging a permutation from $\mathcal{A}$ with a permutation from $\mathcal{B}$.

Note that if $\mathcal{A}$ and $\mathcal{B}$ are both permutation classes, then $\mathcal{A} \oplus \mathcal{B}$, $\mathcal{A} \ominus \mathcal{B}$, $\mathcal{A}[\mathcal{B}]$ and $\mathcal{A} \odot \mathcal{B}$ are permutation classes as well.

A permutation class $\mathcal{C}$ is said to be *sum-closed* (or *skew-closed* or *inflation-closed*) if $\mathcal{C} \oplus \mathcal{C} = \mathcal{C}$ (or $\mathcal{C} \ominus \mathcal{C} = \mathcal{C}$ or $\mathcal{C}[\mathcal{C}] = \mathcal{C}$, respectively).

Notable permutation classes. Certain permutation classes will appear repeatedly in our survey. We present their list now and fix the notation for them.

- Let $\mathbb{1}$ denote the class $\{1\}$, the class containing the single permutation of size 1.

- Let $\mathsf{I} = \mathrm{Av}(21)$ and $\mathsf{D} = \mathrm{Av}(12)$ be the classes containing all the increasing permutations and all the decreasing permutations, respectively.

- Let L be the class $\mathrm{Av}(312, 231)$, whose elements are known as *layered permutations*. Layered permutations are precisely the permutations that can be expressed as a direct sum of decreasing permutations, that is, $\mathsf{L} = \mathsf{I}[\mathsf{D}]$.

- $\overline{\mathsf{L}}$ is the class $\mathrm{Av}(132, 213)$ of all the complements of layered permutations, i.e., the skew sums of increasing sequences. These are known as *co-layered* permutations.

- Sep is the class $\mathrm{Av}(2413, 3142)$, whose elements are known as *separable* permutations. Separable permutations are the smallest infinite class that is both sum-closed and skew-closed.

- $\mathcal{S}$ is the class of all permutations.

2 Ramsey-type properties of permutation classes

2.1 Amalgamation and Ramseyness

Let us first look at the permutation classes that satisfy amalgamation and Ramsey properties in their strongest form. Cameron [19] has obtained the complete characterization of the Fraïssé classes of permutations.

Theorem 2.1 ([19]) *There are six nonempty Fraïssé classes of permutations, namely*

- *the class $\mathbb{1} = \{1\}$,*

- *the class I of increasing permutations,*

- *the class D of decreasing permutations,*

- *the class* L *of layered permutations,*

- *the class* $\overline{\text{L}}$ *of co-layered permutations, and*

- *the class* $\mathcal{S}$ *of all permutations.*

Proof Let us sketch the basic idea of Cameron's argument.

It is easy to check that the six classes listed in Theorem 2.1 are indeed Fraïssé classes, so let us focus on showing that no other Fraïssé class of permutations exists. Let $\mathcal{C}$ be a Fraïssé class. Let us first consider which permutations of size two belong to $\mathcal{C}$.

If $\mathcal{C}$ has no permutation of size two, then clearly $\mathcal{C} = \mathbb{1}$. Suppose that $\mathcal{C}$ has a single permutation of size two, say $12 \in \mathcal{C}$. By amalgamation, $\mathcal{C}$ then contains any increasing permutation, i.e., $\mathsf{I} \subseteq \mathcal{C}$. On the other hand, $\mathcal{C}$ avoids 21, so $\mathcal{C} \subseteq \text{Av}(21) = \mathsf{I}$, and we conclude that $\mathcal{C} = \mathsf{I}$. Symmetrically, if $\mathcal{C}$ contains 21 but not 12, then $\mathcal{C} = \mathsf{D}$.

Suppose that $\mathcal{C}$ contains both 12 and 21, and therefore $\mathcal{C}$ contains all increasing and decreasing permutations. Consider an amalgamation of 12 with 21 identifying the element '1' of 12 with the element '1' of 21. Any such amalgamation contains either the permutation 213 or the permutation 312. Symmetrically, by considering the other possible amalgamations of 12 and 21, we see that $\mathcal{C}$ contains at least one of 312 and 132, at least one of 132 and 231, and at least one of 231 and 213. Consequently, $\mathcal{C}$ contains either both of $\{213, 132\}$ or both of $\{231, 312\}$.

Of the four permutations $\{213, 312, 132, 231\}$, how many belong to $\mathcal{C}$? We just saw that $\mathcal{C}$ contains at least two of them, so suppose first that $\mathcal{C}$ contains 213 and 132, while avoiding 231 and 312. Since $\mathcal{C}$ contains 213, we observe that for every π in $\mathcal{C}$, the permutation $\pi \oplus 1$ is also in $\mathcal{C}$: to see this, amalgamate π with 213 by identifying the largest element of π with the '2' of 213 and the rightmost element of π with the '1' of 213. By a symmetric argument, since $\mathcal{C}$ contains 132, $\mathcal{C}$ is closed under the operation $\pi \mapsto 1 \oplus \pi$. From this, it is easy to see that $\mathcal{C}$ is sum-closed.

Since $\mathcal{C}$ is sum-closed and contains all decreasing permutations, $\mathcal{C}$ contains all layered permutations. On the other hand, $\mathcal{C}$ avoids 231 and 312, which are exactly the basis of L, and therefore $\mathcal{C} = \mathsf{L}$. Symmetrically, if $\mathcal{C}$ contains 231 and 312, but not 213 and 132, then $\mathcal{C} = \overline{\mathsf{L}}$.

Suppose now that $\mathcal{C}$ contains at least three permutations from the set $\{213, 312, 132, 231\}$, say $\{213, 312, 132\} \subseteq \mathcal{C}$. We will show that $\mathcal{C}$ must contain all permutations.

By the above arguments, $\mathcal{C}$ is sum-closed. It therefore contains $1 \oplus 312 = 1423$ and $312 \oplus 1 = 3124$. By amalgamating the '123' in 1423 with the '124' in 3124, we obtain 31524, which contains 3142, and hence also

231. We may then easily verify that C contains all permutations of size at most 4 and is skew-closed.

We will show that for each n, C contains all permutations of length n. For $n \leq 4$, we already know this, so fix $n > 4$ and suppose C contains all permutations of size less than n. Let $\pi = \pi(1)\pi(2)\cdots\pi(n)$ be a permutation of size n. Let σ be the permutation of size $n-1$ order-isomorphic to $\pi(1)\pi(2)\cdots\pi(n-1)$. By assumption, σ is in C. If $\pi(n) = 1$ or $\pi(n) = n$, then $\pi = \sigma \ominus 1$ or $\pi = \sigma \oplus 1$, and π is in C since we have seen that C is sum-closed and skew-closed. Suppose $1 < \pi(n) < n$, and fix indices i and j such that $\pi(i) = \pi(n) - 1$ and $\pi(j) = \pi(n) + 1$. Let τ be the subpermutation of π induced by the elements $\pi(i)$, $\pi(j)$, $\pi(n-1)$ and $\pi(n)$ (note that $\pi(n-1)$ may coincide with $\pi(i)$ or $\pi(j)$). By amalgamating σ and τ through the (at most) three elements corresponding to $\pi(i)$, $\pi(j)$ and $\pi(n-1)$, we obtain π, showing that π is in C. This shows that $C = S$, as claimed. $\qquad\square$

We remark that in the above proof, rather than using the full strength of the amalgamation property, we only needed to know that the class in question is α-amalgamable for permutations α of size at most three. Actually, the above argument slightly deviates from Cameron's original proof, precisely in order to keep the size of the amalgamated part bounded.

Since every Ramsey class with the joint embedding property must be a Fraïssé class, the search for possible Ramsey permutation classes can be restricted to the classes listed in Theorem 2.1. Böttcher and Foniok [17] and independently Sokić [68] have shown that all these classes are Ramsey.

Theorem 2.2 ([17, 68]) *All the Fraïssé permutation classes are Ramsey.*

Notice that the Ramsey property of the classes I and D is equivalent to the classical finite Ramsey theorem. The Ramsey property for L (and, by symmetry, for $\overline{\text{L}}$) can be deduced from the following generalization of the Ramsey theorem, by Graham, Rothschild and Spencer [31].

Theorem 2.3 ([31]) *For every $t, c, n \in \mathbf{N}$ and every t-tuple $(k_1, \ldots, k_t) \in \mathbf{N}^t$ there is a $N \in \mathbf{N}$ such that for every t tuple of sets $X_1, \ldots, X_t$ with $|X_1| = |X_2| = \cdots = |X_t| = N$ and for every coloring of the set*

$$\binom{X_1}{k_1} \times \binom{X_2}{k_2} \times \cdots \times \binom{X_t}{k_t}$$

by c colors, we can find n-element subsets $Y_1 \subseteq X_1, \ldots, Y_t \subseteq X_t$ such that the elements of the set

$$\binom{Y_1}{k_1} \times \binom{Y_2}{k_2} \times \cdots \times \binom{Y_t}{k_t}$$

all receive the same color.

The most challenging part in the proof of Theorem 2.2 is to show that the class S is Ramsey. Böttcher and Foniok achieve this by adapting a technique originally used by Neetil and Rödl [57] in the study of Ramsey properties of graphs, while Sokić, who states his results in the more general setting of hereditary classes of partial orders, uses a different argument.

An open problem, raised by Cameron [19], is to extend the characterization of Fraïssé and Ramsey permutation classes to classes of higher-dimensional permutations; here a d-dimensional permutation is a relational structure with d relations, all of which are linear orders.

Problem 2.4 *What are the Fraïssé classes of d-dimensional permutations, for a given $d \geq 3$? And which of these Fraïssé classes are Ramsey?*

We note that Sokić [68, Theorem 78] (see also [69, Theorem 10]) proved that the class of all d-dimensional permutations is a Ramsey class for any $d \geq 1$. A different proof of this result was obtained by Solecki and Zhao [70].

2.2 Weaker notions of amalgamation

As we have seen, there are only five infinite Fraïssé classes of permutations. This motivates the study of weaker, and hopefully more interesting, Ramsey-type properties.

Let us first look at classes that satisfy a weaker version of amalgamation. For an integer k, let us say that a permutation class C is k-*amalgamable* if it is α-amalgamable for every $\alpha \in C$ of size at most k.

The argument of Cameron, in the version outlined in Section 2.1, shows that all the 3-amalgamable permutation classes are amalgamable. Therefore, to get genuinely new properties, we need to look at k-amalgamable classes for $k \leq 2$.

Atomicity. If we were to consider structures of size 0, then 0-amalgamation would be a synonym for the joint embedding property. Permutation classes that satisfy the joint embedding property are commonly referred to as *atomic* classes. From the general results of Fraïssé [29] on relational structures, one may obtain several equivalent characterizations of atomicity.

Proposition 2.5 ([29], [7, Theorem 1.2]) *The following properties are equivalent for a permutation class C:*

- $\mathcal{C}$ *is atomic,*

- $\mathcal{C}$ *cannot be obtained as a union of two of its proper subclasses,*

- $\mathcal{C}$ *is the age of a finite or infinite permutation,*

- $\mathcal{C}$ *contains an increasing sequence* $\alpha_1 \leq \alpha_2 \leq \cdots$ *of permutations, such that every permutation in* $\mathcal{C}$ *is contained in some* α_n.

Decompositions into atomic classes. Trivially, any permutation class can be written as a (possibly infinite) union of its atomic subclasses: we simply take $\mathcal{C} = \bigcup_{\alpha \in \mathcal{C}} \mathsf{Age}(\alpha)$. However, there are in general many different ways to express a given class as a union of atomic classes. Can we perhaps find an expression that would be 'canonical', in a suitable sense? This question was addressed by Murphy [53], who considered two distinct ways of decomposing a class into a union of atomic subclasses.

Firstly, Murphy considered *maximal decompositions* of a class $\mathcal{C}$, which are expressions of the form $\mathcal{C} = \bigcup_{i \in I} \mathcal{A}_i$ where each $\mathcal{A}_i$ is a maximal atomic subclass of $\mathcal{C}$. Murphy showed, by an application of Zorn's lemma, that such a decomposition always exists, but unfortunately, it is in general not unique. We remark that although some classes have uncountably many maximal atomic subclasses ([53, Proposition 178] gives an example), each permutation class can be expressed as a union of at most countably many maximal atomic subclasses, since it is enough to consider for each $\gamma \in \mathcal{C}$ a maximal atomic subclass of $\mathcal{C}$ containing γ.

Another way to decompose a class $\mathcal{C}$ is to consider the *irredundant decomposition*, which is an expression $\mathcal{C} = \bigcup_{i \in I} \mathcal{A}_i$ where each $\mathcal{A}_i$ is an atomic subclass of $\mathcal{C}$, and with the additional property that for every proper subset $J \subsetneq I$, the union $\bigcup_{j \in J} \mathcal{A}_j$ does not equal $\mathcal{C}$. Notice that an irredundant decomposition necessarily involves at most countably many classes $\mathcal{A}_i$, since by irredundance, each $\mathcal{A}_i$ contains a permutation $\gamma_i \in \mathcal{C}$ which is not contained in $\bigcup_{j \in I \setminus \{i\}} \mathcal{A}_j$. As Murphy pointed out, an irredundant decomposition of $\mathcal{C}$, if it exists, is unique, but unfortunately some classes have no irredundant decomposition.

Overall, there does not seem to be a notion of 'canonical' decomposition that could be applied in full generality.

Natural classes. As we pointed out above, every atomic permutation class $\mathcal{C}$ is the age of a (not necessarily unique, possibly infinite) permutation. Recall that an infinite permutation on a vertex set X is a relational structure with two relations $<_H$ and $<_V$, both determining linear orders of X. One might hope to gain an insight into the structure of $\mathcal{C}$ by looking at the structure of the two linear orders $(X, <_H)$ and $(X, <_V)$.

A permutation class $\mathcal{C}$ is *natural* if is the age of an infinite permutation $\Gamma = (X, <_H, <_V)$, where both $(X, <_H)$ and $(X, <_V)$ are isomorphic to $(\mathbf{N}, <)$, i.e., to the set of positive integers with their natural ordering. Equivalently, one can view a natural class as the set of finite subpermutations of a bijection $\Gamma \colon \mathbf{N} \to \mathbf{N}$.

Natural classes were studied by Atkinson, Murphy and Rukuc [7], by Murphy [53], and by Huczynska and Rukuc [36]. It turns out that the structure of natural classes is a lot more restricted than the structure of general atomic classes. Moreover, as shown by Murphy [53, Theorem 207], there is an algorithm which for a given finite set of permutations F decides whether the class $\mathrm{Av}(F)$ is natural. It is not known whether the corresponding decision problem for atomic classes is decidable.

Problem 2.6 *Given a finite set F of permutations, is there an algorithm to determine whether $\mathrm{Av}(F)$ is atomic?*

Note that the above problem is trivial when $|F| = 1$, since any principal class $\mathrm{Av}(\pi)$ is sum-closed or skew-closed, and hence atomic.

Molecular and well-quasi-ordered classes. Following [37], let us call a class *molecular* if it is a finite union of atomic classes. Molecular classes can be characterized by the following weakening of the joint embedding property.

Proposition 2.7 *For a permutation class $\mathcal{C}$ and an arbitrary $k \in \mathbf{N}$, the class $\mathcal{C}$ is a union of k atomic classes if and only if among every $k+1$ distinct permutations $\gamma_1, \ldots, \gamma_{k+1} \in \mathcal{C}$ there are at least two distinct permutations γ_i and γ_j that have a joint embedding in $\mathcal{C}$. Moreover, $\mathcal{C}$ is not molecular if and only if it contains an infinite subset of permutations no two of which have a joint embedding in $\mathcal{C}$.*

This proposition is not specific to permutations, it generalizes straightforwardly to arbitrary classes of relational structures. In fact, the proposition is direct corollary of a result that can be stated in the setting of arbitrary partially ordered sets.

Let $(P, \preccurlyeq)$ be a partially ordered set. An element $z \in P$ is a *common upper bound* of elements $x, y \in P$ if $x \preccurlyeq z$ and $y \preccurlyeq z$. A subset $I \subseteq P$ is an *ideal* if

- I is a down-set, that is, for every $x \in I$ and $y \in P$, $y \preccurlyeq x$ implies $y \in I$, and

- I is up-directed, that is, every two elements of I have a common upper bound in I.

Notice that when a permutation class $\mathcal{C}$ is considered as a poset with the partial order determined by the containment relation, then ideals of $\mathcal{C}$ are precisely the atomic subclasses of $\mathcal{C}$. Proposition 2.7 is then a special case of the following Dilworth-type result.

Theorem 2.8 ([37, Theorem 1.5]) *For a poset $(P, \preccurlyeq)$ and an arbitrary $k \in \mathbf{N}$, the poset P is a union of k ideals if and only if among every $k + 1$ distinct elements $x_1, \ldots, x_{k+1} \in P$ there are at least two distinct elements x_i and x_j that have a common upper bound in P. Moreover, P is not a union of finitely many ideals if and only if it contains an infinite subset of elements no two of which have a common upper bound in P.*

Proposition 2.7 implies that any non-molecular class contains an infinite antichain. The converse, of course, does not hold: for instance, the class of all permutations is atomic, and therefore also molecular, even though it contains infinite antichains.

The fact that every permutation class without an infinite antichain is molecular was first pointed out by Atkinson, Murphy and Rukuc [6]. Permutation classes not containing any infinite antichain are usually referred to as *well-quasi-ordered* (or sometimes well-partially-ordered) classes. As observed by Atkinson, Murphy and Rukuc [6], this property can be characterized in several equivalent ways.

Lemma 2.9 ([6]) *The following properties are equivalent for a permutation class $\mathcal{C}$:*

- *$\mathcal{C}$ is well-quasi-ordered.*

- *There is no infinite strictly decreasing chain $\mathcal{C} = \mathcal{C}_0 \supsetneq \mathcal{C}_1 \supsetneq \mathcal{C}_2 \supsetneq \cdots$ of subclasses of $\mathcal{C}$.*

- *$\mathcal{C}$ has only countably many subclasses.*

- *Every nonempty set of subclasses of $\mathcal{C}$ has a minimal element.*

Moreover, if $\mathcal{C}$ is a finitely based class, then $\mathcal{C}$ is well-quasi-ordered if and only if each subclass of $\mathcal{C}$ is finitely based.

Let us mention a conjecture relating well-quasi-ordering with algebraicity of generating functions. A permutation class $\mathcal{C}$ is said to be *strongly algebraic* if every subclass of $\mathcal{C}$ has an algebraic generating function over the ring of polynomials with rational coefficients. Note that a class $\mathcal{C}$ that is not well-quasi-ordered cannot be strongly algebraic, since one can find in $\mathcal{C}$ uncountably many subclasses with distinct generating functions, while there are only countably many algebraic power series. Vatter [73] conjectured that the converse holds as well.

Conjecture 2.10 ([73, Conjecture 3.4]) *A permutation class C is well-quasi-ordered if and only if C is strongly algebraic.*

2.3 1-amalgamation, 2-amalgamation, juxtaposition

Recall that a class C is k-amalgamable if it is α-amalgamable for every $\alpha \in C$ of size at most k. With $k = 1$, this notation may seem ambiguous, since 1 may also refer to the unique permutation of size 1. Fortunately, the two corresponding meanings of '1-amalgamable' are identical, so may talk of 1-amalgamable permutation classes without ambiguity.

We know, from the general properties of relational structures mentioned in the introduction, that every unsplittable permutation class is 1-amalgamable. As we shall see in Chapter 3, there are infinitely many unsplittable classes, and hence also infinitely many 1-amalgamable classes.

It is quite hard to find examples of classes that are simultaneously splittable and 1-amalgamable. Indeed, until recently, it was not known whether any such classes exist at all. However, Opler [59] has shown that the class $\mathrm{Av}(1342, 1423)$ is both splittable and 1-amalgamable.

Problem 2.11 *Can you find infinitely many permutation classes that are 1-amalgamable and splittable?*

Vatter [73] has pointed out that every unsplittable class is sum-closed or skew-closed. This property actually applies more generally to any 1-amalgamable class.

Proposition 2.12 *Each infinite 1-amalgamable class is sum-closed or skew-closed.*

Proof For contradiction, let C be an infinite 1-amalgamable class that is neither sum-closed nor skew-closed. This means that among the basis elements of C there is at least one sum-decomposable permutation $\pi = \pi' \oplus \pi''$ as well as at least one skew-decomposable permutation $\sigma = \sigma' \ominus \sigma''$.

Suppose first that both π' and π'' have size at least two. Then $\pi' \oplus 1$ and $1 \oplus \pi''$ are both in C since they are proper subpermutations of the basis element π. However, by amalgamating the rightmost element of $\pi' \oplus 1$ with the leftmost element of $1 \oplus \pi''$, we obtain the permutation $\pi' \oplus 1 \oplus \pi''$, which is not in C. This contradicts the 1-amalgamation property of C.

We conclude that at least one of π', π'' has size 1, and by the same argument, at least one of σ', σ'' has size 1. Suppose then that $\pi = \pi' \oplus 1$

and $\sigma = \sigma' \ominus 1$, with the other cases being symmetric. Note that we may assume that π' and σ' both have size at least two, otherwise $\mathcal{C}$ would be a finite class, contrary to our assumptions. Moreover, we may similarly deduce that π' does not have the form $\pi''' \oplus 1$, because if it had such form, we could obtain the basis element π by amalgamating π' with the permutation 12.

Let $\tau \in \mathcal{C}$ be an amalgamation of π' and σ' obtained by identifying the largest element of π' with the smallest element of σ'. Thus, τ is a union of a copy of π' (which we call the *bottom part* of τ) and a copy of σ' (the *top part* of τ), with the two parts sharing exactly one element. The shared element is not the rightmost element of τ – recall that π' does not have the form $\pi''' \oplus 1$, so its rightmost element differs from its largest element.

Let $\tau(n)$ be the rightmost element of τ. If $\tau(n)$ is in the top part of τ, then the bottom part together with $\tau(n)$ forms a copy of $\pi = \pi' \oplus 1$, and if $\tau(n)$ is in the bottom part, we find a copy of $\sigma = \sigma' \ominus 1$. In both cases we get a contradiction, since τ belongs to $\mathcal{C}$.   □

Vatter has shown that the classes that are sum-closed or skew-closed can be characterized by another unsplittability-type property. To state this precisely, we need more terminology. Let π and σ be permutations of size m and n, respectively. A permutation τ of size $m + n$ is a *horizontal juxtaposition* of π and σ, if the leftmost m elements of τ are order isomorphic to π and the remaining n elements are order-isomorphic to σ. A vertical juxtaposition is defined analogously. For two permutation classes $\mathcal{A}$ and $\mathcal{B}$, their *horizontal juxtaposition* is the class of all permutations that belong to $\mathcal{A} \cup \mathcal{B}$ or are obtainable as a horizontal juxtaposition of a permutation from $\mathcal{A}$ with a permutation from $\mathcal{B}$. Vertical juxtapositions are defined analogously. Let us say that a permutation class is *juxtaposable* if it is contained in a vertical or horizontal juxtaposition of two of its proper subclasses. Juxtaposable classes were characterized by Vatter [73] as follows.

Proposition 2.13 ([73, Proposition 4.22]) *A permutation class is sum-closed or skew-closed if and only if it is not juxtaposable.*

We remark that for atomic permutation classes, the property of being juxtaposable is equivalent to the property of being *grid reducible*. The concept of grid reducible classes was introduced by Vatter [72], and turned out to be extremely useful in the analysis of permutation classes of small growth rates. We will not introduce grid reducibility here, and instead refer the reader to the comprehensive survey in [73, Chapter 4].

Combining Propositions 2.12 and 2.13, we see that the 1-amalgamation property is sandwiched between two unsplittability-type properties.

Corollary 2.14 *For an infinite permutation class C, the following impli-cations hold: C is unsplittable $\implies$ C is 1-amalgamable $\implies$ C is not juxtaposable.*

Neither of the two implications in Corollary 2.14 can be replaced by an equivalence. We have already pointed out Opler's example of split-table 1-amalgamable class. It is much easier to find examples of non-juxtaposable classes that are not 1-amalgamable: take, e.g., any class of the form $\mathrm{Av}(\pi \oplus \sigma)$ where π and σ have size at least two. Such a class is not juxtaposable (any principal class is sum-closed or skew-closed, and hence not juxtaposable), but it is not 1-amalgamable either, since it has no amalgamation identifying the rightmost element of $\pi \oplus 1$ with the leftmost element of $1 \oplus \sigma$.

Let us now briefly mention 2-amalgamable classes. It can be shown, without much difficulty, that the class $\mathrm{Av}(213)$ and its symmetries are 2-amalgamable. Also, the class Sep of separable permutations can be shown to be 2-amalgamable. These examples demonstrate that there are classes which are 2-amalgamable but not 3-amalgamable (recall that every 3-amalgamable permutation class is amalgamable and Ramsey).

By following a similar reasoning as in the proof of Theorem 2.1, it can be shown that the only 2-amalgamable subclasses of Sep are Sep itself, $\mathrm{Av}(213)$ and its symmetries, and the Fraïssé classes $\mathbb{1}$, I, D, L and $\overline{\mathsf{L}}$. In contrast, it is known [4] that there are infinitely many unsplittable sub-classes of Sep. We thus have infinitely many examples of unsplittable (and hence 1-amalgamable) classes which are not 2-amalgamable. However, there is nothing known about 2-amalgamable classes not contained in Sep.

Problem 2.15 *Are there any 2-amalgamable classes not contained in Sep, apart from the class of all permutations? Are there infinitely many 2-amalgamable classes?*

3   Splittability of permutation classes

Among the Ramsey-type properties considered in connection with per-mutation classes, unsplittability seems to have recently gathered most in-terest. In this chapter, we will present an overview of various aspects of unsplittability of permutation classes.

Recall that $\mathcal{A} \odot \mathcal{B}$ denotes the class of permutations obtainable by merging an element of $\mathcal{A}$ with an element of $\mathcal{B}$. A *splitting* of a permutation class $\mathcal{C}$ is a finite multiset $\{\mathcal{A}_1, \ldots, \mathcal{A}_k\}$ of permutation classes such that $\mathcal{C} \subseteq \mathcal{A}_1 \odot \mathcal{A}_2 \odot \cdots \odot \mathcal{A}_k$. The classes $\mathcal{A}_i$ are the *parts* of the splitting. A splitting $\{\mathcal{A}_1, \ldots, \mathcal{A}_k\}$ of a class $\mathcal{C}$ is *nontrivial* if $\mathcal{C}$ is not a subclass of any $\mathcal{A}_i$, and the splitting is *irredundant* if no proper submultiset of $\{\mathcal{A}_1, \ldots, \mathcal{A}_k\}$ is a splitting.

One can easily observe that for a permutation class $\mathcal{C}$, the following properties are all equivalent:

a) $\mathcal{C}$ has no nontrivial splitting into two parts.

b) $\mathcal{C}$ has no splitting into two parts, with both parts being proper subclasses of $\mathcal{C}$.

c) $\mathcal{C}$ has no splitting of the form $\{Av(\sigma), Av(\pi)\}$, with σ and π being two permutations from $\mathcal{C}$.

d) For any two permutations σ and π in $\mathcal{C}$, there is a permutation $\rho \in \mathcal{C}$ such that every two-coloring of the vertices of ρ by red and blue contains a red copy of σ or a blue copy of π.

a') $\mathcal{C}$ has no nontrivial splitting.

b') $\mathcal{C}$ has no splitting all of whose parts are proper subclasses of $\mathcal{C}$.

c') $\mathcal{C}$ has no splitting of the form $\{Av(\sigma_1), Av(\sigma_2), \ldots, Av(\sigma_k)\}$, with $\sigma_1, \ldots, \sigma_k$ being permutations from $\mathcal{C}$.

d') For any k permutations $\sigma_1, \ldots, \sigma_k$ in $\mathcal{C}$, there is a permutation $\rho \in \mathcal{C}$ such that every k-coloring of the vertices of ρ by colors $c_1, \ldots, c_k$ contains, for some i, a monochromatic copy of σ_i in color c_i.

A permutation class is unsplittable if it satisfies any (and hence all) of these eight properties. We may easily observe that every unsplittable class is atomic.

3.1 Examples of splittable classes

What examples of splittable classes do we know about? Some examples are trivial: for instance, every finite class is clearly splittable, except for the trivial class $\mathbb{1} = \{1\}$. Also, as we pointed out already, every non-atomic permutation class is splittable as well. Another easy example is the class $Av(123)$ of the permutations having no increasing subsequence of length three. It is well-known that such permutations are precisely the permutations merged from at most two decreasing sequences, or in other words, $Av(123) = Av(12) \odot Av(12)$. The argument can, of course, be extended to any class of the form $Av(123 \cdots k)$ with $k \geq 3$.

To find further examples of splittable classes, we need less trivial criteria. One such criterion is provided by the next theorem of Jelínek and Valtr.

Theorem 3.1 ([39, Proposition 3.2]) *For any three nonempty permutations α, β and γ, we have*

$$\mathrm{Av}(\alpha \oplus \beta \oplus \gamma) \subseteq \mathrm{Av}(\alpha \oplus \beta) \odot \mathrm{Av}(\beta \oplus \gamma).$$

In particular, the class $\mathrm{Av}(\alpha \oplus \beta \oplus \gamma)$ is splittable.

Notice that the theorem can be seen as a generalization of the fact that $\mathrm{Av}(123) \subseteq \mathrm{Av}(12) \odot \mathrm{Av}(12)$. In this particular example, we actually have $\mathrm{Av}(123) = \mathrm{Av}(12) \odot \mathrm{Av}(12)$. However, in the statement of the theorem, the inclusion cannot be replaced by equality, as can be easily seen by considering less trivial examples.

An even more general family of examples of splittable classes is given by the next result.

Theorem 3.2 ([39, Theorem 3.1]) *If π is a sum-decomposable permutation other than 12, 213 or 132, then the class $\mathrm{Av}(\pi)$ is splittable.*

As we will later see, the classes $\mathrm{Av}(12)$, $\mathrm{Av}(213)$ and $\mathrm{Av}(132)$ are all unsplittable, therefore the three exceptions in Theorem 3.2 are necessary.

The proof of Theorem 3.2 given in [39] is quite technical and we will not attempt to sketch it here.

The proof is constructive in the sense that it provides, for a given π, an explicit construction of a sequence of patterns $\sigma_1, \ldots, \sigma_k \in \mathrm{Av}(\pi)$ such that $\mathrm{Av}(\pi) \subseteq \mathrm{Av}(\sigma_1) \odot \cdots \odot \mathrm{Av}(\sigma_k)$. Unfortunately, the patterns σ_i produced by this construction are quite large, and the splittings obtained in this way do not seem to yield much insight into the structure of the class $\mathrm{Av}(\pi)$.

Consider, for instance, the class $\mathrm{Av}(1423)$. By following the ideas of the proof of Theorem 3.2, we will obtain a splitting

$$\mathrm{Av}(1423) \subseteq \mathrm{Av}(15234) \odot \mathrm{Av}(15234) \odot \mathrm{Av}(52314) \odot \mathrm{Av}(52314).$$

It is unlikely that this is the best way to split this particular class, but there is nothing better known.

In contrast, Theorem 3.1 provides a simple and explicit splittings which have turned out to be useful in attacking difficult enumerative problems. We will mention some of these applications later, in Section 3.3.

There is one more splittability result worth mentioning, which, while not providing new examples of splittable classes, can at least serve to exhibit explicit splittings with simple structure. For a multiset M and integer q, let $q * M$ denote the multiset obtained from M by increasing the multiplicity of each element q times.

Proposition 3.3 ([38, Theorem 3.15]) *Let π be a sum-indecomposable permutation of size n. Suppose that the class $\mathrm{Av}(\pi)$ has a splitting*

$$\{\mathrm{Av}(\pi_1), \ldots, \mathrm{Av}(\pi_k)\},$$

where $\pi_1, \ldots, \pi_k$ are sum-indecomposable permutations. Then the class $\mathrm{Av}(1 \oplus \pi)$ has the splitting

$$q * \{\mathrm{Av}(1 \oplus \pi_1), \ldots, \mathrm{Av}(1 \oplus \pi_k)\},$$

for some $q \leq 16^n$.

As an example of the application of Proposition 3.3, take $\pi = 321$. As we know, $\mathrm{Av}(321)$ has the splitting $\{\mathrm{Av}(21), \mathrm{Av}(21)\}$. By the proposition, the class $\mathrm{Av}(1 \oplus \pi) = \mathrm{Av}(1432)$ then has the splitting

$$q * \{\mathrm{Av}(132), \mathrm{Av}(132)\} = 2q * \{\mathrm{Av}(132)\},$$

with $q \leq 16^3$. In this example, the upper bound on q is far from tight; it is in fact known that $\mathrm{Av}(1432)$ admits the splitting $5 * \{\mathrm{Av}(132)\}$, and the constant 5 is optimal. We shall return to this example in Section 3.4. Even in the general case, the upper bound on q given in the statement of Proposition 3.3 is probably far from optimal.

Problem 3.4 *Can the exponential upper bound on q in Proposition 3.3 be improved to a sub-exponential or polynomial one?*

In Section 3.4, we will see that for some choice of π and of $\pi_1, \ldots, \pi_k$, the optimum value of q may be as large as $\Omega(n \log n)$.

3.2 Examples of unsplittable classes

To prove that a given class is splittable, an obvious approach is to describe an explicit splitting. But how can we prove that a class $\mathcal{C}$ is unsplittable? Perhaps the most natural approach is to show that for every two permutations σ and π in $\mathcal{C}$, there is a permutation τ whose every red-blue coloring contains a red copy of σ or a blue copy of π. This is equivalent to saying that for every σ and π in $\mathcal{C}$, the set $\mathcal{C} \setminus (\mathrm{Av}(\sigma) \odot \mathrm{Av}(\pi))$ is nonempty, since the set $\mathrm{Av}(\sigma) \odot \mathrm{Av}(\pi)$ contains precisely the permutations that admit a red-blue coloring where the red part avoids σ and the blue part avoids π.

Let $\mathcal{R}_{\mathcal{C}}(\sigma, \pi)$ denote the set $\mathcal{C} \setminus (\mathrm{Av}(\sigma) \odot \mathrm{Av}(\pi))$. To construct an explicit $\tau \in \mathcal{R}_{\mathcal{C}}(\sigma, \pi)$, we typically use suitable closure properties of the class $\mathcal{C}$. The following easy proposition provides a motivating example.

Proposition 3.5 ([39]) *Every inflation-closed class is unsplittable.*

Proof Let $\mathcal{C}$ be inflation-closed. Pick $\sigma, \pi \in \mathcal{C}$, and consider the permutation $\tau = \sigma[\pi]$. We easily observe that every red-blue coloring of τ has a red copy of σ or a blue copy of π, i.e., τ is in $\mathcal{R}_{\mathcal{C}}(\sigma, \pi)$. $\square$

A class is inflation-closed if and only if all its basis elements are simple. The above proposition thus implies, in particular, that for every simple permutation ρ, the class $\mathrm{Av}(\rho)$ is unsplittable.

Another result demonstrating the close relationship between inflation and unsplittability is the following proposition, whose proof is based on a similar idea as the proof of Proposition 3.5, and we omit it.

Proposition 3.6 ([4]) *If $\mathcal{C}$ and $\mathcal{D}$ are unsplittable classes, then $\mathcal{C}[\mathcal{D}]$ is unsplittable as well.*

This proposition implies, for instance, that the class L of layered permutations is unsplittable. To see this, note that L equals I[D] (recall that I $= \mathrm{Av}(21)$ and D $= \mathrm{Av}(12)$), and that I and D are both unsplittable by Proposition 3.5.

To deal with a class $\mathcal{C}$ that is not inflation-closed, one typically needs a combination of several weaker closure properties of $\mathcal{C}$. In such situations, it is often convenient to use the notion of unavoidable permutations, which we now introduce. Let $\mathcal{C}$ be a permutation class. A permutation $\pi \in \mathcal{C}$ is *unavoidable* in $\mathcal{C}$, if in every irredundant splitting of $\mathcal{C}$, each part contains the permutation π. Equivalently, π is unavoidable in $\mathcal{C}$ if and only if for any $\rho \in \mathcal{C}$, the set $\mathcal{R}_{\mathcal{C}}(\pi, \rho)$ is nonempty.

Let $\mathcal{U}_{\mathcal{C}}$ be the set of unavoidable permutations of $\mathcal{C}$. We call $\mathcal{U}_{\mathcal{C}}$ the *unavoidable core* of $\mathcal{C}$.

Trivially, the set $\mathcal{U}_{\mathcal{C}}$ is hereditary, that is, it is a permutation class. Equally obviously, $\mathcal{U}_{\mathcal{C}}$ is equal to $\mathcal{C}$ if and only if $\mathcal{C}$ is unsplittable. More generally, in any irredundant splitting of $\mathcal{C}$, all the parts contain $\mathcal{U}_{\mathcal{C}}$ as a subclass.

What makes the unavoidable core useful in proving unsplittability of a class $\mathcal{C}$ is the fact that various closure properties $\mathcal{C}$ are reflected by analogous closure properties of $\mathcal{U}_{\mathcal{C}}$, as the next proposition shows.

Proposition 3.7 ([39]) *Let $\mathcal{C}$ be a permutation class, and let X be a set of permutations. If $\mathcal{C}[X] \subseteq \mathcal{C}$ then $\mathcal{U}_{\mathcal{C}}[X] \subseteq \mathcal{U}_{\mathcal{C}}$, and if $X[\mathcal{C}] \subseteq \mathcal{C}$ then $X[\mathcal{U}_{\mathcal{C}}] \subseteq \mathcal{U}_{\mathcal{C}}$.*

As an example of applying this proposition, consider again the class L of layered permutations. This class is sum-closed, i.e., 12[L] $\subseteq$ L. Moreover, it satisfies $L[21] \subseteq L$. It follows that its unavoidable core $\mathcal{U}_L$ satisfies these two closure properties as well. However, since L is the smallest class with these two properties, we get $\mathcal{U}_L = L$, and hence L is unsplittable.

Apart from the unsplittable examples obtained from the previous general propositions, it is also known the class $\mathrm{Av}(132)$ is unsplittable [39], and by symmetry, $\mathrm{Av}(213)$, $\mathrm{Av}(231)$ and $\mathrm{Av}(312)$ are unsplittable as well.

Despite the various necessary and sufficient conditions for splittability mentioned in Sections 3.1 and 3.2, we are still far away from understanding which permutation classes are splittable, even when we restrict ourselves to principal classes. We have seen that $\mathrm{Av}(\pi)$ is splittable whenever π is sum-decomposable of size at least four, and of course, by symmetry, an analogous result holds for skew-decomposable π as well. On the other hand, $\mathrm{Av}(\pi)$ is unsplittable whenever π is simple or π is one of 132, 213, 231 or 312.

Problem 3.8 *Which permutation classes are splittable? Given a finite basis F of a class $C = \mathrm{Av}(F)$, can we determine algorithmically whether C is splittable? Can we at least do this when C is a principal class?*

As shown by Albert, Atkinson and Klazar [3], there are asymptotically $\frac{n!}{e^2}\left(1 - \frac{4}{n} + O(n^{-2})\right)$ simple permutations of size n. Therefore, for a random permutation π of size n, the probability that $\mathrm{Av}(\pi)$ is unsplittable is at least $\frac{1}{e^2} - o(1)$ as n grows. We have no nontrivial upper bound for this probability.

Problem 3.9 *For a random permutation π, what is the asymptotic probability that $\mathrm{Av}(\pi)$ is unsplittable? Is it unsplittable almost surely?*

Vatter [74] has shown that any class C that avoids at least one layered permutation and at least one co-layered permutation can be split into finitely many copies of $I = \mathrm{Av}(21)$ and $D = \mathrm{Av}(12)$. This means, in particular, that are no unsplittable proper subclasses of L or $\bar{L}$, except for I, D or $\mathbb{1}$. More generally, Albert and Jelínek [4] show that the only unsplittable proper subclasses of the class Sep of separable permutations are the classes obtainable by iterated inflations from $\mathbb{1}$, I, $\mathrm{Av}(132)$ and their symmetries. Thus, as far as subclasses of Sep are concerned, splittability is essentially fully understood.

Extending this understanding beyond the class of separable permutations remains an open problem. Note that any inflation-closed class other than $\mathbb{1}$, I or D contains Sep as a subclass.

Problem 3.10 *Find an example of an unsplittable class which neither contains nor is contained in the class Sep of separable permutations.*

The concept of unavoidable core is not only useful when proving un-splittability of a class, but it may also help to restrict the possible splittings of a class that is known to be splittable. It would therefore help to understand the cores of the splittable classes for which no 'nice' splittings are currently known, such as Av(1423).

Problem 3.11 *What is the unavoidable core of* Av(1423)*? Does it contain any of the permutations 1342, 2413 or 3142?*

3.3 Splittability and enumeration

In the previous sections, we have seen an overview of theoretic results about splittability. Let us now look at some of the motivations for this concept, that is, let us look at examples where splittability turned out to be useful in solving other combinatorial problems.

For a permutation class $\mathcal{C}$, let $\mathcal{C}_n$ denote the set of permutations of size n in $\mathcal{C}$. Finding the cardinality of $\mathcal{C}_n$ is one of the oldest and most basic problems in the study of permutation classes. The enumeration is easy enough to obtain when the permutations in the given class have a nice structure, e.g., when $\mathcal{C}$ avoids a small pattern. Thus, Knuth [45] has shown that Av(132) and Av(123) are enumerated by Catalan numbers, and later on, exact enumeration was obtained for classes such as Av(1234) (Gessel [30]) or Av(1342) (Bóna [12]).

Unfortunately, we are typically not able to give an exact formula enumerating a given class, and must look for asymptotic estimates instead. A crucial result of Marcus and Tardos shows that $|\mathcal{C}_n|$ is at most exponential.

Theorem 3.12 ([52]) *For every proper permutation class $\mathcal{C}$ there is a constant K such that $\mathcal{C}$ contains at most K^n permutations of size n.*

The first goal when enumerating a class $\mathcal{C}$ is then to obtain estimates for the optimum value of the constant K from the Marcus–Tardos Theorem. We therefore look at *upper growth-rate* $\overline{\mathrm{gr}}(\mathcal{C})$ and *lower growth-rate* $\underline{\mathrm{gr}}(\mathcal{C})$, defined by

$$\overline{\mathrm{gr}}(\mathcal{C}) = \limsup_{n \to \infty} \sqrt[n]{|\mathcal{C}_n|} \qquad \text{and} \qquad \underline{\mathrm{gr}}(\mathcal{C}) = \liminf_{n \to \infty} \sqrt[n]{|\mathcal{C}_n|}.$$

When $\overline{\mathrm{gr}}(\mathcal{C}) = \underline{\mathrm{gr}}(\mathcal{C})$ for a class $\mathcal{C}$, we speak simply of the *growth rate* of $\mathcal{C}$, which we denote by $\mathrm{gr}(\mathcal{C})$. It is conjectured (e.g., by Vatter [73, Conjecture 1.1]) that every permutation class has a growth rate.

For every principal permutation class $\mathcal{C} = \mathrm{Av}(\pi)$ the growth rate is known to exist, and is usually called the *Stanley–Wilf limit* of the pattern π, denoted by $SW(\pi)$.

Splittability of a class $\mathcal{C}$ can help in proving upper bounds for the growth rate of $\mathcal{C}$ in an intuitive way: if $\mathcal{C}$ can be split into a small number of parts with each part being 'small', then $\mathcal{C}$ itself cannot be too large. A more precise statement of this idea is presented in the next proposition, which, in various versions and special instances, appeared independently in the works of Bóna [13, 14] and Albert [2].

Proposition 3.13 *Suppose that* $\mathcal{A}$, $\mathcal{B}$ *and* $\mathcal{C}$ *are permutation classes and that* $\mathcal{C} \subseteq \mathcal{A} \odot \mathcal{B}$. *Then*

$$\sqrt{\overline{\mathrm{gr}}(\mathcal{C})} \leq \sqrt{\overline{\mathrm{gr}}(\mathcal{A})} + \sqrt{\overline{\mathrm{gr}}(\mathcal{B})}.$$

The straightforward proof of this proposition can be found, e.g., in the survey of Vatter [73, Proposition 3.5].

The first (implicit) application of a splittability argument for enumeration purposes seems to appear in the work of Bóna [13], who proved, for any permutation π, the identity

$$\sqrt{SW(1 \oplus 1 \oplus \pi)} = 1 + \sqrt{SW(1 \oplus \pi)},$$

and later generalized it as follows

Theorem 3.14 ([14, Theorem 4.2]) *For any two permutations* σ *and* π, *we have*

$$\sqrt{SW(\sigma \oplus 1 \oplus \pi)} = \sqrt{SW(\sigma \oplus 1)} + \sqrt{SW(1 \oplus \pi)}.$$

Note that the inequality $\sqrt{SW(\sigma \oplus 1 \oplus \pi)} \leq \sqrt{SW(\sigma \oplus 1)} + \sqrt{SW(1 \oplus \pi)}$ follows from Theorem 3.1 and Proposition 3.13. Bóna was able to also prove the opposite inequality, by a combinatorial argument which cannot be applied in the more general setting of Theorem 3.1.

Another step in this line of research was made by Claesson, Jelínek and Steingrímsson [22], who proved a slightly weaker version of Theorem 3.1, namely that

$$\mathrm{Av}(\alpha \oplus (\beta \ominus 1) \oplus \gamma) \subseteq \mathrm{Av}(\alpha \oplus (\beta \ominus 1)) \odot \mathrm{Av}((\beta \ominus 1) \oplus \gamma).$$

Combining this with Proposition 3.13, they were able to deduce the following result.

Theorem 3.15 $SW(1324) \leq 16$. *Moreover, for an arbitrary layered permutation* π *of size* k, $SW(\pi) \leq 4k^2$.

The upper bound on $SW(1324)$ was then improved by Bóna [15, 16] by a refinement of the splitting argument, to the currently best known upper bound $SW(1324) \leq 13.76$. Despite these efforts, the precise value of $SW(1324)$ is still unknown, with the currently best known lower bound, due to Bevan [10], being equal to 9.81. The pattern 1324, and its symmetric copy 4231, are the only patterns of size at most four whose Stanley–Wilf limits are not known.

The relationship between splitting and enumeration is a topic of ongoing research. An example of a recent result in this area is the following theorem of Albert et al. [5].

Theorem 3.16 *If $\mathcal{A}$ and $\mathcal{B}$ are permutation classes, and if each of them is sum-closed or skew-closed, then*

$$\sqrt{\mathrm{gr}(\mathcal{A} \odot \mathcal{B})} = \sqrt{\mathrm{gr}(\mathcal{A})} + \sqrt{\mathrm{gr}(\mathcal{B})}.$$

The study of Stanley–Wilf limits and growth rates remains one of the main topics in the area of combinatorics of permutations. For more information about the major recent advances in this area, we refer the reader to the recent survey of Vatter [73].

3.4 Splittability and χ-boundedness

There is a surprising connection between splittability of permutations and χ-boundedness of certain graph classes.

The *chromatic number* $\chi(G)$ of a graph G is the smallest integer k such that the vertices of G can be colored by k colors, with no two adjacent vertices receiving the same color. The *clique-number* $\omega(G)$ of G is the number of vertices in the largest complete subgraph of G. Recall that a class of graphs $\mathcal{G}$ is χ-*bounded* if there is a function $f \colon \mathbf{N} \to \mathbf{N}$ such that every graph $G \in \mathcal{G}$ with $\omega(G) < k$ satisfies $\chi(G) \leq f(k)$.

One area of graph theory where χ-boundedness plays an important role is the study of geometric intersection graphs. Let $\mathcal{X} = \{X_1, X_2, \ldots, X_n\}$ be a system of sets. The *intersection graph* of the set system $\mathcal{X}$ is the graph G whose vertices correspond to the sets of the set system $\mathcal{X}$ and two vertices are adjacent if and only if the two corresponding sets have nonempty intersection. The set system $\mathcal{X}$ is then an *intersection representation* of G.

Many important graph classes can be characterized by admitting intersection representations whose sets belong to a restricted family, often defined in geometric terms. The class of graphs that we will be interested in are the so-called circle graphs. A graph G is a *circle graph* if G is the intersection graph of a system of chords inscribed into a circle.

Circle graphs are also known as *interval overlap graphs*. This is because a graph G on a vertex set $\{v_1, v_2, \ldots, v_n\}$ is a circle graph if and only if there is a system of intervals $I_1, I_2, \ldots, I_n$ on the real line with the property that v_i is adjacent to v_j if and only if the intervals I_i and I_j overlap, i.e., they are neither disjoint nor in inclusion.

There is an alternative, purely combinatorial characterization of circle graphs. Suppose that S is a sequence over an alphabet $V = \{v_1, v_2, \ldots, v_n\}$, in which each symbol v_i appears exactly twice. We say that v_i and v_j alternate in S if they form a subsequence $v_i v_j v_i v_j$ or $v_j v_i v_j v_i$. Let A_S be the *alternation graph* of S, which is the graph on the vertex set V in which two vertices are adjacent if and only if they alternate in S. It is not hard to observe that A_S is a circle graph, and every circle graph is isomorphic to A_S for a suitable S. This shows, by the way, that circle graphs are a special case of the so-called *word-representable* graphs, which are a topic of a survey by Kitaev and Lozin [44].

It is known, by a result of Gyárfás [34, 35], that circle graphs are χ-bounded. However, it is still an open problem to find an asymptotically tight relationship between $\chi(G)$ and $\omega(G)$ for this class. Let $f(k)$ denote the smallest number c such that each circle graph G not having a clique of size k has chromatic number at most c.

Gyárfás proved that $f(k) \leq k^2 2^k (2^k - 2)$. This bound was subsequently improved by Kostochka and Kratochvíl [48] to $f(k) \leq 50 \cdot 2^k - O(k)$, and the currently best known upper bound is due to erný [25], and is equal to $f(k) \leq 21 \cdot 2^k - O(k)$. On the other hand, Kostochka [47] has shown that $f(k) \geq \Omega(k \log k)$, and this remains the best known lower bound.

Not only is there a large gap between the asymptotic bounds on $f(k)$, but also finding values of $f(k)$ for fixed small k is challenging. Trivially, $f(2) = 1$. We know that $f(3) = 5$, by the combined results of Kostochka [46], who proved $f(3) \leq 5$, and Ageev [1], who constructed a triangle free circle graph on 220 vertices with chromatic number 5. Nenashev [54] proved that $f(4) \leq 30$.

We will show that the problem of estimating the values of $f(k)$ is related to splittability of permutation classes. Let δ_k denote the decreasing permutation of length k, that is, $\delta_k = k(k-1)\cdots 1$, and let π_k be the permutation $1 \oplus \delta_k$. Recalling Proposition 3.3 from page 293, we see that the class $\text{Av}(\pi_k)$ has a splitting into some number of copies of $\text{Av}(132)$. The relationship to circle graphs is explained by the next proposition.

Proposition 3.17 ([38, Proposition 3.20]) *For any k, the number $f(k)$ is equal to the smallest integer c such that $\text{Av}(\pi_k)$ has a splitting into c copies of $\text{Av}(132)$.*

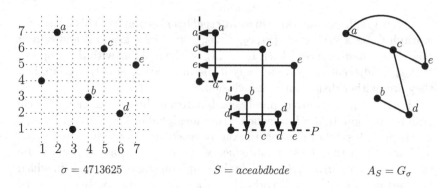

$$\sigma = 4713625 \qquad\qquad S = aceabdbcde \qquad\qquad A_S = G_\sigma$$

Figure 2: Example showing that G_σ is a circle graph. Left: consider $\sigma = 4713625$, and label its non-LR-minima by a, b, c, d, e from left to right. Center: into the diagram of σ, draw the lattice path P whose bottom-left corners are the LR-minima (dashed). Consider the horizontal and vertical projections of the non-LR-minima onto P: these projections form the sequence $S = aceabdbcde$ on P. Right: the alternation graph of S is precisely the graph G_σ.

In view of this proposition, we may regard Proposition 3.3 as a generalization of the χ-boundedness result for circle graphs.

The basic idea of the proof of Proposition 3.17 is not too difficult. Let $\sigma = \sigma(1) \cdots \sigma(n)$ be a permutation. Let us say that $\sigma(i)$ *covers* $\sigma(j)$ if $i < j$ and $\sigma(i) < \sigma(j)$. Recall that an element $\sigma(i)$ that is not covered by any other element of σ is a left-to-right minimum of σ, or just LR-minimum for short. Let V be the set of elements in σ that are not LR-minima.

Let G_σ be the graph on the vertex V in which two vertices $\sigma(a), \sigma(b) \in V$ with $a < b$ are adjacent if and only if $\sigma(a) > \sigma(b)$ and there is a LR-minimum $\sigma(c)$ which covers both $\sigma(a)$ and $\sigma(b)$. Note that in any occurrence of 132 in σ, the two rightmost elements are connected by an edge of G_σ, and conversely, if $\sigma(a)$ and $\sigma(b)$ are adjacent in σ, then there is a (not necessarily unique) occurrence of 132 in σ whose two rightmost elements are $\sigma(a)$ and $\sigma(b)$.

Note that the graph G_σ has a clique of size k if and only if σ contains π_k. Note also that if the graph G_σ has a proper coloring with c colors, then we may extend this coloring arbitrarily to all the elements of σ, and each color will correspond to a 132-avoiding permutation. In particular, σ can be split into at most $\chi(G_\sigma)$ permutations from Av(132).

The key point of the proof is that the graph G_σ is a circle graph, and that every circle graph can be obtained this way. Rather than proving this formally, we refer to the example in Figure 2.

Our previous arguments show that every π_k-avoiding permutation can be merged from at most $f(k)$ permutations avoiding 132. To show that $f(k)$ 132-avoiding permutations are sometimes needed would require a further slightly technical argument, which we omit.

Problem 3.18 *Improve the estimates* $\Omega(k \log k) \leq f(k) \leq O(2^k)$.

The hope is that a better bound for $f(k)$ could then be generalized to a better bound in the more general setting of Proposition 3.3. Vice versa, a better understanding of permutation splitting may lead to an improved bound in Proposition 3.3, which may improve, as a special case, the upper bound on $f(k)$.

4 Ramsey numbers

So far, we have only looked at Ramsey-theoretic problems from the qualitative point of view, that is, we merely wanted to determine the existence of a suitable 'large object' whose every coloring contained a pre-scribed monochromatic substructure. In this last chapter of our survey, we will investigate the quantitative side of the theory, that is, we will look at the problems of determining the minimum size of these 'large objects'.

When dealing with problems of this kind, it is convenient to represent a permutation $\pi = \pi(1) \cdots \pi(n)$ as a *permutation matrix* P_π, which is a $\{0, 1\}$-matrix of size $n \times n$ that contains a 1-entry in column i and row j if and only if $\pi(i) = j$.

We say that a $\{0, 1\}$-matrix M *contains* a $\{0, 1\}$-matrix P if P can be created from M by deleting rows, deleting columns, and changing 1-entries into 0-entries. Observe that a permutation π contains a permutation σ if and only if the matrix P_π contains the matrix P_σ. Thus, the containment poset of permutations can be seen as a restriction of the containment poset of $\{0, 1\}$-matrices.

The poset of $\{0, 1\}$-matrices can be further generalized to ordered graphs. An *ordered graph* is a pair $G^{\prec} = (G, \prec)$, where G is a (simple, loopless, undirected) graph, and $\prec$ is a linear order of the vertices of G. An ordered graph $(G, \prec_G)$ with vertex set V_G is a *subgraph* of an ordered graph $(H, \prec_H)$ with vertex set V_H if there is a function $f \colon V_G \to V_H$ with these properties:

- f is order-preserving, that is, for any two vertices $x, y \in V_G$, if $x \prec_G y$ then $f(x) \prec_H f(y)$.

- f is edge-preserving, that is, if x and y are adjacent vertices of G, then $f(x)$ and $f(y)$ are adjacent in H.

A $\{0,1\}$-matrix M with m rows and n columns can be represented by an ordered graph $G_M^{\prec} = (G_M, \prec)$ with vertex set

$$r_1 \prec r_2 \prec \cdots \prec r_m \prec c_1 \prec c_2 \prec \cdots \prec c_n,$$

where r_j is adjacent to c_i in G_M if and only if M has a 1-entry in row j and column i, and $G_M^{\prec}$ has no edge with both endpoints in $\{r_1, \ldots, r_m\}$ or both endpoints in $\{c_1, \ldots, c_n\}$.

Clearly, if a $\{0,1\}$-matrix M contains a $\{0,1\}$-matrix P, then $G_M^{\prec}$ contains $G_P \prec$ as a subgraph. Note, though, that the converse may fail to hold in the case when M or P has empty rows or columns, since in such situation the graph $G_M^{\prec}$ does not determine the corresponding matrix uniquely.

If P_π is the permutation matrix of a permutation π, we write $G_\pi^{\prec}$ instead of $G_{P_\pi}^{\prec}$.

We let J_n denote the matrix with n rows and n columns and with all entries equal to 1, and we let $K_n^{\prec}$ be the complete ordered graph on n vertices.

We may now formally define the notion of Ramsey number, which is the central topic of this chapter. For a permutation π, its *Ramsey number* $R(\pi)$ is the smallest integer N such that there is a permutation τ of size N with the property that every 2-coloring of the entries of τ creates a monochromatic copy of π.

Similarly, for a $\{0,1\}$-matrix M, its Ramsey number $R(M)$ is the smallest N such that in every 2-coloring of the entries of the matrix J_N, we will find, as a submatrix, a copy of M whose 1-entries all have the same color. For an ordered graph $G^{\prec}$, the Ramsey number $R(G^{\prec})$ is the smallest N such that every 2-coloring of the edges of $K_N^{\prec}$ contains a monochromatic copy of $G^{\prec}$ as a subgraph, and for an unordered graph G, the Ramsey number $R(G)$ is defined analogously.

Observation 4.1 *Let π be a permutation, let P_π be its permutation matrix, and let $G_\pi^{\prec}$ be the ordered graph representing P_π. We then have the inequalities*

$$\frac{1}{2} R(G_\pi^{\prec}) \leq R(P_\pi) \leq R(\pi).$$

Let π be a permutation of size n. How large and how small can $R(\pi)$ be? We have an obvious lower bound $2n - 1 \leq R(P_\pi) \leq R(\pi)$, since a matrix with at most $2n - 2$ rows can be two-colored in such a way that each color appears in at most $n - 1$ distinct rows. This lower bound is tight for some patterns π. For instance, if π is the identity permutation

$123 \cdots n$, then $R(P_\pi) = R(\pi) = 2n - 1$, since any two-coloring of the identity permutation of size $2n - 1$ has a monochromatic copy of π.

A more challenging problem is to investigate the largest possible value of $R(\pi)$ or $R(P_\pi)$. Let us start with an easy upper bound, which, in the matrix setting, has already been pointed out by Conlon, Fox, Lee and Sudakov [23].

Proposition 4.2 *For any permutation π of size n, we have $R(\pi) \leq n^2$, and therefore also $R(P_\pi) \leq n^2$. Moreover, in every red-blue coloring of the entries of J_{n^2} we can find either a blue copy of J_n or a red copy of the permutation matrix of an arbitrary permutation of size n.*

Proof To see that $R(\pi) \leq n^2$, notice that any two-coloring of the permutation $\tau = \pi[\pi]$ has a monochromatic copy of π.

To prove the second part, suppose we are given a 2-coloring of J_{n^2}. Let us partition J_{n^2} into n^2 contiguous blocks, each of size $n \times n$. If one of the blocks is entirely blue, we have a blue copy of J_n. If not, then every block has a red entry, and we can easily find a red copy of the permutation matrix of an arbitrary permutation of size n. □

Given the easy proof of Proposition 4.2, one might expect that the upper bound n^2 can be further significantly improved. However, even an improvement by a constant factor appears to be surprisingly challenging.

Problem 4.3 *Is there a constant $\varepsilon > 0$ such that for every n large enough and every permutation π of size n, we have $R(\pi) < (1 - \varepsilon)n^2$? Do we even have $R(\pi) = o(n^2)$ as n goes to infinity?*

The bound of n^2 in Proposition 4.2 can only be improved by at most a polylogarithmic factor, due to the following result by Balko, Jelínek and Valtr.

Theorem 4.4 ([9]) *There is a constant $c > 0$ such that for almost every permutation π of size n we have*

$$R(G_\pi^{\prec}) \geq c \frac{n^2}{\log^2 n}.$$

Theorem 4.4 is a slight improvement of an earlier result of Conlon, Fox, Lee and Sudakov [23], who found an explicit construction of a permutation π of size n satisfying $R(G_\pi^{\prec}) \geq \Omega\left(\frac{n^2}{\log^2 n \log \log n}\right)$.

Beyond these bounds, not much is known about the Ramsey numbers of permutations or of $\{0, 1\}$-matrices. We may hope that this area will receive more attention in future. As a possible source of inspiration, let us mention several notable results dealing with the closely related areas of graphs and ordered graphs.

For our purposes, the Ramsey numbers of sparse graphs appear particularly relevant. A graph G is called d-*degenerate* if any subgraph of G has a vertex of degree at most d.

Burr and Erdős [18] conjectured that for every d, every d-degenerate unordered graph on n vertices has Ramsey number at most $O_d(n)$. For graphs of maximum degree d, this was proved by Chvátal, Rödl, Szemerédi and Trotter [21], and the conjecture in full generality was recently proved by Lee [50].

The Burr–Erdős conjecture cannot be generalized to ordered graphs, not even to ordered graphs of maximum degree 1. This already follows from Theorem 4.4, which shows that a 1-regular ordered graph (i.e., an ordered matching) can have almost quadratic Ramsey number. In fact, as shown by Conlon, Fox, Lee and Sudakov [23], and independently by Balko, Cibulka, Král and Kynčl [8], the Ramsey number of an ordered matching can be even larger.

Theorem 4.5 ([23, 8]) *Let M_n be the unordered matching with $2n$ vertices and n edges. The is a linear order $\prec$ on the vertex set of M_n such that the ordered graph $M_n^{\prec} = (M_n, \prec)$ satisfies*

$$R(M_n^{\prec}) \geq n^{\Omega(\log n / \log \log n)}.$$

On the other hand, for every linear order $\prec$, we have

$$R(M_n^{\prec}) \leq n^{\lceil \log n \rceil}.$$

Actually, the Ramsey number of a bounded-degree ordered graph $G^{\prec}$ may be superlinear even if we choose the ordering $\prec$ that minimizes $R(G^{\prec})$. This follows from the following result of Balko, Jelínek and Valtr.

Theorem 4.6 ([9]) *For every n there is a 3-regular unordered graph G on n vertices such that for every linear ordering $\prec$ of the vertices of G, the ordered graph $G^{\prec} = (G, \prec)$ satisfies*

$$R(G^{\prec}) \geq \Omega\left(\frac{n^{7/6}}{\log n \log \log n}\right).$$

On the other hand, every 2-regular graph G on n vertices has an ordering $\prec$ such that the Ramsey number of $(G, \prec)$ is linear in n.

An important parameter of an ordered graph $G^{\prec} = (G, \prec)$ is the *interval chromatic number* $\chi_I(G^{\prec})$, defined as the smallest integer c such that the vertices of G can be partitioned into c independent sets each of which is an interval in the ordering $\prec$. Note that for a $\{0,1\}$-matrix M, we have $\chi_I(G_M^{\prec}) \leq 2$. A general upper bound for the Ramsey number of degenerate graphs in terms of χ_I was obtained by Conlon, Fox, Lee and Sudakov [23].

Theorem 4.7 ([23]) *There is a constant c such that every ordered d-degenerate graph $G^{\prec}$ on n vertices with $\chi_I(G^{\prec}) = \chi$ satisfies*

$$R(G^{\prec}) \leq n^{cd \log \chi}.$$

A comprehensive review of the current state of graph Ramsey theory can be found in the survey paper of Conlon, Fox and Sudakov [24].

It is an open problem to obtain similar results for permutations and their permutation matrices, or more generally for sparse matrices.

Problem 4.8 *For which permutations π is $R(\pi)$ linear? For which $\{0,1\}$-matrices M of shape $n \times n$ is $R(M)$ linear in n? What is the largest value of $R(M)$, for an $n \times n$ $\{0,1\}$-matrix M with at most d 1-entries in each row? And what about matrices with at most d 1-entries in each row and column?*

Apart from that, it would be useful to clarify the relationship between the various Ramsey numbers that we associated to a given permutation.

Problem 4.9 *How much can the values of $R(\pi)$, $R(P_\pi)$ and $R(G_\pi^{\prec})$ differ from each other?*

We remark that a somewhat similar problem of establishing relationships between extremal functions of matrices and their corresponding ordered graphs was addressed by Pach and Tardos [60].

We also note that the notion of submatrix, which we used to define a partial order on $\{0,1\}$-matrices, is not the only way to define a partial order on $\{0,1\}$-matrices. A promising alternative is the *interval minor* order, used to great effect by Fox [28] in his work on Stanley–Wilf limits, and implicitly also by Guillemot and Marx [32]. For permutation matrices, the two partial orders coincide, and therefore they both generalize the permutation containment.

Acknowledgement

I gratefully acknowledge the financial support from the Neuron Foundation for Support of Science. I also greatly appreciate the insightful comments of the anonymous reviewer of an earlier version of this manuscript.

References

[1] A. A. Ageev. A triangle-free circle graph with chromatic number 5. *Discrete Math.*, 152(1–3):295–298, 1996.

[2] M. H. Albert. On the length of the longest subsequence avoiding an arbitrary pattern in a random permutation. *Random Structures Algorithms*, 31(2):227–238, 2007.

[3] M. H. Albert, M. D. Atkinson, and M. Klazar. The enumeration of simple permutations. *J. Integer Seq.*, 6, 2003. Article 03.4.4, 18 pages.

[4] M. H. Albert and V. Jelínek. Unsplittable classes of separable permutations. *Electron. J. Combin.*, 23(2):#P2.49, 2016.

[5] M. H. Albert, J. Pantone, and V. Vatter. On the growth of merges and staircases of permutation classes. arXiv:1608.06969, 2016.

[6] M. D. Atkinson, M. M. Murphy, and N. Ruškuc. Partially well-ordered closed sets of permutations. *Order*, 19(2):101–113, 2002.

[7] M. D. Atkinson, M. M. Murphy, and N. Ruškuc. Pattern avoidance classes and subpermutations. *Electron. J. Combin.*, 12:#R60, 2005.

[8] M. Balko, J. Cibulka, K. Král, and J. Kynčl. Ramsey numbers of ordered graphs. *Electron. Notes Discrete Math.*, 49:419–424, 2015.

[9] M. Balko, V. Jelínek, and P. Valtr. On ordered Ramsey numbers of bounded-degree graphs. arXiv:1606.05628, 2016.

[10] D. Bevan. Permutations avoiding 1324 and patterns in Łukasiewicz paths. *J. Lond. Math. Soc.*, 92(1):105–122, 2015.

[11] M. Bodirsky. Ramsey classes: Examples and constructions. *Surveys in Combinatorics 2015*, (424):1, 2015.

[12] M. Bóna. Exact enumeration of 1342-avoiding permutations: A close link with labeled trees and planar maps. *J. Combin. Theory Ser. A*, 80(2):257–272, 1997.

[13] M. Bóna. The limit of a Stanley–Wilf sequence is not always rational, and layered patterns beat monotone patterns. *J. Combin. Theory Ser. A*, 110(2):223–235, 2005.

[14] M. Bóna. New records in Stanley–Wilf limits. *European J. Combin.*, 28(1):75–85, 2007.

[15] M. Bóna. A new upper bound for 1324-avoiding permutations. *Combin. Probab. Comput.*, 23(5):717–724, 2014.

[16] M. Bóna. A new record for 1324-avoiding permutations. *Eur. J. Math.*, 1(1):198–206, 2015.

[17] J. Böttcher and J. Foniok. Ramsey properties of permutations. *Electron. J. Combin.*, 20(1):#P2, 2013.

[18] S. A. Burr and P. Erdős. On the magnitude of generalized Ramsey numbers for graphs. In *Infinite and Finite Sets, Vol. 1 (Keszthely 1973)*, volume 10 of *Colloq. Math. Soc. János Bolyai*, pages 214–240. North-Holland, Amsterdam, 1975.

[19] P. J. Cameron. Homogeneous permutations. *Electron. J. Combin.*, 9(2):#R2, 2002.

[20] G. L. Cherlin. *The classification of countable homogeneous directed graphs and countable homogeneous n-tournaments*, volume 621 of *Memoirs of the American Mathematical Society*. American Mathematical Soc., 1998.

[21] V. Chvátal, V. Rödl, E. Szemerédi, and W. T. Trotter. The Ramsey number of a graph with bounded maximum degree. *J. Combin. Theory Ser. B*, 34(3):239–243, 1983.

[22] A. Claesson, V. Jelínek, and E. Steingrímsson. Upper bounds for the Stanley–Wilf limit of 1324 and other layered patterns. *J. Combin. Theory Ser. A*, 119:1680–1691, 2012.

[23] D. Conlon, J. Fox, C. Lee, and B. Sudakov. Ordered Ramsey numbers. *J. Combin. Theory Ser. B*, 122:353–383, 2017.

[24] D. Conlon, J. Fox, and B. Sudakov. Recent developments in graph Ramsey theory. *Surveys in Combinatorics 2015*, 424:49–118, 2015.

[25] J. Černý. Coloring circle graphs. *Electron. Notes Discrete Math.*, 29(0):457–461, 2007.

[26] M. El-Zahar and N. W. Sauer. Ramsey-type properties of relational structures. *Discrete Math.*, 94(1):1 – 10, 1991.

[27] J. Folkman. Graphs with monochromatic complete subgraphs in every edge coloring. *SIAM J. Appl. Math.*, 18(1):19–24, 1970.

[28] J. Fox. Stanley–Wilf limits are typically exponential. arXiv:1310.8378, 2013.

[29] R. Fraïssé. Sur l'extension aux relations de quelques propriétés des ordres. *Ann. Sci. Ec. Norm. Supér.*, 71(4):363–388, 1954.

[30] I. Gessel. Symmetric functions and P-recursiveness. *J. Combin. Theory Ser. A*, 53(2):257–285, 1990.

[31] R. L. Graham, B. L. Rothschild, and J. H. Spencer. *Ramsey theory.* John Wiley & Sons, 1990.

[32] S. Guillemot and D. Marx. Finding small patterns in permutations in linear time. In *Proceedings of the Twenty-Fifth Annual ACM-SIAM Symposium on Discrete Algorithms*, pages 82–101. ACM, New York, 2014.

[33] A. Gyárfás. On Ramsey covering numbers. In *Infinite and Finite Sets (Keszthely 1973)*, volume 10 of *Colloq. Math. Soc. János Bolyai*, pages 801–816. North-Holland, Amsterdam, 1975.

[34] A. Gyárfás. On the chromatic number of multiple interval graphs and overlap graphs. *Discrete Math.*, 55(2):161–166, 1985.

[35] A. Gyárfás. Corrigendum. *Discrete Math.*, 62(3):333, 1986.

[36] S. Huczynska and N. Ruškuc. Pattern classes of permutations via bijections between linearly ordered sets. *European J. Combin.*, 29(1):118–139, 2008.

[37] V. Jelínek and M. Klazar. Embedding dualities for set partitions and for relational structures. *European J. Combin.*, 32(7):1084–1096, 2011.

[38] V. Jelínek and P. Valtr. Splittings and Ramsey properties of permutation classes. arXiv:1307.0027, 2013.

[39] V. Jelínek and P. Valtr. Splittings and Ramsey properties of permutation classes. *Adv. Appl. Math.*, 63:41–67, 2015.

[40] A. S. Kechris, V. G. Pestov, and S. Todorcevic. Fraïssé limits, Ramsey theory, and topological dynamics of automorphism groups. *Geom. Funct. Anal.*, 15(1):106–189, 2005.

[41] H. A. Kierstead. Classes of graphs that are not vertex Ramsey. *SIAM J. Discrete Math.*, 10(3):373–380, 1997.

[42] H. A. Kierstead and Y. Zhu. Classes of graphs that exclude a tree and a clique and are not vertex Ramsey. *Combinatorica*, 16(4):493–504, 1996.

[43] H. A. Kierstead and Y. Zhu. Radius three trees in graphs with large chromatic number. *SIAM J. Discrete Math.*, 17(4):571–581, 2004.

[44] S. Kitaev and V. Lozin. *Words and graphs*. Springer, 2015.

[45] D. E. Knuth. *The Art of Computer Programming, Volume 1 (3rd Ed.): Fundamental Algorithms*. Addison Wesley Longman Publishing Co., Inc., Redwood City, CA, USA, 1997.

[46] A. Kostochka. Upper bounds on the chromatic number of graphs. *Trudy Instituta Matematiki (Novosibirsk)*, 10:204–226, 1988. (In Russian).

[47] A. Kostochka. Coloring intersection graphs of geometric figures with a given clique number. In János Pach, editor, *Towards a Theory of Geometric Graphs*, volume 342 of *Contemporary Mathematics*, pages 127–138. Amer. Math. Soc., 2004.

[48] A. Kostochka and J. Kratochvíl. Covering and coloring polygon-circle graphs. *Discrete Math.*, 163(1–3):299–305, 1997.

[49] A. H. Lachlan and R. E. Woodrow. Countable ultrahomogeneous undirected graphs. *Trans. Amer. Math. Soc.*, pages 51–94, 1980.

[50] C. Lee. Ramsey numbers of degenerate graphs. arXiv:1505.04773, 2015. To appear in Annals of Mathematics.

[51] D. Macpherson. A survey of homogeneous structures. *Discrete Math.*, 311(15):1599–1634, 2011.

[52] A. Marcus and G. Tardos. Excluded permutation matrices and the Stanley–Wilf conjecture. *J. Combin. Theory Ser. A*, 107(1):153–160, 2004.

[53] M. M. Murphy. *Restricted permutations, antichains, atomic classes, and stack sorting*. PhD thesis, University of St. Andrews, 2002.

[54] G. V. Nenashev. An upper bound on the chromatic number of a circle graph without K_4. *J. Math. Sci. (N.Y.)*, 184(5):629–633, 2012.

[55] J. Nešetřil and V. Rödl. Partitions of vertices. *Commentationes Mathematicae Universitatis Carolinae*, 17(1):85–95, 1976.

[56] J. Nešetřil and V. Rödl. On Ramsey graphs without cycles of short odd lengths. *Commentationes Mathematicae Universitatis Carolinae*, 20(3):565–582, 1979.

[57] J. Nešetřil and V. Rödl. Simple proof of the existence of restricted Ramsey graphs by means of a partite construction. *Combinatorica*, 1(2):199–202, 1981.

[58] J. Nešetřil. Ramsey classes and homogeneous structures. *Combin. Probab. Comput.*, 14(1):171–189, January 2005.

[59] M. Opler. Personal communication, 2016.

[60] J. Pach and G. Tardos. Forbidden paths and cycles in ordered graphs and matrices. *Israel J. Math.*, 155(1):359–380, 2006.

[61] F. P. Ramsey. On a problem of formal logic. *Proc. Lond. Math. Soc.*, s2-30(1):264–286, 1930.

[62] V. Rödl and N. Sauer. The Ramsey property for families of graphs which exclude a given graph. *Canad. J. Math.*, 44:1050–1060, 1992.

[63] V. Rödl, N. Sauer, and X. Zhu. Ramsey families which exclude a graph. *Combinatorica*, 15(4):589–596, 1995.

[64] N. Sauer. Vertex partition problems. In D. Miklós, V. T. Sós, and T. Szőnyi, editors, *Combinatorics, Paul Erdős is eighty, Vol. 1*, pages 361–377. János Bolyai Mathematical Society, 1993.

[65] N. Sauer. On the Ramsey property of families of graphs. *Trans. Amer. Math. Soc.*, 347(3):785–833, 1995.

[66] N. Sauer. Age and weak indivisibility. *European J. Combin.*, 37:24–31, 2014.

[67] J. H. Schmerl. Countable homogeneous partially ordered sets. *Algebra Universalis*, 9(1):317–321, 1979.

[68] M. Sokić. *Ramsey Property of Posets and Related Structures*. PhD thesis, University of Toronto, 2010.

[69] M. Sokić. Ramsey property, ultrametric spaces, finite posets, and universal minimal flows. *Israel J. Math.*, 194(2):609–640, 2013.

[70] S. Solecki and M. Zhao. A Ramsey theorem for partial orders with linear extensions. *European J. Combin.*, 60:21–30, 2017.

[71] D. P. Sumner. Subtrees of a graph and chromatic number. In G. Chartrand, Y. Alavi, D. L. Goldsmith, L. Lesniak-Foster, and D. L. Lick, editors, *The theory and applications of graphs (Kalamazoo, MI, 1980)*, pages 557–576. John Wiley and Sons Dordrecht and Boston, 1981.

[72] V. Vatter. Small permutation classes. *Proc. Lond. Math. Soc.*, 103(5):879–921, 2011.

[73] V. Vatter. Permutation classes. In *Handbook of enumerative combinatorics*, Discrete Math. Appl. (Boca Raton), page 753833. CRC Press, Boca Raton, FL, 2015.

[74] V. Vatter. An Erdős–Hajnal analogue for permutation classes. *Discrete Math. Theor. Comput. Sci.*, 8(2):#4, 2016.

[75] A. Zucker. Topological dynamics of automorphism groups, ultrafilter combinatorics, and the Generic Point Problem. *Trans. Amer. Math. Soc.*, 368(9):6715–6740, 2016.

Computer Science Institute
Charles University in Prague
Malostranské námstí 25, Praha 1, 118 00, Czech Republic
jelinek@iuuk.mff.cuni.cz

Monotone cellular automata

Robert Morris

Abstract

Cellular automata are interacting particle systems whose update rules are local and homogeneous. Since their introduction by
von Neumann almost 50 years ago, many particular such systems
have been investigated, but no general theory has been developed
for their study, and for many simple examples surprisingly little is
known. Understanding the rules that govern their typical global
behaviour is an important and challenging problem in statistical
physics, probability theory and combinatorics.

In this survey we will consider the behaviour of a particular
(large) family of *monotone* cellular automata – those which can
naturally be embedded in d-dimensional space – with random initial
conditions. For example, in the case where a site updates (from
inactive to active) if at least r of its neighbours are already active,
these models are known as *bootstrap percolation*, and have been
extensively studied for various specific underlying graphs.

Our aims are threefold: to provide a relatively gentle introduction to some of the key techniques in the area; to describe some
dramatic recent progress relating to an extremely general class of
models, which were introduced by Bollobás, Smith and Uzzell; and
to discuss applications of these techniques to models in statistical physics, such as the Glauber dynamics of the Ising model, the
abelian sandpile, and kinetically constrained spin models.

1   Introduction

Consider the following deterministic process on a graph G: at each
time step, a healthy vertex becomes *infected* if at least r of its neighbours
are already infected, while infected vertices stay infected forever. Thus,
writing A_t for the set of infected vertices at time t, we have

$$A_{t+1} = A_t \cup \big\{ v \in V(G) \, : \, |N(v) \cap A_t| \geqslant r \big\}$$

for each $t \geqslant 0$. This process, which is called *r-neighbour bootstrap percolation*, was introduced in 1979 by Chalupa, Leath and Reich [23], who
were motivated by the Glauber dynamics of the Ising model on $\mathbb{Z}^d$ (see
Section 5). In this context, the following basic question is of fundamental
importance: What is the typical behaviour of the bootstrap process when
the initial set of infected vertices is chosen randomly?

312

In order to state this question more precisely, let us write $A = A_0$ for the set of vertices infected at time $t = 0$, and define $\mathbb{P}_p$ to be the probability distribution obtained by placing each vertex of G in A independently at random with probability p. (We say that a subset $A \subseteq V(G)$ is p-*random* if it is chosen according to $\mathbb{P}_p$.) The *closure* of A is denoted

$$[A] := \bigcup_{t \geqslant 0} A_t$$

and we say that the set A *percolates* if $[A] = V(G)$. The *critical probability* for r-neighbour bootstrap percolation on G is then defined to be

$$p_c(G, r) := \inf \left\{ p \, : \, \mathbb{P}_p\big([A] = V(G)\big) \geqslant 1/2 \right\}.$$

We remark that the event $\big\{[A] = V(G)\big\}$ typically has a sharp threshold[1], and so the choice of the constant $1/2$ is not important.

The r-neighbour bootstrap process has been studied on a wide range of graphs, including trees [23, 10, 15], the Erdős-Rényi random graph [42] and the random regular graph [11]. In this survey, however, we will restrict our attention to a particular class of processes: roughly speaking, those that can naturally be embedded in d-dimensional Euclidean space. As we will see, this leads to an extremely rich and complex class of models, but we will nevertheless be able to make significant progress towards understanding the behaviour of such processes.

The structure of this survey is as follows. As a warm-up, we will begin (in Sections 2–4) by discussing in some detail an important specific class of models: the two-neighbour process on the two-dimensional torus $\mathbb{Z}_n^2$, and its (significantly more challenging) generalization, the r-neighbour process on $\mathbb{Z}_n^d$. We will also (in Section 5) give a couple of applications of these results in a more complex setting: the zero-temperature Glauber dynamics of the Ising model on $\mathbb{Z}^d$.

After this extended introduction, we will be ready to expand our horizons and consider more general classes of processes. We will begin (in Section 6) by discussing a few instructive examples, in order to build up our intuition; we will then proceed (in Sections 7 and 8) to the main topic of this survey: a series of recent breakthroughs that provide a very precise characterization of monotone cellular automata in two dimensions. Finally, in Sections 9 and 10, we will discuss possible future directions, including a conjectured characterization in higher dimensions, and some related models including nucleation and growth, kinetically constrained spin models, and the abelian sandpile.

---

[1] In the sense that if $p < \big(1 + o(1)\big)p_c(G, r)$ then $\mathbb{P}_p\big([A] = V(G)\big) = o(1)$, whereas if $p > \big(1 + o(1)\big)p_c(G, r)$ then $\mathbb{P}_p\big([A] = V(G)\big) = 1 - o(1)$.

2   The two-neighbour model on $\mathbb{Z}_n^2$

How many *randomly chosen* infected vertices of the graph $\mathbb{Z}_n^2$ are needed to (typically) result in the eventual infection of the entire torus? On the one hand, it is easy to see that there exists a set A of size $n - 1$ such that $[A] = \mathbb{Z}_n^2$ (for example, one could take all but one element of a diagonal line); on the other, if the elements of A are typically far apart, it is hard to see how they could help one another.

The first step in developing our intuition is the following theorem, first proved by van Enter [29] in 1987.

Theorem 2.1 $p_c(\mathbb{Z}_n^2, 2) \to 0$ *as* $n \to \infty$.

Proof The proof has two steps: we first find a large infected square, and then (using sprinkling to maintain independence) show that this square is likely to grow and infect the entire torus. To be precise, for each constant $\varepsilon > 0$ we will show that, if $p = 2\varepsilon$, then $\mathbb{P}_p([A] = \mathbb{Z}_n^2) \to 1$ as $n \to \infty$. Indeed, let A_1 and A_2 be (independent) ε-random subsets of $\mathbb{Z}_n^2$, so that (via a trivial coupling) we have $A \supseteq A_1 \cup A_2$, and set $k = \log \log n$. We will prove the following two statements:

1. With high probability A_1 contains a square R of side-length k.

2. With high probability $[A_2 \cup R] = \mathbb{Z}_n^2$.

Indeed, the first claim follows since each $k \times k$ square lies in A_1 with probability $\varepsilon^{k^2} = n^{-o(1)}$, for disjoint squares the corresponding events are independent, and there exists a collection of $n^{2-o(1)}$ disjoint $k \times k$ squares in $\mathbb{Z}_n^2$. To see the second claim, let us suppose (without loss of generality) that $R = [k]^2$, and observe that if $[m]^2 \subseteq [A_2 \cup R]$ and A_2 contains at least one element of each of the sets $\{(m+1, y) : y \in [m]\}$ and $\{(x, m+1) : x \in [m]\}$ then $[m+1]^2 \subseteq [A_2 \cup R]$. It follows that

$$\mathbb{P}([A_2 \cup R] = \mathbb{Z}_n^2) \geqslant 1 - 2 \sum_{m=k}^{n-1} (1 - \varepsilon)^m \to 1$$

as $n \to \infty$, as claimed.   □

The following year, Aizenman and Lebowitz [1] proved the following fundamental theorem, which determines the *threshold* for the event that A percolates.[2]

---

[2]In fact, Aizenman and Lebowitz determined the threshold for the two-neighbour model on $\mathbb{Z}_n^d$ for all $d \geqslant 2$, but for simplicity we shall work in two dimensions. We remark that a second proof of this theorem was obtained independently several years later by Balogh and Pete [11].

Theorem 2.2

$$p_c(\mathbb{Z}_n^2, 2) = \Theta\left(\frac{1}{\log n}\right).$$

The proof of the upper bound in Theorem 2.2 is essentially just a refinement of the proof of Theorem 2.1, the key observation being that in the first step we can replace the event $R \subseteq A_1$ by the (much more likely) event that R is *internally filled* by A_1, that is, if $R \subseteq [A_1 \cap R]$. Indeed, if $R = [k]^2$ and A_1 is p-random, then the probability of this event can be bounded by

$$\mathbb{P}\big(R \subseteq [A_1 \cap R]\big) \geqslant \prod_{m=1}^{k-1} \left(1 - (1-p)^m\right)^2 \geqslant e^{-C/p} \qquad (2.1)$$

for some large constant C, since if each of the sets $\{(m+1, y) : y \in [m]\}$ and $\{(x, m+1) : x \in [m]\}$ contains an element of A_1 then $R \subseteq [A_1 \cap R]$, and these sets are all disjoint. (The second inequality can be proved, for example, by noting that if $m = o(1/p)$ then $1 - (1-p)^m \sim pm$, if $m = \Theta(1/p)$ then $1 - (1-p)^m = \Theta(1)$, and if $m \gg 1/p$ then $(1-p)^m \leqslant e^{-pm}$.)

Now, suppose that $p \geqslant C/\log n$, and let $k = (\log n)^3$ and A_1 be a p-random set. By applying (2.1) to (roughly) $n^2/k^2 \gg n$ disjoint squares, it follows that with high probability there exists a $k \times k$ square R that is internally filled by A_1. As in the proof of Theorem 2.1, if A_2 is also a p-random set then

$$\mathbb{P}\big([A_2 \cup R] = \mathbb{Z}_n^2\big) \geqslant 1 - 2 \sum_{m=k}^{n-1} (1-p)^m \geqslant 1 - 2ne^{-pk} \to 1$$

as $n \to \infty$, so $p_c \leqslant 2C/\log n$, as required.

For the lower bound, the key tool is the following deterministic lemma, which allows us to identify the "bottleneck" that prevents A from percolating. For technical reasons (see the proofs of the lemmas below), it will be slightly simpler to work on the grid $[n]^2$; however, the proof can be easily extended to the torus $\mathbb{Z}_n^2$ without any additional ideas. Let $\phi(R)$ denote the semi-perimeter of a rectangle $R \subseteq [n]^2$.

Lemma 2.3 (The Aizenman–Lebowitz lemma) *Let R be a rectangle that is internally filled by a set $A \subseteq [n]^2$. For every $1 \leqslant k \leqslant \phi(R)$, there exists an internally filled rectangle $S \subseteq R$ with $k \leqslant \phi(S) < 2k$.*

We remark that variants of this lemma will play a crucial role in almost all of the proofs discussed below. The lemma is a straightforward consequence of the following process, which is simply a convenient reordering of the order in which we infect the sites of $[A] \setminus A$.

Definition 2.4 (The rectangles process)
Let $A = \{x_1, \ldots, x_m\} \subseteq [n]^2$, and consider the collection $\{R_1, \ldots, R_m\}$, where $R_j = \{x_j\}$ for each $j \in [m]$. Now repeat the following steps until STOP:

1. If there exist two rectangles R_i and R_j in the current collection at distance at most two from one another, then choose such a pair, remove them from the collection, and replace them by $[R_i \cup R_j]$.

2. If there do not exist such a pair of rectangles, then STOP.

Before going any further, let's record a couple of simple but important observations for future reference. We leave the (easy) proofs to the reader.

Lemma 2.5 *Let R and S be rectangles in $[n]^2$ at distance at most two from one another. Then $[R \cup S]$ is a rectangle, and moreover*

$$\phi\big([R \cup S]\big) \leqslant \phi(R) + \phi(S).$$

Moreover, if R and S are internally filled, then $[R \cup S]$ is internally filled.

Note that, by Lemma 2.5, every rectangle that appears in the rectangles process applied to A is internally filled by A.

For each $A \subseteq [n]^2$, let us write $\langle A \rangle = \{R_1, \ldots, R_k\}$ for the collection of rectangles obtained by applying the rectangles process to A; we call $\langle A \rangle$ the *span* of A. The next lemma is also just an observation.

Lemma 2.6 *For each $A \subseteq [n]^2$, we have*

$$[A] = \bigcup_{S \in \langle A \rangle} S.$$

In particular, if R is internally filled by A, then $\langle A \cap R \rangle = \{R\}$.

Proof Each rectangle that appears in the rectangles process applied to A is internally filled by A, and so each $S \in \langle A \rangle$ is clearly contained in $[A]$. On the other hand, suppose (for a contradiction) that there exists a vertex of $[A]$ is that is not contained in some rectangle in $\langle A \rangle$, and consider the first such vertex to be infected. This vertex (v, say) has at least two previously infected neighbours, which (by minimality) must be contained in (one or two) rectangles in $\langle A \rangle$. But these cannot be different rectangles, since the distance between them is at most two, and if they are the same rectangle then they also contain v, which is the desired contradiction.

Finally, note that, since the rectangles in $\langle A \rangle$ are at pairwise distance at least three, if they cover R then they must consist of a single rectangle. $\qquad\square$

We can now easily deduce Lemma 2.3.

Proof of Lemma 2.3 If R is internally filled by $A \subseteq [n]^2$, then we have $R \in \langle A \cap R \rangle$, by Lemma 2.6. Let S be the first rectangle that appears in the rectangles process with $\phi(S) \geqslant k$. Note that S is internally filled by A, and that $S = [R_i \cup R_j]$ for some rectangles with $\phi(R_i), \phi(R_j) < k$. Hence, by Lemma 2.5, we have $\phi(S) < 2k$, as required. $\qquad\square$

The final tool we need to prove the lower bound in Theorem 2.2 is the following lemma, which also follows easily from the rectangles process.

Lemma 2.7 *Let $R \subseteq [n]^2$ be a rectangle. If R is internally filled by A, then*

$$|A \cap R| \geqslant \frac{\phi(R)}{2}.$$

Proof We use induction on $\phi(R)$, noting that if $|R| = 1$ then $\phi(R) = 2$, so the base case holds trivially. To prove the induction step, we will show that there exist two *disjointly* internally filled rectangles S_1 and S_2 such that

$$\phi(S_1), \phi(S_2) < \phi(R) \qquad \text{and} \qquad [S_1 \cup S_2] = R.$$

That is, there exist disjoint sets $A_1, A_2 \subseteq A$ with $S_1 = [A_1]$ and $S_2 = [A_2]$.

To prove this, observe first that at every step of the rectangles process all of the rectangles in the current collection are disjointly internally filled. (Proof: this is true at the beginning, and is maintained by uniting two rectangles.) Thus, if we run the rectangles process for the set $A \cap R$, stopping just before the rectangle R appears for the first time, we obtain two rectangles S_1 and S_2 as required.

Now, by Lemma 2.5 and the induction hypothesis, it follows that

$$|A \cap R| \geqslant |A_1| + |A_2| \geqslant \frac{\phi(S_1)}{2} + \frac{\phi(S_2)}{2} \geqslant \frac{\phi(R)}{2},$$

as required. $\qquad\square$

This is in fact not the simplest proof of Lemma 2.7, but it has the significant advantage of being fairly "robust"; in particular, it motivates the strategy we will use later in more complex contexts.

Exercise: Find an alternative (one line!) proof of Lemma 2.7.

We are now ready to prove Theorem 2.2.

Proof of Theorem 2.2 The proof of the upper bound was outlined above, so we will prove only the lower bound. Suppose that $A \subseteq [n]^2$ is such that $[A] = [n]^2$. By Lemma 2.3, there exists an internally filled rectangle R with

$$\log n \leqslant \phi(R) \leqslant 2 \log n.$$

Let X denote the random variable that counts the number of such rectangles when A is a p-random subset of $[n]^2$, and observe that

$$\mathbb{E}[X] \leqslant \sum_{k=\log n}^{2 \log n} n^3 \binom{k^2}{k/2} p^{k/2} \leqslant \sum_{k=\log n}^{2 \log n} n^3 (2ekp)^{k/2} \leqslant \frac{1}{n} \qquad (2.2)$$

if $1/p > e^{12} \log n$, by Lemma 2.7, since there are at most n^3 rectangles in $[n]^2$ with semi-perimeter k, and each has area at most k^2. It follows from Markov's inequality that a p-random set $A \subseteq [n]^2$ percolates with probability at most $1/n$, and hence

$$p_c([n]^2, 2) \geqslant \frac{1}{e^{12} \log n},$$

as required.   □

The proof of Aizenman and Lebowitz can be extended (in a relatively straightforward manner) to determine the threshold for the two-neighbour model in d dimensions (i.e., on the grid $[n]^d$ or the torus $\mathbb{Z}_n^d$). We leave this generalization as an exercise for the reader (see also Section 3).

Exercise: Prove that

$$p_c(\mathbb{Z}_n^d, 2) = \Theta\left(\frac{1}{\log n}\right)^{d-1}$$

for every $d \geqslant 2$.

2.1 A sharp threshold for the two-neighbour process on $\mathbb{Z}_n^2$

The next major breakthrough in the study of the two-neighbour process was made by Holroyd [39] in 2003, who determined the following sharp threshold in two dimensions.

Theorem 2.8

$$p_c(\mathbb{Z}_n^2, 2) = \left(\frac{\pi^2}{18} + o(1)\right) \frac{1}{\log n}.$$

The main technical innovation in Holroyd's proof, which has had an extremely significant impact on the subsequent development of the area, is the concept of a *hierarchy*. Roughly speaking, this is obtained by recording a (cleverly-chosen) subset of the rectangles that appear in the rectangles process. This subset has the following properties:

(a) There are at most $n^{o(1)}$ possible hierarchies of a rectangle R of semi-perimeter $O(\log n)$ (this is useful as we will take a union bound over hierarchies when bounding the probability that R is internally filled).

(b) Each hierarchy is unlikely to occur (at least as unlikely as that corresponding to the simplest possible kind of growth, in which a single droplet finds at each step a single infected site on its boundary).

To be slightly more precise, whenever two 'large' rectangles combine we will record the event, and if a droplet grows for some time without meeting another 'large' rectangle, then we will periodically record its latest size and position. Much more precisely, we have the following technical definition.

Definition 2.9 Let R be a rectangle. A *hierarchy* $\mathcal{H}$ for R consists of a directed rooted tree $G_{\mathcal{H}}$, with all of its edges directed downwards (i.e., away from the root v_0), and for each vertex $v \in V(G_{\mathcal{H}})$ an associated rectangle D_v, such that the following conditions are satisfied:

(i) the root vertex is associated to R, i.e., $D_{v_0} = R$;

(ii) each vertex has out-degree at most 2;

(iii) if $v \in N(u)$ then $D_v \subseteq D_u$;

(iv) if $N(u) = \{v, w\}$ then $D_u = [D_v \cup D_w]$,

where $N(u)$ denotes the out-neighbourhood of u in $G_{\mathcal{H}}$.

Given $s, t > 0$, we say that $\mathcal{H}$ is (s,t)-*good* if it satisfies the following conditions for each $u \in V(G_{\mathcal{H}})$:

(v) u is a leaf (i.e, $|N(u)| = 0$) if and only if $\phi(D_u) \leqslant s$;

(vi) if $N(u) = \{v\}$ and $|N(v)| = 1$ then

$$t \leqslant \phi(D_u) - \phi(D_v) \leqslant 2t;$$

(vii) if $N(u) = \{v\}$ and $|N(v)| \neq 1$ then $\phi(D_u) - \phi(D_v) \leqslant 2t$;

(viii) if $N(u) = \{v, w\}$ then $\phi(D_u) - \phi(D_v) \geqslant t$.

Finally, $\mathcal{H}$ is *satisfied* by A if the following events all occur *disjointly*:

(ix) if v is a leaf then D_v is internally filled by A;

(x) if $N(u) = \{v\}$ then $D_u = [D_v \cup (D_u \cap A)]$.

That is, there exists a family of disjoint subsets of A, one for each of these events, such that each guarantees that the corresponding event occurs.

Let us denote by $\mathcal{H}_R(s,t)$ the family of (s,t)-good hierarchies for a rectangle R. The following lemma is a straightforward consequence of the rectangles process (as described above), so we will leave the details of the proof to the reader.

Lemma 2.10 *Let R be a rectangle that is internally filled by $A \subseteq \mathbb{Z}_n^2$. Then, for every $s \geqslant t \geqslant 1$, there exists an (s,t)-good hierarchy $\mathcal{H} \in \mathcal{H}_R(s,t)$ that is satisfied by A.*

In order to deduce a bound on the probability that a rectangle is internally filled, we will need the following fundamental inequality of van den Berg and Kesten [12]. Given two increasing[3] events $E, F \subseteq \{0,1\}^{\mathbb{Z}_n^2}$, we write $E \circ F$ for the event that E and F occur disjointly.

Lemma 2.11 (The van den Berg-Kesten inequality)
If E and F are increasing events then

$$\mathbb{P}_p(E \circ F) \leqslant \mathbb{P}_p(E) \cdot \mathbb{P}_p(F).$$

We remark that the assumption in Lemma 2.11 that E and F are increasing is in fact not necessary for the conclusion to hold, as was conjectured by van den Berg and Kesten [12], and later proved by Reimer [53].

Let us write $L(\mathcal{H})$ for the set of leaves of $G_\mathcal{H}$ (we will refer to the rectangles corresponding to these vertices as the *seeds* of $\mathcal{H}$), and $\prod_{u \to v}$ for the product over all pairs $\{u, v\} \subseteq V(G_\mathcal{H})$ such that $N(u) = \{v\}$ (we will refer to these as *steps*). Also, given rectangles $R \subseteq R'$, define

$$I(R) = \{[A \cap R] = R\} \quad \text{and} \quad \Delta(R, R') := \{R' = [R \cup (R' \cap A)]\},$$

and note that these events correspond to R being internally filled by A, and R' being internally filled by $A \cup R$, respectively, cf. conditions (ix) and (x) of Definition 2.9.

We are now ready to state and prove Holroyd's fundamental bound on the probability that a rectangle is internally filled by a p-random set A.

---

[3]An event $E \subseteq \{0,1\}^{\mathbb{Z}_n^2}$ is *increasing* if $A \in E$ and $A \subseteq B$ implies $B \in E$.

Lemma 2.12 *Let $R \subseteq \mathbb{Z}_n^2$ be a rectangle. For every $s \geqslant t \geqslant 1$, we have*

$$\mathbb{P}_p\big(I(R)\big) \leqslant \sum_{\mathcal{H} \in \mathcal{H}_R(s,t)} \bigg(\prod_{u \in L(\mathcal{H})} \mathbb{P}_p\big(I(D_u)\big) \bigg) \bigg(\prod_{u \to v} \mathbb{P}_p\big(\Delta(D_v, D_u)\big) \bigg).$$

(2.3)

Proof If R is internally filled by A then, by Lemma 2.10, there must exist an (s,t)-good hierarchy $\mathcal{H} \in \mathcal{H}_R(s,t)$ that is satisfied by A. By Definition 2.9, if $\mathcal{H}$ is satisfied by A then the events $I(D_u)$ (for each $u \in L(\mathcal{H})$) and $\Delta(D_v, D_u)$ (for each pair $\{u, v\}$ such that $N(u) = \{v\}$) all occur disjointly. The lemma now follows by applying the van den Berg-Kesten inequality to this family of events. $\qquad\square$

We will apply Lemma 2.12 with $s = \varepsilon \log n$ and $t = \varepsilon^2 \log n$, for some sufficiently small constant $\varepsilon > 0$. This choice will allow us to obtain sufficiently strong bounds on the probabilities of the events $I(D_u)$ and $\Delta(D_v, D_u)$ on the right-hand side of (2.3), whilst keeping the total number of hierarchies relatively small. Indeed, we have the following bound on the size of $\mathcal{H}_R(s,t)$.

Lemma 2.13 *If $\phi(R) = O(\log n)$ and $t = \Omega(\log n)$, then*

$$|\mathcal{H}_R(s,t)| = (\log n)^{O(1)}.$$

Proof This follows since if $\mathcal{H} \in \mathcal{H}_R(s,t)$ then the height of the tree $G_{\mathcal{H}}$ is bounded, and hence $|V(G_{\mathcal{H}})| = O(1)$. Since each rectangle in the set $\{R_u : u \in V(G_{\mathcal{H}})\}$ is contained in R, it follows that we have only $(\log n)^{O(1)}$ choices for each, and so the claimed bound follows. $\qquad\square$

We also easily obtain the following bound on the probability that a seed is internally filled, cf. (2.2).

Lemma 2.14 *If R is a rectangle and $k = \lceil \phi(R)/2 \rceil$, then*

$$\mathbb{P}_p\big(I(R)\big) \leqslant (ekp)^k.$$

Proof By Lemma 2.7, if R is internally filled by A then $|A \cap R| \geqslant k$. Since the area of R is at most k^2, it follows that

$$\mathbb{P}_p\big(I(R)\big) \leqslant \binom{k^2}{k} p^k \leqslant (ekp)^k,$$

as claimed. $\qquad\square$

It follows immediately from Lemmas 2.12, 2.13 and 2.14 that the expected number of satisfied hierarchies in $\mathcal{H}_R(s,t)$ such that the sum of the semi-perimeters of the seeds is at least $\delta \log n$ is at most

$$(\log n)^{O(1)} \left(ep \cdot \varepsilon \log n\right)^{\delta \log n} \leqslant \frac{1}{n^3}$$

if $p \leqslant 1/\log n$ and $\varepsilon = \varepsilon(\delta) > 0$ is sufficiently small. Since there are only $n^{2+o(1)}$ rectangles in $\mathbb{Z}_n^2$ of semi-perimeter $O(\log n)$, this will be sufficient for our planned application of Markov's inequality.

We may therefore consider from now on only hierarchies with relatively few seeds; however, since doing so is rather more difficult (and technical), we will give only a brief sketch of the necessary ideas. The basic idea is to compare the steps of $\mathcal{H}$ to those in an 'ideal' hierarchy with one large seed of semi-perimeter

$$z(\mathcal{H}) := \sum_{u \in L(\mathcal{H})} \phi(D_u).$$

Since growth becomes easier as the droplet gets larger, it is natural to expect (and possible to prove) that the product

$$\prod_{u \to v} \mathbb{P}_p\big(\Delta(D_v, D_u)\big)$$

will be (asymptotically) minimized (over those $\mathcal{H} \in \mathcal{H}_R(s,t)$ with a given value of $z(\mathcal{H})$) by taking a hierarchy of this form. Moreover, the 'easiest' way for such a droplet to grow is 'along the diagonal', i.e., as an approximate square. The basic fact that allows one to show this can be paraphrased as follows:

"A rectangle is crossed by A if and only if it has no double gap."

More precisely, let us write ∂S for the set of sites directly to the left of a rectangle S, and say that S *is crossed by* A if $S \subseteq [\partial S \cup (A \cap S)]$. Then S is crossed by A if and only if there does not exist a 'double gap', i.e., a consecutive pair of columns of S neither or which contains an element of A. We leave the proof of the following lemma as an exercise for the reader.

Lemma 2.15 *Let S be an $a \times b$ rectangle. Then*

$$\beta(u)^a \leqslant \mathbb{P}_p\big(S \text{ is crossed by } A\big) \leqslant \beta(u)^{a-1},$$

where $u = 1 - (1-p)^b$ and $\beta(u) = \big(u + \sqrt{u(4-3u)}\big)/2$.

Moreover, the choice of s and t (with $t = \varepsilon s$) allows one to approximate the event $\Delta(D_v, D_u)$ by the intersection of four 'crossing' events (to the right, left, up and down), since the 'corner regions' are relatively small, and therefore are unlikely to contain many elements of A.

It follows (after a fair amount of work) that one can bound the probability that a hierarchy $\mathcal{H} \in \mathcal{H}_R(s, t)$ is satisfied by A by

$$\prod_{u \to v} \mathbb{P}_p\big(\Delta(D_v, D_u)\big) \leqslant \prod_{b=z(\mathcal{H})}^{\phi(R)/2} \beta\big(u(b)\big)^{2-\delta},$$

where $u(b) = 1 - (1 - p)^b$ and $\delta > 0$ is an arbitrarily small[4] constant. After some (non-trivial) integration (see [39, Section 2], and also [41]), this implies, for every $\gamma > 0$, the bound

$$\mathbb{P}_p\big(\mathcal{H} \text{ is satisfied by } A\big) \leqslant \exp\left(-\frac{\pi^2 - \gamma}{9p}\right)$$

for every $\mathcal{H} \in \mathcal{H}_R(s, t)$ with $z(\mathcal{H}) \leqslant \delta \log n$, if $\phi(R) \geqslant C \log n$ for some sufficiently large constant $C = C(\gamma)$, and if $\delta = \delta(\gamma) > 0$ was chosen sufficiently small. Thus, if

$$p \leqslant \left(\frac{\pi^2}{18} - \gamma\right) \frac{1}{\log n},$$

then we obtain

$$\mathbb{P}_p\big(I(R)\big) \leqslant (\log n)^{O(1)} \exp\left(-\frac{\pi^2 - \gamma}{9p}\right) \leqslant n^{-2-\gamma}. \tag{2.4}$$

Finally, applying Lemma 2.3 (cf. the proof of Theorem 2.2), if $[A] = \mathbb{Z}_n^2$ then there exists an internally filled rectangle R with

$$C \log n \leqslant \phi(R) \leqslant 2C \log n,$$

where $C = C(\gamma)$ is the sufficiently large constant chosen above. By (2.4) (and Markov's inequality) this has probability at most $n^{-\gamma + o(1)}$, and hence

$$p_c(\mathbb{Z}_n^2, 2) \geqslant \left(\frac{\pi^2}{18} - \gamma\right) \frac{1}{\log n},$$

for every $\gamma > 0$, as required. This completes our sketch of Holroyd's proof of the lower bound in Theorem 2.8; we leave the upper bound to the reader.

---

[4]To prove this for a given δ we must take $\varepsilon = \varepsilon(\delta) > 0$ sufficiently small.

Exercise: Use Lemma 2.15 and the fact (see [39, Section 2]) that

$$\int_0^\infty -\log\left(\beta(1-e^{-z})\right)\,\mathrm{d}z = \frac{\pi^2}{18}$$

to prove that

$$p_c(\mathbb{Z}_n^2, 2) \leqslant \left(\frac{\pi^2}{18} + o(1)\right)\frac{1}{\log n}$$

as $n \to \infty$.

Much more precise bounds on $p_c(\mathbb{Z}_n^2, 2)$ have since been obtained by Gravner and Holroyd [35], Gravner, Holroyd and Morris [36], and Morris [46], culminating in the following (unpublished) theorem.

Theorem 2.16

$$p_c(\mathbb{Z}_n^2, 2) = \frac{\pi^2}{18\log n} - \frac{\Theta(1)}{(\log n)^{3/2}}.$$

The proof of Theorem 2.16 requires a significantly more complicated notion of hierarchy than that described above (requiring the use of Reimer's theorem instead of the van den Berg–Kesten inequality), as well as extremely precise bounds on the probability various types of 'steps'. We refer the reader to [46] for (most of) the details.

3   The r-neighbour model on $\mathbb{Z}_n^d$

The methods of the previous section can be generalized without too much difficulty to the setting of $\mathbb{Z}_n^d$ when $r = 2$. However, when $r \geqslant 3$ things fall apart very quickly, as the closed sets are no longer (unions of) rectangles. The first breakthrough was made by Schonmann [54] in 1992, who proved the following generalization of van Enter's theorem.

Theorem 3.1 *If $d \geqslant r \geqslant 1$, then*

$$p_c(\mathbb{Z}_n^d, r) \to 0$$

as $n \to \infty$.

Note that if $r > d$ then $p_c(\mathbb{Z}_n^d, r) = 1 - o(1)$, since a translate of $\{1,2\}^d$ that contains no element of A will remain uninfected forever (cf. the discussion of 'subcritical update rules' in Sections 7 and 9, below).

Since $p_c(\mathbb{Z}_n^d, r)$ is an increasing function of r, we may assume in the proof of Theorem 3.1 that $r = d$. The key observation is that the d-neighbour process on the side of an infected droplet $D = \{1, \ldots, m\}^d$ is 'dominated' by a $(d-1)$-neighbour process on $\{1, \ldots, m\}^{d-1}$. To be more precise, suppose that D is internally filled, and note that each element of the set $S = \{m+1\} \times \{1, \ldots, m\}^{d-1}$ has exactly one neighbour in D, and thus requires only $d-1$ additional infected neighbours inside S in order to become infected. Hence, if we can obtain a sufficiently strong lower bound on the probability of percolation in the $(d-1)$-neighbour process on $\{1, \ldots, m\}^{d-1}$ for all sufficiently large $m \in \mathbb{N}$, then we will be able to deduce a lower bound on the probability of percolation in the d-neighbour process on $\mathbb{Z}_n^d$. In other words, we will use induction on d.

Let us write $I_r(R) = \{R \subseteq [A \cap R]\}$ for the event that a set of vertices $R \subseteq \mathbb{Z}_n^d$ is internally filled in the r-neighbour process on $\mathbb{Z}_n^d$. The induction hypothesis is as follows.

Lemma 3.2 *For each $p > 0$ and $d \geqslant 1$, there exists $\varepsilon = \varepsilon(d, p) > 0$ such that the following holds. If $R = \{1, \ldots, m\}^d$, then*

$$\mathbb{P}_p\big(I_d(R)\big) \geqslant 1 - e^{-\varepsilon m}$$

for all sufficiently large $m \in \mathbb{N}$.

Proof When $d = 1$ the result holds with $\varepsilon(1, p) = p$, since R is internally filled with probability $1 - (1-p)^m$. So let $d \geqslant 2$ and assume that the claim holds for all smaller values of d. The first step is to show that

$$\mathbb{P}_p\big(I_d(R)\big) \to 1$$

as $m \to \infty$. To do so, as in the proof of Theorem 2.1, we consider two independent $(p/2)$-random subsets A_1 and A_2 of R, find a large droplet $D \subseteq A_1$, and use the induction hypothesis to show that $[D \cup A_2] = R$ with high probability. Indeed, if $D = \{1, \ldots, m_0\}^d$ for some sufficiently large m_0 then, by the induction hypothesis, the probability that it fails to grow one step in each direction decreases exponentially in its side-length. Thus

$$\mathbb{P}\big([D \cup A_2] = R\big) \geqslant 1 - 2d \sum_{k=m_0}^{m-1} e^{-\varepsilon k} \to 1$$

as $m_0 \to \infty$, as claimed.

Now, let $\delta > 0$ be sufficiently small, and fix k such that if $S = \{1, \ldots, k\}^d$ then $\mathbb{P}\big(I_d(S)\big) \geqslant 1 - \delta$. Now, by tiling R with copies of S, it follows from standard results in percolation theory that with probability at least $1 - e^{-\varepsilon m}$, all components of $R \setminus [A \cap R]$ have diameter at most $m/4$. However, it is easy to see that any such component must in fact be empty, and so the lemma (and hence also the theorem) follows. $\qquad \square$

The proof above can be modified to give the following upper bound:

$$p_c(\mathbb{Z}_n^d, r) \leqslant \left(\frac{\Theta(1)}{\log_{(r-1)} n} \right)^{d-r+1}, \qquad (3.1)$$

where $\log_{(r)} n = \log \left(\log_{(r-1)} n \right)$ denotes an r-times iterated logarithm. In a major breakthrough, a matching lower bound was obtained by Cerf and Cirillo [19] (in the case $d = r = 3$), and by Cerf and Manzo [20] (for all $d \geqslant r \geqslant 3$). As noted above, the case $r = 2$ was resolved by Aizenman and Lebowitz [1], while the case $r = 1$ is trivial.

Theorem 3.3 *If $d \geqslant r \geqslant 1$, then*

$$p_c(\mathbb{Z}_n^d, r) = \left(\frac{\Theta(1)}{\log_{(r-1)} n} \right)^{d-r+1}.$$

The basic idea behind the proof of Theorem 3.3 is as follows: there must exist an internally filled component of diameter about $\log n$, and this implies that a box[5] of side-length about $\log n$ must be 'internally spanned'. To be precise, we make the following important definition.

Definition 3.4 A box $R \subseteq \mathbb{Z}_n^d$ is *internally spanned* by A if there exists a connected set $S \subseteq [A \cap R]$ such that R is the smallest box containing S.

Note that if S is a connected subset of $\mathbb{Z}_n^d$ such that $[A \cap S] = S$, then the smallest box containing S is internally spanned by A. This simple observation allows one to prove the following analogue of the Aizenman-Lebowitz lemma (Lemma 2.3) when $r \geqslant 3$. Let us write $\mathrm{diam}(S)$ for the L^∞-diameter of a connected set $S \subseteq \mathbb{Z}_n^d$, i.e., the maximum side-length of the smallest box containing S.

Lemma 3.5 *Let R be a box in $\mathbb{Z}_n^d$, and suppose that R is internally spanned. For every $1 \leqslant k \leqslant \mathrm{diam}(R)$, there exists an internally spanned box $Q \subseteq R$ with $k \leqslant \mathrm{diam}(Q) \leqslant 2k$.*

To prove this lemma, we replace the rectangles process of Section 2 by the following "components process". Since we will later need to use this process in other settings, we define it on a general (finite) graph G.

---

[5]That is, a set of the form $[a_1, b_1] \times \cdots \times [a_d, b_d]$.

Definition 3.6 (The components process) Given a finite graph G, let $A = \{x_1, \ldots, x_m\} \subseteq V(G)$, and set $\mathcal{S} := \{S_1, \ldots, S_m\}$, where $S_j = \{x_j\}$ for each $j \in [m]$. Now repeat the following steps until STOP:

1. If there exists a family of $2 \leqslant s \leqslant r$ sets $\{S_{i_1}, \ldots, S_{i_s}\} \subseteq \mathcal{S}$ such that[6]
$$[S_{i_1} \cup \cdots \cup S_{i_s}]$$
   is connected in G, then choose a minimal such family, remove them from $\mathcal{S}$, and replace them by $[S_{i_1} \cup \cdots \cup S_{i_s}]$.

2. If there does not exist such a family of sets in $\mathcal{S}$, then STOP.

Observe that all members of $\mathcal{S}$ (at all points of the process) are closed under the (r-neighbour) bootstrap process, and that
$$A \subseteq V(\mathcal{S}) := \bigcup_{S \in \mathcal{S}} S \subseteq [A].$$

Therefore, if $V(\mathcal{S})$ is not equal to $[A]$ then it is not closed, i.e., there must exist a vertex v with at least r neighbours in $V(\mathcal{S})$. Since the closure of (the union of) the corresponding elements of $\mathcal{S}$ is connected (via v), it follows that the process does not stop until $V(\mathcal{S}) = [A]$.

Now, for the graph $G = \mathbb{Z}_n^d$, it is moreover straightforward to show that if $[S_{i_1} \cup \cdots \cup S_{i_s}]$ is connected (and minimal) then
$$\text{diam}\big([S_{i_1} \cup \cdots \cup S_{i_s}]\big) \leqslant 2 \cdot \max_{i \in [s]} \big\{\text{diam}(S_i)\big\} + 2,$$

since boxes are closed under the r-neighbour bootstrap process on $\mathbb{Z}_n^d$. Lemma 3.5 now follows easily (cf. the proof of Lemma 2.3), so we leave the details as an exercise for the reader.

Exercise: Write down a complete proof of Lemma 3.5.

In order to prove Theorem 3.3, it remains to bound the probability that a box of side-length k is internally spanned by a p-random set A. For simplicity, we will only sketch the proof in the case $d = r = 3$ (the general case is similar, but the details are somewhat more complicated). Let us denote by $I^\times(R)$ the event that R is internally spanned by A.

Lemma 3.7 There exists $c > 0$ such that the following holds if
$$p \leqslant \frac{c}{\log \log n}.$$

---

[6]The closure here refers to the r-neighbour bootstrap process on G.

Fix $\delta > 0$, and let $n \in \mathbb{N}$ be sufficiently large. If $R \subseteq \mathbb{Z}_n^3$ is a box of diameter $k \leqslant \log n$, then

$$\mathbb{P}_p\big(I^\times(R)\big) \leqslant \delta^k.$$

Proof The key idea is to partition R into $k \times k \times 2$ 'slices', where we assume (for simplicity) that $R = \{1, \dots, k\}^3$. Each vertex has at most one neighbour in a different slice, and we may therefore couple the 3-neighbour process on R with $k/2$ independent 2-neighbour processes on the slices. Since

$$p \leqslant \frac{c}{\log \log n} \leqslant \frac{c}{\log k},$$

it follows from the proof of Aizenman and Lebowitz (see Section 2) that the closure in a given slice contains a component of size larger than $\log k$ with probability at most $1/k$. Moreover, the same method implies that the expected number of vertices v that are contained in the same internally filled component of a given vertex u (only counting those components of diameter at most $\log k$) is $o(1)$. More precisely, if we define, for vertices u and v in a slice X,

$u \sim v \;\Leftrightarrow\;$ there exists a component $Y \subseteq X$, with $u, v \in Y$ and

$\qquad\qquad$ $\mathrm{diam}(Y) \leqslant \log k$, such that Y is internally spanned

$\qquad\qquad$ by A in the 2-neighbour bootstrap process on X,

then the expected size of $\{v \in X : u \sim v\}$ is $o(1)$ for each $v \in X$.

Now, in order to bound the probability that R is internally spanned, we first condition on the subset Z of slices with a component of size larger than $\log k$. We then count the expected number of 'minimal' paths between each consecutive pair in Z, where each step of the path corresponds either to moving between vertices u and v in the same slice with $u \sim v$, or between neighbouring vertices in adjacent slices. By minimality, different steps in the same slice occur disjointly, and so we may use the van den Berg–Kesten inequality (Lemma 2.11) to bound the probability of each such path. The result now follows by some simple counting. $\qquad\square$

For the details, we refer the reader to the original papers [19, 20], or to the more recent applications of the Cerf–Cirillo method in [40, 8, 6].

Exercise: Expand the sketch proof above into a full proof of Lemma 3.7.

It is now easy to deduce Theorem 3.3 (in the case $d = r = 3$) from Lemmas 3.5 and 3.7, cf. the proof of Theorem 2.2.

Proof of Theorem 3.3 The proof of the upper bound was outlined above, so we will prove only the lower bound. If $[A] = \mathbb{Z}_n^3$, then by Lemma 3.5 there exists a box $Q \subseteq \mathbb{Z}_n^3$, with

$$\frac{\log n}{2} \leqslant \operatorname{diam}(Q) \leqslant \log n$$

that is internally spanned by A. By Lemma 3.7, the expected number of such boxes is

$$n^{3+o(1)} \delta^{\log n/2} \to 0$$

as $n \to \infty$, if $\delta = e^{-8}$, say. Hence

$$p_c(\mathbb{Z}_n^3, 3) \geqslant \frac{c}{\log \log n},$$

where $c > 0$ is the constant in Lemma 3.7, as required. □

3.1 A sharp threshold for the r-neighbour process on $\mathbb{Z}_n^d$

Despite the breakthroughs of Cerf and Cirillo [19] and Holroyd [39], the problem of determining a sharp threshold for $p_c(\mathbb{Z}_n^d, r)$ remained open until a few years ago, when it was finally resolved by Balogh, Bollobás, Duminil-Copin and Morris [6, 8].

Theorem 3.8 *For every $d \geqslant r \geqslant 2$, there exists a constant $\lambda(d, r) > 0$ such that*

$$p_c(\mathbb{Z}_n^d, r) = \left(\frac{\lambda(d, r) + o(1)}{\log_{(r-1)} n} \right)^{d-r+1}$$

as $n \to \infty$.

The constant $\lambda(d, r)$ is defined as follows. For each $k \in \mathbb{N}$, let[7]

$$\beta_k(u) = \frac{1}{2} \left(1 - (1 - u)^k + \left(1 + (4u - 2)(1 - u)^k + (1 - u)^{2k} \right)^{1/2} \right),$$

and set $g_k(z) = -\log \left(\beta_k(1 - e^{-z}) \right)$. Then

$$\lambda(d, r) = \int_0^\infty g_{r-1}(z^{d-r+1}) \, dz.$$

This definition, while a little complicated, has the following nice properties:

$$\lambda(d, 2) = \frac{d-1}{2} + o(1) \qquad \text{and} \qquad \lambda(d, d) = \left(\frac{\pi^2}{6} + o(1) \right) \frac{1}{d}$$

---

[7]Note that $\beta_k(u)^2 = (1 - (1 - u)^k) \beta_k(u) + u(1 - u)^k$.

as $d \to \infty$.

The proof of Theorem 3.8 is complicated, but the basic idea (at least when $d = r = 3$) is quite simple: one would like to apply the Cerf–Cirillo method to slices whose width is a large constant C, instead of two. For any fixed C, the two sides of the slice (which are assumed to get help from the adjacent slices) will be able to cooperate enough to affect the constant, but as $C \to \infty$ the effect of this cooperation becomes arbitrarily small. The main technical difficulty is then to extend Holroyd's method to the (significantly) more complicated context of a slice; we remark that proving an analogue of Lemma 2.15 is particularly challenging, and the tools that were developed in [6, 8] in order to resolve this problem have proven extremely useful in other contexts, see Section 8. For the details of the proof, we refer the interested reader to [6].

4   Bootstrap in high dimensions

In Sections 2 and 3 we studied the r-neighbour bootstrap process on $\mathbb{Z}_n^d$ with d and r fixed, and n sufficiently large. It is natural to ask how large n needs to be for these results to hold, and (similarly) to ask for bounds on $p_c(\mathbb{Z}_n^d, r)$ when n and r are allowed to be arbitrary functions of $d \to \infty$. This question was first asked by Balogh and Bollobás [5], who determined $p_c(Q_d, 2)$ up to a constant factor, where $Q_d = \mathbb{Z}_2^d$ is the d-dimensional hypercube. The sharp threshold was determined several years later by Balogh, Bollobás and Morris [9], who proved that

$$\left(1 + \frac{\log d}{\sqrt{d}}\right) \frac{16\lambda}{d^2} 2^{-2\sqrt{d}} \leqslant p_c(Q_d, 2) \leqslant \left(1 + \frac{5(\log d)^2}{\sqrt{d}}\right) \frac{16\lambda}{d^2} 2^{-2\sqrt{d}}$$

for all sufficiently large $d \in \mathbb{N}$, where $\lambda \approx 1.17$ is the smallest positive root of the equation

$$\sum_{k=0}^{\infty} \frac{(-1)^k \lambda^k}{2^{k^2 - k} k!} = 0.$$

They also obtained the following sharp threshold whenever $d \gg \log n \gg 1$.

Theorem 4.1 *If $d \gg \log n \gg 1$, then*

$$p_c(\mathbb{Z}_n^d, 2) = \frac{4\lambda + o(1)}{d^2} 2^{-\sqrt{d \log_2 n}}$$

as $d \to \infty$.

The proof of Theorem 4.1 is based on a very precise enumeration of the number of close to minimum-size percolating sets in a subcube of 'critical' size. This is proved using induction on the dimension, using (a higher dimensional analogue of) the rectangles process. Since the analysis is rather involved, we refer the interested reader to [9] for the details.

Theorems 3.8 and 4.1 determine $p_c(\mathbb{Z}_n^d, 2)$ up to a factor of $1 + o(1)$ when $d = O(1)$ and $d \gg \log n$, but no bounds are known for intermediate values of d.

Open problem: Determine $p_c(\mathbb{Z}_n^d, 2)$ when $1 \ll d = O(\log n)$.

It seems likely that the method described in Section 2 could be extended to give reasonable bounds for a relatively slowly-growing function (perhaps $d = \log \log n$), but new ideas are likely to be necessary in order to deal with larger d, and in particular the range $d = \Theta(\log n)$.

As when $d = O(1)$, the r-neighbour bootstrap process becomes much more complicated when $r \geqslant 3$, and very little is known about $p_c(Q_d, r)$ when $r \geqslant 3$ is fixed as $d \to \infty$. In an important breakthrough, however, the *extremal* problem was recently resolved by Morrison and Noel [48]. To be precise, given a graph G and $r \in \mathbb{N}$, define

$$m(G, r) = \min \big\{ |A| : A \subseteq V(G) \text{ percolates in the } r\text{-neighbour process} \big\}.$$

Morrison and Noel proved the following theorem, which confirms (in a strong form) a conjecture of Balogh and Bollobás [5].

Theorem 4.2 *For each $r \in \mathbb{N}$,*

$$m(Q_d, r) = \frac{d^{r-1}}{r!} + \Theta(d^{r-2}).$$

This breakthrough provides hope that it might be possible to make progress on the following conjecture of Balogh, Bollobás and Morris [9].

Conjecture 4.3 *For each $r \in \mathbb{N}$,*

$$p_c(Q_d, r) = \exp\Big(- \Theta\big(d^{1/2^{r-1}}\big)\Big).$$

To motivate this conjecture, let us give a brief sketch of the proof of the upper bound, which proceeds by considering a growing subcube of infected vertices. The key lemma states that

$$\mathbb{P}_p\big(I_r(Q_\ell)\big) \geqslant \exp\Big(- \big(\log 1/p\big)^{2^{r-1}}\Big)$$

for every $\ell \in \mathbb{N}$, and the proof proceeds by induction on r, noting that if Q_ℓ is already infected, then on a neighbouring subcube we may couple with an $(r-1)$-neighbour process. After roughly $(\log 1/p)^{2^{r-2}}$ steps, each of which adds one dimension to the growing cube, we pass the 'bottleneck' and growth becomes easier. The proof now proceeds as usual: we find a large internally filled subcube somewhere in Q_d, and (using sprinkling) show it is likely to grow to infect the entire hypercube.

4.1 Large d and r

For the d-neighbour model, Balogh, Bollobás and Morris [7] proved the following theorem.

Theorem 4.4 *If $d \gg (\log \log n)^2 \log \log \log n$, then*

$$p_c(\mathbb{Z}_n^d, d) \to \frac{1}{2}$$

as $d \to \infty$.

The proof of Theorem 4.4 is completely different from any of those described above. Note that if $p = 1/2 + \varepsilon$ then almost all vertices are infected in the first step, whereas if $p = 1/2 - \varepsilon$ then almost no vertices are infected. For the upper bound, the idea is to show that if a vertex v is not infected after $k \approx \sqrt{d/\log d}$ steps, then there exists a set X of about $(cd/k)^k$ vertices (for some constant $c > 0$) at distance k from v, none of which was infected in the first step. Since for vertices $x, y \in X$ at distance at least three, the events $x \notin A_1$ and $y \notin A_1$ are independent, it is straightforward to bound the probability that such a set exists, and hence the expected number of vertices that are not in A_k.

The proof of the lower bound is more interesting (and the key idea will also be needed in the next section). The main difficulty is that the process can continue for a long time, and this creates long-distance dependencies that complicate the calculations. The solution is to introduce a modified 'more generous' process that infects vertices with slightly fewer than d infected neighbours in the first few steps: if a vertex is not infected during the early 'generous' part of the algorithm, it needs to gain many new infected neighbours in order to be infected later. This allows one to show that, if a vertex v is infected after k steps of the generous process, there must be a set of about $(c'd/k)^k$ vertices at distance k from v, all of which were infected in the first step. See [7, Section 6] for the details.

Combining Theorems 3.3 and 4.4, it follows that (roughly speaking):

$$p_c(\mathbb{Z}_n^d, d) \to \begin{cases} 1/2 & \text{if } n \ll 2^{2^{\sqrt{d}}} \\ 0 & \text{if } n \gg 2^{2^{\cdot^{\cdot^{2}}}}, \text{ a tower of 2s of height } d. \end{cases}$$

It would be very interesting to improve these bounds, and it seems likely that a significant breakthrough will be necessary in order to describe the transition between the two regimes.

Open problem: Determine $p_c(\mathbb{Z}_n^d, d)$ in the range $2^{2^{\sqrt{d}}} \leqslant n \leqslant 2^{2^{\cdot^{\cdot^2}}}$.

The lower bound seems 'softer' to us, and we expect the threshold to lie closer to the upper bound. More precisely, we suspect that the function

$$f(d) := \min \left\{ n \in \mathbb{N} : p_c(\mathbb{Z}_n^d, d) \leqslant 1/4 \right\}$$

grows like a tower of 2s of height $\Theta(d)$.

5   The Glauber dynamics of the Ising model on $\mathbb{Z}^d$

Before introducing the general model that is the main topic of this survey, let us conclude our gentle introduction to the area by describing an application of the results presented above to the Ising model on $\mathbb{Z}^d$. This is an extremely well-studied model of ferromagnetism (see, for example, [43, 52] and the references therein), and provided the original motivation for the introduction of bootstrap percolation in [23].

Definition 5.1 The zero-temperature Glauber dynamics of the Ising model are defined as follows:

(a) At each time $t \geqslant 0$, the system is in a state $\sigma(t) \in \{+, -\}^{\mathbb{Z}^d}$.

(b) Each vertex $x \in \mathbb{Z}^d$ has an independent exponential clock, which rings randomly at rate 1.

(c) If the clock at vertex x rings at time t, then its state $\sigma_x(t) \in \{+, -\}$ resets according to the following rule:

   – if x has at least $d + 1$ neighbours in state $+$, then $\sigma_x(t) := +$.

   – if x has at least $d + 1$ neighbours in state $-$, then $\sigma_x(t) := -$.

   – if x has exactly d neighbours in each state, then $\sigma_x(t)$ is chosen uniformly at random, independently of all other events.

For the sake of brevity, in this section we will refer to this process as simply 'the Glauber dynamics on $\mathbb{Z}^d$'.

We are interested in the long-term behaviour of the dynamics, starting from a randomly chosen initial state. In particular, does the system 'fixate', or do some vertices change state infinitely many times?

In order to state this question more precisely, let us write $\mathbb{P}_p^+$ for the probability distribution obtained by setting $\sigma_x(0) = +$ independently at random with probability p for each $x \in \mathbb{Z}^d$. We say that a vertex $x \in \mathbb{Z}^d$ *fixates* if its state changes only finitely many times, and we say that the dynamics fixates if every vertex fixates. (If the state of every vertex is eventually $+$, then we moreover say that the dynamics *fixates at* $+$.) The critical probability for the zero-temperature Glauber dynamics of the Ising model on $\mathbb{Z}^d$ is then defined as follows:

$$p_c^{\mathrm{Is}}(\mathbb{Z}^d) := \inf \left\{ p \, : \, \mathbb{P}_p^+ \big(\text{the dynamics on } \mathbb{Z}^d \text{ fixates at } +\big) = 1 \right\}.$$

In one dimension, the critical probability was determined by Arratia [2] in 1983, who proved (a significant generalisation of) a conjecture of Erdős and Ney [32] about annihilating random walks on $\mathbb{Z}$. His result implies that, for every $p \in (0, 1)$, every vertex almost surely changes state infinitely many times in the dynamics on $\mathbb{Z}$, and hence that $p_c^{\mathrm{Is}}(\mathbb{Z}) = 1$.

In two or more dimensions, on the other hand, the problem of determining $p_c^{\mathrm{Is}}(\mathbb{Z}^d)$ is wide open. The following (possibly folklore) conjecture is the main topic of this section.

Conjecture 5.2
$$p_c^{\mathrm{Is}}(\mathbb{Z}^d) = \frac{1}{2}$$

for every $d \geqslant 2$.

An important breakthrough on Conjecture 5.2 was achieved by Fontes, Schonmann and Sidoravicius [34], who proved the following theorem.

Theorem 5.3
$$p_c^{\mathrm{Is}}(\mathbb{Z}^d) < 1$$

for every $d \geqslant 2$.

The proof of Theorem 5.3 (which we will sketch below) uses some of the results of Aizenman and Lebowitz [1] described in Section 2. More recently, Morris [45] used the main result of [34], together with the techniques introduced by Balogh, Bollobás and Morris [7] in their work on majority bootstrap percolation on the hypercube, in order to prove the following strengthening in high dimensions.

Theorem 5.4
$$p_c^{\mathrm{Is}}(\mathbb{Z}^d) \to \frac{1}{2}$$

as $d \to \infty$.

The proof of Fontes, Schonmann and Sidoravicius [34] uses multi-scale analysis, which (roughly speaking) means using induction to control the process inside an increasingly-large sequence of boxes. Indeed, consider two scales, n and N, with $N = n^C$ for some large constant C, and tile $\mathbb{Z}^d$ with copies of $[n]^d$ and $[N]^d$. Suppose that we have already found a coupling of the original dynamics with a dynamics that is 'more generous'[8] to $-$, such that, in the coupled dynamics:

(a) Each tile (i.e., copy of $[n]^d$ in our tiling) is entirely $+$ at time t with probability at least $1 - q$, for suitable t and q.

(b) For non-adjacent tiles (i.e., copies of $[n]^d$ in our tiling that do not share a corner), the events described in (a) are independent.

We then aim to find a similar coupling at scale N, except replacing t and q by some (much larger) t' and (much smaller) q'. The key observation is that the evolution of the $-$ spins in $[N]^d$ can be coupled with two-neighbour bootstrap percolation on $[N/n]^d$, and if $q \ll \left(\log(N/n) \right)^{-d+1}$, then, by the results of Aizenman and Lebowitz [1] (cf. the proof of Theorem 2.2), the closure of this bootstrap process contains a component (of $[n]^d$-tiles) of size larger than $\log n$ with probability at most $1/n$. Now, each such component is surrounded (and hence attacked) by $+$ vertices, and it can be shown that the probability that it survives until time n^{2d} (say) is extremely small. Moreover, since information only propagates at rate 1, it is extremely unlikely that non-adjacent $[N]^d$-tiles interact before this time, if $N \geqslant n^{3d}$ (say). Using these facts, it is straightforward to construct the required coupling, with $t' = t + n^{2d}$ and $q' = 1/n$.

The proof of Theorem 5.4 is similar, except we replace the first step of the argument above (which holds trivially if p is sufficiently close to 1) by a more careful analysis, assuming that the dimension d is sufficiently large. More precisely, we tile $\mathbb{Z}^d$ with copies of $[n]^d$, where $n = 2^d$, and couple the dynamics in each tile with a more generous process which satisfies conditions (a) and (b) above. The coupling is constructed in several steps, but the key idea is to run the more generous bootstrap process used to prove Theorem 4.4 for a bounded number of steps on the set of vertices in $[n]^d$ that were initially in state $-$. With (extremely) high probability this process reaches a set X that is closed under majority bootstrap percolation, and (crucially) there are no long-range dependencies between the events $\{x \in X\}$. This allows one to couple the state at time d with a distribution in which each vertex is in state $-$ with super-polynomially

---

[8] That is, at every time t, the set of vertices in state $-$ in the original dynamics is contained in the the the set of vertices in state $-$ in the coupled dynamics.

small probability (in d), and vertices at distance at least 20 have independent states. A similar idea then allows one to turn the remaining $-$ vertices to $+$ by time d^5 (with extremely high probability), and since 2^d is much larger than d^5, it is extremely unlikely that non-adjacent $[n]^d$-tiles interact before this time. Inserting this argument into the proof of Fontes, Schonmann and Sidoravicius, we obtain Theorem 5.4.

5.1   The symmetric setting, $p = 1/2$

Recall that Arratia [2] proved that on $\mathbb{Z}$, every vertex almost surely changes state infinitely many times for every $p \in (0, 1)$. Nanda, Newman and Stein [50] proved that this is also the case on $\mathbb{Z}^2$ when $p = 1/2$.

Theorem 5.5 *If $p = 1/2$ then, in Glauber dynamics on $\mathbb{Z}^2$, almost surely every vertex changes state an infinite number of times.*

Proof It follows by a standard ergodic theory argument that a given row or column contains a site which fixates at $+$ with probability either 0 or 1. If this probability is zero then we are done (by symmetry between $+$ and $-$, and between rows and columns), so let us assume that it occurs almost surely. This implies that there exists an axis-parallel rectangle R such that three of its corners fixate in alternating states. In other words, there exist x, y and z, with x and y in the same column, and x and z in the same row, such that x fixates at $+$, whereas y and z fixate at $-$.

Now, at any time t after x, y and z have fixated, the boundary (in the dual lattice) between $+$ and $-$ must connect two adjacent sides of R, either side of one of x, y or z. But then that vertex will change state in $[t, t+1]$ with positive probability, which is a contradiction.   □

The problem appears to be much harder in higher dimensions.

Open problem: If $p = 1/2$ and $d \geqslant 3$ then, in Glauber dynamics on $\mathbb{Z}^d$, does every vertex almost surely change state an infinite number of times?

6   A few other specific bootstrap processes on $\mathbb{Z}^2$

Before defining the general model of Bollobás, Smith and Uzzell [17], let us provide a gentle warm-up (and some motivation) by considering a few specific bootstrap-like processes in two dimensions. The only difference between the processes discussed below and the two-neighbour model discussed in Section 2 will be the *update rule*: in each case, we need to define under what conditions a vertex is infected in step $t + 1$.

6.1 Modified bootstrap percolation

The simplest variant of the two-neighbour model is a process known as *modified bootstrap percolation*, whose update rule is as follows: a vertex x is infected once it has at least one already-infected horizontal neighbour, and at least one already-infected vertical neighbour. In other words, $x \in A_{t+1}$ if either $x \in A_t$, or $x + X \subseteq A_t$ for some $X \in \mathcal{M}^{(2)}$, where

$$\mathcal{M}^{(2)} = \Big\{ \{(0,1),(1,0)\}, \{(1,0),(0,-1)\}, \{(0,-1),(-1,0)\}, \{(-1,0),(0,1)\} \Big\}.$$

In this model the closed sets are still unions of rectangles, and the proofs of Aizenman and Lebowitz [1] and Holroyd [39] go through with only minor modifications. In fact, the natural generalization of the modified model to $\mathbb{Z}_n^d$ (in which a vertex is infected if it has at least one infected neighbour in each dimension) is significantly simpler to study than the d-neighbour model when $d \geqslant 3$. Indeed, Holroyd [40] was able to determine the sharp threshold in d dimensions, some years before the corresponding result was obtained for the r-neighbour model, by combining the technique of [39] with the method of Cerf and Cirillo [19].

Theorem 6.1

$$p_c\big(\mathbb{Z}_n^d, \mathcal{M}^{(d)}\big) = \left(\frac{\pi^2}{6} + o(1) \right) \frac{1}{\log_{(d-1)} n}.$$

Note that here we write $p_c\big(\mathbb{Z}_n^d, \mathcal{M}^{(d)}\big)$ for the critical probability of the modified model in d dimensions, see Definition 7.1, below.[9]

Exercise: Verify that the proof of Theorem 2.2 also implies that

$$p_c\big(\mathbb{Z}_n^2, \mathcal{M}^{(2)}\big) = \frac{\Theta(1)}{\log n}.$$

6.2 Anisotropic bootstrap percolation

A significantly more challenging variant, which exhibits rather different behaviour, is the following so-called *anisotropic* model, which was first studied by Gravner and Griffeath [37]: The update rule in this setting is as follows: a vertex x is infected once there are at least three infected vertices in the set

$$x + \big\{ (-2,0), (-1,0), (0,1), (0,-1), (1,0), (2,0) \big\}.$$

---

[9]To be precise, $\mathcal{M}^{(d)}$ denotes the family of sets that contain exactly one element of the pair $\{e_i, -e_i\}$ for each $i \in [d]$, and no other elements, where $e_i = (0, \ldots, 0, 1, 0, \ldots, 0)$ is the ith basis vector.

The threshold for this model was determined by van Enter and Hulshof [31], and the sharp threshold by Duminil-Copin and van Enter [28], who proved the following theorem.

Theorem 6.2

$$p_c(\mathbb{Z}_n^2, \mathcal{A}) = \left(\frac{1}{12} + o(1)\right) \frac{(\log\log n)^2}{\log n}.$$

Here we write $p_c(\mathbb{Z}_n^2, \mathcal{A})$ for the critical probability in the anisotropic model.[10] Note that in the anisotropic model the critical probability has increased by a factor of $(\log\log n)^2$ (compared with the two-neighbour model); this is because the typical growth of a droplet in the anisotropic model is (in a certain sense) *asymptotically one-dimensional*, unlike in the case of the two-neighbour model. This is an important point, so let us spend a little time discussing the proof of the following weaker result which, as mentioned above, was proved by van Enter and Hulshof [31]:

$$p_c(\mathbb{Z}_n^2, \mathcal{A}) = \Theta\left(\frac{(\log\log n)^2}{\log n}\right). \tag{6.1}$$

Let us begin with the upper bound. Let R_0 be a rectangle of height $a = p^{-1}\log(1/p)$ and width $b = p^{-2}$, and note that if A contains the two left-most columns of R_0, and at least one vertex in every column, then it is internally filled by A. This has probability at least

$$p^{2a}\left(1 - (1-p)^a\right)^b \geqslant p^{2a}(1-p)^b \geqslant p^{2a}e^{-2/p}.$$

Next, let $R_1 \supseteq R_0$ be a rectangle of height $5a$ and width b, and note that if A contains two adjacent vertices of each row of R_1, then $[R_0 \cup (A \cap R_1)] = R_1$. This has probability at least

$$\left(1 - (1-p^2)^{b/2}\right)^{5a} \geqslant e^{-10a}.$$

Now, let $R_2 \supseteq R_1$ be a rectangle of height $5a$ and width b^2, and note that if A contains a vertex of each column of R_2, then $[R_1 \cup (A \cap R_2)] = R_2$. This has probability at least

$$\left(1 - (1-p)^{5a}\right)^{b^2} \geqslant \left(1 - p^5\right)^{b^2} \geqslant \frac{1}{2}.$$

It follows that if

$$p \geqslant \frac{2(\log\log n)^2}{\log n}$$

---

[10]This matches the notation introduced in the next section if $\mathcal{A}$ denotes the collection of subsets of $\{(-2,0), (-1,0), (0,1), (0,-1), (1,0), (2,0)\}$ of size three.

then with high probability there exist at least

$$\frac{n^2}{5ab^2} \cdot p^{2a} e^{-2/p-10a-1} \geqslant n^2 \exp\left(-\frac{3}{p}\left(\log \frac{1}{p} \right)^2 \right) \gg 1$$

internally filled translates of R_2 in $\mathbb{Z}_n^2$. Finally, if every row of $b^2 = p^{-4}$ consecutive vertices contains at least two adjacent elements of A, and every column of $\mathbb{Z}_n^2$ intersects A, then any internally filled translate of R_2 grows to infect the whole of $\mathbb{Z}_n^2$. Since both of these events occur with high probability, this proves that $p_c(\mathbb{Z}_n^2, A) = O((\log \log n)^2 / \log n)$.

The alert reader will have noticed several points in the above argument that were not sharp, and indeed a more careful (and somewhat more complicated) proof along the same lines gives the upper bound in Theorem 6.2; we leave the details as an exercise for the reader.

Exercise: By modifying the argument above, show that

$$p_c(\mathbb{Z}_n^2, A) \leqslant \left(\frac{1}{12} + o(1) \right) \frac{(\log \log n)^2}{\log n}.$$

To prove the lower bound in (6.1), we apply the components process (see Definition 3.6) with G the graph on $\mathbb{Z}_n^2$ with edge set

$$E(G) = \big\{ xy : |x_1 - y_1| + 2 \cdot |x_2 - y_2| \leqslant 2 \big\}.$$

Using this process, one can show that if A percolates, then there must exist an internally spanned[11] rectangle R with either

(a) height at most $\delta p^{-1} \log(1/p)$ and width at least $p^{-3/2}$, or

(b) height at least $\delta p^{-1} \log(1/p)$ and width at most $3p^{-3/2}$,

for some (small) $\delta > 0$. (Indeed, simply run the components process until we find an internally spanned rectangle with width between $p^{-3/2}$ and $3p^{-3/2}$.) Now, if R is internally spanned, then every three consecutive columns of R must contain at least one element of A, and every pair of consecutive rows of R must contain at least two elements of A that are within distance two in G. Thus, setting $a = \delta p^{-1} \log(1/p)$ and $b = p^{-3/2}$, the probability that R is internally spanned is at most

$$\max \left\{ (1 - (1-p)^{3a})^{b/3}, (30p^2 b)^{a/2} \right\} \leqslant \exp\left(-\frac{\delta}{5p}\left(\log \frac{1}{p} \right)^2 \right).$$

---

[11]For the anisotropic model, we say that a box $R \subseteq \mathbb{Z}_n^d$ is *internally spanned* by A if there exists a set $S \subseteq [A \cap R]$ that is connected in the graph G, such that R is the smallest box containing S, cf. Definition 3.4.

Since there are at most n^3 rectangles R in $\mathbb{Z}_n^2$ that satisfy either (a) or (b) above, it now follows, by Markov's inequality, that A percolates with probability at most

$$n^3 \exp\left(-\frac{\delta}{5p}\left(\log\frac{1}{p}\right)^2 \right) \leqslant \frac{1}{n}$$

if $p \leqslant \delta^2(\log\log n)^2/\log n$, which implies the lower bound in (6.1).

The lower bound in Theorem 6.2 is significantly more difficult, and follows by adapting the technique of Holroyd [39] to the anisotropic setting. We refer the interested reader to [28, 31] for the details.

Finally, we note that the threshold for a class of anisotropic models in three dimensions was determined by van Enter and Fey [30]. More precisely, they considered the r-neighbour process with (a, b, c)-neighbourhood

$$\{(x,0,0) : \pm x \in [a]\} \cup \{(0,y,0) : \pm y \in [b]\} \cup \{(0,0,z) : \pm z \in [c]\}$$

and $r = a + b + c$. They showed that if $a = b \leqslant c$, then

$$p_c\big(\mathbb{Z}_n^3, \mathcal{A}(a,b,c)\big) = \Theta\left(\frac{1}{\log\log n} \right)^{1/a},$$

whereas if $a < b \leqslant c$, then

$$p_c\big(\mathbb{Z}_n^3, \mathcal{A}(a,b,c)\big) = \Theta\left(\frac{(\log\log\log n)^2}{\log\log n} \right)^{1/a},$$

where we write $\mathcal{A}(a,b,c)$ for the update family of the model. It would be interesting to determine a sharp threshold for this class of models, and more generally for all $c + 1 \leqslant r \leqslant a + b + c$.

6.3 The Duarte model

Our final example is the simplest example of a class of models that cause the greatest difficulty in the general setting of the next two sections: those with 'drift'. Its update rule is as follows: a vertex x is infected once there are at least two infected vertices in the set

$$x + \big\{(-1,0), (0,1), (0,-1)\big\}.$$

This model was first introduced by Duarte [26] in 1988, and first studied rigorously a few years later by Mountford [49], who determined the threshold for percolation. The following sharp threshold was determined only very recently, by Bollobás, Duminil-Copin, Morris and Smith [14].

Theorem 6.3

$$p_c(\mathbb{Z}_n^2, \mathcal{D}) = \left(\frac{1}{8} + o(1)\right) \frac{(\log\log n)^2}{\log n}.$$

Here we write $p_c(\mathbb{Z}_n^2, \mathcal{D})$ for the critical probability in the Duarte model.[12] The threshold is therefore the same (up to a constant factor) as in the anisotropic model, and the typical growth of a droplet in the anisotropic model is again asymptotically one-dimensional. However, in the Duarte model there is a significant difference: growth only occurs in one direction ('to the right'), and a typical droplet is not a rectangle, but a non-polygonal convex shape that is taller at the right end than the left.

To illustrate these points, let us briefly discuss how to modify the proof of (6.1) above to deduce the following weaker bounds, which were first proved by Mountford [49]:

$$p_c(\mathbb{Z}_n^2, \mathcal{D}) = \Theta\left(\frac{(\log\log n)^2}{\log n}\right). \tag{6.2}$$

The proof of the upper bound is very similar to that given above, the main difference being that for the droplet to grow vertically, it is not sufficient that A contains two adjacent vertices of each row. Instead, we need to find a sequence of (single) elements of A, one in each pair of consecutive rows, which form an 'increasing' sequence (meaning their x-coordinates are an increasing function of their y-coordinates). We will typically be able to find such a sequence of length roughly p times the current length of the droplet, and we thus obtain a vertical line of this height in $[A]$. This vertical line can then be used to continue growth to the right, which can then be used to grow upwards, and so on.

To be more precise, let R_0 be a rectangle of height $a = 4p^{-1}\log(1/p)$ and width $b = p^{-5}$, and note that if A contains the left-most column of R_0, and at least one vertex in every column, then it is internally filled by A. This has probability at least

$$p^a\left(1 - (1-p)^a\right)^b \geqslant p^a(1 - p^4)^b \geqslant p^a e^{-2/p}.$$

Next, let $R_1 \supseteq R_0$ be a rectangle of height p^{-3} and width b, and note that if A contains an increasing sequence of vertices as described above, then the right-hand side of R_1 is contained in $[R_0 \cup (A \cap R_1)]$. This occurs with high probability, since $pb \gg p^{-3}$. Finally, if every row or columns of p^{-3} consecutive vertices intersects A, then any translate of the right-hand

---

[12]This matches the notation introduced in the next section if $\mathcal{D}$ denotes the collection of subsets of $\{(-1,0),(0,1),(0,-1)\}$ of size two.

side of R_1 grows to infect the whole of $\mathbb{Z}_n^2$. Since this occurs with high probability, is is easy to deduce that $p_c(\mathbb{Z}_n^2, \mathcal{D}) = O((\log \log n)^2 / \log n)$.

Once again, a more careful (and somewhat more complicated) proof along the same lines gives the upper bound in Theorem 6.3; we again leave the details as an exercise for the reader.

Exercise: By modifying the argument above, show that

$$p_c(\mathbb{Z}_n^2, \mathcal{D}) \leqslant \left(\frac{1}{8} + o(1) \right) \frac{(\log \log n)^2}{\log n}.$$

The proof of the lower bound in (6.2) is significantly more complicated than that of the lower bound in (6.1); we will try to explain the new difficulties that arise, and give an idea of how they may be overcome.

The first problem we run into – the fact that the interaction between vertices is no longer symmetric, cf. Definition 3.6 – is easy to overcome in this particular setting: we simply apply a slight modification of the components process in which we take the closure in the Duarte model, but define connectivity according to the graph G on $\mathbb{Z}_n^2$ with edge set

$$E(G) = \{ xy : |x_1 - y_1| \leqslant 1 \text{ and } |x_1 - y_1| + |x_2 - y_2| \leqslant 2 \}.$$

Using this process, one can show (as before) that if A percolates, then there must exist an internally spanned rectangle R with either

(a) height at most $\delta p^{-1} \log(1/p)$ and width at least $p^{-4/3}$, or

(b) height at least $\delta p^{-1} \log(1/p)$ and width at most $3p^{-4/3}$,

for some (small) $\delta > 0$. In the first case we are done, since every column of an internally spanned rectangle must intersect A, and so a rectangle satisfying condition (a) is internally spanned with probability at most

$$\left(1 - (1-p)^a \right)^b \leqslant \exp\left(-p^{-4/3+2\delta} \right),$$

where $a = \delta p^{-1} \log(1/p)$ and $b = p^{-4/3}$. However, it is not so easy to bound the probability that a rectangle R satisfying condition (b) is internally spanned, as even if $A \cap R$ does not contain a large 'increasing' sequence as in the proof of the upper bound, it can still be internally spanned (for example, via the cooperation of many small internally spanned droplets).

This obstacle is overcome in different ways in the papers [13, 14, 49]; we will describe the method of [13], specialised to the setting of the Duarte model, where most of the difficulties encountered for general update rules do not occur. We first partition the droplet into strips of height $p^{-2/3}$, and observe that each of these must either be 'half-crossed' from above, or from below. Next, we partition (most of) each half-strip into constant width strips that are rotated slightly, so that the right-hand end of each is about $p^{-1/2}$ higher than its left-hand end. The crucial observation is now that each constant width strip must either contain a 'cluster' of elements of A, close enough together that they can cooperate, or must intersect a large internally spanned 'saver' droplet, and moreover all of these events occur disjointly. Since sites in a cluster must lie within distance $p^{-1/2}$ of one another, this allows us to prove a bound on the probability that R satisfying condition (b) is internally spanned of the form

$$p^{\delta a} \leqslant \exp\left(-\frac{\delta^2}{p}\left(\log\frac{1}{p}\right)^2 \right),$$

where $a = \delta p^{-1}\log(1/p)$. The claimed lower bound in (6.2) now follows as in Section 6.2.

The proof of the sharp threshold for the Duarte model requires two important additional innovations, which in turn create a large number of additional technical difficulties. These are the 'method of iterated hierarchies', and the use of non-polygonal droplets. The former technique was introduced in [13] in order to determine the threshold for general 'critical' update families (see Sections 7 and 8, below), and involves applying Holroyd's method of hierarchies (see Section 2.1) at various different scales. We deduce sharp bounds at a given scale by using (even sharper) bounds at a smaller scale in order to control the probability that a saver droplet is internally spanned. At sufficiently small scales, it is relatively easy to obtain extremely strong bounds using (a suitably modified version of) the method of Aizenman and Lebowitz, see Section 2. However, one cannot obtain sharp bounds using rectangular droplets; instead, one needs to define a droplet as follows.

Definition 6.4 Given $\varepsilon > 0$ and $p > 0$, a *Duarte region* $D^* \subseteq \mathbb{R}^2$ is a set of the form

$$D^* = (a, b) + \{(x, y) \in \mathbb{R}^2 : 0 \leqslant x \leqslant w, \ |y| \leqslant f(x)\}, \qquad (6.3)$$

for some $a, b, w \in \mathbb{R}$, where $f\colon [0, \infty) \to [0, \infty)$ is the function

$$f(x) := \frac{1}{p}\log\left(1 + \frac{\varepsilon^2 px}{\log 1/p}\right).$$

A *Duarte droplet* $D \subseteq \mathbb{Z}^2$ is the intersection of a Duarte region with $\mathbb{Z}^2$.

With this crucial definition in hand, it is possible (though rather nontrivial) to carry out the proof outlined above; see [14] for the details.

7   General cellular automata: the BSU model

We now make a significant transition: from considering specific update rules one at a time, to dealing with large families of models simultaneously. As a consequence, many of the definitions become somewhat more technical, and the processes can be more difficult to visualise; nevertheless, the approach we take is surprisingly simple and natural.

Let us begin by defining a large class of d-dimensional monotone cellular automata, introduced by Bollobás, Smith and Uzzell [17].

Definition 7.1 (The $\mathcal{U}$-bootstrap process) Let $\mathcal{U} = \{X_1, \ldots, X_m\}$ be an arbitrary finite collection of finite subsets of $\mathbb{Z}^d \setminus \{0\}$. The $\mathcal{U}$-*bootstrap process* on the d-dimensional torus $\mathbb{Z}_n^d$ is defined as follows: given a set $A \subseteq \mathbb{Z}_n^d$ of initially *infected* sites, set $A_0 = A$, and define, for each $t \geqslant 0$,

$$A_{t+1} = A_t \cup \{x \in \mathbb{Z}_n^d : x + X \subseteq A_t \text{ for some } X \in \mathcal{U}\}.$$

The $\mathcal{U}$-bootstrap process on $\mathbb{Z}^d$ is defined in an analogous way.

Let $[A]_\mathcal{U} = \bigcup_{t \geqslant 0} A_t$ denote the *closure* of A under the $\mathcal{U}$-bootstrap process, and define the *critical probability* of the $\mathcal{U}$-bootstrap process on $\mathbb{Z}_n^d$ to be

$$p_c(\mathbb{Z}_n^d, \mathcal{U}) := \inf \left\{ p : \mathbb{P}_p([A]_\mathcal{U} = \mathbb{Z}_n^d) \geqslant 1/2 \right\}.$$

We will refer to $\mathcal{U}$ as the *update family* of the process. Note that each of the processes considered above is a $\mathcal{U}$-bootstrap process for some update family $\mathcal{U}$. For example, the r-neighbour model on $\mathbb{Z}_n^d$ is obtained by setting $\mathcal{U} = \mathcal{N}_r^d$, the family of r-subsets of the $2d$ nearest neighbours of the origin in $\mathbb{Z}^d$, and the Duarte model is obtained by setting $\mathcal{U} = \mathcal{D}$, the collection of subsets of $\{(-1,0), (0,1), (0,-1)\}$ of size two.

One of the key insights of Bollobás, Smith and Uzzell [17] was that the typical global behaviour of a two-dimensional $\mathcal{U}$-bootstrap process acting on random initial sets should be determined by the action of the process on discrete half-spaces. (The same also turns out to be true in higher dimensions, but there the situation is significantly more complicated.) For the rest of this section, let us therefore restrict ourselves to $\mathbb{Z}^2$.

Definition 7.2 (The set of stable directions) For each $u \in S^1$, let

$$\mathbb{H}_u := \left\{ x \in \mathbb{Z}^2 : \langle x, u \rangle < 0 \right\}$$

denote the discrete half-plane whose boundary is perpendicular to u.

We say that u is a *stable direction* if

$$[\mathbb{H}_u]_\mathcal{U} = \mathbb{H}_u.$$

and we denote by $\mathcal{S} = \mathcal{S}(\mathcal{U}) \subseteq S^1$ the collection of stable directions.

It is easy to determine the stable set of an update family: simply remove from S^1 the (possibly empty) open interval of directions that are destabilised by X, for each $X \in \mathcal{U}$. In particular, note that $\mathcal{S}(\mathcal{N}_1^2)$ is empty and $\mathcal{S}(\mathcal{N}_3^2) = S^1$, while $\mathcal{S}(\mathcal{N}_2^2)$ consists of four isolated points, and $\mathcal{S}(\mathcal{D})$ consists of a closed semicircle and an isolated point. We remark that $\mathcal{S}(\mathcal{U})$ always consists of a finite collection of closed intervals, some of which may be isolated points, all of whose endpoints are rational.

Let $\mathcal{C}$ denote the collection of open semicircles in S^1. The following key definition is due to Bollobás, Smith and Uzzell [17].

Definition 7.3 (Critical and super/subcritical update families)
We say that a two-dimensional update family $\mathcal{U}$ is:

- *supercritical* if there exists $C \in \mathcal{C}$ that is disjoint from $\mathcal{S}$,

- *critical* if there exists $C \in \mathcal{C}$ that has finite intersection with $\mathcal{S}$, and if every $C \in \mathcal{C}$ has non-empty intersection with $\mathcal{S}$,

- *subcritical* if every $C \in \mathcal{C}$ has infinite intersection with $\mathcal{S}$.

Note that this is a partition of the two-dimensional update families.

The importance of Definition 7.3 is demonstrated by the following theorem. Parts (a) and (b) were proved by Bollobás, Smith and Uzzell [17], and part (c) was proved by Balister, Bollobás, Przykucki and Smith [4].

Theorem 7.4 *Let $\mathcal{U}$ be a two-dimensional update family.*

(a) *if $\mathcal{U}$ is supercritical then $p_c(\mathbb{Z}_n^2, \mathcal{U}) = n^{-\Theta(1)}$.*

(b) *if $\mathcal{U}$ is critical then $p_c(\mathbb{Z}_n^2, \mathcal{U}) = (\log n)^{-\Theta(1)}$.*

(c) *if $\mathcal{U}$ is subcritical then $\liminf\limits_{n \to \infty} p_c(\mathbb{Z}_n^2, \mathcal{U}) > 0$.*

It is perhaps difficult to convey to the reader how surprising it is that such a simple and beautiful characterization could be proved in such extraordinary generality. The proof of Theorem 7.4 is quite complicated, and so we will give only a brief outline of some of the key ideas.

7.1   Supercritical update families

The only supercritical model we have come across in the sections above is the 1-neighbour process, which is somewhat trivial. This could easily lead one to suspect that all supercritical models are 'easy', but this would be mistaken: they exhibit an *extremely* rich variety of complex behaviours, and even proving the relatively weak bound claimed in Theorem 7.4(a) will require some important new ideas. Nevertheless, this is easiest part of the proof, and will serve as a useful warm-up.

Recall that if $\mathcal{U}$ is supercritical, then there exists an open semicircle C in S^1 that contains no stable direction. Choose such a semicircle with rational endpoints, and let u^+ denote its midpoint. For simplicity we will rotate $\mathbb{Z}_n^2$ so that $u^+ = (1,0)$, i.e., u^+ points to the right. We claim that there exists a constant $k = k(\mathcal{U})$ such that the closure of the continuous square[13] $R = [-k, k]^2$ contains all vertices in the (minimal) horizontal strip containing R. If $p \geqslant n^{-1/2k^2}$ (say), then with high probability A contains such a square in every translate of this strip, so this will be sufficient to prove Theorem 7.4(a).

To prove the claim, it will suffice to show that the column L directly to the right of R lies in the closure of R. Since u^+ is unstable, it follows that there exists a rule $X_1 \in \mathcal{U}$ that is entirely contained in $\mathbb{H}_{u^+}$, and this allows us to infect all but a constant (depending on the size of $|X_1|$) number of vertices of L, which lie at the top and bottom. Infecting the last few vertices is rather more complicated, and requires the entire semicircle C to be unstable; roughly speaking, the idea is as follows. First, using the rule X_1 we can construct a 'pyramid' of infected sites on the right-hand side of R. The sides of this pyramid correspond to the endpoints of the open interval that is destabilised by X_1, and so (since C is entirely unstable) there must exist another rule $X_2 \in \mathcal{U}$ that destabilises the sides of the pyramid. This gives us another pyramid with slightly steeper sides, and by repeating this process sufficiently many times, we eventually infect the whole of L.

In order to make the above sketch precise we will need an important idea from [17], which also turns out to be extremely useful when proving upper bounds for critical update families. Define a set

$$\mathcal{Q} := \bigcup_{X \in \mathcal{U}} \bigcup_{x \in X} \{u \in S^1 : \langle u, x \rangle = 0\} \tag{7.1}$$

of *quasi-stable* directions by taking the two unit vectors u and $-u$ perpendicular to x (considered as a vector) for every site $x \in X$ and every

---

[13]By this, we mean the set of vertices of the rotated $\mathbb{Z}_n^2$ that lie in $[-k, k]^2$.

update rule $X \in \mathcal{U}$. The following lemma allows one to control the growth of droplets whose sides are all perpendicular to members of $\mathcal{Q}$.

Lemma 7.5 *If u, v are consecutive elements of the set $\mathcal{S}(\mathcal{U}) \cup \mathcal{Q}$, then there exists an update rule X such that*

$$X \subseteq \left(\mathbb{H}_u \cup \ell_u\right) \cap \left(\mathbb{H}_v \cup \ell_v\right),$$

where $\ell_u = \left\{x \in \mathbb{Z}^2 : \langle x, u \rangle = 0\right\}$.

Once $\mathcal{Q}$ has been defined as above, the proof of Lemma 7.5 is actually quite straightforward (see Figure 1)[14]. Indeed, let $w \in S^1$ be a direction between u and v, and note that since w is not stable, there exists an update rule $X \subseteq \mathbb{H}_w$. Suppose the conclusion of the lemma fails; then without loss of generality there exists $x \in X$ such that $\langle x, v \rangle < 0$ and $\langle x, u \rangle > 0$. But this implies that there exists $w' \in S^1$ perpendicular to x with w' between u and v, contradicting the construction of $\mathcal{Q}$.

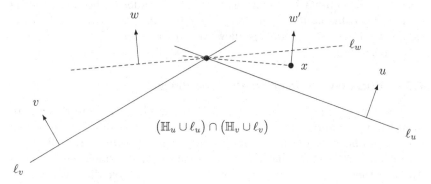

Figure 1: The proof of Lemma 7.5.

Now, it follows from Lemma 7.5 that a 'frontier' consisting of (reasonably long) intervals perpendicular to the elements of $\mathcal{Q} \cap C$ can be extended (under the action of $\mathcal{U}$) by one line in each direction, and hence L is contained in the closure of the rectangle R, as claimed. We refer the reader to [17, Sections 5 and 7] for the details.

7.2 Critical update families: the upper bound

The upper bound in Theorem 7.4(*b*) again uses quasi-stable directions, but also requires one or two extra important ideas. Let us begin by stating

---

[14]All figures used in this paper are by Paul Smith, and are reproduced here with his kind permission. We remark that Figure 1 originally appeared as Figure 4 in [17].

the key lemma in the proof of the upper bound for critical families.

Lemma 7.6 *Let $u \in S^1$ be an isolated point of $\mathcal{S}(\mathcal{U})$. Then there exists a finite set $Z \subseteq \ell_u$ such that $\ell_u \subseteq [\mathbb{H}_u \cup Z]_{\mathcal{U}}$.*

Proof It is sufficient to show that a sufficiently long interval $Z \subseteq \ell_u$ grows by one in both directions. To see that it grows to the right, consider a direction v a little to the right of u, and observe that v is unstable if it is chosen sufficiently close to u. Now let $X \in \mathcal{U}$ be a rule which destabilises v, and observe that if v is chosen sufficiently close to u then $X \setminus \mathbb{H}_u$ is contained in Z. Therefore Z grows to the right, and by symmetry also to the left, as required.   □

The proof is now almost identical to that outlined in Section 7.1, the only difference being that the rectangle R must have size at least p^{-a}, where a is sufficiently large so that for every $u \in \mathcal{S}(\mathcal{U}) \cap C$, there exists a set Z as in Lemma 7.6 of size at most $a - 2$. Indeed, once our growing droplet D is this large, we are very likely to find a translate of Z on the side of D at each step, and this allows one to show that D grows to fill an entire horizontal strip (with very high probability), as before.

7.3   Critical update families: the lower bound

The lower bound in Theorem 7.4(*b*) is significantly more difficult to prove than the upper bounds sketched above, and required the introduction of another important concept: the *covering algorithm*. To define this algorithm, observe first that if $\mathcal{U}$ is a critical two-dimensional update family, then there must exist a set $\mathcal{T}$ of three or four stable directions which intersects every open semicircle in S^1. We will control the process using $\mathcal{T}$-droplets, defined as follows:

Definition 7.7 A $\mathcal{T}$-*droplet* is a non-empty set of the form

$$D = \bigcap_{u \in \mathcal{T}} (\mathbb{H}_u + a_u)$$

for some collection $\{a_u \in \mathbb{Z}^2 : u \in \mathcal{T}\}$.

The covering algorithm replaces the rectangles process of Section 2, with $\mathcal{T}$-droplets playing the role of rectangles. The first step is to choose a sufficiently large constant $\kappa = \kappa(\mathcal{U})$, fix a $\mathcal{T}$-droplet $\hat{D}$ of diameter roughly κ, and place a copy of $\hat{D}$ (arbitrarily) on each element of A. Now, at each step, if two $\mathcal{T}$-droplets in the current collection are within distance κ^2 of one another, then remove them from the collection, and replace them by the smallest $\mathcal{T}$-droplet containing both. If κ is chosen sufficiently large, then one can prove that the final collection of $\mathcal{T}$-droplets cover $[A]_\mathcal{U}$.

If a $\mathcal{T}$-droplet occurs at some point in the covering algorithm, then let us say that it is *covered* by A. Also, let us write $\text{diam}(D)$ for the diameter of a $\mathcal{T}$-droplet D, i.e., the maximum distance between two points in D. Using the covering algorithm, one can prove the following analogues of Lemmas 2.3 and 2.7; we leave the details to the reader.[15]

Lemma 7.8 (Aizenman–Lebowitz lemma for covered droplets)
If $[A]_\mathcal{U} = \mathbb{Z}_n^2$, then for every $\kappa^3 \leqslant k \leqslant n$, there exists a covered $\mathcal{T}$-droplet $D \subseteq \mathbb{Z}_n^2$ with $k \leqslant \text{diam}(D) \leqslant 3k$.

Lemma 7.9 (Extremal lemma for covered droplets) *There exists a constant $\varepsilon = \varepsilon(\mathcal{U}) > 0$ such that*

$$|D \cap A| \geqslant \varepsilon \cdot \text{diam}(D)$$

for every covered $\mathcal{T}$-droplet D.

Once we have these two lemmas, it is relatively straightforward to deduce the claimed lower bound, using the method of Aizenman and Lebowitz [1].

Proof of the lower bound in Theorem 7.4(*b*) Suppose that $A \subseteq \mathbb{Z}_n^2$ is such that $[A]_\mathcal{U} = \mathbb{Z}_n^2$. By Lemma 7.8, there exists a covered $\mathcal{T}$-droplet D with

$$\frac{\log n}{\varepsilon} \leqslant \text{diam}(D) \leqslant \frac{3 \log n}{\varepsilon}.$$

Let X denote the random variable that counts the number of such $\mathcal{T}$-droplets when A is a p-random subset of $\mathbb{Z}_n^2$, and observe that

$$\mathbb{E}[X] \leqslant \sum_{k=\varepsilon^{-1}\log n}^{3\varepsilon^{-1}\log n} n^3 \binom{k^2}{\varepsilon k} p^{\varepsilon k} \leqslant \sum_{k=\varepsilon^{-1}\log n}^{3\varepsilon^{-1}\log n} n^3 \left(\frac{ekp}{\varepsilon}\right)^{\varepsilon k} \leqslant \frac{1}{n}$$

if $p < \varepsilon^3 / \log n$, by Lemma 7.9, since the number of $\mathcal{T}$-droplets in $\mathbb{Z}_n^2$ with semi-perimeter k is $n^{2+o(1)}$, and each has area at most k^2. It follows

---

[15]Note that in both lemmas we implicitly assume that $\mathcal{U}$ is a critical two-dimensional update family, so that there exists a set $\mathcal{T} \subseteq \mathcal{S}(\mathcal{U})$ and (sufficiently large) constant $\kappa > 0$ as described above.

from Markov's inequality that a p-random set $A \subseteq \mathbb{Z}_n^2$ percolates with probability at most $1/n$, and hence

$$p_c(\mathbb{Z}_n^2, \mathcal{U}) \geqslant \frac{\varepsilon^3}{\log n},$$

as required.   □

7.4 Subcritical update families

When there is no 'easy' direction in which to grow, the behaviour of the $\mathcal{U}$-bootstrap process changes dramatically, and the proof of Theorem 7.4(c) more closely resembles that of Theorem 5.3 than the lower bounds on $p_c(\mathbb{Z}_n^2, \mathcal{U})$ that we have seen so far. Indeed, subcritical processes behave much more like classical models from percolation theory than bootstrap processes, and the main result of Balister, Bollobás, Przykucki and Smith [4] is actually the following theorem.

Theorem 7.10 *Let $\mathcal{U}$ be a subcritical two-dimensional update family. If $p > 0$ is sufficiently small, and A is a p-random subset of $\mathbb{Z}^2$, then almost surely $[A]_\mathcal{U}$ contains no infinite component.*

If $\mathcal{U}$ is a subcritical two-dimensional update family, then there must exist a set $\mathcal{T}$ consisting of three stable *intervals* in S^1, such that $\mathcal{T}$ intersects every open semicircle in S^1 (cf. Section 7.3). The proof of Theorem 7.10 is via multi-scale analysis, using $\mathcal{T}$-droplets to control the process.

The basic idea is to partition $\mathbb{Z}^2$ into squares of side-length $n(i)$ for each $i \in \mathbb{N}$, where $n(1) < n(2) < \cdots$ is a suitable increasing sequence of 'scales', and to define (at each scale) a notion of a 'bad' square, so that good squares at scale $i + 1$ contain only small, well-separated clusters of bad squares at scale i. (When $i = 1$, a square is good if and only if it contains no element of A.) The key step is then to cover each cluster of bad squares with a $\mathcal{T}$-droplet whose sides are sufficiently far from any other bad square. This is done by first placing a triangle (whose sides correspond to points in the interior of $\mathcal{T}$) on top of the cluster, and then 'locally adjusting' the sides of the triangle so as to avoid bad squares that happen to be a little too close. (The corners of the initial triangle, the definition of a bad square, and the scales $n(i)$, all need to be chosen rather carefully for this to be possible.) We refer the reader to [4] for the details.

Finally, we remark that the authors of [4] posed a number of extremely interesting (and challenging) open problems; see [4, Section 7].

8 Critical update rules in two dimensions

We saw in the previous section that two-dimensional bootstrap processes have poly-logarithmic thresholds if, and only if, they are critical. These are therefore (in some sense) the 'correct' generalization of the classical two-neighbour process, and of the anisotropic and Duarte models discussed in Section 6. In this section, we will see how to determine the threshold for *every* process in this family.

To be precise, we will define a parameter $\alpha = \alpha(\mathcal{U}) \in \mathbb{N}$, and a bi-partition of the set of critical update families into 'balanced' and 'unbalanced' families, such that the following 'universality' theorem of Bollobás, Duminil-Copin, Morris and Smith [13] holds.

Theorem 8.1 *Let $\mathcal{U}$ be a critical two-dimensional update family.*

(a) *If $\mathcal{U}$ is balanced, then*

$$p_c\big(\mathbb{Z}_n^2, \mathcal{U}\big) = \Theta\left(\frac{1}{\log n}\right)^{1/\alpha}.$$

(b) *If $\mathcal{U}$ is unbalanced, then*

$$p_c\big(\mathbb{Z}_n^2, \mathcal{U}\big) = \Theta\left(\frac{(\log\log n)^2}{\log n}\right)^{1/\alpha}.$$

Roughly speaking, the parameter α is determined by the 'difficulty' of growth in the 'easiest' direction, where the difficulty of a stable direction $u \in \mathcal{S}(\mathcal{U})$ is the number of 'nearby' infected sites needed for $\mathbb{H}_u$ to infect the line ℓ_u, and the difficulty of growth in direction u is the maximum difficulty of a stable direction in the open semicircle centred at u. Even more roughly speaking, a critical two-dimensional update family $\mathcal{U}$ is 'balanced' if growth under the $\mathcal{U}$-bootstrap process is completely two-dimensional (like the two-neighbour process) and unbalanced if it is asymptotically one-dimensional (like the anisotropic and Duarte models).

To define these concepts precisely, we will need some extra terminology. Let $\mathcal{Q}_1 \subseteq S^1$ denote the set of rational directions on the circle, and for each $u \in \mathcal{Q}_1$, let ℓ_u^+ (resp. ℓ_u^-) be the (infinite) subset of the line ℓ_u consisting of the sites to the right (resp. left) of the origin as one looks in the direction of u. Now, let $\alpha_\mathcal{U}^+(u)$ (resp. $\alpha_\mathcal{U}^-(u)$) denote the minimum (possibly infinite) cardinality of a set $Z \subseteq \mathbb{Z}^2$ such that the set $[\mathbb{H}_u \cup Z]_\mathcal{U}$ contains infinitely many sites of ℓ_u^+ (resp. ℓ_u^-).

Definition 8.2 Given $u \in \mathcal{Q}_1$, the *difficulty* of u (with respect to $\mathcal{U}$) is[16]

$$\alpha(u) := \begin{cases} \min\left\{\alpha_{\mathcal{U}}^+(u), \alpha_{\mathcal{U}}^-(u)\right\} & \text{if } \alpha_{\mathcal{U}}^+(u) < \infty \text{ and } \alpha_{\mathcal{U}}^-(u) < \infty \\ \infty & \text{otherwise.} \end{cases}$$

We define the *difficulty* of $\mathcal{U}$ to be

$$\alpha := \min_{C \in \mathcal{C}} \max_{u \in C} \alpha(u), \tag{8.1}$$

where (as before) $\mathcal{C}$ denotes the collection of open semicircles of S^1.

We remark that $\alpha(u) = 0$ if and only if u is unstable, and that

$$\alpha(u) < \infty \qquad \Leftrightarrow \qquad u \text{ is unstable or isolated in } \mathcal{S}(\mathcal{U}).$$

Indeed, it follows from Lemma 7.6 that if u is an isolated point of $\mathcal{S}(\mathcal{U})$ then $\alpha(u) < \infty$. It is moreover not hard to show that if u is not an isolated point of $\mathcal{S}(\mathcal{U})$ then $\alpha(u) = \infty$ (consider a stable direction v a little to one side of u, and observe that $\mathbb{H}_u \cup \mathbb{H}_v$ is stable).

It follows (cf. Lemma 7.6) that a direction u has finite difficulty if, and only if, there exists a finite set Z of sites that, such that $\ell_u \subseteq [\mathbb{H}_u \cup Z]_{\mathcal{U}}$. Moreover, the difficulty is at least k if every such Z contains at least k infected sites within a bounded distance of one another (see Lemma 8.4, below). If the open semicircle centred at u contains no direction of difficulty greater than k, then it is possible for a 'critical droplet' of infected sites to grow in direction u without ever finding more than k infected sites close together.

We can now define what it means for an update family to be balanced.

Definition 8.3 A critical update family $\mathcal{U}$ is *balanced* if there exists a closed semicircle C such that $\alpha(u) \leqslant \alpha$ for all $u \in C$. It is said to be *unbalanced* otherwise.

As noted above, it turns out that growth under the action of balanced critical families is completely two-dimensional, while that for unbalanced critical families is asymptotically one-dimensional. Despite this fact, it turns out that analyzing the $\mathcal{U}$-bootstrap process when $\mathcal{U}$ is unbalanced is significantly more difficult than when it is balanced; the most problematic class of all are the (Duarte-like) 'drift' models, see below.

---

[16]In order to slightly simplify the notation, and since the update family $\mathcal{U}$ will always be clear from the context, we will drop the dependence of the difficulty on $\mathcal{U}$.

8.1 Upper bounds

The upper bounds in Theorem 8.1 follow from a more careful application of the method of quasi-stable directions (see Sections 7.1 and 7.2), combined (in the case of unbalanced families) with the method used in Section 6.3 to bound the critical probability of the Duarte model from above. The main new tool is the following sharp version of Lemma 7.6.

Lemma 8.4 *For every $u \in S^1$, there exists a set $Z \subseteq \mathbb{Z}^2$ of size $\alpha(u)$ such that*
$$\ell_u \subseteq \big[\mathbb{H}_u \cup (Z + a_1 + k_1 b) \cup \cdots \cup (Z + a_r + k_r b)\big]_{\mathcal{U}}$$
for some $r \in \mathbb{N}$ and $a_1, \ldots, a_r, b \in \ell_u$, and every $k_1, \ldots, k_r \in \mathbb{Z}$.

In words, this says that a bounded number of voracious sets are sufficient (together with $\mathbb{H}_u$) to infect ℓ_u, and moreover we may choose any suitable translation of each voracious set. To prove Lemma 8.4, suppose (without loss of generality) that $\alpha_{\mathcal{U}}^+(u) = \alpha(u)$, and observe that (by definition) there exists a set $Z \subseteq \mathbb{Z}^2$ of size $\alpha(u)$ such that $[\mathbb{H}_u \cup Z]_{\mathcal{U}}$ contains infinitely many sites of ℓ_u^+. One next needs to show (exercise) that, given such a set Z, there exists an infinite arithmetic progression in $[\mathbb{H}_u \cup Z]_{\mathcal{U}}$, and hence deduce that a bounded collection of translates of Z (as in the statement of the lemma) are sufficient to infect all of ℓ_u^+. Finally, assuming that $\alpha(u) < \infty$ (otherwise the lemma is trivial), it follows from Lemma 7.6 that $[\mathbb{H}_u \cup \ell_u^+]_{\mathcal{U}} = \ell_u$. We leave the details to the reader.

In order to deduce the claimed upper bounds, we again use the set $\mathcal{Q}$ of quasi-stable directions defined in (7.1). Suppose first that $\mathcal{U}$ is balanced, and observe that there exists a closed arc C in S^1, strictly containing a closed semicircle, such that $\alpha(u) \leqslant \alpha$ for every $u \in C$. We set
$$\mathcal{T} := \big(\mathcal{S}(\mathcal{U}) \cup \mathcal{Q} \cup \{w_1, w_2\}\big) \cap C$$
where w_1 and w_2 are the endpoints of C, and consider $\mathcal{T}$-droplets.

By Lemma 8.4, a $\mathcal{T}$-droplet D grows by one step in direction $u \in \mathcal{T} \setminus \{w_1, w_2\}$ (see Figure 2) with probability at least
$$\big(1 - (1 - p^\alpha)^{\Omega(m)}\big)^{O(1)},$$
where m is the length of the side of D corresponding to u. The proof now follows the usual argument, noting that the 'critical scale' of our droplet (above which growth becomes likely) is therefore of order $p^{-\alpha}$.

If $\mathcal{U}$ is unbalanced, on the other hand, then we replace the closed arc C above by a closed semicircle, and repeat the proof, except starting with a rectangular droplet of height $O\big(p^{-\alpha} \log(1/p)\big)$, and alternately growing

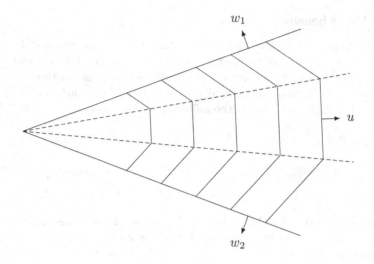

Figure 2: The growth of a droplet when the update family is balanced.

in the 'easy' and 'hard' directions, cf. Section 6.3. We refer the reader to [13, Section 5] for the details of the proof.

Exercise: Expand the above sketch to a complete proof of the upper bound in Theorem 8.1.

8.2 Balanced update families

The proof of the lower bound for balanced update families is again not all that much more difficult than the lower bound of Bollobás, Smith and Uzzell [17]. The key new idea is to modify the covering algorithm (see Section 7.3) by only covering 'clusters' of α nearby sites. The crucial observation is that, by the definition of α, the remaining sites do not contribute significantly to the growth of a droplet.

Observe first that, by (8.1), if $\mathcal{U}$ is a critical two-dimensional update family then there must exist a set $\mathcal{T}$ of three or four stable directions, each of difficulty at least α, which intersects every open semicircle in S^1. Choose a sufficiently large constant $\kappa = \kappa(\mathcal{U})$, and say that two vertices are *strongly connected* if the distance between them is at most κ. Now, define an α-*cluster* to be any strongly connected set of α vertices.

The α-covering algorithm is similar to the covering algorithm, but instead of placing a copy of the $\mathcal{T}$-droplet $\hat{D}$ on each element of A, we place one only on each α-cluster in A.

Definition 8.5 (The α-covering algorithm) Given $K \subseteq A$, choose a maximal collection $\mathcal{B}$ of disjoint α-clusters in K, and a collection $\mathcal{D}$ of copies of a fixed, sufficiently large $\mathcal{T}$-droplet $\hat{D}$, such that the elements of $\mathcal{D}$ cover the elements of $\mathcal{B}$. Now repeat the following steps until STOP:

1. If there are two droplets D and D' in the current collection and an $x \in \mathbb{Z}^2$ such that the set

$$D \cup D' \cup (x + \hat{D})$$

   is strongly connected, then remove them from the collection, and replace them by the smallest $\mathcal{T}$-droplet containing $D \cup D'$.

2. If there do not exist such a pair of droplets, then STOP.

We call the final collection of $\mathcal{T}$-droplets an α-cover of K.

We will also say that a droplet D is α-*covered* if the single droplet $\mathcal{D} = \{D\}$ is an α-cover of $D \cap A$, i.e., a possible output of the α-covering algorithm. The following two key lemmas follow easily from the algorithm, as in Section 7.3; we again leave the details to the reader.

Lemma 8.6 (Aizenman–Lebowitz lemma for α-covered droplets)
Let D be an α-covered droplet. Then for every $1 \ll k \leqslant \mathrm{diam}(D)$, there exists an α-covered droplet $D' \subseteq D$ such that $k \leqslant \mathrm{diam}(D') \leqslant 3k$.

Lemma 8.7 (Extremal lemma for α-covered droplets)
There exists a constant $\varepsilon = \varepsilon(\mathcal{U}) > 0$ such that every α-covered $\mathcal{T}$-droplet D contains at least $\varepsilon \cdot \mathrm{diam}(D)$ disjoint α-clusters of elements of A.

It remains to show that an α-cover $\mathcal{D}$ of a set K is a reasonable approximation of the closure $[K]_\mathcal{U}$. The basic idea is simple: since all α-clusters are contained in some droplet of $\mathcal{D}$, the remaining 'dust' of $K \setminus (D_1 \cup \cdots \cup D_k)$ should contribute only 'locally' to the set of eventually infected sites. The following lemma makes this notion precise.

Lemma 8.8 *There exists a constant $\rho = \rho(\mathcal{U}) > 0$ such that if $\mathcal{D}$ is an α-cover of a set K, then every vertex in $[K]_\mathcal{U}$ is either contained in an element of $\mathcal{D}$, or lies within distance ρ of some element of K.*

To prove the lemma, we partition the 'dust'

$$Y := K \setminus X, \qquad \text{where} \qquad X := \bigcup_{D \in \mathcal{D}} D,$$

into a collection $Y_1, \ldots, Y_s$ of maximal strongly connected components. We then show that, since each strongly connected component has size at most $\alpha - 1$, and each element of $\mathcal{T}$ has difficulty at least α, each Y_i can (with the help of X) only infect a bounded number of extra vertices. Since κ was chosen sufficiently large, this means that different strongly connected components are too far apart to cooperate, and the lemma follows. We refer the reader to [13, Section 6.1] for the details.

Proof of the lower bound in Theorem 8.1(a) Suppose that $A \subseteq \mathbb{Z}_n^2$ is such that $[A]_{\mathcal{U}} = \mathbb{Z}_n^2$. By Lemmas 8.6 and 8.8, there exists an α-covered $\mathcal{T}$-droplet D with

$$\log n \leqslant \operatorname{diam}(D) \leqslant 3 \log n.$$

Let X denote the random variable that counts the number of such $\mathcal{T}$-droplets when A is a p-random subset of $\mathbb{Z}_n^2$, and observe that there are $n^{2+o(1)}$ $\mathcal{T}$-droplets in $\mathbb{Z}_n^2$ with semi-perimeter k, and each contains at most Ck^2 (not necessarily infected) α-clusters for some constant $C = C(\mathcal{U}) > 0$. Thus, by Lemma 8.7, we have

$$\mathbb{E}[X] \leqslant \sum_{k=\log n}^{3 \log n} n^3 \binom{Ck^2}{\varepsilon k} p^{\varepsilon \alpha k} \leqslant \sum_{k=\log n}^{3 \log n} n^3 \left(C^2 k p^\alpha\right)^{\varepsilon k} \leqslant \frac{1}{n}$$

if $p^\alpha \log n$ is sufficiently small. It follows from Markov's inequality that a p-random set $A \subseteq \mathbb{Z}_n^2$ percolates with probability at most $1/n$, and hence

$$p_c(\mathbb{Z}_n^2, \mathcal{U}) = \Omega\left(\frac{1}{\log n}\right)^{1/\alpha},$$

as required. $\square$

For a particular class of balanced models, a sharp threshold is known. Given a two-dimensional update family $\mathcal{U}$, let us say that $\mathcal{U}$ is a *threshold rule* if it consist of the the r-subsets of some neighbourhood $N \subseteq \mathbb{Z}^2$ of the origin. Moreover, let us say that N is a symmetric star-neighbourhood if $x \in N$ implies that $-x \in N$, and moreover that every vertex of $\mathbb{Z}^2$ on the straight line between x and $-x$ is in N. The following theorem was proved by Duminil-Copin and Holroyd [27].

Theorem 8.9 *Let $\mathcal{U}$ be a balanced critical two-dimensional update family. If $\mathcal{U}$ is a threshold rule with a symmetric star-neighbourhood, then*

$$p_c(\mathbb{Z}_n^2, \mathcal{U}) = \left(\frac{\lambda + o(1)}{\log n}\right)^{1/\alpha}$$

for some constant $\lambda = \lambda(\mathcal{U}) > 0$.

The constant $\lambda(\mathcal{U})$ can be computed as the solution of a variational problem, see [27, Section 4] for the details. It is a very interesting (and likely very challenging) open problem to generalise Theorem 8.9 to non-symmetric and unbalanced update families.

8.3 Unbalanced update families

Finally we arrive at the real challenge: proving the lower bound in Theorem 8.1 for unbalanced update families. The proof is significantly more difficult than any of those described above, and we will only be able to give a very rough sketch of the main ideas.

The first key observation is that if $\mathcal{U}$ is an unbalanced critical two-dimensional update family, then there exist a pair of opposite directions $\{u, -u\}$ that both have difficulty at least $\alpha + 1$. We set

$$\mathcal{T} = \{u, -u, v, v'\},$$

where v and v' are stable directions in opposite semicircles with endpoints $\{u, -u\}$, each of difficulty at least α. Observe that the pair $\{u, -u\}$ exists by Definition 8.3, and the pair $\{v, v'\}$ exists by (8.1). Let us rotate our perspective so that u is vertical, and write $h(D)$ and $w(D)$ for (respectively) the height and width of an $\mathcal{T}$-droplet D.

As in the balanced case, we will need a suitable variant of the rectangles process; however, instead of the α-covering algorithm (which fails to capture the non-linear geometry of unbalanced models), we use the following variant of the components process introduced in Section 3. As above, we say that two vertices are strongly connected if the distance between them is at most κ, where $\kappa = \kappa(\mathcal{U})$ is a sufficiently large constant.

Definition 8.10 (The spanning algorithm) Let $K = \{x_1, \ldots, x_m\}$, and set $\mathcal{S} := \{S_1, \ldots, S_m\}$, where $S_j = \{x_j\}$ for each $j \in [m]$. Now repeat the following steps until STOP:

1. If there exist a pair of sets $\{S_1, S_2\} \subseteq \mathcal{S}$ such that

$$[S_1]_{\mathcal{U}} \cup [S_2]_{\mathcal{U}}$$

is strongly connected in G, then remove S_1 and S_2 from $\mathcal{S}$, and replace them by $S_1 \cup S_2$.

2. If there do not exist such a family of sets in $\mathcal{S}$, then STOP.

The *span* of K is defined to be

$$\langle K \rangle = \{D(S) : S \in \mathcal{S}\},$$

where $D(S)$ denotes the smallest $\mathcal{T}$-droplet containing S, and $\mathcal{S}$ is the final collection of sets in the spanning algorithm. We would like to say that a $\mathcal{T}$-droplet D is internally spanned by A if $D \in \langle D \cap A \rangle$. This turns out to be true if we modify Definition 3.4 as follows.

Definition 8.11 A $\mathcal{T}$-droplet D is *internally spanned* by A if there exists a strongly connected set $S \subseteq [D \cap A]_\mathcal{U}$ such that $D = D(S)$. We write $I^\times(D)$ for the event that D is internally spanned.

Fix a small constant $\varepsilon = \varepsilon(\mathcal{U}) > 0$. We say that a $\mathcal{T}$-droplet is *critical* if either of the following conditions hold:

(a) $w(D) \leqslant p^{-\alpha - 1/5}$ and $\varepsilon p^{-\alpha} \log(1/p) \leqslant h(D) \leqslant 3\varepsilon p^{-\alpha} \log(1/p)$.

(b) $p^{-\alpha - 1/5} \leqslant w(D) \leqslant 3p^{-\alpha - 1/5}$ and $h(D) \leqslant \varepsilon p^{-\alpha} \log(1/p)$.

The spanning algorithm allows us to prove the following Aizenman–Lebowitz lemma for internally spanned $\mathcal{T}$-droplets.

Lemma 8.12 (Aizenman–Lebowitz lemma for critical $\mathcal{T}$-droplets)
If $[A]_\mathcal{U} = \mathbb{Z}_n^2$, then there exists an internally spanned critical $\mathcal{T}$-droplet.

Note that the reason we need both types of critical droplet is that the spanning algorithm cannot control the width and the height of the critical droplet simultaneously. We choose the constant ε so that a critical $\mathcal{T}$-droplet is not overwhelmingly likely to grow sideways (that is, perpendicular to u); in particular, if the probability of growing one step is roughly $1 - p^\varepsilon$, then the width $p^{-\alpha - 1/5}$ is large enough so that the probability that a critical droplet of type (b) is internally spanned should be sufficiently small to compensate for the number of choices for the initial rectangle. Unfortunately, however, if $\mathcal{U}$ is not symmetric then there is no easy way to prove this, and we need to use the method of hierarchies. In fact, when $\alpha > 1$ the situation is much worse, and we need a more complicated method, which we call the *method of iterated hierarchies*.

In order to motivate this method, let us first see why a straightforward application of the method of hierarchies cannot work. The first key reason is that the extremal lemma we obtain via the spanning algorithm is much less powerful than that given by the α-covering algorithm.

Lemma 8.13 (Extremal lemma for internally spanned $\mathcal{T}$-droplets)
There exists a constant $\varepsilon = \varepsilon(\mathcal{U}) > 0$ such that every internally spanned $\mathcal{T}$-droplet D contains at least $\varepsilon \cdot \max\{h(D), w(D)\}$ elements of A.

This lemma only implies a non-trivial bound on the probability that D is internally spanned if D has either height or width at most p^{-1}, whereas our critical droplets have height and width larger than $p^{-\alpha}$. The second main obstacle is that when bounding the probability of a 'step' of the hierarchy, we need to bound the probability of the existence of an internally spanned 'saver' droplet (cf. Section 6.3). However, if the seeds and steps both have size only p^{-1} then the number of possible hierarchies will be much too large for our application of the union bound.

Thus, in order to apply the method of hierarchies, we first need to give strong bounds on the probability that a seed or saver droplet of height and width roughly $p^{-\alpha}$ is internally spanned. We prove this in stages, building up from the result for much smaller droplets (which we prove using Lemma 8.13). More precisely, we use the following induction hypothesis:

Definition 8.14 For each $\beta_1, \beta_2 \in \mathbb{N}$, we say that $\mathrm{IH}(\beta_1, \beta_2)$ holds if the following statement is true for some (small, fixed) constant $\eta > 0$:

There exists $\delta = \delta(\beta_1 + \beta_2) > 0$ such that

$$\mathbb{P}_p\big(I^\times(D)\big) \leqslant p^{\delta \max\{w(D),\, h(D)\}}$$

for every $\mathcal{T}$-droplet D such that

$$w(D) \leqslant p^{-\beta_1(1-2\eta)-\eta} \quad \text{and} \quad h(D) \leqslant p^{-\beta_2(1-2\eta)-\eta}. \tag{8.2}$$

The bounds we need on the probability that a seed or a saver droplet is internally spanned are given by the following lemma.

Lemma 8.15 *The assertions* $\mathrm{IH}(\alpha + 1, \alpha)$ *and* $\mathrm{IH}(\alpha, \alpha + 1)$ *both hold.*

We prove Lemma 8.15 by induction on $\beta_1 + \beta_2$. The base case, $\beta_1 = \beta_2 = 1$, follows easily from Lemma 8.13; indeed, the probability that D contains at least $k := \varepsilon \cdot \max\{h(D), w(D)\}$ elements of A is at most

$$\binom{O(k^2)}{k} p^k \leqslant O(pk)^k \leqslant p^{\eta k/2}$$

if $k \leqslant p^{-1+\eta}$, so $\mathrm{IH}(1,1)$ holds, as claimed.

The specific induction statements that were proved in [13] are:

$$\mathrm{IH}(\beta, \beta) \;\Rightarrow\; \mathrm{IH}(\beta + 1, \beta) \qquad \text{for all } 1 \leqslant \beta \leqslant \alpha;$$
$$\mathrm{IH}(\beta, \beta) \;\Rightarrow\; \mathrm{IH}(\beta, \beta + 1) \qquad \text{for all } 1 \leqslant \beta \leqslant \alpha;$$
$$\big(\mathrm{IH}(\beta + 1, \beta) \wedge \mathrm{IH}(\beta, \beta + 1)\big) \;\Rightarrow\; \mathrm{IH}(\beta + 1, \beta + 1) \quad \text{for all } 1 \leqslant \beta \leqslant \alpha - 1.$$

The proof of these statements depends on whether, for the $\mathcal{T}$-droplet D in question, $h(D) \geqslant w(D)$ or vice-versa. This is because we have a pair $\{u, -u\}$ of opposite stable directions, so we may partition the droplet D into many smaller sub-droplets of the same width, and bound the probability that each is vertically crossed (possibly with help from above and below) independently, since these events depend on disjoint sets of infected sites. This turns out to be a good idea if $h(D) \geqslant w(D)$; otherwise, we need to use the spanning algorithm to construct an (s,t)-good hierarchy for an internally spanned $\mathcal{T}$-droplet D (cf. Section 2.1) with $s = t \approx p^{-\beta(1-2\eta)-\eta}$. Note that in this latter case, we can use the induction hypothesis to bound the probability that a seed is internally spanned.

In either case, it remains to bound the probability that it is possible to 'cross' a parallelogram of sites from one side to the other with 'help' from one of the sides in the form of an infected half-plane. We obtain bounds for the probabilities of crossings by showing that, to a certain level of precision, the most likely way these events could occur is via the droplet (or half-plane) advancing row-by-row, rather than via the merging of many smaller droplets. The hardest part of the proof is controlling vertical crossings in the case of models 'with drift' (that is, when one of u and $-u$ has infinite difficulty). As in Section 6.3, the trick is to rotate ones view slightly (by an angle of $p^{1-\eta}$), and show that each constant width (rotated) strip must either contain a reasonably large 'cluster' of elements of A, or must intersect a large internally spanned 'saver' droplet, and moreover all of these events occur disjointly. To prove this, we need introduce yet another algorithm for approximating the closure of a set of sites (the 'iceberg algorithm'). Since the details are rather complicated, we refer the interested reader to [13, Sections 6.3 and 8.3].

9   A general conjecture in higher dimensions

Perhaps the most important open problem on monotone cellular automata is to extend Theorem 7.4 to higher dimensions. In this section we will state a conjecture as to the correct form of this generalization.

Fix an integer $d \geqslant 2$ and let $\mathcal{U}$ be a d-dimensional update family. Let

$$\mathbb{H}_u^d := \left\{ x \in \mathbb{Z}^d : \langle x, u \rangle < 0 \right\}$$

denote the discrete d-dimensional half-space with normal $u \in S^{d-1}$, and define the set of stable directions to be

$$\mathcal{S} = \mathcal{S}(\mathcal{U}) = \left\{ u \in S^{d-1} : [\mathbb{H}_u^d]_{\mathcal{U}} = \mathbb{H}_u^d \right\}.$$

Let μ denote Lebesgue measure on S^{d-1} (this is usually called 'spherical measure'), and let $\mathcal{C}^d$ denote the collection of open hemispheres in S^{d-1}.

Definition 9.1 We say that a d-dimensional update family is:

- *supercritical* if there exists $C \in \mathcal{C}^d$ that is disjoint from $\mathcal{S}$,

- *critical* if there exists $C \in \mathcal{C}^d$ such that $\mu(C \cap \mathcal{S}) = 0$, and every $C \in \mathcal{C}^d$ has non-empty intersection with $\mathcal{S}$,

- *subcritical* if $\mu(C \cap \mathcal{S}) > 0$ for every $C \in \mathcal{C}^d$.

Note that, as in two dimensions, this trichotomy depends only on the stable set $\mathcal{S}$. The following conjecture was made by Bollobás, Duminil-Copin, Morris and Smith [13].

Conjecture 9.2 *Let $\mathcal{U}$ be a d-dimensional update family.*

(a) *If $\mathcal{U}$ is supercritical then $p_c(\mathbb{Z}_n^d, \mathcal{U}) = n^{-\Theta(1)}$.*

(b) *If $\mathcal{U}$ is critical then there exists $r = r(\mathcal{U}) \in \{2, \dots, d\}$ such that*

$$p_c(\mathbb{Z}_n^d, \mathcal{U}) = \left(\frac{1}{\log_{(r-1)} n} \right)^{\Theta(1)}, \qquad (9.1)$$

where $\log_{(r-1)}$ denotes an $(r-1)$-times iterated logarithm.

(c) *If $\mathcal{U}$ is subcritical then $\liminf\limits_{n \to \infty} p_c(\mathbb{Z}_n^d, \mathcal{U}) > 0$.*

The intuition behind this conjecture is as follows. Each direction $u \in S^{d-1}$ should have a 'difficulty' $r(u) \in \{0, \dots, d\}$ that depends on the intersection of $\mathcal{S}$ with its neighbourhood; this should correspond to the r in (9.1) when applied to the $(d-1)$-dimensional process induced on the side of the half-space $\mathbb{H}_u^d$ (if the direction is unstable then $r = 0$, and if the induced process is super-/subcritical then $r \in \{1, d\}$). Now, if there exists an open hemisphere in which all directions have difficulty at most $r - 1$, then the difficulty of $\mathcal{U}$ should be at most r (cf. the proof of (3.1)); if not, then it should be at least $r + 1$ (cf. the proof of Lemma 3.7).

10 Some related models

We finish this survey by briefly discussing some applications (and potential future applications) of the results and techniques described above to three more complicated models: kinetically constrained spin models, a model of nucleation and growth, and the abelian sandpile.

10.1   Kinetically constrained spin models

Suppose we modify the update rule in $\mathcal{U}$-bootstrap percolation so that the process is no longer monotone, i.e., so that infected sites may become uninfected? For example, if we randomly (and independently) resample the state (infected or uninfected) of a vertex v whenever $v + X$ is entirely infected for some $X \in \mathcal{U}$, we obtain an important class of models of the 'liquid-glass transition' known as 'kinetically constrained spin models'. The following general family of models was introduced by Cancrini, Martinelli, Roberto and Toninelli [18].

Definition 10.1 Let $\mathcal{U}$ be a finite collection of finite subsets of $\mathbb{Z}^d \setminus \{0\}$, and let $p \in (0,1)$. The $\mathcal{U}$-*kinetically constrained spin model* on $\mathbb{Z}^d$ with density p is defined as follows:

(a) Each vertex has an independent exponential clock which rings randomly at rate 1.

(b) If the clock at vertex v rings at (continuous) time $t \geqslant 0$, and $v + X \subseteq A_t$ for some $X \in \mathcal{U}$, where $A_t \subseteq \mathbb{Z}^d$ is the set of infected vertices at time t, then v becomes infected with probability p, and healthy with probability $1 - p$, independently of all other events.

Note that if the set of infected vertices at time $t = 0$ percolates in the $\mathcal{U}$-bootstrap process, then the set of infected vertices at all later times also has this property. The model can therefore be thought of as a biased random walk on the family of percolating sets. Let us choose the infected sites at time $t = 0$ to be p-random; the distribution is therefore also given by $\mathbb{P}_p$ at every later time t, though the distributions at different times are (of course) not independent of one another.

Let us write
$$\tau(\mathbb{Z}^d, \mathcal{U}) := \inf \{t \geqslant 0 : \mathbf{0} \in A_t\},$$

for the (random) time at which the origin is first infected. It was pointed out in [18] that the time taken for the $\mathcal{U}$-bootstrap process to infect the origin provides a lower bound on $\tau(\mathbb{Z}^d, \mathcal{U})$. In particular, the following theorem is an immediate consequence of Theorem 8.1.

Theorem 10.2 *Let $\mathcal{U}$ be a critical two-dimensional update family. There exists a constant $c = c(\mathcal{U}) > 0$ such that the following holds with high probability as $p \to 0$.*

(a) *If $\mathcal{U}$ is balanced, then*

$$\tau(\mathbb{Z}^2, \mathcal{U}) \geqslant \exp\left(cp^{-\alpha}\right).$$

(b) *If $\mathcal{U}$ is unbalanced, then*

$$\tau(\mathbb{Z}^2, \mathcal{U}) \geqslant \exp\left(cp^{-\alpha}\big(\log(1/p)\big)^2\right).$$

Cancrini, Martinelli, Roberto and Toninelli [18] introduced a powerful and general analytic technique that allows one to prove upper bounds on $\tau(\mathbb{Z}^d, \mathcal{U})$ for many specific update families. There is some hope (see [44, 47]) that their ideas, combined with the techniques introduced in [13, 17], might be sufficient to prove an almost-matching upper bound for a large collection of critical two-dimensional update families.

10.2 Nucleation and growth

We next discuss a model that was first studied by Dehghanpour and Schonmann [24, 25], who used it to study the metastable behavior of the kinetic Ising model on $\mathbb{Z}^2$ with a small magnetic field and vanishing temperature. (Their results were recently generalized to higher dimensions by Cerf and Manzo [21, 22].) The following general (monotone) version of this model was recently introduced in [16].

Definition 10.3 Let $\mathcal{U}_1, \ldots, \mathcal{U}_r$ be d-dimensional update families, and let $1 \leqslant k_1(n) \leqslant \ldots \leqslant k_r(n) = n$ be functions. In the $(\mathcal{U}_1, \ldots, \mathcal{U}_r)$-*nucleation and growth model*, each vertex $v \in \mathbb{Z}^d$ is initially uninfected, but becomes (permanently) infected at rate $k_\ell(n)/n$ at each time $t \geqslant 0$, where

$$\ell = \ell(v, t) := \max\big\{j \in [r] : v + X \subseteq A_t \text{ for some } X \in \mathcal{U}_j\big\},$$

where A_t is the set of infected sites at time t, and at rate $1/n$ otherwise.

Similarly to the previous subsection, let us write

$$\tau(\mathbb{Z}^d; \mathcal{U}_1, \ldots, \mathcal{U}_r) := \inf\big\{t \geqslant 0 : \mathbf{0} \in A_t\big\},$$

for the (random) time at which the origin is first infected.

Dehghanpour and Schonmann [24] studied the case $d = r = 2$, with $\mathcal{U}_1 = \mathcal{N}_1^2$ and $\mathcal{U}_2 = \mathcal{N}_2^2$ (recall that $\mathcal{N}_r^d$ denotes the family of r-subsets of the $2d$ nearest neighbours of the origin in $\mathbb{Z}^d$), proving that

$$\tau(\mathbb{Z}^2; \mathcal{N}_1^2, \mathcal{N}_2^2) = \begin{cases} \left(\dfrac{n}{k}\right)^{1+o(1)} & \text{if } k \leqslant \sqrt{n} \\[4mm] \left(\dfrac{n^2}{k}\right)^{1/3+o(1)} & \text{if } k \geqslant \sqrt{n} \end{cases} \qquad (10.1)$$

with high probability as $n \to \infty$. The following more precise bounds were proved recently by Bollobás, Griffiths, Morris, Rolla and Smith [16].

Theorem 10.4 *With high probability as $n \to \infty$, the following hold.*

(a) *If $k \ll \log n$ then*

$$\tau\left(\mathbb{Z}^2; \mathcal{N}_1^2, \mathcal{N}_2^2\right) = \left(\frac{\pi^2}{18} + o(1)\right) \frac{n}{\log n}.$$

(b) *If $\log n \ll k \ll \sqrt{n}(\log n)^2$ then*

$$\tau\left(\mathbb{Z}^2; \mathcal{N}_1^2, \mathcal{N}_2^2\right) = \left(\frac{1}{4} + o(1)\right) \frac{n}{k} \log\left(\frac{k}{\log n}\right).$$

(c) *If $\sqrt{n}(\log n)^2 \ll k \ll n$ then*

$$\tau\left(\mathbb{Z}^2; \mathcal{N}_1^2, \mathcal{N}_2^2\right) = \Theta\left(\frac{n^2}{k \log(n/k)}\right)^{1/3}.$$

It would be very interesting to generalize these results to other update families. In particular, we expect the following problem to already be very challenging.

Open problem: Determine $\tau\left(\mathbb{Z}^2; \mathcal{U}_1, \mathcal{U}_2\right)$ up to a constant factor whenever $\mathcal{U}_1$ is supercritical and $\mathcal{U}_2$ is critical.

Cerf and Manzo [21] determined $\tau\left(\mathbb{Z}^d; \mathcal{N}_1^d, \ldots, \mathcal{N}_d^d\right)$ (with high probability) up to a factor of $1 + o(1)$ in the exponent, cf. (10.1). Generalizing their result to arbitrary collections of d-dimensional update families is an important (and likely extremely difficult) problem.

10.3 The abelian sandpile

The final model we would like to briefly discuss was first introduced almost 30 years ago by Bak, Tang and Wiesenfeld [3] as an example of so-called 'self-organised criticality'. In the model, grains of sand are placed on each vertex of a graph G; if there are at least $d(v)$ particles on vertex v, then the pile 'topples' by sending one grain to each of its neighbours. The process is 'abelian' in the sense that the order of toppling does not affect the final configuration. We are interested in this model on the graph $\mathbb{Z}^d$, and with a random initial configuration.

Definition 10.5 Given a function $A: \mathbb{Z}^d \to \mathbb{Z}$, the *abelian sandpile* on $\mathbb{Z}^d$ with initial state A is defined as follows:

(a) At time $t = 0$, place $A(v)$ grains of sand on v for each $v \in \mathbb{Z}^d$.

(b) For each $t \in \mathbb{N}$, do the following at time t: for each vertex $v \in \mathbb{Z}^d$ with at least $2d$ grains of sand at time $t - 1$, remove $2d$ grains from v and place one at each nearest neighbour of v.

Let us say that A *percolates on* $\mathbb{Z}^d$ if every site topples infinitely many times in the abelian sandpile on $\mathbb{Z}^d$ starting from A, and define

$$\lambda_c^S(\mathbb{Z}^d) := \inf \left\{ \lambda \, : \, \mathbb{P}_\lambda \big(A \text{ percolates on } \mathbb{Z}^d \big) = 1 \right\},$$

where $\mathbb{P}_\lambda$ denotes the probability measure in which each $A(v)$ is chosen according to an independent Poisson random variable of mean λ.

The problem of determining $\lambda_c^S(\mathbb{Z}^d)$ was introduced over 15 years ago by Dickman, Muñoz, Vespagnani and Zapperi [55], but the best known bounds (see e.g. [33]) are only

$$d \leqslant \lambda_c^S(\mathbb{Z}^d) \leqslant 2d - 1.$$

Note that there exists an initial distribution with density $2d-1$ which does not percolate (take $A(v)$ to be constant), and distributions with density arbitrarily close to d which do percolate (take $A(v) = 2d$ with probability ε, and $A(v) = d$ otherwise, and apply Theorem 3.1). Thus, to improve the bounds above, we must use some properties of the Poisson distribution other than its mean.

Morris [47] proposed the following approach to this problem in high dimensions. Let us label a vertex v as 'infected' if $A(v) \geqslant 2d$, as 'vulnerable' if $d < A(v) < 2d$, and as 'removed' if $A(v) \leqslant d$. Note that infected vertices topple immediately, and that vulnerable vertices with at least $d - 1$ neighbours that topple at some point will also topple. The vertices that eventually topple therefore contain the closure of the infected set of vertices in $(d - 1)$-neighbour bootstrap percolation on the subgraph of $\mathbb{Z}^d$ induced by the non-removed vertices. Note that if d is large and $\lambda = (1 + \varepsilon)d$, then only a few vertices will be removed, and even fewer vertices will be infected.

The coupling above motivates the following problem, which was first studied by Gravner and McDonald [38]. Let us write $\mathbb{Z}^d(q)$ for the random induced subgraph of $\mathbb{Z}^d$ obtained by removing each vertex independently with probability q, and say that a set $A \subseteq \mathbb{Z}^d(q)$ *percolates in the r-neighbour polluted bootstrap process* if the closure of A in the r-neighbour bootstrap process on $\mathbb{Z}^d(q)$ contains an infinite component. Define

$$p_c^\infty\big(\mathbb{Z}^d(q), r\big) := \inf \Big\{ p \, : \, \mathbb{P}_p\big(A \text{ percolates in the } r\text{-neighbour}$$

$$\text{polluted bootstrap process} \big) \geqslant 1/2 \Big\}$$

for each $d \geqslant r \geqslant 2$. It was proved by Gravner and McDonald [38] that

$$p_c^\infty\big(\mathbb{Z}^2(q), 2\big) = \Theta\big(\sqrt{q}\big)$$

almost surely as $q \to 0$, and a Cerf–Cirillo-type argument (see Section 3) can be used to show that moreover $p_c^\infty\big(\mathbb{Z}^d(q), d\big) > 0$ for every $d \in \mathbb{N}$ and $q > 0$. On the other hand, the following conjecture was stated in [47].

Conjecture 10.6 *For each $d > r \geqslant 1$, there exists $q_0(d, r) > 0$ such that*

$$p_c^\infty\big(\mathbb{Z}^d(q), r\big) = 0$$

almost surely for every $0 < q < q_0(d, r)$.

Note that when $r = 1$ we can take $q_0(d, 1) = 1 - p_c^{\mathrm{site}}(\mathbb{Z}^d)$, but the conjecture is open (and seems to be very difficult) for every $d > r \geqslant 2$. Conjecture 10.6 motivates the following conjecture, also stated in [47].

Conjecture 10.7

$$\frac{\lambda_c^S(\mathbb{Z}^d)}{d} \to 1$$

as $d \to \infty$.

The extension of these conjectures to more general update rules is yet another fascinating, and likely very hard, open problem.

Acknowledgements

The author would like to thank Béla Bollobás for introducing him to bootstrap percolation, and for his encouragement and support over many years. He would also like to thank his frequent collaborators, Józsi Balogh, Béla Bollobás, Hugo Duminil-Copin and Paul Smith, for many interesting conversations over the years, and for their many important contributions to the area, only a few of which we have had space to describe; Paul Smith for Figures 1 and 2, and for several helpful discussions during the preparation of this survey; and Teeradej Kittipassorn and the anonymous referee for a number of helpful comments on the manuscript.

References

[1] M. Aizenman and J.L. Lebowitz, Metastability effects in bootstrap percolation, *J. Phys. A.*, **21** (1988), 3801–3813.

[2] R. Arratia, Site recurrence for annihilating random walks on $\mathbb{Z}^d$, *Ann. Prob.*, **11** (1983), 706–713.

[3] P. Bak, C. Tang and K. Wiesenfeld, Self-organized criticality: an explanation of $1/f$ noise, *Phys. Rev. Letters*, **59** (1987), 381–384.

[4] P. Balister, B. Bollobás, M.J. Przykucki and P.J. Smith, Subcritical $\mathcal{U}$-bootstrap percolation models have non-trivial phase transitions, *Trans. Amer. Math. Soc.*, **368** (2016), 7385–7411.

[5] J. Balogh and B. Bollobás, Bootstrap percolation on the hypercube, *Prob. Theory Rel. Fields*, **134** (2006), 624–648.

[6] J. Balogh, B. Bollobás, H. Duminil-Copin and R. Morris, The sharp threshold for bootstrap percolation in all dimensions, *Trans. Amer. Math. Soc.*, **364** (2012), 2667–2701.

[7] J. Balogh, B. Bollobás and R. Morris, Majority bootstrap percolation on the hypercube, *Combin. Probab. Computing*, **18** (2009), 17–51.

[8] J. Balogh, B. Bollobás and R. Morris, Bootstrap percolation in three dimensions, *Ann. Prob.*, **37** (2009), 1329–1380.

[9] J. Balogh, B. Bollobás and R. Morris, Bootstrap percolation in high dimensions, *Combin. Probab. Computing*, **19** (2010), 643–692.

[10] J. Balogh, Y. Peres and G. Pete, Bootstrap percolation on infinite trees and non-amenable groups, *Combin. Prob. Computing*, **15** (2006), 715–730.

[11] J. Balogh and B. Pittel, Bootstrap percolation on the random regular graph, *Random Structures Algorithms*, **30** (2007), 257–286.

[12] J. van den Berg and H. Kesten, Inequalities with applications to percolation and reliability, *J. Appl. Prob.*, **22** (1985), 556–569.

[13] B. Bollobás, H. Duminil-Copin, R. Morris and P. Smith, Universality of two-dimensional critical cellular automata, *Proc. London Math. Soc.*, to appear, arXiv:1406.6680.

[14] B. Bollobás, H. Duminil-Copin, R. Morris and P. Smith, The sharp threshold for the Duarte model, *Ann. Prob.*, to appear.

[15] B. Bollobás, K. Gunderson, C. Holmgren, S. Janson and M. Przykucki, Bootstrap percolation on Galton-Watson trees, *Electron. J. Prob.*, **19** (2014), 1–27.

[16] B. Bollobás, S. Griffiths, R. Morris, L. Rolla and P. Smith, Nucleation and growth in two dimensions, submitted, arXiv:1508.06267.

[17] B. Bollobás, P.J. Smith and A.J. Uzzell, Monotone cellular automata in a random environment, *Combin. Probab. Computing*, **24** (2015), 687–722.

[18] N. Cancrini, F. Martinelli, C. Roberto and C. Toninelli, Kinetically constrained spin models, *Prob. Theory Rel. Fields*, **140** (2008), 459–504.

[19] R. Cerf and E.N.M. Cirillo, Finite size scaling in three-dimensional bootstrap percolation, *Ann. Prob.*, **27** (1999), 1837–1850.

[20] R. Cerf and F. Manzo, The threshold regime of finite volume bootstrap percolation, *Stochastic Proc. Appl.*, **101** (2002), 69–82.

[21] R. Cerf and F. Manzo, A d-dimensional nucleation and growth model, *Prob. Theory Rel. Fields*, **155** (2013), 427–449.

[22] R. Cerf and F. Manzo, Nucleation and growth for the Ising model in d dimensions at very low temperatures, *Ann. Prob.*, **41** (2013), 3697–3785.

[23] J. Chalupa, P.L. Leath and G.R. Reich, Bootstrap percolation on a Bethe latice, *J. Phys. C.*, **12** (1979), L31–L35.

[24] P. Dehghanpour and R.H. Schonmann, A nucleation-and-growth model, *Prob. Theory Rel. Fields*, **107** (1997), 123–135.

[25] P. Dehghanpour and R.H. Schonmann, Metropolis dynamics relaxation via nucleation and growth, *Comm. Math. Phys.*, **188** (1997), 89–119.

[26] A.M.S. Duarte, Simulation of a cellular automaton with an oriented bootstrap rule, *Phys. A*, **157** (1989), 1075–1079.

[27] H. Duminil-Copin and A. Holroyd, Sharp metastability for threshold growth models, manuscript, available at http://www.unige.ch/~duminil.

[28] H. Duminil-Copin and A.C.D. van Enter, Finite volume Bootstrap Percolation with balanced threshold rules on $\mathbb{Z}^2$, *Ann. Prob.*, **41** (2013), 1218–1242.

[29] A.C.D. van Enter, Proof of Straley's argument for bootstrap percolation, *J. Stat. Phys.*, **48** (1987),943–945.

[30] A.C.D. van Enter and A. Fey, Metastability threshold for anisotropic bootstrap percolation in three dimensions, *J. Stat. Phys.*, **147** (2012), 97–112.

[31] A.C.D. van Enter and W.J.T. Hulshof, Finite-size effects for anisotropic bootstrap percolation: logarithmic corrections, *J. Stat. Phys.*, **128** (2007), 1383–1389.

[32] P. Erdős and P. Ney, Some Problems on Random Intervals and Annihilating Particles, *Ann. Prob.*, **2** (1974), 828–839.

[33] A. Fey, R. Meester and F. Redig, Stabilizability and percolation in the infinite volume sandpile model, *Ann. Prob.*, **37** (2009), 654–675.

[34] L.R. Fontes, R.H. Schonmann and V. Sidoravicius, Stretched Exponential Fixation in Stochastic Ising Models at Zero Temperature, *Commun. Math. Phys.*, **228** (2002), 495–518.

[35] J. Gravner and A.E. Holroyd, Slow convergence in bootstrap percolation, *Ann. Appl. Prob.*, **18** (2008), 909–928.

[36] J. Gravner, A.E. Holroyd and R. Morris, A sharper threshold for bootstrap percolation in two dimensions, *Prob. Theory Rel. Fields*, **153** (2012), 1–23.

[37] J. Gravner and D. Griffeath, Scaling laws for a class of critical cellular automaton growth rules, In: Random Walks (Budapest, 1998), *Bolyai Soc. Math. Stud.*, **9** (1999), 167–186.

[38] J. Gravner and E. McDonald, Bootstrap percolation in a polluted environment, *J. Stat. Phys.* **87** (1997), 915–927.

[39] A. Holroyd, Sharp Metastability Threshold for Two-Dimensional Bootstrap Percolation, *Prob. Theory Rel. Fields*, **125** (2003), 195–224.

[40] A. Holroyd, The metastability threshold for modified bootstrap percolation in d dimensions, *Electron. J. Prob.*, **11** (2006), 418–433.

[41] A.E. Holroyd, T.M. Liggett and D. Romik, Integrals, partitions, and cellular automata, *Trans. Amer. Math. Soc.*, **356** (2004), 3349–3368.

[42] S. Janson, T. Łuczak, T. Turova and T. Vallier, Bootstrap percolation on the random graph $G(n, p)$, *Ann. Appl. Prob.*, **22** (2012), 1989–2047.

[43] F. Martinelli, Lectures on Glauber dynamics for discrete spin models, Lectures on Probability Theory and Statistics, Springer Lecture Notes in Mathematics, **1717** (1998), 93–191.

[44] F. Martinelli and C. Toninelli, Towards a universality picture for the relaxation to equilibrium of kinetically constrained models, arXiv:1701.00107.

[45] R. Morris, Zero-temperature Glauber dynamics on $\mathbb{Z}^d$, *Prob. Theory Rel. Fields*, **149** (2011), 417–434.

[46] R. Morris, The second term for bootstrap percolation in two dimensions, manuscript, available at http://w3.impa.br/~rob/.

[47] R. Morris, Bootstrap percolation and other automata, *European J. Combin.*, to appear.

[48] N. Morrison and J.A. Noel, Extremal Bounds for Bootstrap Percolation in the Hypercube, arXiv:1506.04686.

[49] T.S. Mountford, Critical length for semi-oriented bootstrap percolation, *Stochastic Process. Appl.*, **56** (1995), 185–205.

[50] S. Nanda, C.M. Newman and D. Stein, Dynamics of Ising spin systems at zero temperature, In *On Dobrushin's way (From Probability Theory to Statistical Mechanics)*, eds. R. Minlos, S. Shlosman and Y. Suhov, *Am. Math. Soc. Transl.*, **198** (2000), 183–194.

[51] J. von Neumann, Theory of Self-Reproducing Automata. Univ. Illinois Press, Urbana, 1966.

[52] C.M. Newman and D. Stein, Zero-temperature dynamics of Ising spin systems following a deep quench: results and open problems, *Physica A*, **279** (2000), 159–168.

[53] D. Reimer, Proof of the van den Berg–Kesten Conjecture, *Combin. Prob. Computing*, **9** (2000), 27–32.

[54] R.H. Schonmann, On the behaviour of some cellular automata related to bootstrap percolation, *Ann. Prob.*, **20** (1992), 174–193.

[55] A. Vespagnani, R. Dickman, M. Muñoz and S. Zapperi, Absorbing-state phase transitions in fixed-energy sandpiles, *Phys. Rev. E*, **62** (2000), 45–64.

IMPA
Estrada Dona Castorina, 110
Jardim Botânico
Rio de Janeiro, Brazil
rob@impa.br

Robustness of graph properties

Benny Sudakov

Abstract

A typical result in graph theory says that a graph G, satisfying certain conditions, has some property $\mathcal{P}$. Once such a theorem is established, it is natural to ask how strongly G satisfies $\mathcal{P}$. Can one strengthen the result by showing that G possesses $\mathcal{P}$ in a robust way? What measures of robustness can one utilize? In this survey, we discuss various measures that can be used to study robustness of graph properties, illustrating them with examples.

1 Introduction

Let G be a graph and $\mathcal{P}$ a graph property. Many results in graph theory are of the form "under certain conditions, G has property $\mathcal{P}$". Once such a result is established, it is natural to ask how strongly does G possess $\mathcal{P}$? In other words, we want to determine the *robustness* of G with respect to $\mathcal{P}$. Recently, there has been increasing interest in the study of robustness of graph properties, aiming to strengthen classical results in extremal and probabilistic combinatorics. The goal of this paper is to discuss several such results and to use them to illustrate various measures that can be used to study the robustness of graph properties.

The property which we consider frequently in this survey is Hamiltonicity, which we use as a motivating example. A *Hamilton cycle* in a graph is a cycle which passes through every vertex of the graph, and a graph is *Hamiltonian* if it contains a Hamilton cycle. Hamiltonicity is one of the most central notions in graph theory which has been intensively studied by numerous researchers. The problem of deciding the Hamiltonicity of a graph is one of the NP-complete problems that Karp listed in his seminal paper [53], and accordingly, one cannot hope for a simple classification of such graphs. Nonetheless, there are many results deriving conditions that are sufficient to establish Hamiltonicity. For example, a classical result proved by Dirac in 1952 (see, e.g., [27, Theorem 10.1.1]) asserts that every graph on $n \geq 3$ vertices of minimum degree at least $\frac{n}{2}$ is Hamiltonian. In this context, we say that a graph is a *Dirac graph* if it has minimum degree at least $\frac{n}{2}$. Note that the bound $\frac{n}{2}$ is tight, as can be seen by the following two examples (in both n is odd): the first one is a graph obtained by taking two complete graphs of order $\frac{n+1}{2}$ sharing one vertex, and the second one is the complete bipartite graph with parts of sizes $\frac{n+1}{2}$ and

$\frac{n-1}{2}$. Both graphs have n vertices and minimum degree $\frac{n-1}{2}$, but are not Hamiltonian.

Dirac's theorem is one of the most influential results in the study of Hamiltonicity, and by now many related results are known (see, e.g., [17]). It is therefore very natural to try to strengthen this theorem, by asking whether Dirac graphs are robustly Hamiltonian. It turns out that there are several ways to answer this question using different measures of robustness. For example, one could try to show that a Dirac graph has many Hamilton cycles or that it contains several edge-disjoint Hamilton cycles. One can also study whether Maker can win a Hamiltonicity game played on the edges of a Dirac graph. Another natural question concerns the Hamiltonicity of random subgraphs of Dirac graphs. We can also put some restrictions on the pairs of edges of a Dirac graph and consider whether there are Hamilton cycles which do not contain a pair of conflicting edges. Note that the answer to each of these questions defines in some sense a measure of robustness of a Dirac graph with respect to Hamiltonicity.

Another measure of how strongly a graph satisfies some property is the so-called *resilience* of a property. A graph property is called *monotone increasing* if it is closed under the addition of edges. Roughly speaking, for a monotone increasing graph property, the resilience quantifies the robustness in terms of the number of edges one must delete from G, locally or globally, in order to destroy the property $\mathcal{P}$. Resilience was recently studied extensively in the context of random graphs. The most commonly used model of random graphs, sometimes even synonymous with the term "random graph", is the so called *binomial random graph* $G(n, p)$.

The random graph $G(n, p)$ denotes the probability space whose elements are graphs on a vertex set $[n] = \{1, \ldots, n\}$, and where each pair of vertices forms an edge, randomly and independently, with probability p. Here, p is a positive real not greater than one, which can depend on n. Abusing notation slightly, we denote a graph drawn from this distribution by $G(n, p)$. It is well known (and easy to prove) that this distribution is concentrated on the graphs with roughly $\binom{n}{2} p$ edges. We say that the random graph $G(n, p)$ possesses a graph property $\mathcal{P}$ *asymptotically almost surely*, or a.a.s. for short, if the probability that $G(n, p)$ satisfies $\mathcal{P}$ tends to 1 as the number of vertices n tends to infinity. It is well known (see [52]) that a monotone property $\mathcal{P}$ has a *threshold* p_0 in a sense that if $p \gg p_0$, then the random graph a.a.s. satisfies $\mathcal{P}$, while if $p \ll p_0$, then $G(n, p)$ a.a.s. does not satisfy it. The study of random graphs, introduced in the seminal paper of Erdős and Rényi [31], has experienced spectacular growth in the last fifty years, with hundreds of papers and several monographs (see, e.g., [14, 52, 42]) devoted to the subject. For almost any interesting graph property, we now understand quite well when the random graph $G(n,p)$ typically has this property. It is therefore interesting to obtain robust versions of classical results on random graphs.

In the subsequent sections of this paper, we discuss the robustness notions we mentioned above in more detail and present several extensions of well-known theorems using these notions. It is of course impossible to cover all the known robustness results in one survey, and therefore the choice of results we present is inevitably subjective. Yet we hope to describe enough examples from this fascinating area to appeal to many researchers in extremal and probabilistic combinatorics, and to motivate further study of the subject.

2 Many copies

Given a graph G, let $h(G)$ denote the number of distinct Hamilton cycles in G. The above mentioned theorem of Dirac states that $h(G) > 0$ for every n-vertex graph with minimum degree at least $n/2$. One obvious way to strengthen this theorem is to show that every Dirac graph contains several Hamilton cycles. This was achieved by Sárközy, Selkow and Szemerédi [88], who proved that every Dirac graph G contains not only one but at least $c^n n!$ Hamilton cycles for some small positive constant c. They also conjectured that c can be improved to $1/2 - o(1)$. This is best possible, since there are Dirac graphs with at most $(1/2 + o(1))^n n!$ Hamilton cycles. To see this, consider the random graph $G(n,p)$ with $1/2 < p = 1/2 + o(1)$. Indeed, in this case, with high probability the minimum degree $\delta(G(n,p)) = pn + o(n) \geq n/2$ and the expected number of Hamilton cycles is $p^n(n-1)!$.

The conjecture of Sárközy, Selkow and Szemerédi was proved in a remarkable paper of Cuckler and Kahn [25]. In fact, Cuckler and Kahn proved the following stronger result.

Theorem 2.1 *For every Dirac graph G on n vertices with minimum degree $\delta(G)$,*

$$h(G) \geq \left(\frac{\delta(G)}{e}\right)^n (1 - o(1))^n .$$

By considering a random graph with edge probability $p = \delta(G)/n$, one can easily check that this estimate is tight up to a multiplicative factor $(1 - o(1))^n$. The proof of Cuckler and Kahn uses a self-avoiding random walk on G, in which the next vertex is chosen from the yet unvisited neighbors of the current vertex according to a very cleverly chosen distribution. The authors show that even after walking for at least $n - o(n)$ steps, one can use the remaining vertices to close it into a Hamilton cycle. Since the initial part of the cycle was chosen randomly, this process produces many distinct Hamilton cycles. The details of how to choose the correct edge-weights and the proof that the above strategy works are quite involved and will not be discussed here.

A different approach to this problem was proposed in [38]. This approach is based on the standard estimates for the permanent of a matrix (the famous Minc conjecture, established by Brégman [19], and Van der Waerden conjecture, established by Egorychev [29] and Falikman [35]). Let S_n be the set of all permutations of the set $[n]$. The *permanent* of an $n \times n$ matrix A is defined as $per(A) = \sum_{\sigma \in S_n} \prod_{i=1}^{n} A_{i\sigma(i)}$. Note that every permutation $\sigma \in S_n$ has a cycle representation, which is unique up to the order of cycles. When A is the 0-1 adjacency matrix of a graph, every non-zero summand in the permanent is 1 and corresponds to a (≤ 2)-factor (consisting of cycles and single edges). Thus, the permanent of A counts the number of such factors. A 0-1 matrix A is called r-*regular* if it contains exactly r 1's in every row and column. Given such a matrix, the two above mentioned estimates on the permanent show that

$$r^n \frac{n!}{n^n} \leq per(A) \leq (r!)^{n/r}, \qquad (2.1)$$

which is asymptotically $(1 - o(1))^n (r/e)^n$.

Let G be a graph on n vertices which contains an r-factor (that is, an r-regular spanning subgraph) with r linear in n. Using estimates (2.1) one can show that G contains many 2-factors with few (at most $n^{1-\delta}$ for some constant $\delta > 0$) cycles. These 2-factors are then converted into many Hamilton cycles using rotation-extension type techniques. For some previous applications of the permanent-based approach to Hamiltonicity problems see, e.g., [3] and [43].

To illustrate this technique we sketch the proof of the following statement, which gives a lower bound on the number of Hamilton cycles in a dense graph G in terms of reg(G), where reg(G) is the maximal even r for which G contains an r-factor.

Proposition 2.2 *Let G be a Dirac graph on n vertices. Then the number of Hamilton cycles in G is at least $\left(\frac{reg(G)}{e} \right)^n (1 - o(1))^n$.*

In particular, for a cn-regular graph G on n vertices and constant $c > 1/2$ we show that $h(G) \geq (c + o(1))^n n!$, which is asymptotically tight. In general, the estimate in Proposition 2.2 is weaker than the result of Cuckler and Kahn. On the other hand, since every Dirac graph contains an r-factor with even r about $n/4$ (see [54]), this bound implies immediately the result of Sárközy, Selkow and Szemerédi mentioned above.

Sketch of proof of Proposition 2.2. Let $r = \text{reg}(G)$ and let $H \subseteq G$ be an r-factor of G. We have that $r = \Theta(n)$ and we can use permanent estimates (see [38]) to show that H has at least $(r/e)^n(1-o(1))^n$ 2-factors with at most $s = n^{1/2+o(1)}$ cycles. For every such factor F, we can turn F into a Hamilton cycle of G by adding and removing at most $O(s)$ edges as follows.

Take a non-Hamilton cycle C in F. By the connectivity of G we can find a vertex $v \in V(C)$ and a vertex $u \in V(G) \backslash V(C)$ for which $vu \in E(G)$. Then by deleting the edge vv^+ from C (v^+ is next vertex after v in the cycle) we get a path P which can be extended by the edge vu. Connecting it to a cycle C' which contains u we obtain a longer path P'. Repeat this argument as long as we can. If there are no edges between the endpoints w, w' of the current path P' and the other cycles from F, then we can close P' into a cycle as in the standard proof of Dirac's theorem. Indeed, both w, w' have at least $n/2$ neighbors in P' and therefore there is a pair of consecutive vertices v, v^+ of P' such that w is adjacent to v^+ and w' is adjacent to v.

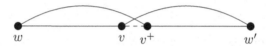

These edges together with P' form a cycle, which is connected to some other cycle in F as we explained above. Note that in each such step we invest at most 4 edge replacements in order to decrease the number of cycles in the factor by 1, and unless the current cycle is a Hamilton cycle, we can always merge two cycles. Therefore, after $O(s)$ edge replacements we get a Hamilton cycle.

In order to complete the proof, note that every Hamilton cycle C we constructed was counted at most $\binom{n}{s}(2s)^{2s} = r^{o(n)}$ times. Indeed, to get a 2-factor F from C, choose s edges of C to delete. This gives at most s paths which need to be turned into a 2-factor by connecting their endpoints. For each endpoint we have at most $2s$ choices of other endpoints to connect it to. Therefore, we obtain at least $(r/e)^n(1-o(1))^n$ Hamilton cycles. $\square$

A *tournament* is an oriented complete graph. It is easy to prove by induction that every tournament has a Hamilton path. Moreover, if the tournament is regular, i.e., if all vertices have the same in/outdegrees, then it also has a Hamilton cycle. Hence, it is logical to ask whether it has many such cycles. Counting Hamilton cycles in tournaments is a very old problem which goes back some seventy years to one of the first applications of the probabilistic method by Szele [91]. He proved that there is a tournament on n vertices which has at least $(n-1)!/2^n$

Hamilton cycles. Alon [3] showed, using permanent estimates, that this result is nearly tight and that every n-vertex tournament has at most $O(n^{3/2}(n-1)!/2^n)$ Hamilton cycles. Thomassen [93] conjectured that the randomness is unnecessary in Szele's result and that in fact every regular tournament contains at least $n^{(1-o(1))n}$ Hamilton cycles. This conjecture was solved by Cuckler [24] using the above mentioned random walk approach. He proved that every regular tournament on n vertices contains at least $\frac{n!}{(2+o(1))^n}$ Hamilton cycles.

Tournaments are a special case of *oriented* graphs, i.e., directed graphs which are obtained by orienting the edges of a simple graph. That is, between every unordered pair of vertices $\{x, y\} \subseteq V(G)$ at most one of the (oriented) edges xy or yx is present. Hamiltonicity problems in oriented graphs are usually much more challenging. Given an oriented graph G, let $\delta^+(G)$ and $\delta^-(G)$ denote the minimum *outdegree* and *indegree* of the vertices in G, respectively. We also use the notation $\delta^{\pm}(G) = \min\{\delta^+(G), \delta^-(G)\}$ and refer to it as the *minimum semi-degree* of G. In the late 70's Thomassen [92] raised the question of determining the minimum semi-degree that ensures the existence of a Hamilton cycle in an oriented graph G. Häggkvist [48] found a construction which gives a lower bound of $\frac{3n-4}{8} - 1$. The problem was only resolved recently by Keevash, Kühn and Osthus [55], who proved an analogue of Dirac's theorem for oriented and large enough graphs.

Theorem 2.3 *For all sufficiently large n, every oriented graph G on n vertices with $\delta^{\pm}(G) \geq \frac{3n-4}{8}$ contains a Hamilton cycle.*

Having obtained such a theorem, it is in the spirit of the above discussion to ask whether one can actually find many Hamilton cycles when $\delta^{\pm}(G) \geq \frac{3n-4}{8}$. It turns out that the permanent-based approach can be used to tackle this question in the case the oriented graph is nearly regular, yielding the following theorem [38].

Theorem 2.4 *Let n be sufficiently large and let G be an oriented graph on n vertices whose in/outdegrees are all $cn \pm o(n)$ for some constant $c > 3/8$. Then $h(G) \geq \left(\frac{(c+o(1))n}{e} \right)^n$.*

The bound on the in/outdegrees in this theorem is tight. This follows from the above-mentioned construction of Häggkvist [48], which shows that there are non-Hamiltonian n-vertex oriented graphs with all in/outdegrees of order $(3/8 - o(1))n$. Since in a regular tournament all in/outdegrees are $\frac{n-1}{2}$, this theorem substantially extends the result of Cuckler [24] mentioned above.

The Hamiltonicity problem was extensively studied for random graphs as well. Early results on the Hamiltonicity of random graphs were proved by Pósa [84] and Korshunov [60]. Improving on these results, Bollobás [12] and Komlós and Szemerédi [59] proved that if $p \geq (\log n + \log \log n + \omega(n))/n$ for any function $\omega(n)$ that goes to infinity together with n, then $G(n, p)$ is a.a.s. Hamiltonian (here and later in the paper all logarithms are natural). The range of p cannot be improved, since if $p \leq (\log n + \log \log n - \omega(n))/n$, then $G(n, p)$ a.a.s. has a vertex of degree at most one.

Once the Hamiltonicity threshold is established, it is natural to try to estimate the number of Hamilton cycles in the random graph. Using linearity of expectation, one can immediately see that the expected number of such cycles in $G(n, p)$ is $\frac{(n-1)!}{2} p^n$. Therefore it is logical to suspect that the actual number of Hamilton cycles in the random graph is typically close to this value, at least when the edge probability is not too small. This was indeed confirmed by Glebov and Krivelevich [45] (also see their paper for the history of the problem and many related results). They proved that for all edge probabilities for which $G(n, p)$ is Hamiltonian, i.e., for $p \geq (\log n + \log \log n + \omega(n))/n$, it has a.a.s. $n! p^n (1 - o(1))^n$ Hamilton cycles.

3   Edge disjoint copies

In the previous section we showed that any Dirac graph contains not only one but exponentially many Hamilton cycles. Another interesting way to extend Dirac's theorem is to show that a minimum degree of at least $n/2$ implies the existence of many edge-disjoint Hamilton cycles. Note that there is no obvious way to deduce such a statement from the results in Section 2. Indeed, a graph might have many Hamilton cycles, all sharing the same small set of edges. The problem of how many edge-disjoint Hamilton cycles one can find in a Dirac graph was first posed by Nash-Williams [81] in 1970. In [82] he proved that any such graph has at least $\frac{5}{224} n$ edge-disjoint Hamilton cycles. He also asked [81, 82] to improve this estimate. Clearly, $\lfloor (n + 1)/4 \rfloor$ is a general upper bound on the number of edge-disjoint Hamilton cycles in a Dirac graph obtained by considering an $n/2$-regular graph, and originally Nash-Williams [81] believed this to be tight.

Babai (see also [81]) found a counterexample to this conjecture. Extending his ideas further, Nash-Williams gave an example of a graph on $n = 8k + 2$ vertices with minimum degree $4k + 1$ and with at most $(n-2)/8$ edge-disjoint Hamilton cycles. He conjectured that this example is tight, that is, any Dirac graph contains at least $(n-2)/8$ edge-disjoint Hamilton cycles. Recall that reg(G) is the maximal even r for which G contains an r-factor. Since a Hamilton cycle is a 2-regular subgraph of G, the maximum number of edge-disjoint Hamilton cycles in G is at most reg(G)/2. In order to study the maximum number of edge-disjoint Hamilton cycles in Dirac graphs, it is therefore natural to ask for the largest even integer r such that every n-vertex graph with minimum degree δ must contain an r-regular spanning subgraph. We denote this function by reg(n, δ). Note that the complete bipartite graph whose parts differ by one vertex shows that reg(n, δ) = 0 for $\delta < n/2$. The case $\delta = n/2$ was solved by Katerinis [54], who showed that the above mentioned example of Nash-Williams is tight and reg($n, n/2$) = $(n-2)/8$. For $\delta > n/2$ it is known (see, e.g., [23]) that

$$\frac{\delta + \sqrt{n(2\delta - n)}}{2} - 1 \le \text{reg}(n, \delta) \le \frac{\delta + \sqrt{n(2\delta - n)}}{2} + 1.$$

After many partial results, the question of Nash-Williams was answered asymptotically by Csaba, Kühn, Lo, Osthus and Treglown [23]. They proved that for large n, every n-vertex Dirac graph G contains at least reg($n, \delta(G)$)/2 edge-disjoint Hamilton cycles. Although this determines the worst-case behavior, for a specific Dirac graph G, reg(G) can be much larger than reg(n, δ). Therefore, it is logical to try and bound the maximum number of edge-disjoint Hamilton cycles in terms of reg(G). This question was raised by Kühn, Lapinskas and Osthus [73], who conjectured the following tight result.

Conjecture 3.1 *Suppose G is a Dirac graph. Then G contains at least reg(G)/2 edge-disjoint Hamilton cycles.*

The following theorem of Ferber, Krivelevich and the author [38] gives an approximate asymptotic version of this conjecture.

Theorem 3.2 *For every $\varepsilon > 0$ and a sufficiently large integer n, the following holds: every graph G on n vertices and with $\delta(G) \ge (1/2 + \varepsilon)n$ contains at least $(1 - \varepsilon)\text{reg}(G)/2$ edge-disjoint Hamilton cycles.*

Sketch of proof. The approach based on the permanent estimates and rotations, described in the previous section, is rather oblivious to the value of reg(G). This makes it well-suited for the proof of the above theorem.

We first construct an auxiliary graph $H \subset G$ with relatively small degree and good expansion properties which we use to perform rotations. More precisely, we claim that for every constant ε and $\alpha \le \varepsilon^2$ there exist some $\beta \ll \varepsilon\alpha$ and a subgraph $H \subset G$ satisfying all of the following: $\delta(H) \ge \varepsilon n/8$ and $G - H$ still has an r-factor G' for an even r with $r \ge$

$(1 - \varepsilon/2)\mathrm{reg}(G)$. Moreover, for every subset of edges E' such that $|E'| \leq \beta n^2$ and $\delta(H - E') \geq \alpha n$ we have that $H - E'$ is connected and every subset $S, |S| \geq \alpha n$ in $H - E'$ has at least $(1/2 + \varepsilon/4)n$ neighbors outside S. This graph can be obtained by randomly choosing $\varepsilon n/16$ 2-factors in G. To prove that it has all the listed above properties we use the permanent estimates from Section 2. Let $S, T \subseteq V(G)$ be two disjoint subsets of sizes $|S| = \alpha n$ and $|T| = \frac{(1-\varepsilon)n}{2}$. It is enough to show that with high probability $|E_H(S,T)| \geq \beta n^2$. Since $\delta(G) \geq (1/2 + \varepsilon)n$, it follows that $d_G(v, T) \geq \varepsilon n/2$ for every $v \in S$. Therefore, $|E_G(S,T)| \geq |S|\varepsilon n/2 = \frac{\varepsilon \cdot \alpha}{2}n^2 \gg \beta n^2$. Using this fact one can bound the probability that $|E_H(S,T)| \geq \beta n^2$ as a ratio of $per(A')/per(A)$, where A is the adjacency matrix G and A' is obtained from A by deleting at least $\frac{\varepsilon \cdot \alpha}{2}n^2 - \beta n^2 \geq \frac{\varepsilon \cdot \alpha}{4}n^2$ ones. The permanent estimates can be used to show that this ratio is exponentially small so that the union bound over all pairs S, T can be applied (see [38] for more details).

Now the proof can be completed similarly to that of Proposition 2.2. We repeatedly find and delete $m = (1 - \varepsilon/2)r/2$ edge-disjoint 2-factors of G', each containing at most $s^* = n^{1/2 + o(1)}$ cycles. Note that by removing such a factor from an r'-regular graph, the resulting graph is $(r' - 2)$-regular, and therefore one can apply the permanent estimates over and over again. Finally we perform rotations using the edges of H to turn every such factor into a Hamilton cycle by replacing at most $O(s^*)$ edges. Moreover, every edge of H which we use is permanently deleted from H. Since the total number of deleted edges in all steps is at most $O(ns^*) = o(n^2)$, the properties of H are not affected. □

The example of a Dirac graph with at most $(n - 2)/8$ edge-disjoint Hamilton cycles is not regular. Accordingly, Nash-Williams conjectured that every d-regular Dirac graph contains $\lfloor d/2 \rfloor$ edge-disjoint Hamilton cycles. Recently, in a remarkable tour de force, Csaba, Kühn, Lo, Osthus and Treglown [23] proved this conjecture for all sufficiently large Dirac graphs.

Recall that a tournament is an orientation of the complete graph. In the previous section we mentioned that every regular tournament has a Hamilton cycle. Does it have many edge-disjoint Hamilton cycles? In principle, since all in/outdegrees are equal it is possible that it contains $(n-1)/2$ edge-disjoint Hamilton cycles. This question was raised by Kelly in 1968, who posed the following striking conjecture.

Conjecture 3.3 *The edges of every regular tournament can be decomposed into Hamilton cycles.*

Recently, this was proved for large tournaments by Kühn and Osthus

[74]. In order to solve the conjecture they developed a very powerful decomposition theorem for the so called *robust outexpanders*. Given an n-vertex directed graph G a ν-*robust outneighborhood* of a vertex-set S is the set of all vertices in G with at least νn inneighbors in S. The graph is called a *robust* (ν, τ)-*outexpander* if the ν-robust outneighborhood of every set S of size $\tau n \leq |S| \leq (1 - \tau)n$ has size at least $|S| + \nu n$. We call an oriented graph r-regular if all its in/outdegrees are equal to r. Using the celebrated Szemerédi's Regularity Lemma (see, e.g., [58] for many other applications of this important tool), Kühn and Osthus [74] proved that a large regular robust outexpander with linear degree has a Hamilton decomposition. One consequence of their result is the following decomposition theorem for regular oriented graphs of high degree, see [75].

Theorem 3.4 *For all sufficiently large n and constant $c > 3/8$, every cn-regular oriented n-vertex graph G has a Hamilton decomposition.*

To prove this theorem, it is enough to simply verify that such oriented graphs are robust outexpanders. The constant $3/8$ is tight, since as we mentioned in Section 2, there are cn-regular oriented n-vertex graphs with $c < 3/8$ which are not Hamiltonian. Since a regular tournament is an $(n - 1)/2$-regular oriented graph, this theorem implies Kelly's conjecture for large tournaments.

An oriented n-vertex graph with minimum semi-degree at least $3n/8$ is Hamiltonian by Theorem 2.3. Can one strengthen this result by showing that such a graph has many edge-disjoint Hamilton cycles? Similarly to the undirected case, for an oriented graph G let reg(G) be the maximum integer r such that G has a spanning r-regular subgraph. If G has t edge-disjoint Hamilton cycles then clearly their union is t-regular and therefore $t \leq$ reg(G). Together with Ferber and Long [40], we believe that this bound is tight.

Conjecture 3.5 *Let n be sufficiently large, and let G be an oriented graph on n vertices with $\delta^{\pm}(G) \geq 3n/8$. Then G contains reg(G) edge-disjoint Hamilton cycles.*

In [40] we prove the following approximate version of this conjecture. If G is an n-vertex oriented graph with $\delta^{\pm}(G) \geq cn$ for some constant $c > 3/8$ and n is sufficiently large, then G contains $(1 - o(1))$reg(G) edge-disjoint Hamilton cycles. The proof is based on various probabilistic arguments and gives a rather short alternative proof of an approximate version of Kelly's conjecture, first established in [76].

Given a regular graph G, denote by $H(G)$ the number of Hamiltonian decompositions of G. We already know that such a G of high degree has a Hamiltonian decomposition, so it is interesting to count the number of decompositions. To estimate the number of Hamilton cycles in G, observe that $per(A_G)$ is an upper bound. Combining this with the permanent estimates explained in (2.1), we see that an r-regular graph has at most $(r!)^{n/r}$ Hamilton cycles. By choosing any Hamilton cycle and deleting it, we obtain an $(r-2)$-regular graph. The number of Hamilton cycles in the new graph can again be bounded from above using (2.1). Continuing this process and multiplying all the estimates for regularities $r, r-2, r-4, \ldots$ we can use Stirling's formula to deduce that $H(G)$ is at most $\left((1+o(1))\frac{r}{e^2}\right)^{nr/2}$. In the case of an oriented r-regular graphs the same arguments gives that the number of Hamilton decompositions is at most $\left((1+o(1))\frac{r}{e^2}\right)^{nr}$. It turns out that some of the tools developed to prove the existence of many edge-disjoint Hamilton cycles can be pushed further to count the decompositions. In particular, for sufficiently dense graphs, both of these estimates have the right order of magnitude. This was established for r-regular n-vertex graphs with $r \geq (1/2 + \epsilon)n$ in [46] and for oriented graphs with $r \geq (3/8 + \epsilon)n$ in [40].

As we explained above, the problem of determining the number of edge-disjoint Hamilton cycles that can be packed into a given graph has a long history and was extensively studied for various classes of graphs. We conclude this section by discussing this question in the context of random graphs. In order to contain s edge-disjoint Hamilton cycles the graph must clearly have minimum degree of at least $2s$. For $G(n, p)$, this happens typically when $p \geq (\log n + (2s-1)\log\log n + \omega(n))/n$. Generalizing such Hamiltonicity result for random graphs, Bollobás and Frieze [16] proved that a minimum degree of $2s$ is indeed a.a.s. sufficient for $G(n, p)$ to contain s edge-disjoint Hamilton cycles. This motivated Frieze and Krivelevich [44] to conjecture that a random graph $G = G(n, p)$ a.a.s. has $\lfloor \delta(G)/2 \rfloor$ edge-disjoint Hamilton cycles for all edge probabilities. This striking conjecture was confirmed in a series of papers by several researches, with two main ranges of edge probabilities covered in [57, 68]. The problems of packing and counting Hamilton cycles in random directed graphs were studied in [39].

4   Random subgraphs and Maker-Breaker games

In this section we discuss two additional, closely related, robust extensions of Dirac's theorem. In both cases, instead of having access to all the edges of a Dirac graph G, we have to find a Hamilton cycle in some very sparse subgraph of G. Moreover, this subgraph is given to us by some random process or by an adversary.

4.1 Random subgraphs

An equivalent way of describing the random graph $G(n,p)$ is to say that it is the probability space of graphs obtained by taking every edge of the complete graph K_n independently with probability p. A variety of questions can be asked when we start with a host graph G other than K_n, and consider the probability space of graphs obtained by taking each one of its edge independently with probability p. We denote this probability space as G_p. If the original graph G has property $\mathcal{P}$, then a natural way to strengthen such a statement is to show that for some $p < 1$ the random subgraph G_p with high probability also has this property. By doing so, one frequently obtains interesting and challenging questions.

For example, it is an easy exercise to prove that a graph G with minimum degree k contains a path and even a cycle with at least $k+1$ vertices. On the other hand, it is more difficult to show that this remains true for a random subgraph G_p of such G when $p < 1$. An even more challenging question is to determine all values of p such that G_p with high probability has a cycle of length at least $k + 1$. This and the related questions were recently studied by several researchers in [65, 86, 47]. In [47] it was proved that for large k and $p \geq \frac{\log k + \log \log k + \omega(k)}{k}$ the random graph G_p a.a.s. contains a cycle of length at least $k + 1$. Since one can take G to be a complete graph on $k+1$ vertices, this result generalizes the classical result on Hamiltonicity of the random graph $G(k,p)$.

In this context, to get a better understanding of the robustness of Dirac's theorem we consider the following question. Let G be an n-vertex graph with minimum degree at least $\frac{n}{2}$. Since Hamiltonicity is a monotone graph property, we know (see [52]) that there exists a threshold p_0 such that if $p \gg p_0$, then G_p is a.a.s. Hamiltonian, and if $p \ll p_0$, then it is a.a.s. not Hamiltonian. What is the Hamiltonicity threshold for G_p, in particular, does G_p stay Hamiltonian for $p \ll 1$? Note that a positive answer to the latter question shows that typically one cannot destroy Hamiltonicity even by randomly removing most of the edges of a Dirac graph. The following theorem, which was proved in [64], answers these questions.

Theorem 4.1 *There exists a positive constant C such that for $p \geq \frac{C \log n}{n}$ and a Dirac graph G on n vertices, the random subgraph G_p is a.a.s. Hamiltonian.*

This theorem establishes the correct order of magnitude of the threshold function for Hamiltonicity of the random subgraph of any Dirac graph. Indeed, if $p \leq (1 + o(1))\frac{\log n}{n}$, then it is easy to see that the graph G_p a.a.s. has isolated vertices.

It is worth comparing this theorem with the robustness results from the two previous sections. Given a Dirac graph G, recall that $h(G)$ is the number of Hamilton cycles in G. Since the expected number of Hamilton cycles in the random subgraph G_p is $p^n h(G)$, Theorem 4.1 implies that $p^n h(G) \geq 1$ for $p \geq \frac{C \log n}{n}$. This shows that $h(G) \geq \left(\frac{n}{C \log n}\right)^n$. We can also partition the edges of a Dirac graph G into $t = C^{-1} n / \log n$ disjoint parts, by putting every edge randomly and independently into one of these parts with probability $p = 1/t$. Every part than behaves like a random subgraph G_p and a.a.s. contains a Hamilton cycle. We have therefore obtained a partition such that most of its parts contain a Hamilton cycle, showing that a Dirac graph has at least $\Omega(n/\log n)$ edge-disjoint Hamilton cycles. The above theorem can thus be used to recover somewhat weaker versions of the results mentioned in Sections 2 and 3.

Sketch of proof of Theorem 4.1. The proof is rather involved, so we only give a very brief overview. Our main tool is Pósa's rotation-extension technique (see [84]), which exploits the expansion property of the graph. Let G be a connected graph and let $P = (v_0, \cdots, v_\ell)$ be a path in G which we want to extend. Suppose that we cannot do this directly, i.e., all neighbors of v_0, v_ℓ lie on P. If G contains an edge of the form (v_0, v_{i+1}) for some i, then $P' = (v_i, \cdots, v_0, v_{i+1}, \cdots, v_\ell)$ forms another path of length ℓ in G. We say that P' is obtained from P by a *rotation* with *fixed endpoint*

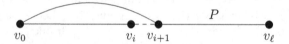

v_ℓ, *pivot point* v_{i+1} and *broken edge* (v_i, v_{i+1}). Note that after performing this rotation, we can now try to extend the new path using edges incident to v_i. We can also close the cycle using the edge (v_i, v_ℓ), if it exists. Since G is connected this will also allow us to extend the path. As we perform more and more rotations, we will get more and more new endpoints and such closing pairs. We employ this rotation-extension technique repeatedly until we can extend the path.

The proof of Theorem 4.1 splits into three different cases, based on the structure of the Dirac graph G. The first two cases apply when G is close to one of the two extremal configurations, i.e., if it is close to a complete bipartite graph or to the union of two disjoint complete graphs. These cases are easier to handle since one can use techniques developed to show Hamiltonicity of random graphs. The third case applies when G has at least cn^2 edges between any two disjoint subsets of size $n/2$ for some constant $c > 0$. In this case one can prove that a random subgraph G_p has typically strong expansion properties and in particular still has many edges between any two disjoint subsets of size $n/2$. Using this expansion property of G_p one can carefully perform rotations of its longest path to obtain a set S_P of linear size with the following property: for every $v \in S_P$ there is a set T_v of size at least $(1/2 + \delta)n$ with constant $\delta > 0$ such that for every $w \in T_v$ there is a path of maximal length from v to w. Since the degree of v in G was at least $n/2$, there are linearly many edges in G between v and T_v. If any of these edges is in G_p we can close the longest path into a cycle, which is a Hamilton cycle by maximality of the initial path and the connectivity of G_p. Since the set S_P has linear size, one can easily show that this a.a.s. occurs for at least one vertex in S_P. $\qquad\square$

4.2 Maker-Breaker games

Let V be a set of elements and $\mathcal{F} \subseteq 2^V$ be a family of subsets of V. A *Maker-Breaker game* involves two players, named Maker and Breaker, who alternately occupy the elements of V, called the *board* of the game. The Breaker makes the first move. The game ends when there are no unoccupied elements of V. Maker wins the game if, in the end, the vertices occupied by Maker contain (as a subset) at least one of the sets in $\mathcal{F}$, the family of *winning sets* of the game. Otherwise Breaker wins.

Chvátal and Erdős [21] were the first to consider biased Maker-Breaker games on the edge set of the complete graph. They realized that such graph games are often "easily" won by Maker when played fairly (that is when Maker and Breaker each claim one element at a time). Thus, for many graph games, it is logical to give Breaker some advantage. In a $(1 : b)$ Maker-Breaker game we follow the same rules as above, except that Breaker claims b elements each round. It is not too difficult to see that, if for some fixed game, Maker can win the $(1 : b)$ game, then Maker can win the $(1 : b')$ game for every $b' < b$. Therefore Maker-Breaker games are *bias monotone*. It is thus natural to consider the *critical bias* of a game, defined as the maximal b_0 such that Maker wins the $(1 : b_0)$ game.

One of the first biased games that Chvátal and Erdős considered in [21] was the Hamiltonicity game played on the edge set of the complete graph. They proved that Maker wins the $(1 : 1)$ game, and that for any fixed positive ε and $b(n) \geq (1 + \varepsilon)\frac{n}{\log n}$, Breaker wins the $(1 : b)$ game for large enough n. Despite many results by various researchers (see [62] for the history), the problem of determining the critical bias of this game was open until recently. It was resolved by Krivelevich [62] who proved that the critical bias is asymptotically $\frac{n}{\log n}$. We refer the reader to [8, 49] for more information on Maker-Breaker games, as well as general positional games.

In the spirit of this survey we would like to strengthen Dirac's theorem from the view point of the Maker-Breaker game. Let G be a Dirac graph and consider the Hamiltonicity Maker-Breaker game played on G. Can Maker win the $(1 : 1)$ game and if yes, what is critical bias? As the step answering this question, the following theorem [64] established the threshold b_0 such that if $b \ll b_0$, then Maker wins, and if $b \gg b_0$, then Breaker wins.

Theorem 4.2 *There exists a constant $c > 0$ such that for $b \leq \frac{cn}{\log n}$ and a Dirac graph G on n vertices, Maker has a winning strategy for the $(1 : b)$ Maker-Breaker Hamiltonicity game on G.*

The theorem implies that the critical bias of this game has order of magnitude $\frac{n}{\log n}$ (since the critical bias is at most $(1 + o(1))\frac{n}{\log n}$ by the result of Chvátal and Erdős mentioned above). Note that in this theorem, once all the elements on the board are claimed, the edge density of Maker's graph is of order of magnitude $\frac{\log n}{n}$ and that this is the same as in Theorem 4.1. This suggests that as in many other Maker-Breaker games, the "probabilistic intuition", a relation between the critical bias and the threshold probability of random graphs holds here (see [8]). In fact, this is not a coincidence, and the proofs of Theorems 4.1 and 4.2 are both done in one unified framework.

As we already mentioned in Section 2, the random graph $G(n, p)$ is a.a.s. Hamiltonian for $p \geq \frac{\log n + \log \log n + \omega(n)}{n}$. Moreover, the Hamiltonicity threshold coincides with the threshold of having minimum degree 2. Can one strengthen these facts using the framework of Maker-Breaker games? Given a graph G with a vertex of degree 3, Breaker can claim at least two edges incident to this vertex. This will leave Maker with a graph containing a vertex of degree 1 (or less), which is not Hamiltonian. Note that even if we allow Maker to move first, it is enough to have two non-adjacent vertices of degree 3 for Breaker to win. This will a.a.s. happen in $G(n, p)$ if $p \leq \frac{\log n + 3 \log \log n - \omega(n)}{n}$. On the other hand, for $p \geq \frac{\log n + 3 \log \log n + \omega(n)}{n}$ the minimum degree of a random graph is a.a.s. 4 so it is natural to believe that Maker can now build a Hamilton cycle. Indeed, such a robust version of the Hamiltonicity of random graphs was proved by Ben-Shimon, Ferber, Hefetz and Krivelevich [9].

Theorem 4.3 *If $p \geq \frac{\log n + 3 \log \log n + \omega(n)}{n}$ then a.a.s. Maker has a winning strategy for the $(1 : 1)$ Hamiltonicity Maker-Breaker game on $G(n, p)$.*

For larger values of the edge probability p this theorem suggests that the $(1 : 1)$ game should be an easy win for Maker. The real question for denser random graphs is therefore to determine the critical bias for the Hamiltonicity game. Extending the above-mentioned result of Krivelevich, Ferber, Glebov, Krivelevich and Naor [36] proved that for $p \gg \frac{\log n}{n}$ a.a.s. the threshold bias for the Hamiltonicity game on $G(n, p)$ is asymptotically $\frac{np}{\log np}$. An interesting topic for future research is to study Hamiltonicity games on random directed graphs.

5 Compatible Hamilton cycles

In this section, we consider the following setting which leads to yet another type of robustness that can be used to study the Hamiltonicity of Dirac graphs and similar questions. Suppose that we are given a Dirac graph G together with a set of restrictions on its edges given to us by an adversary. The adversary can forbid certain pairs of edges to appear together on the Hamilton cycle we are trying to build. Finding a Hamilton cycle satisfying all the imposed restrictions will certainly show some kind of robustness of a Dirac graph with respect to Hamiltonicity. The type of restrictions the adversary can impose is described formally by the following definition.

Definition 5.1 Let $G = (V, E)$ be a graph.

- An *incompatibility system* $\mathcal{F}$ defined over G is a family $\mathcal{F} = \{F_v\}_{v \in V}$ such that for every $v \in V$, F_v is a family of unordered pairs $F_v \subseteq \{\{e, e'\} : e \neq e' \in E, e \cap e' = \{v\}\}$.
- If $\{e, e'\} \in F_v$ for some edges e, e' and a vertex v, then we say that e and e' are *incompatible* in $\mathcal{F}$. Otherwise, they are *compatible* in $\mathcal{F}$. A subgraph $H \subseteq G$ is *compatible* with $\mathcal{F}$, if all its pairs of edges e and e' are compatible.
- For a positive integer Δ, an incompatibility system $\mathcal{F}$ is Δ-*bounded* if for each vertex $v \in V$ and an edge e incident to v, there are at most Δ other edges e' incident to v that are incompatible with e.

This definition is motivated by two concepts in graph theory. First, it generalizes *transition systems* introduced by Kotzig [61] in 1968. In our terminology, a transition system is simply a 1-bounded incompatibility system. Kotzig's work was motivated by a problem of Nash-Williams on cycle coverings of Eulerian graphs (see, e.g., Section 8.7 of [17]).

Incompatibility systems and compatible Hamilton cycles also generalize the concept of *properly colored* Hamilton cycles in edge-colored graphs. A cycle is properly colored if its adjacent edges have distinct colors. The problem of finding properly colored Hamilton cycles in an edge-colored graph was first introduced by Daykin [26]. He asked whether there exists a constant μ such that for large enough n, every edge-coloring of the complete graph K_n in which each vertex is incident to at most μn edges of the same color contains a properly colored Hamilton cycle (we refer to such a coloring as a μn-*bounded edge coloring*). Daykin's question has been answered independently by Bollobás and Erdős [15] with $\mu = \frac{1}{69}$, and by Chen and Daykin [20] with $\mu = \frac{1}{17}$. Bollobás and Erdős further conjectured that all ($\lfloor \frac{n}{2} \rfloor - 1$)-bounded edge colorings of K_n admit a properly colored Hamilton cycle. After several subsequent improvements (see, e.g., [4]), Lo [78] recently settled the conjecture asymptotically, proving that for any positive ε, every ($\frac{1}{2} - \varepsilon)n$-bounded edge coloring of K_n admits a properly colored Hamilton cycle.

Note that a μn-bounded edge coloring obviously defines also a μn-bounded incompatibility system, and thus the question mentioned above can be considered as a special case of the problem of finding compatible Hamilton cycles. However, in general, the restrictions introduced by incompatibility systems need not come from an edge-coloring of graphs, and therefore results on properly colored Hamilton cycles do not necessarily generalize easily to incompatibility systems.

The study of compatible Hamilton cycles in Dirac graphs, although interesting in its own right, is further motivated by the following problem of Häggkvist (see [17, Conjecture 8.40]). In 1988 he conjectured that for every 1-bounded incompatibility system $\mathcal{F}$ over a Dirac graph G, there exists a Hamilton cycle compatible with $\mathcal{F}$. This conjecture can be settled using Theorem 4.2 on the Hamiltonicity Maker-Breaker game played on Dirac graphs. Recall that this theorem asserts the existence of a positive constant β such that Maker has a winning strategy in a $(1 : \beta n / \log n)$ Hamiltonicity Maker-Breaker game played on Dirac graphs. To see how this implies the conjecture, given a graph G and a 1-bounded incompatibility system $\mathcal{F}$, consider a Breaker's strategy claiming, at each turn, the edges that are incompatible with the edge that Maker claimed in the previous turn. This strategy forces Maker's graph to be compatible with $\mathcal{F}$ at all stages. Since Maker has a winning strategy for the $(1 : 2)$ game, we see that there exists a Hamilton cycle compatible with $\mathcal{F}$.

Note that the above analysis gives much more, asserting the existence of a compatible Hamilton cycle for every $\frac{1}{2}\beta n / \log n$-bounded incompatibility system. Is this the best possible result? Can we find a compatible Hamilton cycle for every Δ-bounded system, when Δ is linear in n? The following theorem, proved by Krivelevich, Lee and the author [67], answers these questions.

Theorem 5.2 *There exists a constant $\mu > 0$ such that the following holds for large enough n. For every n-vertex Dirac graph G and a μn-bounded incompatibility system $\mathcal{F}$ defined over G, there exists a Hamilton cycle in G compatible with $\mathcal{F}$.*

This shows that Dirac graphs are very robust against incompatibility systems, i.e., one can find a Hamilton cycle even after forbidding a quadratic number of pairs of edges incident to each vertex from being used together in the cycle. The order of magnitude is best possible since we can forbid all pairs incident to some vertex from being used together to disallow a compatible Hamilton cycle. However, it is not clear what the best possible value of μ is and determining it is an interesting open problem.

The proof in [67] provides the existence of a positive constant μ of approximately 10^{-16} (although no serious attempt was made to optimize it). On the other hand, the following variant of a construction of Bollobás and Erdős [15] shows that μ is at most $\frac{1}{4}$. Let n be an integer of the form $4k - 1$, and let G be an edge-disjoint union of two $\frac{n+1}{4}$-regular graphs G_1 and G_2 on the same n-vertex set. Color the edges of G_1 in red and those of G_2 in blue. Note that G does not contain a properly colored Hamilton cycle since any Hamilton cycle of G is of odd length. Let $\mathcal{F}$ be an incompatibility system defined over G, where incident edges of the same color are incompatible. Then there exists a Hamilton cycle compatible with $\mathcal{F}$ if and only if there exists a properly colored Hamilton cycle. Since there is no properly colored Hamilton cycle, we see that there is no Hamilton cycle compatible with $\mathcal{F}$.

It is also natural to use the notion of compatibility to strengthen classical results on the Hamiltonicity of random graphs. An obvious question here is given a Δ-bounded incompatibility system over $G(n, p)$, when with high probability we can find a compatible Hamilton cycle? Since $G(n, p)$ a.a.s. has no Hamilton cycles for $p \ll \frac{\log n}{n}$, we need to consider values of p above this threshold. Also, since all degrees in our graph are close to np, it is logical to ask whether we can take Δ as large as μnp for some constant $\mu > 0$. The following theorem from [66] summarizes what we know about this problem.

Theorem 5.3 *There exists a positive real μ such that for $p \gg \frac{\log n}{n}$, the graph $G = G(n, p)$ a.a.s. has the following property: for every μnp-bounded incompatibility system defined over G, there exists a compatible Hamilton cycle. Moreover if $p \gg \frac{\log^8 n}{n}$, then we can take $\mu = \left(1 - \frac{1}{\sqrt{2}} - o(1)\right) np$.*

Recalling the above discussion, this result can be seen as an answer to a generalized version of Daykin's question. In fact, Theorem 5.3 generalizes it in two directions. Firstly, we replace properly colored Hamilton cycles by compatible Hamilton cycles and secondly, we replace the complete graph by random graphs $G(n, p)$ for $p \gg \frac{\log n}{n}$ (note that for $p = 1$, the graph $G(n, 1)$ is K_n with probability 1). It is unclear what the best possible value of μ is. The example of Bollobás and Erdős [15] of a $\lfloor \frac{1}{2} n \rfloor$-bounded edge-coloring of K_n with no properly colored Hamilton cycles implies that the optimal value of μ is at most $\frac{1}{2}$, since it provides an upper bound for the case $p = 1$.

Sketch of proof of Theorem 5.3. Let $\mathcal{F}$ be a μnp-bounded incompatibility system over $G = G(n, p)$ for some small, but constant, μ. First we construct an expander graph $R \subset G$ which is compatible with $\mathcal{F}$ and has at most $O(n)$ edges and such that every $X \subset V(R)$ of size at most $n/4$ has at least $2|X|$ neighbors outside X. This graph can be obtained by choosing, randomly with replacement, d (a large constant) edges incident to every vertex in G and showing that the resulting graph has all the desired properties with positive probability. Once we have R, consider a path P in G such that $P \cup R$ is compatible with $\mathcal{F}$ and P is a longest path in $P \cup R$. Using the rotation-extension technique (from the previous section) together with the properties of R, we can rotate both endpoints of P to obtain $\Omega(n^2)$ pairs (u, v) such that we can extend P by adding the edge (u, v). Let us call these pairs *boosters*. Since $G = G(n, p)$ is a random graph, a.a.s. $\Omega(n^2 p)$ of the boosters are actual edges in G. Moreover, since R has only $O(n)$ edges, one can show that one of the boosters is compatible with $P \cup R$. We can therefore repeat this process until we obtain a compatible Hamilton cycle. $\square$

To complete this section, we mention another closely related way to get a robust version of the Hamiltonicity result for random graphs. In an edge-colored graph, we say that a subgraph is *rainbow* if all its edges have distinct colors. There is a vast literature on the branch of Ramsey theory where one seeks rainbow subgraphs in edge-colored graphs. Note that one can easily avoid rainbow copies by using the same color for all edges, and hence in order to find a rainbow subgraph one usually imposes some restrictions on the distribution of colors. Erdős, Simonovits and Sós [34] and Rado [85] developed anti-Ramsey theory where one attempts to determine the maximum number of colors that can be used to color the edges of the complete graph without creating a rainbow copy of a fixed graph. In a different direction, one can try to find a rainbow copy of a target graph by imposing global conditions on the coloring of the host graph. For a real Δ, we say that an edge-coloring of G is *globally Δ-bounded* if each color appears at most Δ times on the edges of G. In 1982, Erdős, Nešetřil and Rödl [30] initiated the study of the problem of finding rainbow subgraphs in a globally Δ-bounded coloring of graphs. One very interesting question of this type is to find sufficient conditions for the existence of a rainbow Hamilton cycle in any globally Δ-bounded coloring. Substantially improving earlier results, Albert, Frieze and Reed [1] proved the existence of a constant $\mu > 0$ for which every globally μn-bounded coloring of K_n (for large enough n) admits a rainbow Hamilton cycle. In fact, they proved a stronger statement, asserting that for all graphs Γ with vertex set $E(K_n)$ (the edge set of the complete graph) and maximum degree at most μn, there exists a Hamilton cycle in K_n which is also an independent set in Γ.

It turns out that the proof technique used in proving Theorem 5.3 can be easily modified to give the following result (see [66]) that extends the above to random graphs.

Theorem 5.4 *There exists a constant $\mu > 0$ such that for $p \gg \frac{\log n}{n}$, the random graph $G = G(n,p)$ a.a.s. has the following property: every globally $\mu n p$-bounded coloring of G contains a rainbow Hamilton cycle.*

Theorem 5.4 is clearly best possible up to the constant μ since one can forbid all rainbow Hamilton cycles in a globally $(1 + o(1))np$-bounded coloring by simply coloring all edges incident to some fixed vertex with the same color.

6 Resilience of graph properties

In this section we discuss the following question. Given a graph G with some property $\mathcal{P}$, how much should one change G in order to destroy $\mathcal{P}$? We call this *the resilience of G with respect to $\mathcal{P}$*. There are two natural kinds of resilience: global and local. It is more convenient to first define these quantities with respect to monotone increasing properties. Recall that $\mathcal{P}$ is monotone increasing if it is preserved under edge addition.

Definition 6.1 Let G be a graph and $\mathcal{P}$ be a monotone increasing property.

- The global resilience of G with respect to $\mathcal{P}$ is the minimum number r such that one can obtain a graph not having $\mathcal{P}$ by deleting r edges from G.

- The local resilience of a graph G with respect to $\mathcal{P}$ is the minimum number r such that one can obtain a graph not having $\mathcal{P}$ by deleting at most r edges at every vertex of G.

The notion of global resilience is not new. In fact, problems about global resilience are popular in extremal graph theory. For example, the celebrated Turán problem can be rephrased as follows. How many edges should one delete from the complete graph K_n to make it H-free for some fixed graph H? The notion of local resilience is more recent and its systematic study was initiated by the author together with Vu [90]. It is motivated by an observation that many interesting properties are easy to destroy by small local changes. For example, to destroy the Hamiltonicity it suffices to delete all edges incident to one vertex, which is much less than the total number of edges in a graph.

If a graph property is not monotone, like containing an induced copy of a fixed graph or having a trivial automorphism group, then we may have to both delete and add edges. This leads to the following more general definition.

Definition 6.2 Given a property $\mathcal{P}$. The global/local resilience of G with respect to $\mathcal{P}$ is the minimum number r such that there is a graph H on $V(G)$ with the total number of edges/maximum degree at most r for which the graph $G \triangle H$ does not have $\mathcal{P}$.

One can observe that there is a certain duality between properties and resilience. If the property is local (such as containing a triangle), then it makes more sense to talk about the global resilience. On the other hand, if the property is global (such as being Hamiltonian), then the local resilience seems to be the right parameter to consider. There are some interesting exceptions to this rule, like chromatic number, which we discuss later in this section.

Using the above notions one can easily generate many questions by choosing some graph with an interesting property and asking for its resilience with respect to this property. For example, asking for the local resilience of the complete graph with respect to Hamiltonicity leads immediately to the celebrated theorem of Dirac, which has been discussed in depth earlier in this paper. Following this approach, in the rest of this section we will revisit several classical theorems on random graphs and discuss the corresponding resilience results.

6.1 Perfect Matching and Hamiltonicity

Let G be a graph on n vertices, where n is even. A perfect matching in G is a set of disjoint edges that covers all n vertices. The threshold for the appearance of such a matching in $G(n, p)$ was determined by Erdős and Rényi already in one of their early papers on random graphs. In

[33] they proved that if $p \geq \frac{\log n + \omega(n)}{n}$, then a.a.s. $G(n,p)$ has a perfect matching. This determines the threshold since for $p \leq \frac{\log n - \omega(n)}{n}$ we a.a.s. have isolated vertices.

To give a lower bound for the local resilience of having a perfect matching consider the following way to destroy this property. Split the vertex set of G into two parts X and Y of size $n/2 + 1$ and $n/2 - 1$ respectively. Then delete all edges inside the set X. Thus X becomes an independent set and it is impossible to match all of its vertices with vertices of Y, since $|Y| < |X|$. In $G(n,p)$ (with p sufficiently large), with high probability all vertices have degree $(1 + o(1))np$. Thus, one would expect that a.a.s. a random graph has an induced subgraph on $n/2 + 1$ vertices whose maximum degree is $(1/2 + o(1))np$. So, the local resilience of $G(n,p)$ with respect to having a perfect matching is a.a.s. at most $(1/2 + o(1))np$.

It was proved in [90] that for $p \gg \log n/n$ this trivial upper bound is actually the truth. Moreover, in this case one can obtain the following rather accurate estimates for error terms. If G' is a subgraph of $G = G(n,p)$ with maximum degree at most $np/2 - 8\sqrt{np \log n}$ then a.a.s. $G - G'$ has a perfect matching. On the other hand, $G(n,p)$ a.a.s. contains a subgraph G'' with maximum degree at most $np/2 + 2\sqrt{np \log n}$ such that $G - G''$ has no perfect matching. The proof of the first part is not very difficult. By randomly splitting the vertices of $G - G'$ into two sets of size $n/2$ we can obtain a bipartite subgraph $H \subset G - G'$ with very good expansion properties. Using these properties one can verify that H satisfies Hall's condition (see [27]), i.e., every subset X on one side has at least $|X|$ neighbors on the other side. The second part follows by analyzing more carefully the degree distribution of the subgraph of $G(n,p)$ on $n/2 + 1$ vertices.

The question of resilience of random graphs with respect to Hamiltonicity is substantially more difficult. Here too one can destroy all Hamilton cycles of $G = G(n,p)$ by splitting the vertex set of G into two parts whose sizes differ by at most two and deleting all the edges inside the larger part. Therefore, similar as above, the local resilience of the random graph with respect to Hamiltonicity is a.a.s. at most $(1/2 + o(1))np$. Note that we can also destroy Hamiltonicity by simply disconnecting the graph. To do so, split the vertex set into two parts whose sizes differ by at most one and delete all edges between them. However in the case of random graphs, this does not change the asymptotics of our lower bound for resilience, since bipartite graph we remove has typically maximum degree $(1/2 + o(1))np$. This suggests that the local resilience of $G(n,p)$ with respect to Hamiltonicity is with high probability $(1/2 + o(1))np$, at least when the random graph is reasonably dense. This was indeed proved for $p \geq \log^4 n/n$ in

[90], where it was conjectured that the value of edge probability p can be decreased all the way to $p \gg \log n / n$. This conjecture was resolved by Lee and Sudakov [77], who proved the following result.

Theorem 6.3 *For every $\varepsilon > 0$, there exist a constant $C = C(\varepsilon)$ such that if $p \geq C \log n / n$, then a.a.s. every subgraph G of $G(n, p)$ with minimum degree at least $(1/2 + \varepsilon)np$ is Hamiltonian.*

Since a complete graph on n vertices is also a random graph $G(n, p)$ with $p = 1$, this theorem can be viewed as a far reaching generalization of Dirac's theorem.

The proof of Theorem 6.3 uses the rotation-extension technique, mentioned in Section 4.1, with a few additional ideas. One is to split the graph G into two graphs G_1, G_2, where the first graph has only small fraction of the edges and will be used to perform rotations and the second graph G_2 will be used to perform extensions. We also partition the longest path P in G_1 into several intervals $I_1, \ldots, I_k$ and show that for most indices j the majority of newly constructed paths have their broken edges outside the interval I_j. Therefore these paths traverse I_j either in the original or the reverse order. This is used to show that, rotating the longest path in G_1, one can obtain a set S_P of $(1/2 + \varepsilon)n$ new endpoints. Furthermore for every $v \in S_P$, rotating the path again and keeping v fixed we can obtain a set T_v of size at least $(1/2 + \varepsilon)n$ such that for all $w \in T_v$ there is a longest path in G_1 starting at v and ending in w. Since G_2 contains most of the edges of G and the minimum degree of G is at least $(1/2 + \varepsilon)np$ one can show that one of these paths can be closed into cycle by using the edges of G_2. Repeating this procedure several times we obtain a Hamilton cycle.

When the edge probability p is close to the threshold for Hamiltonicity, i.e., $p = (1 + \varepsilon) \log n / n$, the random graph $G(n, p)$ has some vertices whose degrees are significantly smaller than np. Therefore in this range of edge probability, one can easily create isolated vertices by deleting $(1/2 + o(1))np$ edges incident to such vertices. Hence in this case, we need to revise the definition of resilience and make the number of edges we allow to delete at a vertex v depending on the degree of v. For further discussion of this regime and known results see e.g., [10] and its references.

Many techniques developed for Hamiltonicity of random graphs rely only on properties of the edge distribution of $G(n, p)$ and therefore can be used to study pseudo-random graphs as well. There are several closely related definitions of pseudo-random graphs (see, e.g., [70] for discussion). Here we use the following one, which is based on spectral properties of such graphs. Consider a graph G on n vertices. Since its adjacency matrix is symmetric, it has n real eigenvalues which we denote by $\lambda_1(G) \geq \lambda_2(G) \geq$

... $\geq \lambda_n(G)$. The quantity $\lambda(G) = \max_{i \geq 2} |\lambda_i(G)|$ is called the *second eigenvalue* of G and plays an important role. We say that G is an (n, d, λ)-*graph* if it is d-regular, has n vertices and $\lambda(G)$ is at most λ.

It is well known (see, e.g., [70]) that if λ is much smaller than the degree d, then G has strong pseudo-random properties, i.e., the edges of G are distributed like in the random graph $G(n, d/n)$. Therefore it is natural to ask whether (similar to $G(n, d/n)$) pseudo-random graphs with $\lambda \ll d$ are Hamiltonian. Such a result was establish by the author together with Krivelevich in [69], where we proved that (n, d, λ)-graphs with $\lambda < d/\log n$ (see [69] for slightly better bound) are Hamiltonian. It would be very interesting to determine the right order of magnitude of the ratio d/λ which already implies Hamiltonicity. Together with Krivelevich [70] we proposed the following conjecture.

Conjecture 6.4 *There exists a positive constant C such that for all sufficiently large n, every (n, d, λ)-graph with $d/\lambda > C$ contains a Hamilton cycle.*

One can further show that when $d \gg \lambda$ the corresponding graph is robustly Hamiltonian. Indeed, together with Vu [90] we proved that if $d/\lambda > \log^2 n$ (actually $\log^{1+\delta} n$ is enough) then the local resilience of such an (n, d, λ)-graph with respect to Hamiltonicity is $(1/2 + o(1))d$. By the above discussion the constant $1/2$ is best possible.

Another interesting open problem is the resilience of random regular graphs with respect to Hamiltonicity. The n-vertex *random d-regular graph*, which is denoted by $G_{n,d}$ (where dn is even), is the uniform probability space of all d-regular graphs on n vertices labeled by the set $[n]$. In this model, one cannot apply the techniques used to study $G(n, p)$ as these two models do not share the same probabilistic properties. Whereas the appearances of edges in $G(n, p)$ are independent, the appearances of edges in $G_{n,d}$ are not. Nevertheless, many results obtained thus far for the random regular graph model $G_{n,d}$ are very similar to the results obtained in $G(n, p)$ with suitable expected degrees, namely, $d = np$. For a detailed discussion of random regular graphs and their properties we refer the interested reader to the excellent survey of Wormald [94].

It was proved by Robinson and Wormald [87] that the random regular graph $G_{n,d}$ is a.a.s. Hamiltonian already for $d \geq 3$. Therefore the question of local resilience of $G_{n,d}$ with respect to this property already makes sense for constant degrees. By the above discussion, it is logical to guess that this resilience should be typically of order $d/2$, at least for large d. More precisely, together with Ben-Shimon and Krivelevich [11] we made the following conjecture.

Conjecture 6.5 *For every $\varepsilon > 0$ there exists an integer $d_0 = d_0(\varepsilon)$ such that, for every fixed integer $d > d_0$, asymptotically almost surely the local resilience of $G_{n,d}$ with respect to Hamiltonicity is at least $(1/2 - \varepsilon)d$ and at most $(1/2 + \varepsilon)d$.*

The upper bound of this conjecture follows easily from the known properties of the edge distribution of $G_{n,d}$ together with constructions mentioned above. So the main task is to prove the lower bound. When $d \gg \log^2 n$ one can prove this conjecture using the above mentioned result from [90] on (n, d, λ)-graphs and the known properties of the second eigenvalue of $G_{n,d}$ (see, e.g., [94, 71]). For smaller values of d this conjecture is still open, although one can show (see [11]) that a.a.s. the local resilience is linear in d.

6.2 Chromatic number

One of the most important parameters of the random graph $G_{n,p}$ is its chromatic number, which we denote by $\chi(G_{n,p})$. Trivially for every graph $\chi(G) \geq |V(G)|/\alpha(G)$, where $\alpha(G)$ denotes the size of a largest independent set in G. It can be easily shown, using first moment computations, that a.a.s. $\alpha(G_{n,p}) \leq 2\log_b(np)$, where $b = 1/(1 - p)$ (all other logarithms in this paper are in the natural base e). This provides a lower bound on the chromatic number of the random graph, showing that with high probability $\chi(G_{n,p}) \geq \frac{n}{2\log_b(np)}$. The problem of determining the asymptotic behavior of $\chi(G_{n,p})$, posed by Erdős and Rényi in the early 60s, stayed for many years as one of the major open questions in the theory of random graphs until its solution by Bollobás [13], using a novel approach based on martingales that enabled him to prove that a.a.s. $\chi(G_{n,p}) = (1 + o(1))\frac{n}{2\log_b(np)}$ for dense random graphs. Later Łuczak [80] showed that this estimate also holds for all values of $p \geq c/n$.

Given two arbitrary graphs G and H it is a folklore result which is easy to prove (see, e.g., [79] Chapter 9) that

$$\chi(G \cup H) \leq \chi(G)\chi(H).$$

Moreover there are pairs of graphs for which the equality holds. Therefore adding a few edges or low degree graph to a graph G sometimes can have a substantial impact on its chromatic number. The question, whether this is the case for random graphs was first posed in [90], where the authors study the resilience of the chromatic number of $G(n, p)$. Unlike for Hamiltonicity, changing the adjacency of few vertices in a graph usually has only a minor effect on its chromatic number. Therefore it is interesting to study both global and local resilience of the chromatic number.

Note that for an arbitrary graph G one can double its chromatic number by choosing an arbitrary subset of $2\chi(G)$ vertices and adding all the missing edges to make this subset a clique. This gives an upper bound $O(\chi^2(G))$ on the global resilience and a bound $O(\chi(G))$ on a local resilience of the chromatic number. Together with Vu [90] we conjectured that this is with high probability tight for relatively dense random graphs (say $p \geq n^{-1+\delta}$ for any fixed $\delta > 0$). Improving the earlier results from [90], Alon and Sudakov [5] proved the following theorem, which makes a substantial progress on this conjecture.

Theorem 6.6 *Let $\varepsilon > 0$ be a fixed constant and let $n^{-1/3+\delta} \leq p \leq 1/2$ for some $\delta > 0$. Then a.a.s.*

1. *for every collection E of $2^{-12}\varepsilon^2 \frac{n^2}{\log_b^2(np)}$ edges the chromatic number of $G_{n,p} \cup E$ is still at most $(1+\varepsilon)\frac{n}{2\log_b(np)}$,*

2. *and for every graph H on n vertices with maximum degree $\Delta(H) \leq 2^{-8}\varepsilon \frac{n}{\log_b(np)\log\log n}$ the chromatic number of $G_{n,p} \cup H$ is still at most $(1+\varepsilon)\frac{n}{2\log_b(np)}$.*

This determines the global resilience of dense random graphs up to a constant factor and leaves only a small multiplicative gap of $\log\log n$ for the local resilience. Both these results show that adding quite large and dense graphs to $G_{n,p}$ with $p \gg n^{-1/3}$ typically has very little impact on its chromatic number. For p below $n^{-1/3}$ much less is known. It was proved in [90] that for every positive integer d and for every $\varepsilon > 0$ there is a constant $c = c(d, \varepsilon)$ such that the following holds. For all $p > c/n$, adding to $G(n, p)$ any graph with maximum degree d a.a.s. cannot increase its chromatic number by a factor of larger than $(1 + \varepsilon)$.

Sketch of proof of Theorem 6.6 (1). Choose $k_0 = (2 - o(1))\log_b^2(np)$ such that the expected number of independent sets of this size in $G(n, p)$ is $\mu = \binom{n}{k_0}(1 - p)^{\binom{k_0}{2}} > n^4$. The expected number of such independent sets containing a given pair of vertices is

$$\mu_0 = \binom{n-2}{k_0-2}(1-p)^{\binom{k_0}{2}} = (1+o(1))\frac{k_0^2}{n^2}.$$

Let $\mathcal{I}$ be a largest collection of independent sets of size k_0 in $G(n,p)$ such that no pair of vertices belongs to more than $4\mu_0$ of these sets. This condition shows that by changing one edge of the random graph we cannot affect the size of $\mathcal{I}$ by more than $4\mu_0$. Therefore using a standard estimate for the tails of martingales we can show that $|\mathcal{I}|$ equals $(1 - o(1))\mathbb{E}[|\mathcal{I}|]$

with probability at least $1 - 2^{-2n}$. Moreover one can also show (see [72])
that $\mathbb{E}[|\mathcal{I}|] = (1 - o(1))\mu$.

Let E be the collection of $2^{-12}\varepsilon^2 \frac{n^2}{\log_b^2(np)}$ edges which was added to
$G(n,p)$. Consider an auxiliary bipartite graph H with parts $\mathcal{I}$ and E in
which an independent set $I \in \mathcal{I}$ is adjacent to an edge $(u,v) \in E$ iff both
vertices u,v belong to I. By the definition of $\mathcal{I}$, every edge $(u,v) \in E$
is contained in at most $4\mu_0$ sets from $\mathcal{I}$. Therefore the number of edges
$e(H)$ is bounded by $4\mu_0 m$. Thus there is an independent set $I \in \mathcal{I}$, whose
degree in H is at most $e(H)/|\mathcal{I}|$. Such I contains at most

$$\frac{e(H)}{|\mathcal{I}|} \leq \frac{4\mu_0|E|}{|\mathcal{I}|} \leq \frac{5\mu_0|E|}{\mu} \leq 6k_0^2\frac{|E|}{n^2} \leq 2^{-7}\varepsilon^2$$

edges from E.

This argument shows that we can find an independent set in $G(n,p)$
which contains no edges from E. Color it by color one, remove it from the
graph and continue this process. One can further prove that, as long as
the number of remaining vertices is larger than $\varepsilon n/\log_b(np)$, we can still
find an independent set I in $G(n,p)$ which has very few edges from E.
Then (by a standard lower bound on the independence number in a graph
with a given number of edges) there is a large independent set in I not
containing any edges of E which we can color by a new color. Using some
simple computations, see [5], we can then show that the total number of
colors used in this process is at most $(1 + O(\varepsilon))\frac{n}{2\log_b(np)}$. Coloring the
remaining $\varepsilon n/\log_b^2(np)$ vertices by additional colors we obtain a coloring
of $G(n,p) \cup E$ into $(1 + O(\varepsilon))\chi(G(n,p))$ colors. $\square$

6.3   Symmetry

An *automorphism* of a graph G is a permutation $\pi : V(G) \rightarrow V(G)$
of the vertices of G such that $(\pi(u), \pi(v))$ is an edge of G if and only if
(u,v) is an edge of G. The collection of all automorphisms of G forms a
group which is denoted by $Aut(G)$. It is clear that for any graph G the
identity belongs to $Aut(G)$. We say that $Aut(G)$ is trivial or equivalently
G is *asymmetric* if $Aut(G)$ does not contain any permutation other than
the identity, otherwise we call G *symmetric*. The automorphism group
was one of the first objects studied by P. Erdős and A. Rényi in their sequence of papers which started the theory of random graphs. In 1963 they
proved that for $p \geq (1+\varepsilon) \log n/n$, $G(n,p)$ is asymptotically almost surely
asymmetric. In fact, Erdős and Rényi studied a more general question
of how much should one change the random graph to have a non-trivial
automorphism. In [32] they proved that if both $1 - p, p \gg \log n/n$, then

a.a.s. we need to add and delete at least $(2 + o(1))np(1 - p)$ edges from $G(n, p)$ to obtain a symmetric graph. This determines the global resilience of $G(n, p)$ with respect to having a trivial automorphism group.

To explain the quantity $(2 + o(1))np(1 - p)$, notice that next to the identity, the simplest permutation is the transposition of two vertices, say u and v. This permutation is an automorphism if u and v have exactly the same neighbors. To achieve this one needs to delete $d(u) + d(v) - 2codeg(u, v)$ edges of G, where $codeg(u, v)$ is the number of common neighbors of these vertices. It shows that in the case of $G(n, p)$, with high probability it is enough to delete $2(1 + o(1))np(1 - p)$ edges to have a symmetric graph. Together with Kim and Vu, the author extended the above theorem of Erdős and Rényi showing that small local changes are not enough to make a random graph symmetric. In [56] they proved the following result.

Theorem 6.7 *If both $1 - p, p \gg \log n / n$, then the local resilience of $G(n, p)$ with respect to being asymmetric is a.a.s. $(1 + o(1))np(1 - p)$.*

It is worth mentioning that the local resilience was not the main object of study in [56], but it was used as a tool to prove a new result. As a corollary of the above theorem (more precisely of its proof), it was shown there that a random regular graph of relatively large degree is asymptotically almost surely non-symmetric, confirming a conjecture of Wormald [94]. The main idea in [56] was as follows. Consider the indicator graph function $I(G)$, where $I(G) = 1$ if G is non-symmetric and 0 otherwise. We want to show that with high probability $I(G) = 1$ where G is a random regular graph. One may want to view this statement as a sharp concentration result, namely, $I(G)$ is a.a.s. close to its mean. However, it is impossible to prove a sharp concentration result for a random variable having only two values close to each other. The idea here is to "blow up" $I(G)$ using the notion of local resilience. Instead of $I(G)$ we used a function $D(G)$ which (roughly speaking) equals the local resilience of G with respect to being non-symmetric. This function is zero if G is symmetric and rather large otherwise. This gives us room to show that $D(G)$ is strongly concentrated around a large positive value, and from this we can conclude that asymptotically almost surely the random regular graph is non-symmetric.

6.4 Further results

There are some additional results on resilience of random graphs whose details we will not discuss in this survey. For convenience of the reader we conclude this section with a short list of such results together with relevant references.

- In the past few years, there has been some considerable success in extending classical results in extremal combinatorics to sparse random settings due to the breakthroughs made independently by Conlon and Gowers [22] and Schacht [89]. In particular, they proved Turán-type theorems for random graphs, establishing the global resilience with respect to containing a fixed non-bipartite graph H.

- Resilience of $G(n,p)$ with respect to *pancyclicity*, which is a property of containing cycles of all lengths from 3 to n, was established in [63].

- In [6] the authors studied the resilience of the property of containing given almost spanning tree of bounded degree.

- Resilience of a random directed graph with respect to Hamiltonicity was studied in [50, 41].

- In [28] the authors studied the resilience of $G(n,p)$ with respect to containing cycle of a given linear length.

- An H-factor in a graph G is a collection of vertex disjoint copies of a fixed graph H covering all vertices of G. The resilience with respect to containing such a factor was studied in [51, 7]. More generally, the resilience with respect to containing almost spanning and spanning bounded degree graphs was studied in [18, 83, 2].

- In [37] the authors made a very interesting connection between the local resilience in random graphs and winning strategies in biased Maker-Breaker games.

7 Conclusion

Although we have made an effort to provide a systematic coverage of recent robust extensions of various classical results in graph theory and random graphs, there are certainly quite a few of them that were left out of this survey, due to the limitations of space and time (and of the author's energy). Still, we would like to believe that we have presented enough examples demonstrating how one can revisit known results and use various measures of robustness to ask new and interesting questions. Therefore we hope that this survey will motivate future research on the subject.

Acknowledgements

Research supported by SNSF grant 200021-149111. We would like to thank F. Dräxler, A. Ferber, N. Kamčev, M. Krivelevich, M. Kwan, J. Noel, A. Pokrovskiy and the referee for helpful comments.

References

[1] M. Albert, A. Frieze, and B. Reed, Multicoloured Hamilton cycles. *Electron. J. Combin.* **2**, R10.

[2] P. Allen, J. Böttcher, J. Ehrenmüller and A. Taraz, The bandwidth theorem in sparse graphs, preprint.

[3] N. Alon, The maximum number of Hamiltonian paths in tournaments, *Combinatorica* 10 (1990), 319–324.

[4] N. Alon and G. Gutin, Properly colored Hamilton cycles in edge-colored complete graphs, *Random Struct. Algor.* **11** (1997), 179–186.

[5] N. Alon and B. Sudakov, Increasing the chromatic number of a random graph, *Journal of Combinatorics* 1 (2010), 345–356.

[6] J. Balogh, B. Csaba and W. Samotij, Local resilience of almost spanning trees in random graphs, *Random Structures and Algorithms* 38 (2011), 121–139.

[7] J. Balogh, C. Lee and W. Samotij, Corrádi and Hajnal's theorem for sparse random graphs, *Combin. Probab. Comput.* 21 (2012), 23–55.

[8] J. Beck, *Combinatorial Games*, Cambridge University Press, Cambridge (2008).

[9] S. Ben-Shimon, A. Ferber, D. Hefetz and M. Krivelevich, Hitting time results for Maker-Breaker games , *Random Structures and Algorithms* 41 (2012), 23–46.

[10] S. Ben-Shimon, M. Krivelevich and B. Sudakov, On the resilience of Hamiltonicity and optimal packing of Hamilton cycles in random graphs, *SIAM J. Discrete Math.* 25 (2011), 1176–1193.

[11] S. Ben-Shimon, M. Krivelevich and B. Sudakov, Local resilience and Hamiltonicity Maker-Breaker games in random regular graphs, *Combinatorics, Probability and Computing* 20 (2011), 173–211.

[12] B. Bollobás, The evolution of sparse graphs, in *Graph Theory and Combinatorics*, Academic Press, New York (1984), 35–57.

[13] B. Bollobás, The chromatic number of random graphs, *Combinatorica* 8 (1988), 49–55.

[14] B. Bollobás, *Random Graphs*, 2nd ed., Cambridge University Press, Cambridge (2001).

[15] B. Bollobás and P. Erdös, Alternating hamiltonian cycles, *Israel J. Math.* **23** (1976), 126–131.

[16] B. Bollobás and A. Frieze, On matchings and Hamiltonian cycles in random graphs, in: *Random graphs '83*, North-Holland Math. Stud. 118, North-Holland, Amsterdam, 1985, 23–46.

[17] J. Bondy, Basic Graph Theory: Paths and Circuits, in: *Handbook of Combinatorics Vol. 1* (ed. by R. Graham, M. Grötschel and L. Lovász), North-Holland, Amsterdam, 1995, 3–110.

[18] J. Böttcher, Y. Kohayakawa and A. Taraz, Almost spanning subgraphs of random graphs after adversarial edge removal, *Combinatorics, Probability and Computing* 22 (2013), 639–683.

[19] L. M. Brégman, Some properties of non-negative matrices and their permanents, *Sov. Mat. Dokl.* 14 (1973), 945–949.

[20] C. Chen and D. Daykin, Graphs with Hamiltonian cycles having adjacent lines different colors, *J. Combin. Theory Ser. B* **21** (1976), 135–139.

[21] V. Chvátal and P. Erdős, Biased positional games, *Annals of Discrete Math.* 2 (1978), 221–228.

[22] D. Conlon and T. Gowers, Combinatorial theorems in sparse random sets, *Annals of Math.* 184 (2016), 367–454.

[23] B. Csaba, D. Kühn, A. Lo, D. Osthus and A. Treglown, Proof of the 1-factorization and Hamilton decomposition conjectures, *Memoirs of the American Mathematical Society* 244 (2016), monograph 1154, 170 pages.

[24] B. Cuckler, Hamilton cycles in regular tournaments, *Combinatorics, Probability and Computing* 16 (2007), 239–249.

[25] B. Cuckler and J. Kahn, Hamiltonian cycles in Dirac graphs, *Combinatorica* 29 (2009), 299–326.

[26] D. E. Daykin, Graphs with cycles having adjacent lines different colors, *J. Combin. Theory Ser. B* **20** (1976), 149–152.

[27] R. Diestel, *Graph theory*, 3rd ed., Springer-Verlag, Berlin, 2005.

[28] D. Dellamonica, Y. Kohayakawa, M. Marciniszyn and A. Steger, On the resilience of long cycles in random graphs, *Electronic J. Combin.* 15 (2008), R32.

[29] G. Egorychev, The solution of the Van der Waerden problem for permanents, *Dokl. Akad. Nauk SSSR* 258 (1981), 1041–1044.

[30] P. Erdős, J. Nešetřil and V. Rödl, On some problems related to partitions of edges of a graph, in: *Graphs and other combinatorial topics* (Prague, 1982), Teubner-Texte Math. 59, Teubner, Leipzig, 1983, 54–63.

[31] P. Erdős and A. Rényi, On the evolution of random graphs, *Magyar Tud. Akad. Mat. Kutató Int. Közl.* 5 (1960), 17–61.

[32] P. Erdős and A. Rényi, Asymmetric graphs, *Acta Math. Acad. Sci. Hung.* 14 (1963), 295–315.

[33] P. Erdős and A. Rényi, On the existence of a factor of degree one of a connected random graph, *Acta Math. Acad. Sci. Hungar.* 17 (1966), 359–368.

[34] P. Erdős, M. Simonovits and V. Sós, Anti-Ramsey theorems, in: *Infinite and finite sets, Vol. 11, edited by A. Hajnal, R Rado and V. Sós*, Colloq. Math. Soc. János Bolyai 10, North-Holland, Amsterdam (1975), 633–643.

[35] D. Falikman, A proof of the Van der Waerden problem for permanents of a doubly stochastic matrix, *Mat. Zametki* 29 (1981), 931–938.

[36] A. Ferber, R. Glebov, M. Krivelevich and A. Naor, Biased games on random boards, *Random Structures and Algorithms* 46 (2015), 651–676.

[37] A. Ferber, M. Krivelevich and H. Naves, Generating random graphs in biased maker-breaker games *Random Structures Algorithms* 47 (2015), 615–634.

[38] A. Ferber, M. Krivelevich and B. Sudakov, Counting and packing Hamilton cycles in dense graphs and oriented graphs, *J. Combinatorial Theory Ser. B*, to appear.

[39] A. Ferber, G. Kronenberg and E. Long, Packing, Counting and Covering Hamilton cycles in random directed graphs, *Israel J. Mathematics*, to appear.

[40] A. Ferber, E. Long and B. Sudakov, Counting Hamiltonian decompositions of oriented graphs, arXiv:1609.09550.

[41] A. Ferber, R. Nenadov, A. Noever, U. Peter and N. Škorić, Robust hamiltonicity of random directed graphs, *Proceedings of 26th ACM-SIAM SODA*, ACM Press (2015), 1752–1758.

[42] A. Frieze and M. Karoński, *Introduction to Random Graphs*, Cambridge University Press, 2016.

[43] A. Frieze and M. Krivelevich, On packing Hamilton cycles in ε-regular graphs, *J. Combinatorial Theory Ser. B* 94 (2005), 159–172.

[44] A. Frieze and M. Krivelevich, On two Hamilton cycle problems in random graphs, *Israel J. of Mathematics* 166 (2008), 221–234.

[45] R. Glebov and M. Krivelevich, On the number of Hamilton cycles in sparse random graphs, *SIAM Journal on Discrete Math.* 27 (2013), 27–42.

[46] R. Glebov, Z. Luria and B. Sudakov, The number of Hamiltonian decompositions of regular graphs, *Israel J. of Mathematics*, to appear.

[47] R. Glebov, H. Naves and B. Sudakov, The threshold probability for long cycles, *Combinatorics, Probability and Computing*, to appear.

[48] R. Häggkvist, Hamilton cycles in oriented graphs, *Combinatorics, Probability and Computing* 2 (1993), 25–32.

[49] D. Hefetz, M. Krivelevich, M. Stojakovic and T. Szabó, *Positional Games*, Birkhäuser Basel, 2014.

[50] D. Hefetz, A. Steger and B. Sudakov, Random directed graphs are robustly Hamiltonian, *Random Structures and Algorithms* 49 (2016), 345–362.

[51] H. Huang, C. Lee and B. Sudakov, Bandwidth theorem for random graphs, *J. Combin. Theory Ser. B* 102 (2012), 14–37.

[52] S. Janson, T. Łuczak and A. Ruciński, *Random graphs*, Wiley, New York, 2000.

[53] R. Karp, Reducibility among combinatorial problems, in: *Complexity of Computer Computations*, New York: Plenum (1972), 85–103.

[54] P. Katerinis, Minimum degree of a graph and the existence of k-factors, *Proc. Indian Acad. Sci. Math. Sci.* 94 (1985), 123–127.

[55] P. Keevash, D. Kühn and D. Osthus, An exact minimum degree condition for Hamilton cycles in oriented graphs, *J. London Math. Soc.* 79 (2009), 144–166.

[56] J.H. Kim, B. Sudakov and V. Vu, On asymmetry of random graphs and random regular graphs, *Random Structures and Algorithms* 21 (2002), 216–224.

[57] F. Knox, D. Kühn and D. Osthus, Edge-disjoint Hamilton cycles in random graphs, *Random Structures and Algorithms* 46 (2015), 397–445.

[58] J. Komlós and M. Simonovits, Szemerédi's regularity lemma and its applications in graph theory, in *Combinatorics, Paul Erdős is eighty*, Vol. 2, Bolyai Soc. Math. Stud. 2, Budapest, 1996, 295–352.

[59] J. Komlós and E. Szemerédi, Limit distribution for the existence of Hamiltonian cycles in a random graph, *Discrete Mathematics* **43** (1983), 55–63.

[60] A. Korshunov, Solution of a problem of Erdős and Rényi on Hamilton cycles in non-oriented graphs, *Soviet Math. Dokl,* **17** (1976), 760–764.

[61] A. Kotzig, Moves without forbidden transitions in a graph, *Matematický časopis* **18** (1968), 76–80.

[62] M. Krivelevich, The critical bias for the Hamiltonicity game is $(1 + o(1))n/\ln n$, *Journal of the American Mathematical Society* 24 (2011), 125–131.

[63] M. Krivelevich, C. Lee and B. Sudakov, Resilient pancyclicity of random and pseudorandom graphs, *SIAM J. Discrete Math.* 24 (2010), 1–16.

[64] M. Krivelevich, C. Lee and B. Sudakov, Robust Hamiltonicity of Dirac graphs, *Transactions of the Amer. Math. Soc.* 366 (2014), 3095–3130.

[65] M. Krivelevich, C. Lee and B. Sudakov, Long paths and cycles in random subgraphs of graphs with large minimum degree, *Random Structures and Algorithms* 46 (2015), 320–345.

[66] M. Krivelevich, C. Lee and B. Sudakov, Compatible Hamilton cycles in random graphs, *Random Structures and Algorithms* 49 (2016), 533–557.

[67] M. Krivelevich, C. Lee and B. Sudakov, Compatible Hamilton cycles in Dirac graphs, *Combinatorica*, to appear.

[68] M. Krivelevich and W. Samotij, Optimal packings of Hamilton cycles in sparse random graphs, *SIAM J. Discrete Mathematics* 26 (2012), 964–982.

[69] M. Krivelevich and B. Sudakov, Sparse pseudo-random graphs are Hamiltonian, *J. Graph Theory* 42 (2003), 17–33.

[70] M. Krivelevich and B. Sudakov, Pseudo-random graphs, in: *More Sets, Graphs and Numbers*, Bolyai Society Mathematical Studies 15, Springer, 2006, 199–262.

[71] M. Krivelevich, B. Sudakov, V. Vu and N. Wormald, Random regular graphs of high degree, *Random Structures and Algorithms* 18 (2001), 346–363.

[72] M. Krivelevich, B. Sudakov, V. Vu and N. Wormald, On the probability of independent sets in random graphs, *Random Structures and Algorithms* 22 (2003), 1–14.

[73] D. Kühn, J. Lapinskas and D. Osthus, Optimal packings of Hamilton cycles in graphs of high minimum degree, *Combinatorics, Probability, Computing* 22 (2013), 394–416.

[74] D. Kühn and D. Osthus, Hamilton decompositions of regular expanders: a proof of Kelly's conjecture for large tournaments, *Advances in Mathematics* 237 (2013), 62–146.

[75] D. Kühn and D. Osthus, Hamilton decompositions of regular expanders: applications, *J. Combinatorial Theory Series B* 104 (2014), 1–27.

[76] D. Kühn, D. Osthus and A. Treglown, Hamilton decompositions of regular tournaments, *Proceedings of the London Mathematical Society* 101 (2010), 303–335.

[77] C. Lee and B. Sudakov, Dirac's theorem for random graphs, *Random Structures and Algorithms* 41 (2012), 293–305.

[78] A. Lo, Properly coloured Hamiltonian cycles in edge-coloured complete graphs, *Combinatorica*, to appear.

[79] L. Lovász, *Combinatorial problems and exercises*, 2nd edition, AMS Chelsea Publishing, 2007.

[80] T. Łuczak, *The chromatic number of random graphs*, Combinatorica 11 (1991), 45–54.

[81] C. Nash-Williams, Hamiltonian lines in graphs whose vertices have sufficiently large valencies, in: *Combinatorial Theory and Its Applications*, III, North-Holland, 1970, 813–819.

[82] C. Nash-Williams, Edge-disjoint Hamiltonian circuits in graphs with vertices of large valency, in: *Studies in Pure Mathematics*, Academic Press, London, 1971, 157–183.

[83] A. Noever and A. Steger, Local resilience for squares of almost spanning cycles in sparse random graphs, preprint.

[84] L. Pósa, Hamiltonian circuits in random graphs, *Discrete Mathematics*, **14** (1976), 359–364.

[85] R. Rado, Anti-Ramsey theorems, in: *Infinite and finite sets, Vol. 11,* edited by A. Hajnal, R. Rado and V. Sós, Colloq. Math. Soc. János Bolyai 10, North-Holland, Amsterdam (1975), 1159–1168.

[86] O. Riordan, Long cycles in random subgraphs of graphs with large minimum degree *Random Structures and Algorithms* 45 (2014), 764–767.

[87] R. Robinson, and N. Wormald, Almost all regular graphs are Hamiltonian *Random Structures and Algorithms* 5 (1994), 363–374.

[88] G. Sárközy, S. Selkow and E. Szemerédi, On the number of Hamiltonian cycles in Dirac graphs, *Discrete Math.* 265 (2003), 237–250.

[89] M. Schacht, Extremal results for random discrete structures, *Annals of Math.* 184 (2016), 333–365.

[90] B. Sudakov and V. Vu, Local resilience of graphs, *Random Structures and Algorithms* 33 (2008), 409–433.

[91] T. Szele, Kombinatorikai vizsgalatok az iranyitott teljes graffal, *Mt. Fiz. Lapok* 50 (1943), 223–256.

[92] C. Thomassen, Long cycles in digraphs with constraints on the degrees, in: *Surveys in Combinatorics* (B. Bollobás ed.), London Math. Soc. Lecture Notes 38, Cambridge University Press, 1979, 211–228.

[93] C. Thomassen, Hamilton circuits in regular tournaments, *Annals of Discrete Math.* 27 (1985), 159–162.

[94] N.C. Wormald, Models of random regular graphs, in: *Surveys in Combinatorics*, Cambridge University Press, 1999, 239–298.

Department of Mathematics
ETH
8092 Zurich
Switzerland
benjamin.sudakov@math.ethz.ch

Some Applications of Relative Entropy in Additive Combinatorics

Julia Wolf

Abstract

This survey looks at some recent applications of relative entropy in additive combinatorics. Specifically, we examine to what extent entropy-increment arguments can replace or even outperform more traditional energy-increment strategies or alternative approximation arguments based on the Hahn-Banach theorem.

1 Introduction

Entropy has a long history as a tool in combinatorics. Starting with a well-known estimate for the sum of the first few binomial coefficients, some of the classical applications include Spencer's theorem that six standard deviations suffice, which states that given n finite sets, there exists a two-colouring of the elements such that all sets have discrepancy at most $\ll n^{1/2}$; a proof of the Loomis-Whitney inequality, which gives an upper bound on the volume of an n-dimensional body in Euclidean space in terms of its $(n-1)$-dimensional projections; or Radhakrishnans proof [33] of Bregman's theorem on the maximum permanent of a $0/1$ matrix with given row sums. For a beautiful introduction to these fascinating applications, as well as an extensive annotated bibliography, see [10].

There are other more recent results in additive combinatorics in particular where the concept of entropy has played a crucial role. Notable examples include Fox's improvement [7] of the bounds in the graph removal lemma (see also [30], which appeared in the proof-reading stages of this article); Szegedy's information-theoretic approach [41] to Sidorenko's conjecture (see also the blog post [14] by Gowers); Tao's solution [45] to the Erds discrepancy problem (see the discussion [44] on Tao's blog).

Since the above developments appear to be well captured by discussions online, we shall not cover them in any detail here. Instead we shall focus on a particular strand of recent results in additive combinatorics that could all be described as "approximation theorems" of a certain kind.

The text naturally splits into five parts. To start with we give a very brief introduction to the concept of entropy and its variants, in particular relative entropy (also known as Kullback-Leibler divergence). In Section 3, we state and prove a rather general sparse approximation theorem due to Lee [28]. In Section 4 we show how it can be used to derive a variant of

Chang's theorem [3] on the dimension of the large Fourier spectrum of a set, as well as Bloom's [2] more recent and powerful refinement. We subsequently derive a quadratic decomposition theorem in Section 5, which is the only part of this survey for which we claim any originality. Finally, in Section 6, we discuss, following Vadhan and Zheng [48], how the notion of relative entropy can been used to derive an optimal version of the transference principle (also known as the dense model theorem), which has had far-reaching applications in number theory, graph theory, and theoretical computer science in recent years.

In the latter two applications, at least two alternative approaches exist in the literature. One is an iterative (and in principle) constructive approach based on an ℓ^2 (or energy) increment. The idea behind it, which goes back at least as far as Szemerdi's regularity lemma, was used by Green and Tao in proving the first quadratic decompositions, as well as the original version of their transference principle. As observed by Fox [7], it sometimes turns out to be quantitatively advantageous to aim for entropy increments instead. For a discussion of the different types of increment strategies, see also [42].

An alternative approach to quadratic decompositions was pioneered by Gowers and the author in [15], and was based on the Hahn-Banach theorem (or the duality of linear programming). Gowers also used it to give an alternative proof of the transference principle [13]. While versatile and relatively clean, it suffers the disadvantage of being non-constructive. In the final section we shall see how the notion of relative entropy can be used to give a constructive proof of the Hahn-Banach theorem (in the form of a Min-Max statement from game theory).

The purpose of this article is thus to advertise relative entropy as a powerful tool in arithmetic combinatorics, and to present the above sphere of results in a unified framework.

Acknowledgements. The author would like to thank Thomas Bloom, James Lee, Luka Rimanic, Tom Sanders and Madhur Tulsiani for helpful conversations on various aspects of this text. She is greatly indebted to James Lee for insightful comments on an earlier draft of this manuscript, and to the anonymous referee for numerous suggestions which improved the presentation.

2   A very brief introduction to entropy

Throughout we use the expectation operator $\mathbb{E}_{x \in X}$ for any finite set X to denote the normalised (finite) sum over all elements $x \in X$, that is,

$$\mathbb{E}_{x \in X} g(x) := \frac{1}{|X|} \sum_{x \in X} g(x).$$

For a function $f : X \to \mathbb{C}$, define its L^p norm as $\|f\|_p := (\mathbb{E}_{x \in X} |f(x)|^p)^{1/p}$. We also have an inner product $\langle f, g \rangle := \mathbb{E}_{x \in X} f(x)\overline{g(x)}$, which gives rise to a Hilbert space of real/complex-valued functions on X. We denote the set of measures on X by

$$\Delta_X := \{f : X \to [0, \infty) : \|f\|_1 = 1\}.$$

This family Δ_X contains, in particular, the so-called characteristic measure μ_A for any subset $A \subseteq X$, which is defined for each $x \in X$ by $\mu_A(x) := \alpha^{-1} 1_A(x)$, where 1_A is the characteristic function of the set A, and $\alpha := |A|/|X|$ is its density. (When $A = \emptyset$, then μ_A is identically zero.) The *entropy* of a distribution measures its information content, or the expected amount of surprise upon observing an event in a probability space. In fact, the shape of the entropy function can be derived from a set of natural conditions (see for example [36], Chapter 9).

Definition 2.1 (Entropy) For $f \in \Delta_X$, define the *entropy* of f to be

$$\text{Ent}(f) := \mathbb{E}_{x \in X} f(x) \log f(x).$$

It is not difficult to check that for $f = \mu_A$, the characteristic measure of a subset $A \subseteq X$, we have $\text{Ent}(f) = \log(\alpha^{-1})$. We may also connect this definition to the notion of *Shannon entropy* $H(Y)$ of a random variable Y taking values in X by setting $f(x) = |X|\mathbb{P}[Y = x]$ and observing that

$$H(Y) := \sum_{x \in X} \mathbb{P}[Y = x] \log \frac{1}{\mathbb{P}[Y = x]} = \log |X| - \text{Ent}(f).$$

By Jensen's inequality, $H(Y) \leq \log |X|$, and thus $\text{Ent}(f) \geq 0$. In fact, $\text{Ent}(f)$ can be viewed as the entropy of f relative to the uniform distribution. Relative entropy is also known as Kullback-Leibler divergence [27].

Definition 2.2 (Relative entropy)   For $f, g \in \Delta_X$, we define the *Kullback-Leibler (KL) divergence* or *relative entropy* from f to g by

$$D_{KL}(f\|g) := \mathbb{E}_{x \in X} f(x) \log \frac{f(x)}{g(x)}$$

whenever $\text{supp}(f) \subseteq \text{supp}(g)$.

It is easy to see that $D_{KL}(f\|1) = \text{Ent}(f)$. Again, taking $f(x) = |X|\mathbb{P}[Y = x]$, we can recover a perhaps more familiar formulation for discrete distributions Y and Z taking values in a finite range X, namely

$$D_{KL}(Y\|Z) := \sum_{x \in X} \mathbb{P}[Y = x] \log \frac{\mathbb{P}[Y = x]}{\mathbb{P}[Z = x]},$$

which by convention is $+\infty$ if $\text{supp}(Y) \not\subseteq \text{supp}(Z)$.

It is clear that KL divergence is not in general symmetric and thus not a metric (it does not satisfy the triangle inequality either), but it does satisfy non-negativity, and equals zero only if the distributions are identical.

Kullback-Leibler divergence is a measure of the information gained when one revises one's beliefs from the prior probability distribution Z to the posterior probability distribution Y. In other words, it is the amount of information lost when Z is used to approximate Y.

The notion of relative entropy is related to the total variation $\delta(Y, Z)$ between two random variables Y, Z via *Pinsker's inequality*

$$\delta(Y, Z) \leq \sqrt{\frac{1}{2} D_{KL}(Y\|Z)}.$$

We shall not explicitly make use of this fact, or indeed more advanced entropy concepts such as conditional entropy, in what follows, but since

$$\mathbb{P}[Z \in S] - \delta(Y, Z) \leq \mathbb{P}[Y \in S] \leq \mathbb{P}[Z \in S] + \delta(Y, Z),$$

it supports the intuition that if $D_{KL}(Y\|Z)$ is small and it is rare for Z to lie in some set S, then it is also rare for Y to lie in S.

One reason that entropy increments can perform quantitatively better than energy increments in some applications is that every measure is at most $\log|X|$ from the uniform measure when measured in terms of KL divergence, whereas it can be as far as $|X|^{1/2}$ in the ℓ^2-distance.

3   A sparse approximation theorem

Throughout, $\mathcal{F} \subseteq L^2(X)$ will be a collection of functions $\phi : X \to \mathbb{R}$ satisfying $\|\phi\|_\infty \leq 1$. We define the semi-norm $\|\cdot\|_{\mathcal{F}}$ by

$$\|h\|_{\mathcal{F}} := \sup_{\phi \in \mathcal{F}} |\langle h, \phi \rangle|$$

for any $h \in \Delta_X$.

Definition 3.1 (Generalised Riesz products) Let $d \geq 1$ be an integer. We say a function $R \in L^2(X)$ is a *Riesz $\mathcal{F}$-product of degree at most d* if

$$R(x) = \prod_{i=1}^{d}(1 + \epsilon_i \phi_i(x))$$

for some $\epsilon_1, \ldots, \epsilon_d \in \{-1, 0, 1\}$ and $\phi_1, \ldots, \phi_d \in \mathcal{F}$.

Note that the ϕ_is appearing in the generalised Riesz product are not necessarily distinct, and that by definition, a Riesz product is always non-negative on X.

The main result of this section, due to Lee [28], states that any $f \in \Delta_X$ can be well approximated by a number of low-degree Riesz $\mathcal{F}$-products in the sense that the resulting error has small inner product with every member of the family $\mathcal{F}$.

Theorem 3.2 (Sparse approximation theorem) *For every $0 < \epsilon < e^{-3}$ and $f \in \Delta_X$, there is a $g \in \Delta_X$ such that $\|f - g\|_{\mathcal{F}} \leq \epsilon$ and there is a subset $\mathcal{F}' \subseteq \mathcal{F}$ of size $|\mathcal{F}'| \leq 9\epsilon^{-2} \cdot \mathrm{Ent}(f)$ such that g is a non-negative linear combination of Riesz $\mathcal{F}'$-products of degree at most*

$$d \leq 18\epsilon^{-1} \cdot \mathrm{Ent}(f) + O\left(\frac{\log \epsilon^{-1}}{\log \log \epsilon^{-1}}\right).$$

Proof For some $T > 0$ we shall define a family of functions

$$\{g_t : t \in [0, T]\} \subseteq \Delta_X$$

by setting $g_0 := 1$ and

$$g_t := \frac{\exp(\int_0^t \phi_s ds)}{\mathbb{E} \exp(\int_0^t \phi_s ds)}$$

for $t \in [0, T]$. The maps $s \mapsto \phi_s$ shall be defined to be piecewise constant on a finite sequence of intervals by the following procedure.

Procedure 1 *Having defined the maps $s \mapsto \phi_s$ on intervals $[0, t_1)$, $[t_1, t_2), \ldots, [t_{i-1}, t_i)$ for some $i \in \mathbb{N}$ with $0 < t_1 < t_2 < \cdots < t_i$, we define t_{i+1} and ϕ_s on $[t_i, t_{i+1})$, as follows.*
If there exists $\phi \in \mathcal{F}$ such that

$$|\langle g_{t_i}, \phi \rangle - \langle f, \phi \rangle| > 2\epsilon/3, \tag{3.1}$$

set

$$t_{i+1} := \inf\{t \geq t_i : |\langle g_t, \phi \rangle - \langle f, \phi \rangle| \leq \epsilon/3\},$$

and

$$\phi_s := \text{sign}(\langle f - g_{t_i}, \phi \rangle)\phi$$

for all $s \in [t_i, t_{i+1})$.

If there is no such ϕ satisfying (3.1) at time t_i, set $T := t_i$ and $I := i$.

While this appears to be pulled out of a hat, neither the form of g_t nor the procedure for determining the maps ϕ ought to be surprising. Indeed, the aim here is to minimise $\text{Ent}(g)$ over all $g \in \Delta_X$ for which $\|f - g\|_{\mathcal{F}} \leq \eta$. The Lagrangian for this problem is $\mathcal{L}(g, \lambda) = \text{Ent}(g) - \sum_{\phi \in \mathcal{F}} \lambda_\phi(\langle f - g, \phi \rangle - \eta)$, with solution $\nabla_g \text{Ent}(g) = -\sum_{\phi \in \mathcal{F}} \lambda_\phi \phi$ and hence $g = \exp(\sum_{\phi \in \mathcal{F}} \lambda_\phi \phi)$ (which also explains the name *gradient descent* for this general approach to solving optimisation problems of this kind). Moreover, in order to minimise the correlation of the error term $f - g$ with elements of $\mathcal{F}$, g must absorb precisely those ϕ for which the correlation is large. The latter feature is not dissimilar to other iterative approaches, in particular the widely used ℓ^2 energy increment argument.

Interpreting the Kullback-Leibler divergence $D_{KL}(f\|g_t)$ as the amount of information lost when g_t is used to approximate f, we can expect said divergence to decrease over the course of the argument, starting from $\text{Ent}(f)$, which is the amount of information lost when we approximate f by the uniform distribution. As a result, we have the following bounds on the stopping time T and the number of intervals I.

Claim 3.3 *The procedure above stops at time $T \leq 3\epsilon^{-1} \cdot \text{Ent}(f)$.*

Proof of Claim 3.3: For $t \in [0, T)$, we compute

$$\frac{d}{dt} D_{KL}(f\|g_t) = \frac{d}{dt} \mathbb{E}[f \log \frac{f}{g_t}] = -\mathbb{E}(f/g_t)\frac{d}{dt}g_t,$$

and evaluating the derivative of g_t with respect to t as $g_t(\phi_t - \langle g_t, \phi_t \rangle)$ we obtain, after some rearranging, that

$$\frac{d}{dt} D_{KL}(f\|g_t) = \langle \phi_t, g_t - f \rangle.$$

By the set-up, for $t \in [t_i, t_{i+1})$, $\phi_t = \text{sign}(\langle f - g_{t_i}, \phi \rangle)\phi$ for some ϕ such that $|\langle f - g_t, \phi \rangle| \geq \epsilon/3$ for all $t \in [t_i, t_{i+1})$, so that

$$\frac{d}{dt} D_{KL}(f\|g_t) = \langle \phi_t, g_t - f \rangle = -\text{sign}(\langle g_t - f, \phi \rangle)\langle \phi, g_t - f \rangle \leq -\epsilon/3.$$

Now since $D_{KL}(f\|g_0) = \text{Ent}(f)$ and $D_{KL}(f\|g_t) \geq 0$ for all t, by the Mean-Value theorem the procedure must terminate in time T satisfying $0 \leq \text{Ent}(f) - \epsilon/3 \cdot T$. $\square$

Claim 3.4 *The procedure above uses $I \leq 9\epsilon^{-2} \cdot \text{Ent}(f)$ intervals.*

Proof of Claim 3.4: We shall show that $t_i - t_{i-1} \geq \epsilon/3$ for all $i \leq I$, from which it follows that the number of intervals I satisfies $I \leq T/(\epsilon/3)$, so that the required bound follows from Claim 3.3.

Fix an interval $[t_{i-1}, t_i)$ for some $i \leq I$, and for simplicity write $\phi = \phi_{t_{i-1}}$, which by definition equals ϕ_t for all $t \in [t_{i-1}, t_i)$. Then, re-using our previous calculation of the derivative of g_t, we have for $t \in [t_{i-1}, t_i)$ that

$$\frac{d}{dt}\langle \phi, g_t \rangle = -\langle \phi, g_t(\phi_t - \langle g_t, \phi_t \rangle) \rangle = -\langle \phi^2, g_t \rangle + \langle \phi, g_t \rangle^2 \geq -\|\phi\|_\infty^2 \|g_t\|_1 \geq -1.$$

Since, by definition of ϕ and t_i,

$$\langle \phi, g_{t_{i-1}} \rangle - \langle \phi, g_{t_i} \rangle = \langle \phi, f - g_{t_{i-1}} \rangle - \langle \phi, f - g_{t_i} \rangle \geq 2\epsilon/3 - \epsilon/3 = \epsilon/3,$$

we obtain, again by the Mean-Value theorem, the desired conclusion that $t_i - t_{i-1} \geq \epsilon/3$. $\qquad\square$

Note that by virtue of the construction we have $g_T \in \Delta_X$ and $\|f - g_T\|_\mathcal{F} \leq 2\epsilon/3$, so the function g_T would be an ideal approximant if it could be described in the appropriate form. In order to achieve this, we shall need to truncate the exponential in its definition. Let $\psi_t := \int_0^t (1 + \phi_s) ds$ and $p_d(x) := \sum_{j=0}^d \frac{x^j}{j!}$.

Claim 3.5 *There exists an integer d such that the function*

$$g := \frac{p_d(\psi_T)}{\mathbb{E}p_d(\psi_T)} \in \Delta_X$$

satisfies $\|g - g_T\|_1 \leq \epsilon/3$ and $\|g - g_T\|_\mathcal{F} \leq \epsilon/3$.

Proof of Claim 3.5: Notice that

$$g_T = \frac{\exp(\int_0^T \phi_s ds)}{\mathbb{E}\exp(\int_0^T \phi_s ds)} = \frac{\exp(\int_0^T (1 + \phi_s) ds)}{\mathbb{E}\exp(\int_0^T (1 + \phi_s) ds)} = \frac{\exp(\psi_T)}{\mathbb{E}\exp(\psi_T)}.$$

Set

$$d := 6T + O\left(\frac{\log \epsilon^{-1}}{\log \log \epsilon^{-1}}\right)$$

so that the remainder in the degree-d Taylor expansion of $\exp(x)$ is bounded via

$$\sup_{x \in [0, 2T]} \frac{|\exp(x) - p_d(x)|}{\exp(x)} \leq \frac{(2T)^{d+1}}{(d+1)!} \leq \epsilon/6.$$

Since $\|\psi_T\|_\infty \le 2T$, it follows that

$$\|p_d(\psi_T) - \exp(\psi_T)\|_1 \le (\epsilon/6)\mathbb{E}\exp(\psi_T) \qquad (3.2)$$

and hence

$$\|g - g_T\|_1 = \left\| \frac{p_d(\psi_T)}{\mathbb{E}p_d(\psi_T)} - \frac{\exp(\psi_T)}{\mathbb{E}\exp(\psi_T)} \right\|_1$$

$$\le \left\| \frac{p_d(\psi_T)}{\mathbb{E}\exp(\psi_T)} - \frac{\exp(\psi_T)}{\mathbb{E}\exp(\psi_T)} \right\|_1 + \left\| \frac{p_d(\psi_T)}{\mathbb{E}p_d(\psi_T)} - \frac{p_d(\psi_T)}{\mathbb{E}\exp(\psi_T)} \right\|_1.$$

The first term is bounded above by $\epsilon/6$ by (3.2), and the second term can be written as

$$\mathbb{E}p_d(\psi_T) \left| \frac{1}{\mathbb{E}p_d(\psi_T)} - \frac{1}{\mathbb{E}\exp(\psi_T)} \right| = \left| 1 - \frac{\mathbb{E}p_d(\psi_T)}{\mathbb{E}\exp(\psi_T)} \right|,$$

Again, it follows from (3.2) that $\mathbb{E}p_d(\psi_T) \ge (1-\epsilon/6)\mathbb{E}\exp(\psi_T)$, so that the second term is also bounded by $\epsilon/6$, and thus $\|g-g_T\|_1 \le \epsilon/3$. Since for all $\phi \in \mathcal{F}$ we have $\|\phi\|_\infty \le 1$, we also have $\|g-g_T\|_{\mathcal{F}} \le \|\phi\|_\infty \|g-g_T\|_1 \le \epsilon/3$.
$\square$

It follows from the construction of g_T and Claim 3.5 that $\|f - g\|_{\mathcal{F}} \le \|f - g_T\|_{\mathcal{F}} + \|g_T - g\|_{\mathcal{F}} \le 2\epsilon/3 + \epsilon/3 = \epsilon$. It remains to show that g is indeed of the form claimed in Theorem 3.2.

Let $\mathcal{F}'$ be the family of functionals ϕ_s encountered in the course of the above argument, i.e. $\mathcal{F}' := \{\phi_s : s \in [0,T)\} \subseteq \mathcal{F}$. By Claim 3.4, $|\mathcal{F}'| = I \le 9\epsilon^{-2}\mathrm{Ent}(f)$. Moreover, ψ_T is by definition of the form $\sum_{\phi \in \mathcal{F}'} \lambda_\phi(1+\phi)$ for some non-negative coefficients λ_ϕ. It follows that $p_d(\psi_T)$, and hence g, is a non-negative linear combination of Riesz $\mathcal{F}'$-products of degree at most d, as claimed. $\square$

4   The structure of the large Fourier spectrum

In this section we shall describe a first application of Theorem 3.2, also due to Lee [28]. Let G be a finite abelian group. The *Fourier transform* $\hat{f} : \hat{G} \to \mathbb{C}$ of a function $f : G \to \mathbb{C}$ is then defined, for each character $\gamma \in \hat{G}$, by the formula

$$\hat{f}(\gamma) := \mathbb{E}_{x \in G} f(x)\gamma(x).$$

We shall sometimes abuse notation and treat characters additively. The *inversion formula* states that with the above definition of the Fourier transform, we can recover f from its Fourier coefficients via the sum

$$f(x) = \sum_{\gamma \in \hat{G}} \hat{f}(\gamma)\gamma(x). \qquad (4.1)$$

while Plancherel's identity asserts that

$$\langle f, g \rangle = \langle \widehat{f}, \widehat{g} \rangle,$$

where

$$\langle f, g \rangle := \mathbb{E}_x f(x)\overline{g(x)} \quad \text{and} \quad \langle \widehat{f}, \widehat{g} \rangle := \sum_{\gamma} \widehat{f}(\gamma)\overline{\widehat{g}(\gamma)}. \tag{4.2}$$

We shall refer to (4.2) as *Parseval's identity* whenever f and g are equal. Note that inner products on physical space are normalised, while those on frequency space are not, and the same convention applies to the $L^p(G)$ and $\ell^p(\widehat{G})$ norms, respectively.

Definition 4.1 Let $\rho > 0$, and let $f : G \to \mathbb{C}$. Define the *ρ-large spectrum* of f to be the set

$$\mathrm{Spec}_\rho(f) := \{\gamma \in \widehat{G} : |\widehat{f}(\gamma)| \geq \rho \|f\|_1\}.$$

We shall also write

$$\mathrm{Spec}_{>0}(f) := \{\gamma \in \widehat{G} : |\widehat{f}(\gamma)| > 0\}.$$

It is not difficult to see that the ρ-large spectrum of a bounded function cannot be too large. Indeed, by Parseval's identity we have

$$\|f\|_2^2 = \|\widehat{f}\|_2^2 \geq \sum_{\gamma \in \mathrm{Spec}_\rho(f)} |\widehat{f}(\gamma)|^2 \geq \rho^2 \|f\|_1^2 |\mathrm{Spec}_\rho(f)|,$$

and hence $|\mathrm{Spec}_\rho(f)| \leq \rho^{-2}\|f\|_2^2/\|f\|_1^2$. In particular, when $f = \mu_A$ is the characteristic measure of a subset $A \subseteq G$ of density $\alpha := |A|/|G|$, then $|\mathrm{Spec}_\rho(\mu_A)| \leq \rho^{-2}\alpha^{-1}$.

In general, this bound is best possible as can be seen by taking A to be a subspace of $\mathbb{F}_2^n$ of fixed codimension, for example. However, it is not the most efficient description of the large Fourier spectrum. In fact, we shall see that the set of large Fourier coefficients of a function is determined by even fewer frequencies, in the following sense.

Definition 4.2 (*s*-covered) Let s be a positive integer. A subset $\Gamma \subseteq \widehat{G}$ is said to be *s-covered* if there exists a subset $S \subseteq \widehat{G}$ of size $|S| \leq s$ such that

$$\Gamma \subseteq \left\{ \sum_{\gamma \in S} \epsilon_\gamma \gamma : \epsilon_\gamma \in \{-1, 0, 1\} \right\}.$$

The following structural result concerning the large spectrum is due to Chang [3]. It has seen a number of important applications in additive combinatorics, for example to Roth's theorem [38], Freiman's theorem [3] and the structure of Boolean functions with small ℓ^1 norm [21], to name just a few.

Theorem 4.3 (Chang's theorem) *Let $A \subseteq G$ be a subset of density α. For every $\rho > 0$, $\mathrm{Spec}_\rho(\mu_A)$ is s-covered with*

$$s \leq 18\rho^{-2} \cdot \log(\alpha^{-1}).$$

This constitutes a logarithmic improvement over the bound arising from Parseval's theorem.

The proof of Chang's theorem in [3, 17] proceeds via Rudin's inequality. A first proof using entropy, restricted to the case $G = \mathbb{F}_2^n$, was later given by Impagliazzo et al. [26]. In fact, it turns out that the two are not unrelated: Friedgut [8] showed very recently how to derive hypercontractivity estimates of Bonami-Beckner type, which generalise Rudin's inequality, using entropy-based arguments.[1]

Green [16] showed that the bound in Chang's theorem as stated is tight up to a constant. However, as we shall see, it is possible to refine Theorem 4.3 in at least two ways which turn out to be advantageous in applications. First, it is possible to relax the notion of covering somewhat; second, we may require only a large subset of the spectrum to be covered.

To examine the first strengthening we need the following (non-standard) definition.

Definition 4.4 ((s,d)-covered) Let s, d be positive integers. We say a subset $\Gamma \subseteq \widehat{G}$ is (s,d)-*covered* if there exists a subset $S \subseteq \widehat{G}$ of size $|S| \leq s$ such that

$$\Gamma \subseteq \left\{ \sum_{\gamma \in S} \epsilon_\gamma \gamma : \epsilon_\gamma \in \mathbb{Z}, \sum_{\gamma \in S} |\epsilon_\gamma| \leq d \right\}.$$

Clearly if $\Gamma \subseteq \widehat{G}$ is s-covered for some $s \in \mathbb{N}$, then it is (s, s)-covered in the sense of Definition 4.4. Also, note that in the popular model setting of $G = \mathbb{F}_2^n$, the two notions of covering introduced above coincide.

In [40] Shekredov proved the following refinement of Chang's theorem.

---

[1]A comprehensive historical account of hypercontractivity estimates can be found in [31].

Theorem 4.5 (Shkredov's theorem) *Let $A \subseteq G$ be a subset of density α. For every $\rho > 0$, $\mathrm{Spec}_\rho(\mu_A)$ is (s, d)-covered with*

$$s \leq 2^{30}\rho^{-2}\log(\alpha^{-1}) \quad and \quad d \leq 8\log(\alpha^{-1}).$$

While the notion of covering is in some sense weaker than that in Chang's original theorem, it turns out that Shkredov's result performs better in certain applications. For example, one obtains a quantitative improvement in Bogolyubov's lemma, which states that for $A \subseteq G = \mathbb{Z}/p\mathbb{Z}$ with p a prime, the iterated sum set $2A - 2A := \{a_1 + a_2 - a_3 - a_4 : a_1, a_2, a_3, a_4 \in A\}$ contains a Bohr set

$$B(K, \delta) := \{x \in G : \|xt/p\| \leq \delta \text{ for all } t \in K\}$$

for some set $K \subseteq \widehat{G}$ of size $|K| \leq 2^{33}\alpha^{-1}\log(\alpha^{-1})$ and width $\delta = (2^8 \log(\alpha^{-1}))^{-1}$, instead of $\delta = \alpha(2^8 \log(\alpha^{-1}))^{-1}$ by an application of Chang's original result.

Following Lee [28] with some small modifications, we shall deduce a slightly weaker version of Theorem 4.5 from Theorem 3.2. Indeed, we were unable to determine at the time of writing whether the full strength of Shkredov's result can be recovered in this way.

Let $\mathcal{F} = \{\Re\gamma, \Im\gamma : \gamma \in \widehat{G}\}$, and note that this is indeed a valid choice for our family of test functions. We first prove an easy lemma which says that the non-zero spectrum of generalised Riesz $\mathcal{F}'$-products for $\mathcal{F}' \subseteq \mathcal{F}$ is efficiently covered.

Lemma 4.6 *Let $\mathcal{F}' \subseteq \mathcal{F} = \{\Re\gamma, \Im\gamma : \gamma \in \widehat{G}\}$, and let $\delta > 0$. Let $R : G \to \mathbb{R}$ be a Riesz $\mathcal{F}'$-product of degree at most d, and let $r : G \to \mathbb{R}$ be any non-negative linear combination of Riesz $\mathcal{F}'$-products of degree at most d such that $r \in \Delta_G$. Then $\mathrm{Spec}_\delta(R)$ is d-covered, and $\mathrm{Spec}_\delta(r)$ is $(2|\mathcal{F}'|, d)$-covered.*

Proof Consider a Riesz $\mathcal{F}'$-product $R = \prod_{i=1}^{d}(1 + \epsilon_i\phi_i)$ of degree at most d, with $\phi_1, \ldots, \phi_d \in \mathcal{F}'$ (not necessarily distinct) and $\epsilon_1, \ldots, \epsilon_d \in \{-1, 0, 1\}$. This means that for each $j = 1, \ldots, d$, ϕ_j is of the form $\Re\gamma_j = \frac{1}{2}(\gamma_j + \overline{\gamma}_j)$ or $\Im\gamma_j = \frac{1}{2}(\gamma_j - \overline{\gamma}_j)$ for some character $\gamma_j \in \widehat{G}$. In particular, whenever $\widehat{R}(\gamma) \neq 0$, by uniqueness of the Fourier expansion γ is a product of at most d elements from the multiset $S' := \{\gamma_1, \ldots, \gamma_d, \overline{\gamma}_1, \ldots, \overline{\gamma}_d\} \subseteq \widehat{G}$. This means that $\mathrm{Spec}_{>0}(R)$ is $(2d, d)$-covered. Note that we can replace a character γ' occurring k times in S' by $\{\gamma', \gamma'^2, \ldots, \gamma'^k, \overline{\gamma}', \overline{\gamma}'^2, \ldots, \overline{\gamma}'^k\}$ to obtain a set S of size at most δ, allowing us to conclude that in fact, $\mathrm{Spec}_{>0}(R)$ is d-covered. Since $\mathrm{Spec}_\delta(R) \subseteq \mathrm{Spec}_{>0}(R)$, the same is obviously true of $\mathrm{Spec}_\delta(R)$.

Suppose now that $r = \sum_i c_i R_i$ is a non-negative linear combination of Riesz $\mathcal{F}'$-products of degree at most d such that $r \in \Delta_G$. Clearly

$$1 = \|r\|_1 = \sum_i c_i \|R_i\|_1$$

and if $\gamma \in \mathrm{Spec}_\delta(r)$, then

$$|\sum_i c_i \widehat{R}_i(\gamma)| = |\widehat{r}(\gamma)| \geq \delta \|r\|_1 = \delta \sum_i c_i \|R_i\|_1,$$

It follows by the triangle inequality and averaging that there exists at least one i such that

$$|\widehat{R}_i(\gamma)| \geq \delta \|R_i\|_1,$$

i.e. $\gamma \in \mathrm{Spec}_\delta(R_i)$, and thus the first part of the argument implies that $\mathrm{Spec}_\delta(r)$ is $(2|\mathcal{F}'|, d)$-covered. $\qquad\square$

Now Theorem 3.2 allows us to approximate any function by a small number of generalised Riesz products, implying that the spectrum of such a function can also be efficiently covered.

Proof of (a weaker version of) Theorem 4.5 using Theorem 3.2:
Let $f = \mu_A$ be the characteristic measure of $A \subseteq G$, whose entropy $\mathrm{Ent}(f)$ equals $\log(\alpha^{-1})$. Set $\epsilon = \rho/4$ in Theorem 3.2 to obtain a function $g \in \Delta_G$ such that $\|f - g\|_\mathcal{F} \leq \rho/4$ and a subset $\mathcal{F}' \subseteq \mathcal{F} = \{\Re\gamma, \Im\gamma : \gamma \in \widehat{G}\}$ of size $|\mathcal{F}'| \leq 144\rho^{-2} \cdot \mathrm{Ent}(f)$ such that g is a non-negative linear combination of Riesz $\mathcal{F}'$-products of degree at most $d \leq 72\rho^{-1}\mathrm{Ent}(f) + O(\log\rho^{-1}/\log\log\rho^{-1})$. Since $\|f - g\|_\mathcal{F} \leq \rho/4$, we have $\mathrm{Spec}_\rho(f) \subseteq \mathrm{Spec}_{\rho/2}(g)$. By Lemma 4.6, $\mathrm{Spec}_{\rho/2}(g)$ is $(2|\mathcal{F}'|, d)$-covered, and hence so is $\mathrm{Spec}_\rho(f)$. $\qquad\square$

This statement implies a version of Bogolyubov's lemma in which the width of the Bohr set is $\delta = \alpha^{1/2}(2^8 \log(\alpha^{-1}))^{-1}$, a slight gain over Chang's original result.

In many applications it is not necessary to be able to cover the entire spectrumobtaining control over a large subset often suffices. This observation was formalised and exploited by Bloom [2], who used his refinement of Chang's theorem (Theorem 4.7 below) to derive the best known bounds in Roth's theorem to date.

Theorem 4.7 (Bloom's theorem) *For every $0 < \rho < e^{-3}$ and $f \in \Delta_G$, there exists a subset $S \subseteq \mathrm{Spec}_\rho(f)$ of size*

$$|S| \geq \frac{\rho}{2}|\mathrm{Spec}_\rho(f)|$$

which is d-covered for

$$d \leq 36\sqrt{2}\rho^{-1} \cdot \text{Ent}(f) + O\left(\frac{\log \rho^{-1}}{\log\log \rho^{-1}}\right).$$

Proof We set $\epsilon = \rho/(2\sqrt{2})$ in Theorem 3.2 to obtain a function $g \in \Delta_G$ such that $\|f - g\|_{\mathcal{F}} \leq \rho/(2\sqrt{2})$, and a subset $\mathcal{F}' \subseteq \mathcal{F}$ of size $|\mathcal{F}'| \leq 72\rho^{-2} \cdot \text{Ent}(f)$ such that

$$g = \sum_{i=1}^{\ell} c_i R_i,$$

where ℓ is a positive integer, the coefficients c_i are positive reals and each R_i is a Riesz $\mathcal{F}'$-product of degree at most

$$d \leq 36\sqrt{2}\rho^{-1} \cdot \text{Ent}(f) + O\left(\frac{\log \rho^{-1}}{\log\log \rho^{-1}}\right).$$

Note that by Lemma 4.6, the non-zero spectrum of every Riesz product R_i is d-covered. Thus if we could show that for some i, $\text{Spec}_{>0}(R_i)$ is also a reasonably large proportion of $\text{Spec}_{\rho}(f)$, we would be done.

We achieve this by choosing an element at random from $\{1, 2, \ldots, \ell\}$, where we choose j with probability $c_j \mathbb{E}R_j$. Note that since $g \in \Delta_G$, we have that the sum of the probabilities equals $\sum_{i=1}^{\ell} c_i \mathbb{E}R_i = \mathbb{E}\sum_{i=1}^{\ell} c_i R_i = \mathbb{E}g = 1$.

Fix $\gamma \in \text{Spec}_{\rho}(f) \subseteq \text{Spec}_{\rho/2}(g)$, where the latter inclusion holds since $\|f - g\|_{\mathcal{F}} \leq \rho/(2\sqrt{2})$. Now

$$\mathbb{E}_j|\langle\gamma, R_j/(\mathbb{E}R_j)\rangle| = \sum_{i=1}^{\ell} c_i \mathbb{E}R_i|\langle\gamma, R_i/(\mathbb{E}R_i)\rangle|$$

$$\geq |\langle\gamma, \sum_{i=1}^{\ell} c_i R_i\rangle|$$

$$= |\langle\gamma, g\rangle|$$

$$\geq \rho/2,$$

where the first expectation is a probabilistic one with respect to the random choice of j from $\{1, 2, \ldots, \ell\}$. It follows from the preceding line that for any fixed $\gamma \in \text{Spec}_{\rho}(f)$

$$\mathbb{P}_j[\widehat{R}_j(\gamma) \neq 0] = \mathbb{P}_j[|\langle\gamma, R_j/(\mathbb{E}R_j)\rangle| > 0] \geq \rho/2,$$

i.e. that any $\gamma \in \text{Spec}_{\rho}(f)$ lies, with probability at least $\rho/2$, in $\text{Spec}_{>0}(R_i)$ for a randomly chosen i, whence $\mathbb{E}_j|\text{Spec}_{>0}(R_j)| \geq (\rho/2)|\text{Spec}_{\rho}(f)|$. We conclude that there exists a choice of j for which the desired inequality $|\mathrm{Spec}_{>0}(R_j)| \geq (\rho/2)|\mathrm{Spec}_\rho(f)|$ holds. We now set $S := \mathrm{Spec}_{>0}(R_i) \subseteq \mathrm{Spec}_\rho(f)$ to obtain the desired conclusion. $\qquad\square$

5   Quadratic decompositions

The usefulness of the Fourier transform lies in the fact that we can decompose any bounded function on G into a weighted sum of characters, which we shall often refer to as *linear phase functions* in the sequel. Indeed, since the characters form an orthonormal basis for the space of functions $G \to \mathbb{C}$, such a decomposition is unique, and gives rise to the inversion formula (4.1). Those phases with small coefficients can often be neglected in applications, but not always.

Starting with the ground-breaking work of Gowers in [11, 12], the idea that bounded functions are well approximated by their large Fourier coefficients has been successfully generalised and led to the development of so-called *higher-order Fourier analysis* over the past decade. A higher-order analogue of the Fourier inversion formula decomposes any bounded function as a weighted sum of higher-degree *polynomial phase functions*, plus a pseudorandom error term that is negligible in a wider range of applications.

The notion of pseudorandomness that will be appropriate here is due to Gowers [11, 12].

Definition 5.1 (U^k norm) Let $k \geq 2$ be an integer. The U^k *norm* of a function $f : G \to \mathbb{C}$ is defined by the formula

$$\|f\|_{U^k}^{2^k} := \mathbb{E}_{x,h_1,\ldots,h_k \in G}\Delta_{h_1}\Delta_{h_2}\ldots\Delta_{h_k}f(x),$$

where $\Delta_h f(x) := f(x)\overline{f(x+h)}$ is to be thought of as a discrete phase derivative.

It is straightforward to verify that

$$\|f\|_{U^2} = \|\widehat{f}\|_4,$$

and hence that as a measure of pseudorandomness the U^2 norm and the size of the Fourier coefficients of f are, at least in a qualitative sense, equivalent. It is also not too difficult to see, but certainly not obvious, that $\|\cdot\|_{U^k}$ is indeed a norm, and that these norms are nested in the sense that for any integer $k \geq 2$,

$$\|f\|_{U^2} \leq \|f\|_{U^3} \leq \cdots \leq \|f\|_{U^k} \leq \|f\|_\infty.$$

We leave the details to the keen reader, who may wish to refer to the book [43] or the lecture notes [20].

The uniformity norms defined above are useful because of the following proposition, again due to Gowers, which states that error terms which are small in U^{k+1} are negligible when it comes to counting linear patterns of complexity k. A useful example of a linear pattern of complexity k which the reader may wish to bear in mind is an arithmetic progression of length $k + 2$, but we shall not define the notion of "complexity" formally here as it is somewhat involved [24].

Proposition 5.2 *Let $\mathcal{L}$ be a linear pattern of complexity at most k, and let $f : G \to \mathbb{C}$ be a function satisfying $\|f\|_\infty \leq 1$. Then*

$$|\mathbb{E}_{x_1,\ldots,x_d} \prod_{i=1}^{m} f(L_i(x_1,\ldots,x_d))| \leq \|f\|_{U^{k+1}}.$$

This proposition implies that the count of 3-term arithmetic progressions (a linear pattern of complexity $k = 1$) in a subset $A \subseteq G$ is controlled by the U^2 norm of the balanced function of A, and hence by the size of its Fourier coefficients. In what follows, we shall restrict our attention to the case $k = 2$, and say that a function is *quadratically uniform* if it is small in the U^3 norm. In particular, Proposition 5.2 implies that any subset $A \subseteq G$ of density α in G contains the expected number of 4-term arithmetic progressions, namely $\alpha^4 |G|^2$, whenever its balanced function is quadratically uniform.

When the balanced function fails to be quadratically uniform, i.e. when its U^3 norm is non-negligible, we can invoke Theorem 5.3 below, which says that such a function must correlate with a *quadratic phase function*, and hence the set A must be somewhat quadratically structured. For simplicity of the exposition we concentrate on the so-called finite-field model setting here, see [18, 49]. Theorem 5.3 as stated is due to Samorodnitsky [37] ($p = 2$) and Green and Tao [22] ($p > 2$), but the sophisticated ideas on which the proofs build already formed the cornerstone of Gowers's proof of Szemerdi's theorem [11].

Theorem 5.3 (U^3 inverse theorem) *Let $\delta > 0$, and let $f : \mathbb{F}_p^n \to \mathbb{C}$ be a function satisfying $\|f\|_\infty \leq 1$ and $\|f\|_{U^3} \geq \delta$. Then there exists a quadratic form q over $\mathbb{F}_p^n$ such that*

$$|\mathbb{E}_x f(x)\omega^{q(x)}| \geq c(\delta),$$

where $\omega := \exp(2\pi i/p)$, for some constant $c(\delta)$ going to 0 as δ tends to 0.

So far we have expressed the dichotomy between quadratic uniformity and quadratic structure as two separate statements about the U^3 norm of a function. It is often convenient to combine these into one, by decomposing the function into a quadratically uniform and a quadratically structure part, as alluded to in the introductory paragraph of this section. Indeed, in the past decade a significant amount of work has been done towards justifying the labels *quadratic* and *higher-order Fourier analysis* by obtaining quadratic and higher-order decompositions that (at least weakly) possess many of the useful properties that the traditional Fourier decomposition exhibits. One of the first explicit quadratic decomposition theorems was provided in [20].

Theorem 5.4 (Green-Tao decomposition) *Let* $f : \mathbb{F}_p^n \to [-1,1]$ *be a function and let* $\delta > 0$. *Then there is a so-called quadratic factor* (B_1, B_2) *of complexity at most* $O_\delta(1)$ *such that*

$$f = \mathbb{E}(f|(B_1, B_2)) + g,$$

where $\|g\|_{U^3} \leq \delta$.

A *quadratic factor* (B_1, B_2) *of complexity* $O_\delta(1)$ is a simultaneous level set of $O_\delta(1)$ linear and quadratic phases, and the projection $\mathbb{E}(f|(B_1, B_2))$ of f onto such a factor is simply f averaged over each of the level sets comprising the factor. For a precise definition, see [20]. The proof of Theorem 5.4 uses an ℓ^2 energy increment and makes crucial use of Theorem 5.3 above.

A quadratic decomposition which more closely resembles that obtained in classical Fourier analysis is due to Gowers and the author [15][2]

Theorem 5.5 (Gowers-W. decomposition) *Let* $f : \mathbb{F}_n^p \to \mathbb{C}$ *be a function such that* $\|f\|_2 \leq 1$. *Then for every* $\epsilon > 0$ *and* $\eta > 0$, *there exists* $M = \exp(O(\log(\eta\epsilon)^{-O(1)}))$ *such that* f *has a decomposition of the form*

$$f = \sum_i \lambda_i \omega^{q_i} + h + \ell,$$

where the q_i *are quadratic forms on* $\mathbb{F}_p^n$, $\|h\|_{U^3} \leq \epsilon$, $\|\ell\|_1 \leq \eta$ *and* $\sum_i |\lambda_i| \leq M$.

Under the so-called Polynomial Freiman-Ruzsa Conjecture (see [18, 49]) one might expect a polynomial tradeoff between the uniformity of the

---

[2]The statement given incorporates Sanders's quantitative improvement on the Bogolyubov lemma in [39], and is thus different from the published version in [15].

function h and the complexity M of the quadratically structured part of the above decomposition. The proof of Theorem 5.5 was based on Theorem 5.3 and the finite-dimensional Hahn-Banach theorem (see also Section 6).

In [47], a different technique from machine learning, known as *boosting*[3], was used to obtain the following variant.

Theorem 5.6 (Tulsiani-W. decomposition) *Let $\mathcal{F}$ be a class of functions $\phi : X \to [-1, 1]$ closed under negation, and let $\epsilon, \delta > 0$ and $B > 1$. Let A be an algorithm which, given oracle access to a function $f : X \to [-B, B]$ satisfying $\|f\| \geq \epsilon$, outputs, with probability at least $1 - \delta$, a function $\phi \in \mathcal{F}$ such that $\langle f, \phi \rangle \geq \eta$ for some $\eta = \eta(\epsilon, B)$. Then there exists an algorithm which, given any function $f : X \to [-1, 1]$, outputs with probability at least $1 - \eta^{-2}\delta$ a decomposition*

$$f = \sum_{i=1}^{k} c_i \phi_i + h + \ell$$

satisfying $k \leq \eta^{-2}$, $\|h\| \leq \epsilon$, $\|\ell\|_1 \leq (2B)^{-1}$ and $\phi_i \in \mathcal{F}$ for all $i = 1, \ldots, k$. The algorithm makes at most k calls to A.

The norm $\| \cdot \|$ can be taken to be $\| \cdot \|_{\mathcal{F}}$, but also the U^3 norm. An explicit probabilistic algorithm A was given in the paper. Note that when such an algorithm is deterministic, the resulting decomposition algorithm is deterministic. One of the advantages of Theorem 5.6 over Theorem 5.5 was that it naturally gave a bound on the number of quadratic phases involved in the decomposition, which the authors had to work much harder to obtain in [15].

As an immediate application of the sparse approximation theorem in Section 3 (Theorem 3.2) we are able to deduce yet another quadratic decomposition theorem.

Theorem 5.7 *Let $f : \mathbb{F}_p^n \to \mathbb{R}^+$ be a function such that $\|f\|_1 \leq 1$, and let $0 < \epsilon < e^{-3}$. Then there exists a function $g : \mathbb{F}_p^n \to \mathbb{R}^+$ such that $\mathbb{E}f = \mathbb{E}g$, and a set $\mathcal{Q}'$ of quadratic forms of size $|\mathcal{Q}'| \leq 9\epsilon^{-2}\mathrm{Ent}(f/\|f\|_1)$ such that f can be written as*

$$f = g + h,$$

where $g = \sum_i \lambda_i \omega^{q_i}$ with real coefficients λ_i, $\|h\|_{\mathcal{Q}} \leq \epsilon$ and each q_i is of the form

$$q_i = \sum_{q' \in Q_i'} q' - \sum_{q'' \in Q_i''} q'',$$

---

[3]In fact, this technique is closely related to the gradient descent approach described in Section 3, so Theorem 5.7 should not come as a surprise.

where $Q_i', Q_i'' \subseteq \mathcal{Q}'$ are (multi)sets whose sizes are bounded above by

$$|Q_i'| + |Q_i''| \le 18\epsilon^{-1}\mathrm{Ent}(f/\|f\|_1) + O\left(\frac{\log(\epsilon^{-1})}{\log\log(\epsilon^{-1})}\right).$$

Proof Let $\mathcal{F} := \{\Re\omega^{q(x)}, \Im\omega^{q(x)} : q \text{ a quadratic form on } \mathbb{F}_p^n\}$, and use Theorem 3.2 applied to $f/\|f\|_1$ to obtain a family $\mathcal{F}' \subseteq \mathcal{F}$, which gives rise to a family of quadratic forms $\mathcal{Q}'$, and a function $g' : \mathbb{F}_p^n \to \mathbb{R}^+$ such that $g' \in \Delta_G$ and g' is a non-negative linear combination of Riesz $\mathcal{Q}'$-products. It is easy to check that $g := g'\|f\|_1$ satisfies the stated conditions. $\square$

Theorem 5.7 could be interpreted as a quadratic Chang-type theorem as it states that the large quadratic Fourier spectrum is spanned by a small number of quadratic phases. But in some respects it is weaker than the above-cited quadratic decomposition theorems: it does not give a bound on the number of terms in the structured part of the decomposition, and the error h is only guaranteed to not have any correlation with a quadratic phase (rather than being small in U^3).

However, both of these drawbacks are superficial. The equality $\|g\|_1 = \mathbb{E}g = \mathbb{E}f = \|f\|_1$ can replace the bound on the L^1 norm of the coefficients in applications. Moreover, using the inverse theorem for the U^3 norm (Theorem 5.3), we can derive the following corollary.

Corollary 5.8 *Let $f : \mathbb{F}_p^n \to \mathbb{R}^+$ be a function such that $\|f\|_1 \le 1$, and let $0 < \epsilon < e^{-3}$. Then there exists a function $g : \mathbb{F}_p^n \to \mathbb{R}^+$ such that $\mathbb{E}f = \mathbb{E}g$ and a set $\mathcal{Q}'$ of quadratic forms of size $|\mathcal{Q}'| \le 18c(\epsilon)^{-2}\mathrm{Ent}(f/\|f\|_1)$ such that f can be written as*

$$f = g + h,$$

where $g = \sum_i \lambda_i \omega^{q_i}$ with real coefficients λ_i, $\|h\|_{U^3} \le \epsilon$ and each q_i is of the form

$$q_i = \sum_{q' \in Q_i'} q' - \sum_{q'' \in Q_i''} q'',$$

where $Q_i', Q_i'' \subseteq \mathcal{Q}'$ are (multi)sets whose sizes are bounded above by

$$|Q_i'| + |Q_i''| \le 26c(\epsilon)^{-1}\mathrm{Ent}(f/\|f\|_1) + O\left(\frac{\log(\epsilon^{-1})}{\log\log(\epsilon^{-1})}\right),$$

and $c(\epsilon)$ is the constant occurring in Theorem 5.3.

Proof Observe that if h has correlation $c(\epsilon)$ with a quadratic form, then it has correlation at least $c(\epsilon)/\sqrt{2}$ with an element of $\mathcal{F}$ as defined in the proof of Theorem 5.7. The result therefore follows from Theorem 5.3.
$\square$

Just as with Theorems 5.4, 5.5 and 5.6, in many applications stronger information is required about the structured part of the decomposition given by Corollary 5.8. For example, we often need the quadratic phases that make up g to have high rank. We shall not pursue such refinements or their applications here.

6   The transference principle

In view of the application described in the previous section, it seems natural to try and apply the entropy-increment method to derive a third type of result in arithmetic combinatorics which is generally accessible by either the ℓ^2 energy increment or the Hahn-Banach approach: the *transference principle* states, informally speaking, that any distribution which is dense with respect to a pseudorandom measure can be well approximated by a genuinely dense distribution. We phrase this principle here in such a way that it naturally fits with the content of the previous sections. As before, let $\mathcal{F}$ be a family of functions $\{\phi : X \to [0,1]\}$, and for every integer $k \geq 1$, define $\mathcal{F}^k$ to be the set of all k-fold products of elements of $\mathcal{F}$.

Theorem 6.1 (Transference principle, I) *For all $\epsilon > 0$, there exists $k = \epsilon^{-O(1)}$ and $\eta = \exp(-\epsilon^{-O(1)})$ such that the following holds.*

For any family $\mathcal{F} = \{\phi : X \to [0,1]\}$ and any measure $\nu : X \to \mathbb{R}^+$ which is η-pseudorandom with respect to $\mathcal{F}^k$ in the sense that $\|\nu - 1_X\|_{\mathcal{F}^k} \leq \eta$, and any function $f : X \to \mathbb{R}^+$ with $0 \leq f \leq \nu$ and $\mathbb{E}f \leq 1$, there exists a bounded function $g : X \to [0,1]$ such that $\mathbb{E}g = \mathbb{E}f$ and $\|f - g\|_{\mathcal{F}} \leq \epsilon$.

Such a statement was first introduced by Green [19] in his work on 3-term arithmetic progressions in the primes, in the case where the family $\mathcal{F}$ consisted of linear characters and ν was a majorant of the primes, based on the set of almost-primes which exhibits all the right pseudorandomness properties. The Fourier-analytic manifestations of the transference principle are well summarised in Prendiville's recent survey [32]. Its true power, however, only came to light a few years later in Green and Tao's celebrated proof that there are arbitrarily long arithmetic progressions in the primes [23]. The principle was made explicit for the first time by Tao and Ziegler in [46], and a little later given a new proof by Gowers [13] and independently Reingold, Trevisan, Tulsiani and Vadhan [35], who introduced the Hahn-Banach (or duality of linear programming) approach which greatly improved the quantitative dependence of the parameters involved. This improved dependence turned out to be crucial for applications in theoretical computer science, where the transference principle is now widely known as the *dense model theorem*. Its applications include but are not limited to leakage-resilient cryptography and connections to computational differential privacy (for references, see [48]). Reingold et al. [35] also investigated the broader implications of their approach: they obtained a new proof of Impagliazzo's hard core set theorem, which states that if a Boolean function is somewhat hard on average, then there must be a subset of inputs (the hard core) on which it is extremely hard, and outside of which it is easy; and a new proof of Frieze and Kannan's weak regularity lemma [9], which allows one to partition the vertex set of any graph into a small number of parts in such a way that the edge density between any two subsets of vertices in the graph is close to the value expected, based on the edge densities between the parts of the partition and the intersection sizes of the vertex subsets within these parts.

To round off this historical account, let us mention that new light was shed on the exact nature of the pseudorandomness conditions needed for the transference principle to work by Conlon, Fox and Zhao [5] (see also the expository note [51] by Zhao), and indeed their expository article [4] is the most compact and up-to-date reference at the time of writing. In fact, their work furnishes a very general regularity and counting lemma in sparse pseudorandom hypergraphs, which allows one to prove analogues of well-known combinatorial theorems such as Ramseys theorem and Turn's theorem relative to certain sparse pseudorandom hypergraphs.

The remainder of this section is structured as follows. First, we give a proof of Theorem 6.1 following [34] which relies on the finite-dimensional Hahn-Banach theorem, presented here in the (equivalent) form of the Min-Max theorem from game theory. In the second half of this section we describe recent work of Vadhan and Zheng [48], who use a relative-entropy approach to prove a more constructive variant of the Min-Max theorem that leads to an asymptotically optimal dense model theorem.

By $\mathrm{Conv}(\mathcal{F})$ we denote the convex hull of $\mathcal{F}$, and by $\mathrm{Avg}_k(\mathcal{F})$ the family of all averages of at most k elements of $\mathcal{F}$. It will also be convenient to define, for any $t \in \mathbb{R}$, the threshold function $\mathrm{Th}_t : \mathbb{R} \to \{0,1\}$ by $\mathrm{Th}_t(x) = 1$ if $x \geq t$, and $\mathrm{Th}_t(x) = 0$ if $x < t$.

Let us now restate Theorem 6.1 in the following fashion.

Theorem 6.2 (Transference principle, II) *Let $\epsilon, \delta > 0$. Let $\mathcal{F} = \{\phi : X \to [0,1]\}$ be a family of functions and $\nu : X \to \mathbb{R}^+$ any measure, and let $f : X \to \mathbb{R}^+$ be any function with $0 \leq f \leq \nu$ and $\mathbb{E}f = \delta$. Suppose that for any bounded function $g : X \to [0,1]$ such that $\mathbb{E}g = \delta$ we have $\|f - g\|_{\mathcal{F}} \geq \epsilon\delta$.*
Then ν is not pseudorandom in the following sense:

1. *There exists an integer* $k = O(\epsilon^{-2}\log(\epsilon^{-1}\delta^{-1}))$, $t \in \mathbb{R}$ *and* $\psi = \mathrm{Th}_t(\overline{\phi})$ *for some* $\overline{\phi} \in \mathrm{Avg}_k(\mathcal{F})$ *such that*

$$\langle \psi, \nu - 1_X \rangle = \Omega(\epsilon\delta).$$

2. *There exists* $k = \mathrm{poly}(\epsilon^{-1}, \delta^{-1})$ *and* $\phi \in \mathcal{F}^k$ *such that*

$$\langle \phi, \nu - 1_X \rangle \geq \exp(-\mathrm{poly}(\epsilon^{-1}, \delta^{-1})).$$

Part (2) immediately implies Theorem 6.1, and is itself derived from (1) by an application of the Bolzano-Weierstrass theorem, which allows us to replace the threshold function with a product of functions from $\mathcal{F}$. We refer the reader to [35] for details of this reduction, and focus on Part (1) below.

The driving force behind the proof of Theorem 6.2 (1) will be the Min-Max theorem from game theory, which asserts, roughly speaking, that in any zero-sum game between two players, if Player 2 can respond to any (potentially randomised) strategy of Player 1 and achieve a payoff of at least c, then Player 2 in fact has a *universal* (randomised) strategy that guarantees such a payoff regardless of Player 1's strategy.

Theorem 6.3 (Min-Max theorem) *Consider a 2-player zero-sum game in which Player 1's set of pure strategies is* $\mathcal{V} \subseteq \Delta_X$ *and Player 2's is* $\mathcal{W}$, *and the expected pay-off to Player 2 is* $\mathbb{E}f(V, W)$ *for some function* $f : X \times \mathcal{W} \to [0, 1]$. *Then*

$$\max_{W \in \mathrm{Conv}(\mathcal{W})} \min_{V \in \mathcal{V}} \mathbb{E}f(V, W) = \min_{V \in \mathrm{Conv}(\mathcal{V})} \max_{W \in \mathcal{W}} \mathbb{E}f(V, W).$$

In particular, there exist mixed strategies $V^* \in \mathrm{Conv}(\mathcal{V})$ *for Player 1 and* $W^* \in \mathrm{Conv}(\mathcal{W})$ *for Player 2, and a value* c *such that*

$$\min_{V \in \mathcal{V}} \mathbb{E}f(V, W^*) = \max_{W \in \mathcal{W}} \mathbb{E}f(V^*, W) = c.$$

Having stated the essential ingredient, we shall now begin a proof of Part (1) of Theorem 6.1 following [35]. First, we give some intuition behind the general argument. Let $\mathcal{H}$ be the set of all functions $g : X \to [0, 1]$ such that $\mathbb{E}g = \delta$. Suppose that for all $h \in \mathcal{H}$, there exists $\phi \in \mathcal{F}$ such that

$$|\langle f - h, \phi \rangle| > \epsilon\delta. \tag{6.1}$$

We must conclude that $\nu - 1_X$ cannot be $\epsilon\delta$-pseudorandom with respect to $\mathcal{F}$. In order to see this, suppose for the purpose of simplifying the argument that there is a pair h, ϕ as above such that $h(x) = 1$ for all $x \in \text{supp}(\phi)$. Then, since $f \leq \nu$ and by the aforementioned property of h,

$$\langle \nu - 1_X, \phi \rangle = \langle \nu, \phi \rangle - \langle 1_X, \phi \rangle \geq \langle f, \phi \rangle - \langle 1_X, \phi \rangle = \langle f, \phi \rangle - \langle h, \phi \rangle = \langle f - h, \phi \rangle, \tag{6.2}$$

which in conjunction with (6.1) says that $\nu - 1_X$ is not $\epsilon\delta$-pseudorandom with respect to $\mathcal{F}$. Since we had assumed h to be a of a special form, this conclusion is not quite valid in general, but we shall see that the general approach can be made to work with a little more effort. In what follows it will be convenient to replace $\mathcal{F}$ by $\overline{\mathcal{F}} := \mathcal{F} \cup (1 - \mathcal{F})$, as this allows us to remove the absolute value in (6.1). Starting with that equation, we make use of the full strength of the Min-Max theorem to establish the existence of a *universal* distinguisher ([35], Claim 2.1).

Lemma 6.4 *There exists a function $\tilde{\phi} \in \text{Conv}(\overline{\mathcal{F}})$ such that for all $h \in \mathcal{H}$, $\langle f - h, \tilde{\phi} \rangle > \epsilon\delta$.*

Proof We consider the following 2-player zero-sum game. The first player picks $g \in V := \delta^{-1}\mathcal{H} \subseteq \Delta_X$, the second picks $\phi \in W := \overline{\mathcal{F}}$. The payoff for Player 1 is $-\langle \phi, \delta^{-1}f - g \rangle$, and the payoff for Player 2 is $\langle \phi, \delta^{-1}f - g \rangle$. Now by the Min-Max theorem (Theorem 6.3), the game has a value c for which Player 1 has an optimal mixed strategy $g^* \in \text{Conv}(\delta^{-1}\mathcal{H})$ and Player 2 has an optimal mixed strategy $\phi^* \in \text{Conv}(\overline{\mathcal{F}})$ in the sense that for all $\phi \in \overline{\mathcal{F}}$

$$\langle \phi, \delta^{-1}f - g^* \rangle \leq c, \tag{6.3}$$

and for all $g \in \delta^{-1}\mathcal{H}$,

$$\langle \phi^*, \delta^{-1}f - g \rangle \geq c \tag{6.4}$$

Since g^* is a distribution over elements of $\delta^{-1}\mathcal{H}$, it is itself in $\delta^{-1}\mathcal{H}$ and hence by hypothesis of the theorem, there exists $\phi \in \overline{\mathcal{F}}$ such that $\langle \phi, \delta^{-1}f - g^* \rangle > \epsilon$. It follows from (6.3) that $c > \epsilon$, and from (6.4) that $\langle \phi^*, \delta^{-1}f - g \rangle \geq c > \epsilon$ for all $g \in \delta^{-1}\mathcal{H}$, or $\langle \phi^*, f - g \rangle > \epsilon\delta$ for all $g \in \mathcal{H}$. $\square$

Note that since ϕ^* was optimal, $\tilde{\phi}$ is a distribution over functions in $\overline{\mathcal{F}}$. It is therefore reasonable to expect to be able to replace this universal distinguisher $\tilde{\phi}$ by an average of a small number of functions in $\mathcal{F}$. To this end, let S be a set of the $\delta|X|$ elements of X for which the value of $\tilde{\phi}$ is largest. Since S has density δ in X, that is $1_S \in \mathcal{H}$, we have $\langle f - 1_S, \tilde{\phi} \rangle > \epsilon\delta$. We shall show that f and 1_S can also be distinguished by a (Boolean) threshold function.

Claim 6.5 *There exists a value of $t \in [\epsilon/2, 1]$ such that*

$$\langle f, \text{Th}_t(\tilde{\phi}) \rangle - \langle 1_S, \text{Th}_{t-\epsilon/2}(\tilde{\phi}) \rangle > \epsilon\delta/2.$$

Proof of Claim 6.5: Note that by definition of the threshold function Th_t,

$$\int_0^1 \mathrm{Th}_t(\tilde{\phi})(x)dt = \tilde{\phi}(x),$$

and thus by hypothesis

$$\int_0^1 \langle f, \mathrm{Th}_t(\tilde{\phi})\rangle dt = \langle f, \tilde{\phi}\rangle > \langle 1_S, \tilde{\phi}\rangle + \epsilon\delta = \int_0^1 \langle 1_S, \mathrm{Th}_t(\tilde{\phi})\rangle dt + \epsilon\delta. \quad (6.5)$$

Suppose the statement of the claim is false, and we have that for all $t \in [\epsilon/2, 1]$,

$$\langle f, \mathrm{Th}_t(\tilde{\phi})\rangle - \langle 1_S, \mathrm{Th}_{t-\epsilon/2}(\tilde{\phi})\rangle \le \epsilon\delta/2.$$

Then

$$\int_0^1 \langle f, \mathrm{Th}_t(\tilde{\phi})\rangle dt = \int_0^{\epsilon/2} \langle f, \mathrm{Th}_t(\tilde{\phi})\rangle dt + \int_{\epsilon/2}^1 \langle f, \mathrm{Th}_t(\tilde{\phi})\rangle dt$$

is less than or equal to

$$\int_0^{\epsilon/2} \langle f, 1_X\rangle dt + \int_{\epsilon/2}^1 (\langle 1_S, \mathrm{Th}_{t-\epsilon/2}(\tilde{\phi})\rangle + \epsilon\delta/2)dt$$

which in turn is bounded above by

$$\epsilon/2 \cdot \delta + \int_0^1 \langle 1_S, \mathrm{Th}_t(\tilde{\phi})\rangle dt + \epsilon\delta/2 \le \int_0^1 \langle 1_S, \mathrm{Th}_t(\tilde{\phi})\rangle dt + \epsilon\delta,$$

contradicting the assumption that $\tilde{\phi}$ is capable of distinguishing f and 1_S in (6.5). □

We shall next show that $\tilde{\phi}$, and its derived Boolean threshold, are also capable of distinguishing ν and 1_X. First, let $t' := t - \epsilon/2$ and observe that the function 1_S equals 1 on $\mathrm{supp}(\mathrm{Th}_{t'}(\tilde{\phi}))$. Indeed, suppose this were not the case and we had $1_S(x) < 1$ for some $x \in \mathrm{supp}(\mathrm{Th}_{t'}(\tilde{\phi}))$. Then since the set S was defined to contain the $\delta|X|$ elements of X for which the value of $\tilde{\phi}$ is largest, we would certainly have $1_S(y) = 0$ for all $y \notin \mathrm{supp}(\mathrm{Th}_{t'}(\tilde{\phi}))$, in which case

$$\langle 1_S, \mathrm{Th}_{t'}(\tilde{\phi})\rangle = \langle 1_S, 1_X\rangle = \delta = \langle f, 1_X\rangle \ge \langle f, \mathrm{Th}_t(\tilde{\phi})\rangle,$$

contradicting Claim 6.5. Note that we are now in possession of the simplifying assumption made in the intuitive outline of the proof, preceding (6.2): we have a function which has density δ in X, namely 1_S, and a threshold function $\text{Th}_{t'}(\tilde{\phi})$ on whose support 1_S equals 1, and which distinguishes f and 1_S. It follows, arguing as in (6.2), that

$$\langle \nu - 1_X, \text{Th}_{t'}(\tilde{\phi}) \rangle = \langle \nu, \text{Th}_{t'}(\tilde{\phi}) \rangle - \langle 1_X, \text{Th}_{t'}(\tilde{\phi}) \rangle \geq \langle f, \text{Th}_{t'}(\tilde{\phi}) \rangle - \langle 1_S, \text{Th}_{t'}(\tilde{\phi}) \rangle,$$

which is bounded below by

$$\langle f, \text{Th}_t(\tilde{\phi}) \rangle - \langle 1_S, \text{Th}_{t'}(\tilde{\phi}) \rangle \geq \epsilon\delta/2$$

by Claim 6.5. So again, the threshold function $\text{Th}_{t'}(\tilde{\phi})$ distinguishes ν and 1_X. The additional slack in the statement will now allow us to replace the threshold function $\text{Th}_{t'}(\tilde{\phi})$ with the threshold of a function that is an average of at most k functions in $\overline{\mathcal{F}}$ (rather than an element of $\text{Conv}(\overline{\mathcal{F}})$).

Lemma 6.6 *Under the assumption of (6.1) and with the earlier value of t, there exists a distinguisher $\overline{\phi} \in \text{Avg}_k(\mathcal{F})$ such that*

$$\langle \text{Th}_{t-2\epsilon/5}(\overline{\phi}), \nu - 1_X \rangle = \Omega(\epsilon\delta).$$

Proof We view $\tilde{\phi}$ as a distribution over functions in $\overline{\mathcal{F}}$. It follows from a Chernoff bound that $\tilde{\phi}$ will be well approximated by the average of a few functions sampled randomly from this distribution. To see this, pick k functions $\phi_1, \phi_2, \ldots, \phi_k$ randomly and independently from $\overline{\mathcal{F}}$ with probability given by the distribution $\tilde{\phi}$. Then for $k = O(\epsilon^{-2}\log(\epsilon^{-1}\delta^{-1}))$, we have that for every fixed element y,

$$\mathbb{P}_{\phi_1,\phi_2,\ldots,\phi_k}\left[\left|\tilde{\phi}(y) - \frac{\phi_1(y) + \phi_2(y) + \cdots + \phi_k(y)}{k}\right| > \epsilon/10\right] \leq \epsilon\delta/100.$$

Therefore for any probability distribution Y, we find that

$$\mathbb{E}_{\phi_1,\phi_2,\ldots,\phi_k}\left[\mathbb{P}_{y \in Y}\left[\left|\tilde{\phi}(y) - \frac{\phi_1(y) + \phi_2(y) + \cdots + \phi_k(y)}{k}\right| > \epsilon/10\right]\right] \leq \epsilon\delta/100,$$

and by Markov's inequality

$$\mathbb{P}_{\phi_1,\phi_2,\ldots,\phi_k}\left[\mathbb{P}_{y \in Y}\left[\left|\tilde{\phi}(y) - \frac{\phi_1(y) + \phi_2(y) + \cdots + \phi_k(y)}{k}\right| > \epsilon/10\right] > \epsilon\delta/10\right] \leq 1/10$$

when y is drawn from any distribution Y. It follows that there exists a choice of $\phi_1, \phi_2, \ldots, \phi_k$ such that, upon letting $\overline{\phi} := \frac{1}{k}\sum_{i=1}^{k}\phi_i$, we have

$$\mathbb{P}_{y \in Y}\left[|\tilde{\phi}(y) - \overline{\phi}(y)| > \epsilon/10\right] \leq \epsilon\delta/10$$

whenever y is chosen according to ν or to 1_X. In particular, this implies that

$$\langle \mathrm{Th}_{t-\epsilon/10}(\overline{\phi}), \nu \rangle \geq \langle \mathrm{Th}_t(\tilde{\phi}), \nu \rangle - \epsilon\delta/5 \geq \langle \mathrm{Th}_t(\tilde{\phi}), f \rangle - \epsilon\delta/5$$

and

$$\langle \mathrm{Th}_{t-2\epsilon/5}(\overline{\phi}), 1_X \rangle \leq \langle \mathrm{Th}_{t-\epsilon/2}(\tilde{\phi}), 1_X \rangle + \epsilon\delta/10 = \langle \mathrm{Th}_{t-\epsilon/2}(\tilde{\phi}), 1_S \rangle + \epsilon\delta/10.$$

This means that

$$\langle \mathrm{Th}_{t-2\epsilon/5}(\overline{\phi}), \nu - 1_X \rangle \geq \langle \mathrm{Th}_{t-\epsilon/10}(\overline{\phi}), \nu \rangle - \langle \mathrm{Th}_{t-2\epsilon/5}(\overline{\phi}), 1_X \rangle$$

is bounded below by

$$\langle \mathrm{Th}_t(\tilde{\phi}), f \rangle - \epsilon\delta/5 - (\langle \mathrm{Th}_{t-\epsilon/2}(\tilde{\phi}), 1_S \rangle + \epsilon\delta/10),$$

which by Claim 6.5 is $\Omega(\epsilon\delta)$. It follows that $\mathrm{Th}_{t-2\epsilon/5}(\overline{\phi})$ distinguishes ν and 1_X, as desired. $\qquad\square$

This concludes the proof of Part (1) of Theorem 6.2. We now turn our attention to the aforementioned work of Vadhan and Zheng [48], as a consequence of which we shall be able to prove a version of Theorem 6.2 in which the parameter k can be taken to be $O(\epsilon^{-2}\log(\delta^{-1}))$. This strengthened version of the dense model theorem had previously been proved by Zhang [50], who also showed that this dependence of k on δ and ϵ is asymptotically optimal (see also [29]).

Vadhan and Zheng's main innovation is a more constructive, uniform[4] Min-Max theorem which we state and prove below. In fact, it is quite likely that an optimal dense model theorem could be obtained by applying the relative-entropy method directly, rather than to the main ingredient in the proof. However, given that the uniform version of the Min-Max theorem has numerous other applications we have preferred to include it as stated in [48].

Theorem 6.7 (Uniform Min-Max theorem) *Let* $m \in \mathbb{R}^+$. *Consider a 2-player zero-sum game in which Player 1's set of pure strategies is* $\mathcal{V} \subseteq \Delta_X$ *and Player 2's is* $\mathcal{W}$, *and the expected pay-off to Player 2 is* $\mathbb{E}f(V, W)$ *for some function* $f : X \times \mathcal{W} \to [-m, m]$.

---

[4]The terminology "uniform" takes its meaning from applications of the Min-Max theorem in cryptography, where one seeks to construct an adversary "uniformly" (see [52]).

Then for all $0 < \epsilon < 1$ there exists an integer S and an algorithm which produces a sequence of strategies $V^{(1)}, V^{(2)}, \ldots, V^{(S)} \in \mathrm{Conv}(\mathcal{V})$ and $W^{(1)}, W^{(2)}, \ldots, W^{(S)} \in \mathcal{W}$ and outputs a mixed strategy $W^* \in \mathrm{Conv}(\mathcal{W})$ such that

$$\mathbb{E}f(V, W^*) \geq \mathrm{avg}_{1 \leq i \leq S} \mathbb{E}f(V^{(i)}, W^{(i)}) - O(m\epsilon).$$

for any Player 1 strategy $V \in \mathcal{V}$, where S satisfies $D_{KL}(V||V^{(1)}) \leq S\epsilon^2$ for all $V \in \mathcal{V}$.

We first show that Theorem 6.7 implies the original Min-Max Theorem.

Proof of Theorem 6.3 using Theorem 6.7: Since $W^{(i)}$ in Theorem 6.7 will be defined to be Player 2's best response to the mixed strategy $V^{(i)}$, we have

$$\mathbb{E}f(V^{(i)}, W^{(i)}) = \max_{W \in \mathcal{W}} \mathbb{E}f(V^{(i)}, W).$$

By Theorem 6.7, there exists a strategy $W^* \in \mathrm{Conv}(\mathcal{W})$ such that

$$\min_{V \in \mathcal{V}} \mathbb{E}f(V, W^*) \geq \mathrm{avg}_{1 \leq i \leq S} \mathbb{E}f(V^{(i)}, W^{(i)}) - O(m\epsilon)$$

$$= \mathrm{avg}_{1 \leq i \leq S} \max_{W \in \mathcal{W}} \mathbb{E}f(V^{(i)}, W) - O(m\epsilon).$$

But $\max_{W \in \mathcal{W}} \mathbb{E}f(V^{(i)}, W) \geq \min_{V \in \mathrm{Conv}(\mathcal{V})} \max_{W \in \mathcal{W}} \mathbb{E}f(V, W)$ for all i, so

$$\max_{W \in \mathrm{Conv}(\mathcal{W})} \min_{V \in \mathcal{V}} \mathbb{E}f(V, W) \geq \min_{V \in \mathcal{V}} \mathbb{E}f(V, W^*)$$

$$\geq \min_{V \in \mathrm{Conv}(\mathcal{V})} \max_{W \in \mathcal{W}} \mathbb{E}f(V, W) - O(m\epsilon).$$

Letting $\epsilon \to 0$ yields the statement of Theorem 6.3, the reverse inequality being trivial. $\square$

The proof of Theorem 6.7 proceeds by an iterative procedure which incrementally decreases the relative entropy of $V \in \mathcal{V}$ using multiplicative weight updates analogous to the proof of Theorem 3.2 in Section 2. Vadhan and Zheng credit the idea to Barak, Hardt and Kale, who used multiplicative weight updates coupled with approximate KL-projections to give a simple, more efficient and uniform proof of the hard-core lemma [1].

Definition 6.8 (KL projection) Let Z be a distribution on X and let $\mathcal{V}$ be any non-empty closed convex set of distributions on X. We say $\widehat{Y} \in \mathcal{V}$ is a *KL projection* of Z onto $\mathcal{V}$ if

$$\widehat{Y} = \arg \min_{Y \in \mathcal{V}} D_{KL}(Y||Z).$$

KL projections satisfy the following Pythagorean property (see [6], Chapter 11).

Proposition 6.9 (Pythagorean property) *Let $\mathcal{V}$ be any non-empty closed convex set of distributions on X, and let $\widehat{Y} \in \mathcal{V}$ be a KL projection of Z onto $\mathcal{V}$. Then for all $y \in \mathcal{V}$,*

$$D_{KL}(Y||\widehat{Y}) + D_{KL}(\widehat{Y}||Z) \leq D_{KL}(Y||Z).$$

It is easy to see that the Pythagorean property implies that the KL projection is unique. However, finding the exact KL projection is often not feasible, so we make do with the following approximate notion.

Definition 6.10 (Approximate KL projection) Let $0 < \eta < 1$, let Z be a distribution on X and let $\mathcal{V}$ be any non-empty closed convex set of distributions on X. We say $\widetilde{Y}$ is an η-*approximate KL projection* of Z onto $\mathcal{V}$ if $\widetilde{Y} \in \mathcal{V}$ and, for all $Y \in \mathcal{V}$,

$$D_{KL}(Y||\widetilde{Y}) \leq D_{KL}(Y||Z) + \eta.$$

Note that if

$$D_{KL}(Y||\widetilde{Y}) \leq D_{KL}(Y||\widehat{Y}) + \eta$$

for all $Y \in \mathcal{V}$, where $\widehat{Y}$ is the (exact) KL projection of Z onto $\mathcal{V}$, then $\widetilde{Y}$ is an η-approximate KL projection of Z onto $\mathcal{V}$. We are now ready to state the iterative procedure leading to a proof of Theorem 6.7.

Procedure 2 *Apply the following iterative procedure.*
Start the algorithm with an arbitrary $V^{(1)} \in \text{Conv}(\mathcal{V})$. For $i = 1, 2, \ldots, S$,

- *let $W^{(i)} \in \mathcal{W}$ be a best Player-2 response to $V^{(i)}$;*

- *let $V^{(i)'}$ be such that*

$$\mathbb{P}[V^{(i)'} = x] \propto \exp(-\epsilon f(x, W^{(i)})/2m) \cdot \mathbb{P}[V^{(i)} = x];$$

- *let $V^{(i+1)}$ be an arbitrary ϵ^2-approximate KL projection of $V^{(i)'}$ onto $\text{Conv}(\mathcal{V})$.*

Output the mixed strategy

$$W^* := \frac{1}{S} \sum_{i=1}^{S} W^{(i)}.$$

The following lemma states that such multiplicative weight updates decrease KL divergence.

Lemma 6.11 *Let A, B be distributions over X and $h : X \to [0, 1]$ be any function. Define a distribution A' such that*

$$\mathbb{P}[A' = x] \propto \exp(\epsilon h(x))\mathbb{P}[A = x]$$

for $0 \leq \epsilon \leq 1$. Then

$$D_{KL}(B||A') \leq D_{KL}(B||A) - (\log e)\epsilon(\mathbb{E}[h(B)] - \mathbb{E}[h(A)] - \epsilon).$$

Proof By definition,

$$D_{KL}(B||A) - D_{KL}(B||A') = \sum_x \mathbb{P}[B = x] \log \frac{\mathbb{P}[A' = x]}{\mathbb{P}[A = x]}$$

$$= \sum_x \mathbb{P}[B = x] \log \frac{e^{\epsilon h(x)}}{\sum_y e^{\epsilon h(y)}\mathbb{P}[A = y]},$$

which equals

$$(\log e) \left(\epsilon \mathbb{E}h(B) - \ln \sum_y e^{\epsilon h(y)}\mathbb{P}[A = y] \right).$$

Bounding the exponential above and below by $e^z \leq 1+z+z^2$ and $1+z \leq e^z$, respectively, we find that

$$D_{KL}(B||A) - D_{KL}(B||A') \geq (\log e) \left(\epsilon\mathbb{E}[h(B)] - \ln(1 + \epsilon\mathbb{E}[h(A)] + \epsilon^2) \right)$$
$$\geq (\log e)\epsilon(\mathbb{E}[h(B)] - \mathbb{E}[h(A)] - \epsilon)$$

as desired. □

Proof of Theorem 6.7: Let S be the least integer such that for any $V \in \mathcal{V}$, $D_{KL}(V||V^{(1)}) \leq S\epsilon^2$. By Lemma 6.11 with $A = V^{(i)}$, $A' = V^{(i)'}$, $B = V$ and $h(x) = -f(x, W^{(i)})/2m$, we have

$$D_{KL}(V||V^{(i)}) - D_{KL}(V||V^{(i)'})$$
$$\geq (\log e)\epsilon \left(\frac{\mathbb{E}f(V^{(i)}, W^{(i)}) - \mathbb{E}f(V, W^{(i)})}{2m} - \epsilon \right).$$

Now since $V^{(i+1)}$ is an ϵ^2-approximate KL projection of $V^{(i)'}$ onto $\mathrm{Conv}(\mathcal{V})$, we have $D_{KL}(V||V^{(i+1)}) \leq D_{KL}(V||V^{(i)'}) + \epsilon^2$ and thus

$$D_{KL}(V||V^{(i)}) - D_{KL}(V||V^{(i+1)})$$
$$\geq (\log e)\epsilon \left(\frac{\mathbb{E}f(V^{(i)}, W^{(i)}) - \mathbb{E}f(V, W^{(i)})}{2m} - \epsilon \right) - \epsilon^2.$$

Summing the telescoping series from 1 to S, we have

$$D_{KL}(V||V^{(1)}) - D_{KL}(V||V^{(S+1)})$$
$$\geq (\log e)\epsilon \sum_{i=1}^{S} \left(\frac{\mathbb{E}f(V^{(i)}, W^{(i)}) - \mathbb{E}f(V, W^{(i)})}{2m} - \epsilon \right) - S\epsilon^2,$$

with the latter expression being equal to

$$(\log e)S\epsilon \left(\frac{\mathrm{avg}_{1 \leq i \leq S}\mathbb{E}f(V^{(i)}, W^{(i)}) - \mathbb{E}f(V, W^*)}{2m} - \epsilon \right) - S\epsilon^2.$$

Since $D_{KL}(V||V^{(S+1)}) \geq 0$, it follows that

$$\frac{\mathrm{avg}_{1 \leq i \leq S}\mathbb{E}f(V^{(i)}, W^{(i)}) - \mathbb{E}f(V, W^*)}{2m} \leq \frac{D_{KL}(V||V^{(1)}) + S\epsilon^2}{(\log e)S\epsilon} + \epsilon = O(\epsilon),$$

which proves the theorem. □

Vadhan and Zheng [48] use the Uniform Min-Max theorem to strengthen and reprove a number of known results in complexity theory. For example, they give a new proof of Impagliazzo's hard core theorem [25] with optimal hard core density and optimal complexity blow-up. They also deduce an optimal weak regularity lemma for graphs of density $o(1)$ ([48], Section 6.2). Here we shall focus on showing how Theorem 6.7 gives a quick proof of the following version of the dense model theorem due to Zhang [50].

Theorem 6.12 (Optimal dense model theorem) *Let $\epsilon, \delta > 0$. Let $\mathcal{F} = \{\phi : X \to [0,1]\}$ be any family of functions and $\nu : X \to \mathbb{R}^+$ any measure, and let $f : X \to \mathbb{R}^+$ be any function with $0 \leq f \leq \nu$ and $\mathbb{E}f = \delta$. Suppose that for any bounded function $g : X \to [0,1]$ such that $\mathbb{E}g = \delta$, we have $\|f - g\|_{\mathcal{F}} \geq \epsilon\delta$.*
Then ν is not pseudorandom in the sense that there exists an integer $k = O(\epsilon^{-2} \log(\delta^{-1}))$, $t \in \mathbb{R}$ and $\psi = \mathrm{Th}_t(\overline{\phi})$ for some $\overline{\phi} \in \mathrm{Avg}_k(\mathcal{F})$ such that

$$\langle \psi, \nu - 1_X \rangle = \Omega(\epsilon\delta).$$

Proof Note that in the proof of Theorem 6.2, we only actually used the inequality

$$c =: \min_{V \in \mathcal{V}} \mathbb{E}f(V, W^*) \geq \max_{W \in \mathcal{W}} \mathbb{E}f(V^*, W),$$

which we now replace with

$$c = \min_{V \in \mathcal{V}} \mathbb{E}f(V, W^*) \geq \text{avg}_{1 \leq i \leq S} \mathbb{E}f(V^{(i)}, W^{(i)}) - O(\epsilon)$$

as a result of applying Theorem 6.7 with $m = 1$, $\mathcal{V} = \delta^{-1}\mathcal{H} \subseteq \Delta_X$ and $\mathcal{W} = \mathcal{F}$. Note that since $\mathcal{V}$ contains the uniform distribution on X, and we may start the algorithm with an arbitrary $V^{(1)} \in \mathcal{V}$, we may choose S to be the least integer such that for all $V \in \mathcal{V}$, $D_{KL}(V || V^{(1)}) = D_{KL}(V || U_X) = \text{Ent}(V) \leq S\epsilon^2$, which means S can be taken to be $O(\epsilon^{-2}\log(\delta^{-1}))$. We proceed almost exactly as in the proof of Lemma 6.4 to obtain our universal distinguisher. Indeed, note that by hypothesis, for every $V \in \mathcal{V}$ there exists $W \in \mathcal{W}$ such that $\mathbb{E}f(V, W) \geq \epsilon$. This is true in particular for every $V^{(i)}$, implying that

$$\text{avg}_{1 \leq i \leq S} \mathbb{E}f(V^{(i)}, W^{(i)}) = \text{avg}_{1 \leq i \leq S} \max_{W \in \mathcal{W}} \mathbb{E}f(V^{(i)}, W) \geq \epsilon,$$

so that as before, $c \geq \epsilon$. We conclude that there exists a universal distinguisher ϕ^* such that for all $g \in \mathcal{H}$,

$$\langle \phi^*, f - g \rangle \geq \epsilon\delta,$$

but by definition of Procedure 2, the distinguisher ϕ^* is already an average of S elements of $\overline{\mathcal{F}}$. $\square$

It would be interesting to determine whether Theorem 6.12 has any direct applications in additive combinatorics. Regardless of the answer to this particular question, there is little doubt that the general approach using relative entropy outlined in this article will find numerous further uses.

References

[1] Boaz Barak, Moritz Hardt, and Satyen Kale. The uniform hardcore lemma via approximate Bregman projections. In *Proceedings of the Twentieth Annual ACM-SIAM Symposium on Discrete Algorithms*, pages 1193–1200. SIAM, Philadelphia, PA, 2009.

[2] Thomas Bloom. A quantitative improvement for Roth's theorem on arithmetic progressions. *J. Lond. Math. Soc. (2)*, 93(3):643–663, 2016.

[3] Mei-Chu Chang. A polynomial bound in Freiman's theorem. *Duke Math. J.*, 113(3):399–419, 2002.

[4] David Conlon, Jacob Fox, and Yufei Zhao. The Green-Tao theorem: an exposition. *EMS Surv. Math. Sci.*, 1(2):249–282, 2014.

[5] David Conlon, Jacob Fox, and Yufei Zhao. A relative Szemerédi theorem. *Geom. Funct. Anal.*, 25(3):733–762, 2015.

[6] Thomas Cover and Joy Thomas. *Elements of information theory.* Wiley Series in Telecommunications. John Wiley & Sons, Inc., New York, New York, USA, 1991.

[7] Jacob Fox. A new proof of the graph removal lemma. *Ann. of Math. (2)*, 174(1):561–579, 2011.

[8] Ehud Friedgut. An information-theoretic proof of a hypercontractive inequality. *arXiv*, 1504.01506, April 2015.

[9] Alan Frieze and Ravi Kannan. Quick approximation to matrices and applications. *Combinatorica*, 19(2):175–220, 1999.

[10] David Galvin. Three tutorial lectures on entropy and counting. *arXiv*, 1406.7872, June 2014.

[11] Timothy Gowers. A new proof of Szemerédi's theorem for arithmetic progressions of length four. *Geom. Funct. Anal.*, 8(3):529–551, 1998.

[12] Timothy Gowers. A new proof of Szemerédi's theorem. *Geom. Funct. Anal.*, 11(3):465–588, 2001.

[13] Timothy Gowers. Decompositions, approximate structure, trans-ference, and the Hahn-Banach theorem. *Bull. Lond. Math. Soc.*, 42(4):573–606, 2010.

[14] Timothy Gowers. Entropy and Sidorenko's conjecture — after Szegedy. *Personal blog*, https://gowers.wordpress.com/2015/11/18/entropy-and-sidorenkos-conjecture-after-szegedy/, November 2015.

[15] Timothy Gowers and Julia Wolf. Linear forms and quadratic uniformity for functions on $\mathbb{F}_p^n$. *Mathematika*, 57(2):215–237, 2012.

[16] Ben Green. Some constructions in the inverse spectral theory of cyclic groups. *Combin. Probab. Comput.*, 12(2):127–138, 2003.

[17] Ben Green. Spectral structure of sets of integers. In *Fourier analysis and convexity*, pages 83–96. Birkhäuser Boston, Boston, MA, 2004.

[18] Ben Green. Finite field models in additive combinatorics. In Bridget S Webb, editor, *Surveys in combinatorics 2005*, pages 1–27. Cambridge Univ. Press, Cambridge, Cambridge, 2005.

[19] Ben Green. Roth's theorem in the primes. *Ann. of Math. (2)*, 161(3):1609–1636, 2005.

[20] Ben Green. Montréal notes on quadratic Fourier analysis. In *Additive combinatorics*, pages 69–102. Amer. Math. Soc., Providence, RI, 2007.

[21] Ben Green and Tom Sanders. Boolean functions with small spectral norm. *Geom. Funct. Anal.*, 18(1):144–162, 2008.

[22] Ben Green and Terence Tao. An inverse theorem for the Gowers $U^3(G)$ norm. *Proc. Edinb. Math. Soc. (2)*, 51(1):73–153, 2008.

[23] Ben Green and Terence Tao. The primes contain arbitrarily long arithmetic progressions. *Ann. of Math. (2)*, 167(2):481–547, 2008.

[24] Ben Green and Terence Tao. Linear equations in primes. *Ann. of Math. (2)*, 171(3):1753–1850, 2010.

[25] Russell Impagliazzo. Hard-core distributions for somewhat hard problems. In *FOCS '07. 48th Annual IEEE Symposium on Foundations of Computer Science, 2007.*, pages 538–545. IEEE, 1995.

[26] Russell Impagliazzo, Cristopher Moore, and Alexander Russell. An entropic proof of Chang's inequality. *SIAM J. Discrete Math.*, 28(1):173–176, 2014.

[27] Solomon Kullback and Richard Leibler. On information and sufficiency. *Ann. Math. Stat.*, 22(1):79–86, 1951.

[28] James Lee. Covering the large spectrum and generalized Riesz products. *arXiv*, 1508.07109, August 2015.

[29] Chi-Jen Lu, Shi-Chun Tsai, and Hsin-Lung Wu. Complexity of hardcore set proofs. *Comput. Complexity*, 20(1):145–171, 2011.

[30] Guy Moshkovitz and Asaf Shapira. A sparse regular approximation lemma. *arXiv*, 1610.02676, October 2016.

[31] Ryan O'Donnell.   Lecture 16:   The hypercontractivity theorem. *Personal website,* https://www.cs.cmu.edu/~odonnell/boolean-analysis/lecture16.pdf.

[32] Sean Prendiville. Four variants of the Fourier-analytic transference principle. *arXiv,* 1509.09200, September 2015.

[33] Jaikumar Radhakrishnan. An entropy proof of Bregman's theorem. *J. Combin. Theory Ser. A,* 77(1):161–164, 1997.

[34] Omer Reingold, Luca Trevisan, Madhur Tulsiani, and Salil Vadhan. Dense subsets of pseudorandom sets. *Electronic Colloquium on Computational Complexity,* 45:1–33, 2008.

[35] Omer Reingold, Luca Trevisan, Madhur Tulsiani, and Salil Vadhan. New proofs of the Green-Tao-Ziegler dense model theorem: an exposition. *arXiv,* 0806.0381, June 2008.

[36] Sheldon Ross. *A first course in probability.* Macmillan Publishing Co., Inc., New York; Collier Macmillan Publishers, London, 1976.

[37] Alex Samorodnitsky.   Low-degree tests at large distances.   In *STOC'07—Proceedings of the 39th Annual ACM Symposium on Theory of Computing,* pages 506–515. ACM, New York, New York, New York, USA, 2007.

[38] Tom Sanders. On Roth's theorem on progressions. *Ann. of Math. (2),* 174(1):619–636, 2011.

[39] Tom Sanders.   On the Bogolyubov–Ruzsa lemma.   *Anal. PDE,* 5(3):627–655, 2012.

[40] Ilya Shkredov. On sets of large exponential sums. *Dokl. Math.,* 74(3):860–864, 2006.

[41] Balazs Szegedy. An information theoretic approach to Sidorenko's conjecture. *arXiv,* 1406.6738, June 2014.

[42] Terence Tao.   Moser's entropy compression argument.   *Personal blog,*   https://terrytao.wordpress.com/2009/08/05/mosers-entropy-compression-argument/, August 2009.

[43] Terence Tao. *Higher order Fourier analysis,* volume 142 of *Graduate Studies in Mathematics.* American Mathematical Society, Providence, RI, 2012.

[44] Terence Tao. Entropy and rare events. *Personal blog*, https://terrytao.wordpress.com/2015/09/20/entropy-and-rare-events/, September 2015.

[45] Terence Tao. The logarithmically averaged Chowla and Elliott conjectures for two-point correlations. *Forum Math. Pi*, 4:e8–36, 2016.

[46] Terence Tao and Tamar Ziegler. The primes contain arbitrarily long polynomial progressions. *Acta Math.*, 201(2):213–305, 2008.

[47] Madhur Tulsiani and Julia Wolf. Quadratic Goldreich-Levin theorems. *SIAM J. Comput.*, 43(2):730–766, 2014.

[48] Salil Vadhan and Jia Zheng. A uniform min-max theorem with applications in cryptography. In *Advances in Cryptology – CRYPTO 2013*, pages 93–110. Springer Berlin Heidelberg, Berlin, Heidelberg, 2013.

[49] Julia Wolf. Finite field models in arithmetic combinatorics – ten years on. *Finite Fields Appl.*, 32:233–274, 2015.

[50] Jiapeng Zhang. On the query complexity for Showing Dense Model. *Electronic Colloquium on Computational Complexity*, 38, 2011.

[51] Yufei Zhao. An arithmetic transference proof of a relative Szemerédi theorem. *arXiv*, 1307.4959, July 2013.

[52] Jia Zheng. *A uniform min-max theorem and characterizations of computational randomness*. PhD thesis, 2014.

School of Mathematics
University of Bristol
Bristol BS8 1TW
UK
julia.wolf@bristol.ac.uk

Printed in the United States
by Baker & Taylor Publisher Services

Printed in the United States
by Baker & Taylor Publisher Services